FISH PHYSIOLOGY

VOLUME IX
Reproduction

Part B
Behavior and Fertility Control

CONTRIBUTORS

S. T. H. CHAN

EDWARD M. DONALDSON

FREDERICK W. GOETZ

GEORGE A. HUNTER

T. J. LAM

N. R. LILEY

N. E. STACEY

JOACHIM STOSS

GARY H. THORGAARD

W. S. B. YEUNG

FISH PHYSIOLOGY

Edited by

W. S. HOAR
DEPARTMENT OF ZOOLOGY
UNIVERSITY OF BRITISH COLUMBIA
VANCOUVER, BRITISH COLUMBIA, CANADA

D. J. RANDALL
DEPARTMENT OF ZOOLOGY
UNIVERSITY OF BRITISH COLUMBIA
VANCOUVER, BRITISH COLUMBIA, CANADA
and

E. M. DONALDSON
WEST VANCOUVER LABORATORY
FISHERIES RESEARCH BRANCH
DEPARTMENT OF FISHERIES AND OCEANS
WEST VANCOUVER, BRITISH COLUMBIA, CANADA

VOLUME IX

Reproduction

Part B

Behavior and Fertility Control

1983

ACADEMIC PRESS
A Subsidiary of Harcourt Brace Jovanovich, Publishers
New York   London
Paris   San Diego   San Francisco   São Paulo   Sydney   Tokyo   Toronto

ACADEMIC PRESS, INC.
111 Fifth Avenue, New York, New York 10003

United Kingdom Edition published by
ACADEMIC PRESS, INC. (LONDON) LTD.
24/28 Oval Road, London NW1 7DX

Library of Congress Cataloging in Publication Data

Hoar, William Stewart, Date
    Fish physiology.

    Includes bibliographies.
    CONTENTS: v. 1. Excretion, ionic regulation, and
metabolism.--v. 2. The endocrine system.--[etc].--
v. 8 Bioenergetics and Growth, edited by W. S. Hoar,
D. J. Randall, and J. R. Brett.--v. 9B Reproduction:
Behavior and Fertility Control, edited by W. S. Hoar,
D. J. Randall, and E. M. Donaldson.
    1. Fishes--Physiology. I. Hoar, W. S.
author. II. Randall, D. J., Date III. Donaldson, E. M.
IV. Title.
QL639.1.H6   597'.01   76-84233
ISBN 0-12-350429-5 (v. 9B)

PRINTED IN THE UNITED STATES OF AMERICA

83 84 85 86   9 8 7 6 5 4 3 2 1

CONTENTS

4. Sex Control and Sex Reversal in Fish under Natural Conditions
S. T. H. Chan and W. S. B. Yeung

5. Hormonal Sex Control and Its Application to Fish Culture
George A. Hunter and Edward M. Donaldson

6. Fish Gamete Preservation and Spermatozoan Physiology
Joachim Stoss

7. Induced Final Maturation, Ovulation, and Spermiation in
    Cultured Fish
    Edward M. Donaldson and George A. Hunter

8. Chromosome Set Manipulation and Sex Control in Fish
    Gary H. Thorgaard

CONTRIBUTORS

Numbers in parentheses indicate the pages on which the authors' contributions begin.

S. T. H. CHAN (171), *Department of Zoology, University of Hong Kong, Hong Kong*

EDWARD M. DONALDSON (223, 351), *West Vancouver Laboratory, Fisheries Research Branch, Department of Fisheries and Oceans, West Vancouver, British Columbia V7V 1N6, Canada*

FREDERICK W. GOETZ (117), *Department of Biology, University of Notre Dame, Notre Dame, Indiana 46556*

GEORGE A. HUNTER (223, 351), *West Vancouver Laboratory, Fisheries Research Branch, Department of Fisheries and Oceans, West Vancouver, British Columbia V7V 1N6, Canada*

T. J. LAM (65), *Department of Zoology, National University of Singapore, Singapore*

N. R. LILEY (1), *Department of Zoology, The University of British Columbia, Vancouver, British Columbia V6T 2A9, Canada*

N. E. STACEY (1), *Department of Zoology, The University of Alberta, Edmonton, Alberta T6C 2E9, Canada*

JOACHIM STOSS (305)*, *West Vancouver Laboratory, Fisheries Research Branch, Department of Fisheries and Oceans, West Vancouver, British Columbia V7V 1N6, Canada*

GARY H. THORGAARD (405), *Program in Genetics and Cell Biology, Washington State University, Pullman, Washington 99163*

W. S. B. YEUNG (171), *Department of Zoology, University of Hong Kong, Hong Kong*

*Present address: Finnmark Landseruksskole, 9850 Rustefjelbma, Norway.

PREFACE

The Preface to Volume I of "Fish Physiology" noted that a six-volume treatise would attempt to review recent advances in selected areas of fish physiology, to relate these advances to the existing body of literature, and to delineate useful areas for future study. The hope expressed at that time was that the series would serve the biologists of the 1970s as its predecessor "The Physiology of Fishes" (M. E. Brown, editor) had served its readers throughout the 1960s. Our general objectives remain, but with Volumes VII (Locomotion) and VIII (Bioenergetics and Growth) the emphasis has been somewhat altered; these later volumes presented in-depth reviews and assessments of current research in selected areas of fish physiology—especially those areas where advances have been particularly rapid during the past decade. In keeping with this concept, we are pleased to add to the series Volume IXA and IXB on fish reproduction.

When Volume III was published in 1969, the physiology of fish reproduction was reviewed in three chapters. The present treatment in two Parts (A and B) attests to the rapid developments in this field. Moreover, Volume IX deals only with selected topics on reproductive physiology, especially the endocrinology, behavior, environment interactions, and fertility-related topics. Several subjects included in Volume III are not reviewed in these volumes (viviparity, for example), whereas others that now merit consideration in separate chapters were not sufficiently developed to require any comment in Volume III (the hypothalamic hormones and hormone receptors, for example). With the exception of Chapter 1, Part A, which is devoted to the Cyclostomes, and Chapter 2, Part A, which is devoted to Chondrichthyes, the books deal with the much more thoroughly studied teleost fishes.

Volume IX reflects the practical importance of studies in fish reproductive physiology. The control of fertility is now a subject of great economic importance in the manipulation of valuable fisheries resources. Many significant advances and future trends in the research on fertility of teleost fishes are evaluated in several chapters of Part B.

Finally, the editors are happy to express their appreciation to all those who devoted their time to this project; the authors are all active research scientists, and in most cases, they had to find the many hours required for writing in an already full program. We are fortunate to have had the pleasant cooperation of the leaders in this rapidly changing area of fish physiology.

W. S. HOAR
D. J. RANDALL
E. M. DONALDSON

CONTENTS OF OTHER VOLUMES

1

HORMONES, PHEROMONES, AND REPRODUCTIVE BEHAVIOR IN FISH

N. R. LILEY
Department of Zoology
University of British Columbia
Vancouver, British Columbia, Canada

N. E. STACEY
Department of Zoology
The University of Alberta
Edmonton, Alberta, Canada

I. INTRODUCTION

Teleosts display a variety of reproductive behaviors (Balon, 1975, 1981; Keenleyside, 1979). At one extreme, breeding individuals within a school simply release gametes freely into the water. In other species, breeding may involve preparation of a nest site, defence of a territory, and elaborate pair

1

formation and mating ceremonies. This may be followed by extended care of the eggs and young by one or both sexes.

The role of hormones in the regulation of various aspects of reproductive behavior has been reported by a number of researchers (Baggerman, 1969; Liley, 1969, 1980; Fiedler, 1974; Stacey, 1981). In this discussion the more recent studies are examined, particularly those dealing mainly with the role of endocrine factors as determinants of reproductive behavior. However, it is important to emphasize the two-way nature of the relationship between the endocrine system and the biological and physical environment. Not only does the endocrine system regulate the behavioral responses necessary for successful reproduction, but it is also responsive to social and other exogenous stimuli. The smooth progression through the reproductive cycle depends on the continuing interplay between the endocrine system and the environment. In effect, behavior provides the link between the organism and its environment. Lam (Chapter 2) and Peter (Chapter 3, Volume 9A, this series) consider in more detail the nature and mechanisms of the influence of physical, biotic, and social factors on the endocrine system.

As in all vertebrates, with the possible exception of the cyclostomes, a fundamental feature of the fish endocrine system is the interdependence of the hypothalamus, pituitary, and gonads: The hypothalamus–pituitary–gonad axis (HPG). One might expect that, as in other vertebrates, the gonadal hormones play a major role in mediating reproductive behavior, either by acting directly on brain structures governing certain behavior patterns, or by acting indirectly to influence behavior through their effects on the development of secondary sexual characteristics. This apparently is the case in male teleosts, but the role of gonadal hormones in females is far from clear. Gonadal hormone secretion is in turn governed by pituitary gonadotropin, and there are claims that, in addition to its action on gonadal growth and steroidogenesis, gonadotropin has a direct effect on certain behaviors. Other pituitary factors, notably prolactin and neurohypophysial hormones, and chemical mediators such as prostaglandins have also been implicated in the causation of certain behaviors.

The nature and identity of the hormones of the HPG have been reviewed elsewhere (Fontaine, 1976; Idler, 1973; Ng and Idler, Chapter 8, Volume 9A, this series; Fostier et al., Chapter 7, Volume 9A, this series). Apart from certain differences in the chemistry of the pituitary hormones, the basic hormonal repertoire of fish is essentially the same as other vertebrate groups. Of particular interest is the fact that recent advances in the identification and measurement of tissue and plasma hormones have made it possible for researchers to describe in considerable detail the relationships between changes in hormone levels and the onset, maintenance, and completion of the various components of reproductive behavior. In effect this information provides the "first line" of evidence regarding which factors may be playing a causal role. Because of the central role of gonadal steroids in the regulation of reproductive behavior in the tetrapods, recent information regarding these hormones in teleosts is considered briefly here.

II. ANNUAL CYCLES IN GONADAL STEROIDS IN RELATION TO THE ONSET AND MAINTENANCE OF REPRODUCTIVE BEHAVIOR

A. Gonadal Steroids in Male Teleosts

Correlations between the annual breeding cycle and cycles in plasma or testicular androgens have been demonstrated in numerous species of teleost: threespine stickleback, *Gasterosteus aculeatus* (Gottfried and van Mullem, 1967), Atlantic salmon, *Salmo salar* (Idler *et al.*, 1971), the goldfish, *Carassius auratus* (Schreck and Hopwood, 1974), plaice, *Pleuronectes platessa* (Wingfield and Grimm, 1977), brown trout, *Salmo trutta* (Billard *et al.*, 1978), rainbow trout, *Salmo gairdneri* (Campbell *et al.*, 1980; Sanchez-Rodriguez *et al.*, 1978; Scott *et al.*, 1980a), and striped mullet, *Mugil cephalus* (Dindo and MacGregor, 1981). Several of these studies demonstrate clearly that a relatively sudden rise in androgen coincides with spermiation, and therefore, presumably, with readiness to display reproductive behavior.

Testosterone and 11-ketotestosterone appear to be the predominant testicular steroids in the teleosts examined. However, it is still not clear whether they play equally important roles in the development of secondary sexual characteristics and reproductive behavior, or whether one of them, ketotestosterone, should be regarded as the major androgen in teleosts.

In the rainbow trout, both testosterone and ketotestosterone increase slowly initially and then more rapidly from July to November; thereafter, testosterone levels decline, but those of ketotestosterone continue to rise during the winter spawning season to a peak in February (Scott *et al.*, 1980a). Spermiation and the acquisition of secondary sexual features (e.g., coloration, kype, watery flesh, aggressive behavior) appear correlated with the period of high levels of ketotestosterone. A similar pattern has been described in the Atlantic salmon (Idler *et al.*, 1971), brook trout, *Salvelinus fontinalis* (Sangalang and Freeman, 1974), and the winter flounder, *Pseudopleuronectes americanus* (Campbell *et al.*, 1976). In each of these studies, ketotestosterone is the gonadal steroid most clearly associated with the onset of breeding activity.

Ketotestosterone has been identified in many, but not all, species examined (Fostier *et al.*, Chapter 7, Volume 9A, this series). A number of studies have confirmed that ketotestosterone has androgenic properties. Treatment of immature sockeye salmon, *Onchorhynchus nerka*, with ketotestosterone induced male sexual coloration, thickening of the skin, elongation of the snout, and spermiation in males (Idler *et al.*, 1961). The effect on appearance was similar but less pronounced in females. Arai (1967) found ketotestosterone to be considerably more effective than testosterone in its androgenic properties when administered to female medakas, *Oryzias latipes*. Yamazaki and Donaldson (1969), Hishida and Kawamoto (1970), and Takahashi (1975) demonstrated the androgenic properties of ketotestosterone in goldfish, the medaka, and the guppy, *Poecilia reticulata*, respectively.

Although testosterone has been demonstrated to be effective as an androgen, it is not clear from the studies on plasma and tissue testosterone that testosterone is the major androgen associated with reproductive morphology and behavior. Wingfield and Grimm (1977) and Dindo and MacGregor (1981) observed a marked increase in plasma testosterone concentration at the time of spawning in *Pleuronectes platessa* and *Mugil cephalus*, respectively. However, Dindo and MacGregor note that, because of cross reactions with the antiserum used, both dihydrotestosterone and ketotestosterone would have additive effects on total "testosterone" levels measured. Scott *et al.* (1980a) proposed that testosterone may be present as an intermediate product in the synthesis of ketotestosterone, or it may play a role in the earlier stages of spermiogenesis. The discovery of testosterone in females of a number of species of fish at levels similar to or exceeding those found in males (references in Scott *et al.*, 1980b; MacGregor *et al.*, 1981) cast some doubt on testosterone as the primary androgen in fish.

Therefore the evidence, which has accumulated over the last few years, strongly suggests that 11-ketotestosterone is the major androgen in many species of teleost fish. Testosterone may be the functional androgen in certain species, but in others its role may be that of a precursor in ketotestosterone synthesis or as an important agent in the earlier stages of gonadal maturation. Unfortunately, in spite of the fact that ketotestosterone was discovered in the 1960s (Idler *et al.*, 1961), there have been remarkably few experimental investigations of the function of ketotestosterone, and only one of these (Kyle, 1982) has been concerned with the effects of this steroid on reproductive behavior.

A number of other steroids have been identified in male teleosts, including low levels of progestins and estrogens (Fostier *et al.*, Chapter 7, Volume 9A, this series). Progestins and corticosteroid concentrations increase during the spawning season in several species. Both types of hormones have been implicated in the final stages of gonadal maturation; however, at this stage, there is no reason to suspect that these hormones play a causal role in the appearance and maintenance of reproductive behavior, although this possibility should be examined.

B. Gonadal Steroids in Female Teleosts

Studies of a number of teleost species have shown that plasma steroids undergo dramatic changes associated with female reproduction. Generally, these studies have followed plasma steroid levels over the course of a reproductive season or annual cycle and, thus, the results rarely provide detailed information as to what changes, if any, might immediately precede the onset of reproductive behavior. Nevertheless, these studies have demonstrated gonadal steroids are carried in the blood, and thereby, have indicated which gonadal steroids at least have the potential to influence female behaviors.

17β-Estradiol has been identified in the plasma of a variety of oviparous teleosts, and it reaches peak levels during the prespawning period in rainbow trout (Whitehead *et al.*, 1978; Billard *et al.*, 1978; Scott *et al.*, 1980b; van Bohemen and Lambert, 1981), brown trout, *Salmo trutta* (Crim and Idler, 1978), Atlantic salmon (Idler *et al.*, 1981), goldfish (Schreck and Hopwood, 1974), common carp, *Cyprinus carpio* (Elefthériou *et al.*, 1968), plaice, *Pleuronectes platessa* (Wingfield and Grimm, 1977), and striped mullet, *Mugil cephalus* (Dindo and MacGregor, 1981). Presumably, increased levels of plasma estradiol during the period of rapid ovarian growth stimulate synthesis and secretion of hepatic vitellogenin (see Ng and Idler, Chapter 8, Volume 9A, this series); however, whether plasma estradiol may also be involved in stimulating female reproductive behaviors is not clear. For example, in salmonids (Whitehead *et al.*, 1978; Scott *et al.*, 1980b; van Bohemen and Lambert, 1981) and plaice (Wingfield and Grimm, 1977), plasma estradiol reaches maximal levels at least 1 month prior to spawning. If the demonstrated latencies (generally several days) of estrogen-induced female sexual behaviors in other vertebrate classes are in any way comparable to what might occur in fishes, and the only relevant study in teleosts (Liley, 1972) suggests this is so, then it seems unlikely that female sexual behaviors in oviparous teleosts would be stimulated directly by the prolonged elevations of plasma estradiol which can precede spawning by several months. Furthermore, it is clear that in some salmonids (Fostier *et al.*, 1978; Jalabert *et al.*, 1978; Scott *et al.*, 1980b), the carp (Elefthériou *et al.*, 1968), the plaice (Wingfield and Grimm, 1977), and striped mullet (Dindo and MacGregor, 1981), plasma levels of estradiol actually decrease prior to the occurrence of ovulation and spawning; this would not be expected if estradiol stimulated female sexual behavior in teleosts as it does in other vertebrates. Although gonadectomy-induced increases in plasma gonadotropin in rainbow trout suggested that the normal preovulatory decrease in plasma estradiol might function to remove negative feedback on the pituitary, and thereby indirectly stimulate spawning, estradiol replacement in ovariectomized females failed to provide clear support for this proposal (Bommelaer *et al.*, 1981).

Testosterone is a major circulating steroid in female rainbow trout (Scott *et al.*, 1980b; Campbell *et al.*, 1980), Atlantic salmon (Stuart-Kregor *et al.*, 1981), winter flounder, *Pseudopleuronectes americanus* (Campbell *et al.*, 1976), plaice (Wingfield and Grimm, 1977), king mackerel, *Scomberomus cavalla* (MacGregor *et al.*, 1981), *Sarotherodon (Tilapia)* aureus*, (Katz and Eckstein, 1974), and cod, *Gadus morhua* (Sangalang and Freeman, 1977). At least in rainbow trout, testosterone is considerably more abundant than estradiol (Scott *et al.*, 1980b). Indeed, it is clear that in the prespawning period of several species (Scott *et al.*, 1980a,b; Stuart-Kregor *et al.*, 1981; Campbell *et al.*, 1976, 1980), plasma testosterone levels in females exceed those of males. In contrast, ketotestosterone, the principal androgenic steroid in teleosts, is usually either undetectable in female plasma (Wingfield and Grimm, 1976), or present in very much lower concentrations than in the male (Simpson and Wright, 1977; Scott *et al.*, 1980a,b; Campbell *et al.*, 1976, 1980); exceptions include *Sarotherodon aureus* (Katz and Eckstein, 1974) and, possibly, Atlantic salmon and sockeye salmon, *Oncorhynchus nerka* (Schmidt and Idler, 1962), in which ketotestosterone has been measured in high concentrations in blood of reproductively mature females. The function of plasma androgens in female teleosts is not understood. A number of possible functions has been suggested, including stimulation (Scott *et al.*, 1980b) or inhibition (Bommelaer *et al.*, 1981) of gonadotropin secretion, stimulation of behavior (Scott *et al.*, 1980b), and, at least in the case of testosterone, serves as a precursor in the formation of other steroids by aromatization (Scott *et al.*, 1980b). Steroid aromatase is known to be present in high concentration in the teleost brain (Callard *et al.*, 1978, 1981).

Other plasma steroids, which are elevated at the time of ovulation and spawning and thus may influence reproductive behaviors, include various progestogens and corticosteroids (see Fostier *et al.*, Chapter 7, Volume 9A, this series).

To date, plasma steroid levels associated with reproduction in female teleosts have been determined only in oviparous species. As discussed in Section V,B, there is evidence that female sexual behavior in at least some of these externally fertilizing species is regulated not by steroids, but by pros-

---

*Mouthbrooding species formerly placed in the genus *Tilapia* are now assigned to *Sarotherodon* (Trewavas, 1973). The revised nomenclature is used throughout this chapter.

taglandin, which may be released into the bloodstream when ovulated eggs are in the ovaries, and then act on the brain to rapidly trigger spawning behavior (Stacey, 1981). However, in at least one teleost species, the ovoviviparous guppy, female sexual behavior is regulated by estrogen (Liley, 1972). This raises the possibility that, in oviparous species, periovulatory changes in plasma steroids may play regulatory roles in various female behaviors associated with reproduction (e.g., migration, pair formation, nest site preparation, territoriality), even if they do not stimulate sexual behavior *per se*.

III. SECONDARY SEXUAL CHARACTERISTICS

Morphological sexual characteristics are intimately involved in behavioral interactions as passive or active signals, and therefore, must be included in any consideration of the role of the endocrine system in the regulation of behavior.

Numerous studies have demonstrated that the development of secondary sexual characteristics is under endocrine control (Yamamoto, 1969; Schreck, 1974; Liley, 1980). Usually, features such as nuptial coloration and pearl organs are temporary and appear only during the breeding season. However, some structural components are permanent, becoming fully developed at the onset of maturity and remaining as sexually dimorphic features even in nonbreeding fish (e.g., gonopodia of poeciliid fishes; the enlarged dorsal fin of Arctic grayling, *Thymallus arcticus*). In most species examined, the sexual characteristics are "male positive" in that it is the male that undergoes the most striking change at maturation, developing from a more femalelike form.

A. Males

Studies involving treatment with androgens with or without castration have demonstrated an endocrine control of nuptial coloration: in threespine and ninespine stickleback, *Gasterosteus aculeatus* and *G. pungitius*, respectively, the minnow, *Phoxinus laevis* (see Yamamoto, 1969), the blue gourami, *Trichogaster trichopterus* (Johns and Liley, 1970), *Sarotherodon macrocephala* (Levy and Aronson, 1955), and *S. mossambicus* (Billy, 1982). Numerous investigations have concerned poeciliids (see Schreck, 1974) in which it has been established that treatment of females with androgen will induce the development of the gonopodium and characteristic male coloration. Pandey (1969a) hypophysectomized adult male guppies and found that the patches of bright lipophores (yellow and red pigment) became faint or disappeared, but the gonopodium remained unaffected. Treatment with methyl testosterone partially restored the lipophore content (Pandey, 1969b) indicating that the pituitary affects coloration indirectly by regulating androgen production in the testes.

The dramatic changes in coloration and body shape characteristic of spawning male Pacific salmon appear to be governed by androgens; males castrated just prior to breeding fail to assume nuptial coloration (Robertson, 1961; McBride *et al.*, 1963). Idler *et al.* (1961) demonstrated that treatment of sockeye salmon, *Oncorhynchus nerka*, with 11-ketotestosterone resulted in the development of species-typical male coloration.

Although it is generally assumed that nuptial coloration has an important role in reproductive behavior and is the product of sexual selection, it has been examined experimentally only in relatively few cases. Haskins *et al.* (1961) and Endler (1980) provide evidence of the influence of male coloration on mate selection by female guppies. Female sticklebacks, *Gasterosteus aculeatus*, evidently select their mates partly on the basis of color (McPhail, 1969; Semler, 1971). It is likely that, in addition to providing species, sex, or even individual identification, in these and perhaps many other species, changes in coloration associated with breeding have what Fernald (1976) refers to as a "behavioral amplifying" effect. Working with the cichlid, *Haplochromis burtoni*, Fernald (1976) found that androgen treatment increased the number of aggressive encounters and caused an increase in the intensity of the black eye bar. Fernald suggests that as aggressive activity increases, the intensity of the eye bar simultaneously serves as a more potent stimulus eliciting agonistic responses from conspecifics.

In addition to their effects upon coloration, androgens may also affect the development and maintenance of diverse morphological structures [e.g., pearl organs and breeding tubercles (Wiley and Collette, 1970; Smith, 1974), and fin modifications such as simple elongation in the dorsal fins of the blue gourami (Johns and Liley, 1970; D. L. Kramer, 1972) and the gobiid fish, *Pterogobius zonoleucus* (Egami, 1959b), extension of the caudal fin into a sword in the swordtail, *Xiphophorus helleri* (Baldwin and Goldin, 1939), and development of the gonopodia from the anal fin in Poeciliids (Yamamoto, 1969; Schreck, 1974; Lindsay, 1974)]. In the medaka, distinctive structures of the male anal fin, teeth, and body shape all appear to be under androgen control (Yamamoto, 1969). Levy and Aronson (1955) demonstrated that the genital papilla of males of *Sarotherodon macrocephala* lengthens under the influence of androgen treatment.

Smith (1974) found that treatment with methyl testosterone induced the formation of breeding tubercles and the mucus secreting dorsal pad in fathead minnows, *Pimephales promelas*. The appearance of this pad normally coincides with the onset of breeding behavior during which the male rubs the dorsal surface against a rock surface that eventually serves as a spawning site. It is suggested that the mucus coating may serve to lubricate the site and prevent damage and/or perhaps assist in attachment of the eggs. Smith (1976a) suggested that a similar, but more widespread, epidermal thickening of several cyprinid species may provide protection during their abrasive spawning behavior. The thickening of the skin and increased mucus production in spawning salmonids may serve a similar function. These changes can be induced in nonspawning fish by androgen treatment (Yamazaki, 1972).

There are other situations in which mucus secretions play a specialized role in breeding. Wai and Hoar (1963) showed that in the stickleback, *Gasterosteus aculeatus*, androgen stimulates secretion of kidney mucus used in gluing during nestbuilding. However, the trophic secretions of discus fish, *Symphyosodon aequifasciata*, appear to be governed by a prolactin-like hormone from the pituitary (Blüm and Fiedler, 1964). Production of mucus used in construction of a bubble nest by anabantids is also believed to be influenced by fish prolactin (Machemer, 1971).

B. Females

There are fewer studies that demonstrate a hormonal dependence of secondary sexual features in female fish—the most striking changes in coloration and morphology associated with breeding occur in the males. Idler *et al.* (1961) found that administration of estradiol to female sockeye salmon a few months before spawning resulted in an acceleration of darkening in coloration, characteristic of spawning fish. However, estradiol was ineffective in restoring the distinctive opercular pattern lost after ovariectomy in females of *Sarotherodon macrocephala* (Aronson and Holz-Tucker, 1947). Furthermore, hormone treatment caused the opercula of intact females to assume a castrate appearance similar to that of immature fish.

In many species of fish the urinogenital papilla is sexually dimorphic and becomes more prominent immediately prior to spawning. In *Sarotherodon macrocephala* the genital papilla became smaller following ovariectomy (Aronson and Holz-Tucker, 1947). Treatment with testosterone or estradiol caused the genital tube to grow rapidly. The possible role of estrogen in the acquisition of the female papilla was demonstrated in another tilapia species, *S. mariae*, by Jensen and Shelton (1979) who found that estrogen treatment of fry for several weeks resulted in the development of males with normal testes but with femalelike urogenital papilla. Alterations in the genitalia caused by sex hormones have also been reported in the tilapias *S. mossambicus* (Clemens and Inslee, 1968) and *S. niloticus* and *S. macrochir* (Jalabert *et al.*, 1974), the medaka (Yamamoto, 1969), and in the bluntnose minnow, *Hyborhynchus notatus* (Ramaswami and Hasler, 1955).

In the European bitterling, *Rhodeus amaurus*, the size of the ovipositor increases considerably immediately prior to spawning. Ball (1960) reviewed the available information and concluded that there is no reason to suppose that normal ovipositor growth is not under ovarian steroid control, but that growth reported under certain experimental conditions was mainly a response to stress. Shirai (1962, 1964) noted a clear correlation between ovipositor size and ovarian condition in Japanese bitterling, *R. ocellatus*, and proposed that a dual mechanism may be involved: an estrogen responsible for long-term growth, and a second factor governing the short-term cyclical changes during the breeding cycle responsible for the rapid lengthening at each spawning episode.

IV. CHEMICAL SIGNALS (PHEROMONES)

Recent studies have indicated that chemical secretions may be important in species, sex, or even individual recognition (Liley, 1982). Those chemical messages which operate in the context of reproduction appear to be "simple", causing arousal and perhaps "rough" orientation. Such signals may serve to initiate reproductive behavior, but subsequent, perhaps more complex, interactions depend on other sensory modalities. Most investigations have revealed pheromones that act as "releasers", i.e., the chemical elicits a more or less immediate response in another individual. Other pheromones have "priming" effects which involve longer term endogenous changes, the behavioral effects of which only become apparent hours or days later.

A. Males

Observations alone have suggested that female lampreys, *Lampetra*, channel catfish, *Ictalurus punctatus*, and glandulocaudine characids are attracted by chemicals emitted by conspecific males at spawning (Roule, 1931; Bailey and Harrison, 1945; Nelson, 1964; Atkins and Fink, 1979). More convincing evidence comes from laboratory studies which demonstrate that females of *Petromyzon marinus* (Teeter, 1980), rainbow trout (Newcombe and Hartman, 1973), several species of *Ictalurus* (Kendle, 1970; Rubec, 1979), a number of belontiids (Lee and Ingersoll, 1979), *Blennius pavo* (Laumen *et al.*, 1974), the black goby, *Gobius jozo* (Colombo *et al.*, 1980), and the threespine stickleback (Golubev and Marusov, 1979) are attracted to the odor of mature conspecific males. Males of *Hypsoblennius* responded to the odor of actively courting males (N.B., females were not eliminated as a possible source of odor) (Losey, 1969), and in the threespine stickleback, males in breeding condition responded by assuming an aggressive posture and retreating from a source of the odor of a male in nuptial coloration (Golubev and Marusov, 1979). This reaction was stronger in males in the second phase of the breeding cycle, defending a nest with eggs, than in the first phase, that of selecting a territory.

Chen and Martinich (1975) observed that introduction of water from a tank containing a male zebra danio, *Brachydanio rerio*, induced ovulation in an isolated female by the following day. However, water from a container which held a large number of danios of both sexes inhibited ovulation. Chemical stimuli from male angelfish, *Pterophyllum scalare*, induced a spawning rate in isolated females similar to that of females paired with males (Chien, 1973). In the last two examples, it is likely that chemical signals exert priming effects, probably through the endocrine system of the female.

Most studies of fish pheromones have indicated the presence of chemicals with a stimulatory role. Experimental studies suggest the existence of an inhibitory pheromone in a number of belontiids: A chemical released by a male is believed to inhibit aggression and nest-building behavior in other males (Rossi, 1969; Baenninger, 1968; Ingersoll et al., 1976).

There are indications that in a number of species, sexual pheromone is produced in the testes or associated structures. In the testes of the male black goby, *Gobius jozo*, Leydig cells, concentrated into a large mesorchial gland, appear to be specialized for the synthesis of 5β-reduced androgen conjugates. Colombo et al. (1980) found that ovulated females, on exposure to synthetic etiocholanolone glucuronide (a major soluble conjugate secreted by the male), were attracted to the source of the chemical and in some cases released their eggs. Nonovulated females were unresponsive. Prominent mesorchial glands with steroidogenic features have been described in other gobiids (references in Colombo et al., 1980). This finding is of particular interest in that it provides a rare example (see Liley, 1982) of a discrete gland apparently specialized for the production of pheromone (other possible functions have not been investigated). Furthermore, the chemical signal itself is a specialized soluble product and not simply a by-product of steroid metabolism. The testicular hormones are quite different chemically from the pheromonal steroids which are synthesized through an independent biosynthetic pathway.

There is little comparable information for other species, although a number of observations suggest that pheromones are produced in the testes and perhaps released with the gonadal products. For example, conspecific females were strongly attracted to water that had previously held ripe males and to water taken downstream of spawning rainbow trout (Newcombe and

Hartman, 1973). An interspecific interaction was noted by Hunter and Hasler (1965) who observed that the redfin shiner, *Notropis umbratilis*, was stimulated to breed by odors discharged during spawning of the green sunfish, *Lepomis cyanellus*. In Pacific herring, *Clupea harengus pallasi*, milt or testis homogenate added to a group of mature conspecifics in the laboratory triggered a dramatic onset of spawning behavior in both sexes (Stacey and Hourston, 1982). Extracts of testis of the pond smelt, *Hypomesus olidus*, stimulated courtship in conspecifics, although this material was not as effective as fluids taken from the ovarian cavity of females (Okada *et al.*, 1978).

The sexual attractant released by male sea lampreys, *Petromyzon marinus*, also appears to be present in the urinogenital fluid (Teeter, 1980). However, the finding that fluid, which contained no visible milt, evoked a female response, but milt alone was ineffective, suggests that the active substance is present in the urine rather than the products of the testes. Rubec (1979) and Richards (1974, 1976) demonstrated that urine and perhaps skin mucus are important sources of chemical signal(s) in a number of Ictalurid catfish. However, according to these studies the chemicals play a role in individual and species discrimination as well as sex recognition, and therefore, may not be involved in a reproductive function.

Male glandulocaudine fish possess distinct glands which Nelson (1964) suggests secrete sexual pheromone. Histochemical investigation of the caudal gland of *Corynopoma riisei* indicates that the product is probably a mucopolysaccharide, with the more plentiful muco- and or glycoproteins acting as a carrier or diluent (Atkins and Fink, 1979). However, it should be noted that at present there is no direct evidence that the secretion functions as a pheromone. A role in reproduction is suggested by the fact that the secretory cells become reduced in isolated males, but enlarge when courtship activity is resumed.

Laumen *et al.* (1974) demonstrated that a pheromone is secreted by appendices of the anal fin spines of mature males of *Blennius pavo*. Experiments in which immature males were injected with mammalian luteinizing hormone (LH) and methyltestosterone led Laumen and co-workers to conclude that the development and function of the glands is under the direct influence of hypophysial gonadotropin. In *Hypsoblennius* also there are anal secretory pads which may be the source of pheromone. However, it was also noted in these species that ejaculation occurred at the same stage of courtship that the pheromone first appeared; this suggests that the pheromone may be produced in the genital system and perhaps released with the milt (Losey, 1979).

The results of studies by Laumen *et al.* (1974) provided the only direct evidence of an endocrine control of pheromone secretion in a male teleost. However, the clear correlation in various species between reproductive maturity and pheromone release and responsiveness to pheromone suggests that an endocrine involvement in the regulation of pheromone production and release will prove to be widespread.

B. Females

Pheromones that attract and stimulate conspecific males have been demonstrated in the frillfin goby, *Bathygobius soporator* (Tavolga, 1956), the sea lamprey, *Petromyzon marinus* (Teeter, 1980), pond smelt, *Hypomesus olidus* (Okada *et al.*, 1978), the loach, *Misgurnus anguillicaudatus* (Honda, 1980b), several species of Ictalurid catfish (Timms and Kleerekoper, 1972; Rubec, 1979), belontiids (Mainardi and Rossi, 1968; Rossi, 1969; Cheal and Davis, 1974; Pollack *et al.*, 1978; Lee and Ingersoll, 1979), poeciliids (Amouriq, 1964; Liley, 1966; Zeiske, 1968; Gandolfi, 1969; Parzefall, 1970, 1973; Crow and Liley, 1979; Meyer and Liley, 1982; Thiessen and Sturdivant, 1977; Brett and Grosse, 1982), rainbow trout (Newcombe and Hartman, 1973; Emanuel and Dodson, 1979; Honda, 1980a), the ayu, *Plecoglossus altivelis* (Honda, 1979), a characid, *Asyntanax mexicanus* (Wilkens, 1972), two species of cichlid, *Haplochromis burtoni* and *Sarotherodon mossambicus* (Crapon de Caprona, 1980; Silverman, 1978), the zebrafish, *Brachydanio rerio* (van den Hurk *et al.*, 1982), the goldfish (Partridge *et al.*, 1976), and the threespine stickleback (Golubev and Marusov, 1979).

In most cases, the female chemical acts as a releaser, which causes a rapid increase in sexual activity and which, in some cases, attracts males either to the female or, under experimental conditions, to the vicinity of water that held females.

An increase in nest building in male belontiids after exposure to water which held mature females (Mainardi and Rossi, 1968; Rossi, 1969; Cheal and Davis, 1974) suggests a priming effect. Similarly, the increase in aggression and courtship by male cichlids, *Haplochromis burtoni*, several days after brief exposure to water which held females (Crapon de Caprona, 1980) points to the existence of priming chemicals the effects of which only become apparent after the stimulus has disappeared.

There have been a number of attempts to identify the source and chemical nature of sexual pheromones in female fish. Work with oviparous species suggests that pheromone is present in the fluids released from the ovaries at the time of ovulation (Emanuel and Dodson, 1979; Honda, 1979, 1980a,b; van den Hurk *et al.*, 1982; Newcombe and Hartman, 1973; Okada *et al.*, 1978; Partridge *et al.*, 1976; Tavolga, 1956; Teeter, 1980). However, it is not clear from any of these studies that the tested ovarian fluids were free of contamination by urine, and it is of interest that Yamazaki and Watanabe

(1979) speculated that the source of an attractant in female goldfish may be the kidney rather than the ovary. This suggestion is based on the observation that treatment with estrogen causes conspicuous changes in the kidneys of hypophysectomized male goldfish. These males also exhibit a female-like attractiveness to other males. Urine is a carrier for chemicals which mediate sex and individual recognition in several species of *Ictalurus* (Rubec, 1979; Richards, 1974, 1976).

In poeciliids, there are indications that pheromone is produced in the ovary. Liley (1966), Parzefall (1973), Crow and Liley (1979), and Brett and Grosse (1982) provided evidence that the production and/or release of pheromone is linked to the gestation cycle: Females showed maximum attractiveness for a brief period shortly after parturition. In experiments with guppies, Meyer and Liley (1982) found that either ovariectomy or hypophysectomy resulted in a loss of pheromone production. Pheromone production was not restored in ovariectomized fish by treatment with estrogen, but it did return after estrogen therapy of hypophysectomized fish (with ovaries intact but regressed). Meyer and Liley conclude that pheromone is produced in the ovary under the control of ovarian hormone.

Amouriq (1965) reported that extract of ovarian tissue of the guppy caused an increase in locomotory activity of conspecific males. A suspension of an estrogen, hexestrol, produced a similar response, leading Amouriq to conclude that the pheromone is probably an ovarian steroid. This conclusion was criticized by Liley (1969) because there was a considerable delay in the onset of the male response to estrogen, suggesting that the hormone may be exerting a more general metabolic effect. Furthermore, Meyer and Liley (1982) could not detect a response by male guppies to a suspension–solution of an estrogen (17β-estradiol) added as one of their test solutions. Nevertheless, there are indications for a number of species that the sex pheromone is ether soluble and, therefore, may be a steroid or a lipid (Partridge *et al.*, 1976; Honda, 1979, 1980a). Van den Hurk *et al.* (1982) prepared extracts of the ovary of the zebrafish, *Brachydanio rerio*, and concluded that the attractant is contained in a steroid–glucuronid fraction. Colombo *et al.* (1982) have recently shown that male guppies and goldfish are attracted to water bearing etiocholanolone-3-glucuronide (i.e., the major 5β-reduced androgen conjugate identified in the mesorchial gland of the male *Gobius jozo*). Rubec (1979) attempted to fractionate and identify the active constituents in the urine of *Ictalurus melas* and concluded that at least two pheromones are present: one lipid and the other proteinaceous. Algranati and Perlmutter (1981) extracted intrasexual attractant from tank water holding zebrafish and identified the active constituent as a cholesterol ester.

Little is known of the control of pheromone production and release in female fish. It is likely that these events are under endocrine control, but this has been examined experimentally in only a small number of cases. In addition to Meyer and Liley's study referred to previously, Yamazaki and Watanabe (1979) were able to induce femalelike attractiveness in hypophysectomized male goldfish by treatment with estrogen.

C. Pheromones: Summary and Discussion

Data from numerous studies point to the existence of some form of chemical mediation in the reproductive behavior of a variety of species of fish. Most experimental investigations simply confirm that a chemical product of one individual elicits an approach or, in some instances, a more specific sexual response in another individual. In the majority of cases the chemical acts as a "releaser"; a few chemicals with "priming" effects have also been detected. For the most part, the understanding of the functional significance of these chemically mediated interactions is severely limited by the restrictive experimental contexts in which they have been investigated. Nevertheless, it is likely that chemical signals do play an important role in reproduction by facilitating orientation and arousal and thereby ensuring both physical and physiological synchronization of potential partners. Perhaps for many species chemical communication occurs mainly in the preliminary phases of reproduction; more complex responses depend on other sensory modalities. Chemical communication is likely to play a major role and is perhaps the "dominant" form of communication among fish that are active at night or in turbid waters.

It should be emphasized that for most of the observed chemically mediated interactions it is not clear whether one is dealing with "pheromonal communication" in the generally accepted sense (see Liley, 1982, for fuller discussion). The questions remain as to whether the chemical observed to evoke a specific response is a "pheromone" (a discrete chemical signal which has evolved as a component in the species' communication system) or, whether the chemical is simply a metabolic product of one individual which elicits responses in conspecifics in much the same way that various abiotic physical and chemical stimuli elicit specific adaptive responses. For example, as a result of the temporal contiguity of ovulation, the release of ovarian metabolic products associated with ovarian maturation, and behavioral receptivity, selection may have favored an enhanced responsiveness in males to the ovarian products of conspecific females, without any corresponding specialization for signaling the male in the female. The male evidently perceives cues which identify a female's condition. However, the unanswered question is: Does the female signal?

Raising this question in no way diminishes the role of such chemicals in reproductive behavior, but simply directs attention to the many unanswered questions regarding chemical "communication." In particular there is a need for more careful investigation of the source and nature of the chemical products affecting behavior. With relatively few exceptions, notably the testicular gland and specialized product of the black goby (Colombo *et al.*, 1980) and the distinctive glands of glandulocaudine males (Atkins and Fink, 1979) and blennies (Laumen *et al.*, 1974), there is little clear evidence of evolutionary specialization in the synthesis and release of the supposed chemical signals.

Even less is known of the mechanisms involved in the regulation of pheromone production and the relationship between supposed pheromones and the endocrine system. Fragmentary and largely circumstantial evidence implicates the endocrine system in both the control of pheromone production and in the maintenance of behavior responsiveness to chemical signals. Furthermore, it is likely that some pheromones may be either hormones themselves or derived endocrine products (Colombo *et al.*, 1980, 1982; van den Hurk *et al.*, 1982).

V. REPRODUCTIVE BEHAVIOR

As already mentioned, patterns of reproductive behavior among teleost fishes range from the simple release of gametes in the proximity of conspecifics to complex sequences which may include defence and preparation of a nest site or territory, pair formation, and spawning. In some groups, fertilization is internal and results in a release of fertilized eggs (*Trachycoristes*, von Ihering, 1937; *Corynopoma*, Kutaygil, 1959), larvae (*Sebastodes*, Moser, 1967), juveniles (*Poeciliidae*, Turner, 1947), or even sexually mature offspring (*Cymatogaster aggregata*, Wiebe, 1968). In many oviparous species, eggs and young may be protected and cared for, and in some cases provided with nourishment.

It is not surprising that such a variety of reproductive behaviors and associated specializations has created problems in terminology. In the following discussion, reproductive behavior is used as a general term to encompass all activities involved in reproduction. Sexual behavior is restricted to any behavioral interaction between the sexes leading to the union of gametes. In the case of externally fertilizing species, it is important to distinguish between prespawning and spawning behaviors. Prespawning behaviors includes sexual activities, often referred to as courtship, involved in the search for, and attraction and excitation of, a potential sexual partner. However, prespawning behavior may also include nonsexual responses such as those concerned with the preparation and defence of a nest site or territory. The term spawning is restricted to those motor patterns by which males and females directly synchronize their behavior to achieve a coordinated release of gametes (e.g., oviposition and release of milt). For internally fertilizing species the terms corresponding to prespawning and spawning are premating or courtship behavior and mating (copulation). Release of eggs or young by internally fertilizing species is referred to as oviposition and parturition, respectively. Parental behavior refers to any postspawning or postmating care of eggs or young.

A. Male Reproductive Behavior

The clear correlation between gonadal cycles, levels of gonadal steroids, and the appearance of reproductive behavior suggests that gonadal hormones play a major causal role in the appearance and development of reproductive behavior. However, in view of the diversity of teleostean reproductive behavior, it should not be surprising if different components of the breeding repertoire prove to be governed by different causal agents. Indeed, a particular source of confusion has been the fact that in many species aggressive behavior occurs as an integral part of reproductive behavior. Aggressive actions may be components in the sexual responses, or a major feature of the defence of territory, nest, brood, or mate. However, aggression also occurs in a number of nonreproductive contexts, including competition for food, space, or status in a dominance hierarchy. Furthermore, females and juveniles may also display aggressive behavior in similar nonreproductive situations. Therefore, one would expect that the causal basis underlying aggressive behavior may vary in different functional contexts. In assessing the behavioral effects of castration or hormone therapy, it is important to identify the nature of the aggressive behavior observed. The persistence of aggressive behavior after castration should not in itself be taken as a reliable indication that reproductive behavior as a whole occurs independently of gonadal hormones.

Investigations have involved a variety of approaches: fish have been treated with gonadal and pituitary hormones, with or without prior gonadectomy. In addition, chemical blocking agents such as antigonadotropins, steroid enzyme inhibitors and steroid antagonists have also been applied, often in combination with other endocrine treatments. For the most part, investigators have sought effects specific to one sex or a particular behavior. In some of the early studies, more general effects of hormone treatments were described (Liley, 1969).

1. Prespawning Behavior

Most hormone-behavior studies have concentrated on species which have elaborate prespawning behavior, e.g., Gasterosteids, Cichlids, Belontiids, and Centrarchids. These are groups in which the male prepares and defends a nest site and may also care for the eggs and young after spawning.

The threespined stickleback, *Gasterosteus aculeatus*, has been the subject of intensive ethological and endocrinological study (reviewed by Wootton, 1976). Several researchers have found that courtship and nest-building behaviors disappear rapidly after castration (Hoar, 1962a,b; Baggerman, 1957, 1966, 1969; Wootton, 1970). Treatment with an antiandrogen, cyproterone acetate, delayed the onset of breeding in winter-condition fish; however, in spring, at the early stages of the breeding cycle, the same treatment caused a reduction in sexual and aggressive behaviors, but nest maintenance was not affected (Rouse *et al.*, 1977). These results suggest that cyproterone acetate may be only weakly antiandrogenic under these conditions; nevertheless, the data provide some confirmation of the role of androgens in the control of sexual behavior.

Replacement therapy is highly effective in restoring secondary sexual characters and reproductive behavior in castrated sticklebacks (Hoar, 1962a,b; Wai and Hoar, 1963). However, it is of interest that the effectiveness of androgen treatment appears to depend on photoperiod: A greater proportion of castrated males maintained under long photoperiod built nests and with a shorter delay after receiving androgen, than males held under short photoperiod (Hoar, 1962b). The apparent refractoriness of androgen-treated fish under short photoperiod led Hoar to suggest that, although reproductive behavior requires gonadal hormone, its full expression occurs only when gonadotropic activity of the pituitary is maintained at a high level by a long photoperiod.

Aggressive behavior appears to be less dependent on the presence of gonadal androgen. Depending on the season and photoperiod conditions, aggressive behavior may persist at a high level after gonadectomy (Hoar, 1962a,b; Baggerman, 1966; Wootton, 1970). Hoar (1962a) and Baggerman (1966) speculated that there is a seasonal shift in the causation of aggressive behavior: before the onset of breeding, aggressive behavior is regulated by increasing levels of pituitary gonadotropin as the fish responds to increasing photoperiod; gradually the mechanism underlying aggressive behavior becomes less sensitive to gonadotropin and becomes increasingly controlled by gonadal hormones. However, it should be noted that in Hoar's investigations (1962,a,b), there was no attempt to distinguish between nonreproductive aggression and aggression in defence of a breeding territory or nest site (see also Wootton, 1970).

There is only limited experimental support for the proposal that the pituitary is directly involved in the causation of aggressive behavior. In experiments designed to measure the effects of mammalian pituitary hormones, only treatment with LH consistently produced an increase in aggressive behavior of males held under short photoperiod (Hoar, 1962a). [Ahsan and Hoar (1963) also found LH to be the most effective mammalian hormone in stimulating gonadal development in immature fish.] Males maintained in long photoperiod and treated with a gonadotropin-blocking agent, methallibure, showed a decrease in aggression (Carew, 1968).

Research on cichlid fish has provided conflicting results. On the one hand, Noble and Kumpf (1936), Aronson (1951), Aronson *et al.* (1960), and Heinrich (1967) claimed that mating and/or nest-digging behaviors persist after castration in *Hemichromis bimaculatus*, *Sarotherodon macrocephala*, *Aequidens latifrons*, and *S. heudeloti* and *S. nilotica*, respectively. In contrast, Reinboth and Rixner (1970) reported that sexual, aggressive, and nest-digging behaviors were abolished or reduced by castration of *Hemihaplochromis multicolor*.

Studies in which fish have been treated with exogenous androgens suggest that gonadal hormones are normally involved in the maintenance of reproductive behavior. Reinboth and Rixner (1970) noted that testosterone therapy restored male coloration and behavior in castrated *Hemihaplochromis multicolor*. Females treated with testosterone acquired male coloration, established territories, dug a pit, and demonstrated sexual behavior. Similarly, females of *Haplochromis burtoni* could not be distinguished from normal males in appearance and behavior after androgen treatment (Wapler-Leong and Reinboth, 1974). Clemens and Inslee (1968) obtained functional sex reversal in genetic females of *Sarotherodon mossambicus* by treating them with methyltestosterone for the first 69 days of life. The sex-reversed fish exhibited male coloration and nest-building behavior when placed with ripe females. This result has been confirmed by Billy (1982) who also demonstrated that maximum sex reversal occurred in groups of fish treated with methyltestosterone in the first 21 days after release from the female's mouth. Billy (1982) examined the behavior of females sex reversed as juveniles or treated with hormone as adults. In the former group, genetic females developed as males and performed the full repertoire of male courtship patterns. However, both males and females receiving androgen early in development were consistently more aggressive as adults compared with untreated males. Females treated for 40 days as adults performed a number of male-typical displays, but were not as responsive to the hormone as those females given a non-sex-reversing treatment as juveniles and then treated as adults. Evidently, an early exposure to exogenous androgen, even though insufficient to cause sex reversal, sensitizes the fish to a subsequent treatment.

Fernald (1976) injected testosterone into intact adult male *Haplochromis burtoni*. Approach and attack directed toward blinded (and therefore unresponsive) target fish of another species increased markedly. There were no significant changes in courtship, nest building, or other activities. Fernald argues that approach is a pivotal behavioral act which, depending on the sex and behavior of the fish approached, is followed by attack (males and juveniles) or courtship (mature females). The increase in approach after testosterone treatment is interpreted as an indication of an increase in sexual motivation: Failure to perform courtship results from the lack of appropriate stimuli in the target fish.

Heiligenberg and Kramer (1972) demonstrated that social isolation results in a decrease in aggressivity of males of *Haplochromis burtoni*. Hannes and Franck (1983) found what appears to be a corresponding decrease in plasma testosterone and corticosteroids in males of the same species isolated for 2 months.

Schwanck (1980) examined the relationship between endocrine condition and territorial aggressive behavior in young males of *Tilapia mariae*. Schwank argued on the basis of the findings of Aronson (1951) and Levy and Aronson (1955) that the size of the genital papilla may be used as an indicator of hormonal state. In a series of paired encounters, the fish with the larger papillae won most of the encounters. Body size was a decisive factor only if genital papillae were equal or nearly equal. These results suggest that aggressive behavior is to some extent governed by endogenous androgen levels. However, as Schwank observes, it is clear that not all aggression and territorial behavior can be regarded as male reproductive and therefore androgen controlled. Juveniles and mature females can behave aggressively and, in some cases, hold territories, perhaps in competition for food.

Several species of Belontiid fish have been examined experimentally, and, as in the studies of cichlids, there have been claims that reproductive behavior does not depend on the continued presence of the testes (Noble and Kumpf, 1936; Forselius, 1957). However, Johns and Liley (1970) found that 11 of 16 castrated male blue gouramis, *Trichogaster trichopterus*, failed to build nests or court females. After treatment with methyltestosterone, these males performed the full range of reproductive behavior including care of the unfertilized eggs resulting from spawning. The remaining untreated castrates built nests and spawned within 7 days of being paired with mature females. Although unable to detect traces of regenerating testes in the latter group of castrates, Johns and Liley concluded that nest-building and sexual behavior were probably maintained by either an unidentified fragment of regenerating testicular tissue or an extragonadal source of androgen. This conclusion was supported by the fact that the dorsal fin of the spawning "castrates" remained long (characteristic of an intact male, and shown to be androgen dependent), whereas the dorsal fins of the nonspawning fish became shorter and rounder and more similar to that of a female.

In an investigation involving the paradise fish, *Macropodus opercularis*, Villars and Davis (1977) observed a marked decline in male sexual behavior 1 week after castration, whereas nest-building activity was unaffected. The decrease in sexual behavior was prevented by treatment with testosterone. One week later the sexual responsiveness of untreated castrates increased, paralleling the regeneration of the testes. Regeneration could be prevented by an antigonadotropin, methallibure, but the antigonadotropin had no effect on intact males. This result confirmed an earlier finding (Davis *et al.*, 1976) that persistence of sexual behavior after treatment with antigonadotropin was probably attributable to a recovery in endogenous hormone production.

Johns and Liley (1970) concluded that nestbuilding is regulated by gonadal hormones. In apparent contrasts, Villars and Davis (1977) found that, unlike sexual behavior, nest building was not affected 1 week after castration. However, as regeneration is known to have occurred within 2 weeks, this result could simply be interpreted as an indication that nest building requires a lower level of endogenous androgen than sexual behavior for its maintenance. That gonadal hormone is involved in the regulation of nest building is confirmed by D.L. Kramer's (1972) observation that treatment with methyltestosterone results in nest building in female blue gouramis. In *Colisa lalia*, a related species, androgen treatment was also reported to induce nest building behavior in females (Forsélius, 1957). Machemer and Fiedler (1965) and Machemer (1971) found that a combination of androgen and prolactin was more effective than methyltestosterone or prolactin alone in causing nest-building behavior in female paradise fish; prolactin probably serves to stimulate the secretion of the mucus which is the important constituent in the foam nest.

The investigations previously cited lead to the conclusion that sexual and nest-building behaviors of Belontiids are regulated by gonadal hormones. Although defence of a nest site was also eliminated by castration (Johns and Liley, 1970), nonterritorial aggressive behavior still occurred. Castrated males placed with intact or other castrate males performed agonistic behavior until a dominance relationship was established. The agonistic behavior of castrates did not obviously differ qualitatively or quantitatively from that of intact males (Johns and Liley, 1970). Villars and Davis (1977) obtained a similar result with paradise fish. Furthermore, it is clear that in certain circumstances, females also perform a full range of aggressive behavior. Davis and Kassel (1975) reported their own and others' observations that in *Macroprodus opercularis* the two sexes share qualitatively similar threat, attack, and submissive behaviors. However, there are quantitative differences, for example, males perform lateral displays more frequently than females. Davis and Kassel (1975) found that aggressive behavior appears in juveniles well before the gonads are functional, and increases in both males and females with the growth and maturation of the testes and ovaries. Sexual differences in aggressive behavior only become apparent in adult fish.

The conclusion that nonterritorial aggressive behavior may be largely independent of gonadal hormonal control is supported by data from studies by Weiss and Coughlin (1979). They could find no difference in the aggressive behavior of fighting fish, *Betta splendens*, when intact fish were compared with either castrates with regenerating testes or with castrates without detectable regeneration.

Smith (1969) examined the effect of castration on the agonistic and reproductive behavior of two species of centrarchid sunfish, *Lepomis megalotis* and *L. gibbosus*. Nest digging came to a halt after castration and was restored by testosterone treatment. These results provide a clear indication that nest building, and perhaps other reproductive behaviors (sexual responses toward females were not examined) are under gonadal hormonal control. Avila and Chiszar (1972, in Henderson and Chiszar, 1977) observed nest-digging and rim-circling behavior by female bluegill sunfish, *L. macrochirus*, after treatment with methyltestosterone. In apparent contradiction to these findings, B. Kramer (1971, 1972, 1973) concluded that in *L. gibbosus* sexual behavior is controlled directly by gonadotropic hormone. Evidence for this comes from two sources. First, sexual behavior persisted in males receiving high doses of an antiandrogen, cyproterone acetate, but declined in males receiving androgen. B. Kramer (1971) proposed that in the latter case the decrease in sexual behavior is a consequence of a reduction in gonadotropin secretion (release) as a result of a negative-feedback effect of androgen (central inhibition). Second, treatment with methallibure, an antigonadotropic agent, suppressed sexual behavior within 5 days, whereas mammalian luteinizing hormone induced a marked increase in sexual activity in males pretreated with methallibure (B. Kramer, 1972, 1973).

Nest-building behavior disappeared completely after cyproterone acetate, but it was not inhibited initially by testosterone (B. Kramer, 1971). After about 1 week of androgen treatment, a decline in digging occurred, which Kramer attributed to a decrease in endogenous androgen resulting from an inhibition induced by exogenous testosterone. A mammalian gonadotropin (LH) stimulated nest-digging behavior in centrally inhibited fish; methallibure caused a decrease (B. Kramer, 1971, 1973). Kramer (1971) concluded that nest digging is stimulated by androgen, but that gonadotropin also has a motivating influence.

In his first experiment, Smith (1969) observed that castrated males of *L. megalotis* and *L. gibbosus* resident in small aquaria and separated by glass partitions maintained high levels of aggressive behavior similar to those of intact males. However, Smith (1969) noted that under those experimental conditions it was impossible to distinguish between territoriality and aggression devoid of topographical reference. In a second experiment in which fish were released into a large pool, none of the castrated males built nests, and there was a marked decrease in aggressive behavior. After testosterone treatment, castrated males built nests scattered throughout the pool instead of rim-to-rim in shallow water as in intact males. Surprisingly, testosterone did not raise aggressive activity in castrates—perhaps because the nests of these fish were more widely spaced than in intact fish. Furthermore, although treatment with human chorionic gonadotropin (HCG) was effective in stimulating nest-building behavior in *Lepomis* males (presumably by stimulating androgen secretion) it did not affect aggressive behavior (Smith, 1970). However, aggressive behavior remained high in males held under short photoperiod provided that the temperature remained high (25°C), but aggressiveness declined when males were kept under a short photoperiod at low temperature (13°C). Smith concluded that in these species, aggressive behavior is not dependent on androgen or gonadotropin levels but is influenced more by water temperature and social conditions.

B. Kramer (1971) arrived at a different conclusion. Because males of *L. gibbosus* treated with the antiandrogen cyproterone acetate remained considerably more aggressive than those receiving testosterone, Kramer suggests that gonadotropin is directly involved in the control of aggression. There was some recovery of aggressive behavior in centrally inhibited males after a further injection of testosterone. Opercular spreads and leading, two behaviors believed to indicate a conflict between aggressive and sexual tendencies, decreased significantly after treatment with methallibure (B. Kramer, 1971, 1973). Injection of testosterone or mammalian LH into fish pretreated with methallibure caused a significant increase in aggressive behavior and opercular spreads. Kramer (1971) concluded that aggressive behavior in the sunfish is regulated by a synergistic action of gonadotropin and androgen.

B. Kramer (1972) also noted that treatment with reserpine produced a rapid increase in aggressive behavior in males, but leading (a sexual behavior) decreased. Because reserpine is believed to deplete the intraneural storage of catecholamines especially norepinephrine, Kramer proposed that, although LH and androgen exert a long-term control of reproductive behavior, the short-term control of aggression and nest-building behavior may be mediated by catecholamines. Chlorpromazine, which inhibits the action of released norepinephrine, depresses both aggressive behavior and nest-building behavior, although sexual responsiveness remains high.

The investigations by Smith (1969, 1970) and B. Kramer (1971, 1972,

1973) emphasized the need to distinguish between aggressive behavior involved in prespawning behavior, in particular nest-site defence, and in nonreproductive aggression. Smith's results indicate that reproductive aggression is under androgen control but that nonreproductive aggression may be affected by a variety of causal factors. Indeed, several authors have noted that, although defence of a nest is limited to sexually mature sunfish males, immature fish of both sexes perform agonistic behaviors (Greenberg, 1947; Hale, 1956). Henderson and Chiszar (1977) explored the effect of size and sex of the resident and of intruder on aggressive behavior of the bluegill sunfish, *L. macrochirus*, in fish previously established in isolation in an aquarium. All fish were in winter-nonbreeding condition. The sex of the intruder or resident had no effect on resident aggressiveness.

In an investigation of the effects of social isolation on aggressive responses of fish in reproductive condition, the sexes did not differ (Chiszar *et al.*, 1976). However, there were seasonal differences which indicate that the effects of social isolation interact with the reproductive condition of the fish: fish captured in November exhibited peak frequencies of social-aggressive responses after 7 days isolation, but fish tested in the breeding season displayed the maximum response after only 1–3 days.

A puzzling result was obtained by Tavolga (1955) who studied the frillfin goby, *Bathygobius soporator*. Castration abolished aggressive behavior but courtship remained. However, gonadectomized males no longer discriminated between males and females, or between gravid and nongravid females, but courted all equally. Spawning behavior of castrated males with gravid females appeared to be normal, and the male brooded infertile eggs resulting from the spawning. Tavolga (1956) suggested that the loss of endogenous androgen may result in an alteration of the olfactory sensitivity mechanism underlying mate discrimination in intact males. Hypophysectomy was followed by complete loss of sexual, territorial, and agonistic behaviors; this indicated that pituitary hormones exert a direct influence on sexual behavior.

Only relatively recently has attention been directed toward species in which reproductive behavior is limited to a brief pairing and spawning (as in many Cyprinids and Characids). One such approach is that of van den Hurk (1977) who explored the endocrine control of reproductive behavior of male zebrafish, *Brachydanio rerio*. Cytochemical investigation of the testes indicated an increased steroid-synthesizing capacity during the prespawning agonistic and courtship stages. Inhibition of steroid synthesis by administration of 17β-estradiol caused a reduction in reproductive activity. Treatment with androgens suppressed steroid synthesis but maintained agonistic and courtship behavior. In an apparent contradiction to this finding, van den Hurke *et al.* (1982) discovered that castration does not eliminate the male sexual response. Further, von den Hurk and co-workers proposed that the disappearance of immunocytochemically demonstrable gonadotropin in the pituitary during the prespawning agonistic stage may indicate a role for gonadotropin in the control of reproductive behavior. R. van den Hurk (personal communication) noted the difficulty of detecting traces of regenerating testes and commented that only those fish in which traces of testicular tissues were not detectable were included in the experiments (5% of all animals castrated).

Studies with goldfish, *Carassius auratus* (Partridge *et al.*, 1976), another species in which reproductive behavior is relatively "simple," indicate that responsiveness to sexual pheromones produced by the female is governed by the male's endocrine state: unspermiated male goldfish failed to respond to the pheromone, but spermiated males (milt may be easily expressed by pressure on the abdomen) did react. However, Partridge and co-workers also noted that spermiated males exhibited an increased sensitivity to food odor, suggesting that perhaps the increased response to a pheromone reflects a general increase in olfactory responsiveness induced by physiological changes associated with spermiation. Work by Goff (1979) provides some support for this suggestion. Recording from the olfactory bulb, Goff found a nonspecific increase in responsiveness to odors in spermiated male goldfish. Earlier investigations by Oshima and Gorbman (1968, 1969) and Hara (1967) established that administration of sex steroids to goldfish augmented the response of the olfactory epithelium to chemical stimulation. The effects induced by sex hormones involved changes in amplitude and patterns of response rather than a change in threshold.

These findings indicate that responsiveness of male goldfish to sexual stimuli is governed by gonadal hormones. Kyle (1982) provided evidence that androgens are also responsible for the maintenance of motor responses involved in courtship and spawning. Kyle treated intact sexually inactive males with a number of steroids and antisteroids. All males implanted systemically with pellets of testosterone spawned within 5 days. Intraperitoneal injections of androgens were far less effective. The proportion of males spawning after injections of testosterone propionate, 11-ketotestosterone, or estradiol benzoate, did not differ significantly from that of control fish, although there was an indication of a stimulatory effect in the case of ketotestosterone.

Kyle (1982) and Kyle *et al.* (1982b) reported a transitory increase in circulating gonadotropin (GtH) levels after exposure to sexual stimuli (either a receptive female or a pair of spawning goldfish). Serum levels of GtH had already increased within 20 min at 20°C, or 1 hr at 14°C, and declined to control levels after 2 hr. The volume of milt that could be hand-stripped also rose after 1- to 24-hr exposure to sexual stimuli. Males separated from spawning pairs by clear glass or perforated partitions failed to demonstrate an increase in GtH or milt volume—a clear indication that contact with spawning fish is essential to the response. The coincidence of the surge in GtH and milt volumes coincides with the onset of courtship and suggests that these events share a common mechanism (Kyle, 1982). A dramatic surge in GtH at the time of spawning has also been detected in males of the white sucker, *Catostomus commersoni* (Mackenzie *et al.*, 1982) taken from natural spawning beds in Alberta.

The question arises then as to whether a surge in GtH plays a causal role in the onset of spawning activity. Numerous studies in which adult fish have been treated with fish and mammalian gonadotropic hormone preparations give no grounds for suggesting that GtH plays a direct causal role in the onset of spawning. It is more likely that GtH (or perhaps releasing factor) acts indirectly, perhaps through the induction of an increase in milt volume which in turn affects neural or endocrine mechanisms, or by mediating other endocrine changes which in turn play a more direct role. Clearly at this stage any interpretation of these findings is highly speculative; nevertheless, it does emphasize the need for more intensive analyses of the short-term endocrine changes associated with the spawning process.

The weakly electric fish *Sternopygus dariensis* emits electric discharges which are sexually dimorphic: The male discharges at a lower frequency than the female (Meyer, 1983). Experimental work indicates that this dimorphism is used in sexual discrimination (Hopkins, 1972). Androgen treatment causes a decrease in discharge frequency in both males, females, and juveniles, suggesting that the naturally occurring lower discharge of males may be the result of endogenous androgen levels. Dihydrotestosterone caused a much greater decrease than testosterone; estradiol was either without effect or in some cases caused a slight increase in discharge frequency. Two other species, one having much less pronounced sexual dimorphism in discharge frequency, the other sexually monomorphic, displayed a decrease in frequency after treatment with androgen. Meyer (1983) proposed that androgens may exert an effect on a medullary pacemaker nucleus which controls the discharge frequency. Meyer and Zakon (1982) also detected a decrease in the "tuning" of the electroreceptors of *Sternopygus*, which paralleled the decrease in discharge frequency after androgen treatment.

Working with the medaka *Oryzias latipes*, Okada and Yamashita (1944, in Yamamoto, 1969) demonstrated that testosterone treatment of females or implantation of a testis results in masculinization and the performance of male behavior, including pursuit of normal females. Yamamoto (1969) confirmed that complete functional sex reversal may be achieved with androgen treatment of genetic female medakas. Unfortunately, there has been no detailed comparison of the behavior of normal males and sex-reversed genetic females.

The ovoviviparous Poeciliids are well known for the conspicuous premating behavior of the males. Persistent courtship is accompanied by frequent mating attempts, few of which are successful (Liley, 1966). Castration of the male platyfish, *Xiphophorus maculatus*, resulted in a significant decrease in (but not disappearance of) all courtship and insemination activities except backing (Chizinsky, 1968). Aggressive nipping also decreased. Pecking, probably not significant in reproduction, remained unaffected. To account for the persistence of sexual behavior, Chizinsky proposed that the expression of sexual behavior in the adult may be relatively independent of gonadal control and is perhaps governed by the forebrain.

Other, less direct, evidence suggests that sexual behavior in poeciliids is governed by gonadal hormones. In particular, androgen-induced sex reversal in female poeciliids has been demonstrated many times (Schreck, 1974; Lindsay, 1974), and numerous researchers have reported that sex-reversed genetic females readily perform male sexual behavior. However, it is of interest that in three studies (Tavolga, 1949; Laskowski, 1954; Clemens *et al.*, 1966) there are indications that masculinized females court less vigorously and are less successful in mating than normal males, suggesting that although androgens are effective in masculinizing females, genetic factors may not be completely overridden. Lindsay (1974) conducted a careful examination of this particular aspect and found that masculinized guppy females performed fewer displays and spent less time displaying than normal males; the frequency of gonopodial contacts (or attempts) remained the same.

Aggressive behavior is also governed to some extent by androgen levels. Noble and Borne (1940) observed that females of *Xiphophorus helleri* treated with testosterone propionate rose in the pecking order until a reversal in sexual behavior occurred. Franck and Hannes (1979) found a positive correlation between serum testosterone levels and intensity of aggression directed toward a smaller opponent behind a transparent barrier. Four weeks of social isolation resulted in a marked decrease in both androgens and corticosteroids, and in aggression to a smaller opponent (Hannes and Franck, 1983). However, isolated males proceeded to much higher levels of aggression when they were confronted with one another in contests for rank-order position. Interestingly, Hannes *et al.* (1983) found a decrease in testosterone levels 20 min after an aggressive encounter, with those of the losers being significantly lower than the levels in the winners. Testosterone increased to a level above that of controls 72 hr after the encounter. Evidently, hormone levels do not return asymptotically to control level, but go through an oscillation which is not complete 72 hr after the encounter. The decrease in androgen following an encounter was accompanied by a dramatic increase in adrenal corticoids. Hannes *et al.* (1983) proposed that the decrease in androgen immediately after an aggressive encounter is the result of stress.

Although the foregoing studies implicate gonadal hormones in the regulation of aggressive behavior in male poeciliids, it is equally clear that aggressive behavior is not completely dependent on gonadal hormones. Chizinsky (1968) noted that aggressive displays persisted, but they occurred at a lower level after castration of male *Platypoecilus maculatus*. Furthermore, females may also perform aggressive behavior and establish dominance (Braddock, 1945; Laskowski, 1954).

2. SPAWNING BEHAVIOR

Although gonadal hormones appear to be involved in the "long-term" maintenance of reproductive behavior including nest preparation and defence, and courtship responses, neurohypophysial hormones have been implicated in the short-term control of the spawning act in a small number of species.

Involvement of neurohypophysial hormone in the spawning behavior of the killifish, *Fundulus heteroclitus*, was first proposed by Wilhelmi *et al.* (1955) when it was discovered that intraperitoneal injections of large doses of mammalian neurohypophysial hormone preparations induced a "spawning reflex response." This response occurs in gonadectomized–hypophysectomized fish of both sexes and is not preceded by pair formation or any distinct prespawning behavior (a prominent feature of "normal" spawning). Comparable results have been obtained with females of *Oryzias* (Egami, 1959a) and *Rhodeus* (Egami and Ishii, 1962) and male and female *Jordanella floridae* (Crawford, 1975). Blüm (1968) observed an expansion of melanophores and the performance of the spawning reflex in immature *Pterophyllum scalare* following injections of reserpine. Blum proposed that these responses were mediated by pituitary hormones, i.e., melanophore stimulating hormone (MSH) and the neurohypophysial hormone ichthyotocin. Neurohypophysial hormones have no apparent effect on several species tested: goldfish (Pickford, in Macey *et al.*, 1974; Stacey, 1977), *Misgurnus fossilis* and *Salmo* (Egami and Ishii, 1962), *Gasterosteus aculeatus* (T. J. Lam and Y. Nagahama, personal communication), and *Heteropneustes fossilis* (Sundararaj and Goswami, 1966). Injections of oxytocin or isotocin failed to stimulate spawning behavior in male sea horses, *Hippocampus hippocampus*, but complete parturition movements were induced by the treatment even though the brood pouch was empty (Fiedler, 1970).

Treatment of *Fundulus* with teleost neurohypophysial hormones, arginine vasotocin and isotocin, confirmed the earlier results obtained with mammalian preparations (Macey *et al.*, 1974) and revealed that arginine vasotocin was the more potent of the two principal components (Pickford and Strecker, 1977).

Macey *et al.* (1974) found that destruction of the nucleus preopticus reduced or eliminated the reflex response to exogenous neurohypophysial hormone preparations. They suggested that the nucleus preopticus is involved in the spawning behavior of the killifish and that neurohypophysial hormones exert their effect by their action on the nucleus preopticus. However, a more recent investigation casts doubt on that interpretation. Peter (1977) and Pickford *et al.* (1980) reported that arginine vasopressin injected directly into the third ventricle of the brain was no more effective in eliciting the reflex response than intraperitoneal injections; this suggests that the hormone exerts its effect through a peripheral action. Peter (1977) concluded that, in view of the large doses normally required to elicit a spawning-reflex response, it appears that the activation of a peripheral receptor by neurohypophysial hormones is probably not part of the normal mechanism for triggering spawning behavior in teleosts. However, the possibility remains that neurohypophysial hormones may be involved via their ability to stimulate oviduct and ovarian smooth muscle in teleosts (Heller, 1972). In this regard, it is perhaps significant that the three species in which neurohypophysial hormones have the most striking effect, *Fundulus*, *Oryzias*, and *Jordanella*, are all killifishes of the family Cyprinodontidae. A characteristic of this group is that during a breeding season they may spawn daily for several days or even weeks; the female deposits relatively few eggs at a time. In *Jordanella*, in which the female "places" the eggs individually or in small groups, this type of oviposition appears to be associated with the presence of a large muscular oviduct (Crawford, 1975). In view of the ability of neurohypophysial hormones to stimulate oviduct and ovarian smooth muscle in teleosts it may be possible that these hormones induce spawning-type responses through their effects on oviduct and ovarian smooth muscle, and perhaps through effects on comparable muscular tissue in the male.

3. PARENTAL BEHAVIOR

Postspawning care of eggs and young is of widespread occurrence among the teleosts. Evidence of a gonadal involvement in the maintenance of parental behavior comes from only two species. Castration of male threespine stickleback early in the parental phase resulted in a decline in fanning, indicating that parental fanning is maintained by a testis hormone (Smith and Hoar, 1967). In the blue gourami, *Trichogaster trichopterus*, presentation of batches of eggs sufficient to induce parental responses in nonspawning males did not evoke parental behavior in females, although a few females did retrieve eggs occasionally (D. L. Kramer, 1972). After treatment with methyltestosterone, several females began to perform parental behavior in response to the presentation of large clutches of eggs.

Following Fiedler's (1962) demonstration that treatment with mammalian prolactin induced parental-type fanning in the wrasse, *Crenilabrus ocellatus*, a number of investigations have suggested that a prolactin-like hormone may be involved in the regulation of parental behavior in fish. Injection of low doses of mammalian prolactin induced parental fanning in the discus fish, *Symphysodon aequifasciatus axelrodi*, and the angel fish, *Pterophyllum scalare* (Blüm, 1974). These behavioral effects reached a maximum 48–72 hr after injection. High doses of prolactin inhibit fanning. Parental fanning was not detected after prolactin treatment in several other species of cichlids tested (*Aequidens latifrons, Cichlasoma severum,* and *Astronotus ocellatus*) (Blüm and Fiedler, 1965), but a decrease in aggression, a reduction in feeding, and a tranquillizing effect were observed. In *Aequidens latifrons* prolactin induced digging, a behavior normally observed during parental care (Blüm, 1974). Injection of L-dopa, which is believed to act as a prolactin-inhibiting factor, caused a reduction in parental fanning by female *Cichlasoma nigrofasciatum* (Fiedler *et al.*, 1979). Similarly, L-dopa and apomorphine suppressed parental calling behavior in pairs of *Hemichromis bimaculatus*.

Blüm and Fiedler (1974) identified prolactin sensitive neurons in the forebrains of *Lepomis gibbosus, Astronotus ocellatus,* and *Tilapia mariae;* species in which parental fanning is present. In contrast, in a number of mouthbrooding tilapias and in goldfish and *Idus idus,* which show no parental care, prolactin had only a slight effect on forebrain activity.

Another HPG component has been implicated in parental behavior. Thyrotropin-releasing hormone (TRH) caused either an increase or a decrease in parental fanning of females of *C. nigrofasciatum;* what occurred depended on the hormonal status of the specimens (Fiedler *et al.*, 1979). Whether this was a direct effect of TRH or mediated through an effect on prolactin secretion is not clear.

Prolactin also stimulates an increase in the number of epidermal mucus secreting cells. This effect is particularly pronounced in the discus fish in which the secretions normally provide supplementary nutrient for the young (Blüm and Fiedler, 1965; Blüm, 1974). These same changes in behavior and mucus secretion were induced in *Symphysodon* by paralactin a "prolactin" of teleostean origin (Blüm, 1974).

Smith and Hoar (1967) were unable to find any evidence of a role for prolactin in the regulation of parental fanning in threespine sticklebacks. However, Molenda and Fiedler (1971) observed that low doses of prolactin caused an increase in fanning in male sticklebacks, but high doses similar to those used by Smith and Hoar inhibited fanning in males with nests.

4. CONCLUSION: HORMONES AND BEHAVIOR IN MALE FISH

Numerous investigations have demonstrated the effectiveness of exogenous androgens in causing the development of secondary sexual characteristics and the appearance of male reproductive behavior in intact or castrated males, juveniles or females. These results leave little doubt that androgens play a primary role in maintaining all aspects of male reproductive behavior including (as appropriate) territorial defence, preparation of a nest site, spawning or mating, and parental care.

Presently, experimental studies do not allow the identification with any certainty of which of the naturally occurring gonadal steroids is the androgen most directly concerned in the regulation of behavior of untreated fish. Only Kyle (1982) has compared the behavioral effectiveness of ketotestosterone with that of testosterone, although a number of researchers (Idler *et al.*, 1961; Arai, 1967) have suggested that ketotestosterone may be more potent in the induction of morphological sexual characteristics.

The results of castration have been highly variable and frequently contradictory. Castration is claimed to eliminate reproductive behavior in some species but not in others. Nevertheless, in view of the consistency of the effectiveness of treatments with exogenous androgens, one may state with conviction that it is premature to accept Fiedler's (1974) conclusion that reproductive behavior in male fish is governed directly by gonadotropic hormone and cannot be eliminated by castration. Experiments involving castration have rarely included adequate checks on the completeness of castration. The work of Villars and Davis (1977), Davis *et al.* (1976), and Weiss and Coughlin (1979) illustrated only too clearly the difficulty of obtaining complete castration and the speed with which endocrinologically functional tissue may regenerate. The appearance of secondary sexual characteristics in a small number of "castrated" blue gouramis which spawned raised the suspicion that undetected testicular material was present (Johns and Liley, 1970). These results emphasize the need for more reliable checks, perhaps by radioimmunoassay, for the persistence of circulating androgen following gonadectomy.

Fiedler's (1974) conclusion regarding the role of gonadotropin in male sexual behavior was also based in part on the results of the application of hormones and inhibitory agents (antiandrogens and antigonadotropins) to intact animals. Interpretation of the results of these studies relies heavily on unsubstantiated conjecture as to the effectiveness of the stimulatory, inhibitory, and negative-feedback effects of such treatments. There is ample evidence to suggest that both the antigonadotropin, methallibure, and the commonly used antiandrogen, cyproterone acetate, are only partially effective as "blocking agents" (Pandey, 1970; Davis *et al.*, 1976; Villars and Davis, 1977; Rouse *et al.*, 1977; Kyle, 1982; Fostier *et al.*, Chapter 7, Volume 9A, this series).

However, it is recognized that not all reports of persistence of sexual behavior after castration may be accounted for by the aforementioned explanations. There may be a process comparable to the "corticalization of function" proposed by Beach (1964) to account for the persistence of sexual behavior after castration in some mammals. Therefore, although gonadal hormones may be essential to the development of reproductive behavior, these activities become less dependent on gonadal function after maturation and experience of breeding. Aronson (1959) and Chizinsky (1968) suggested that this may account for the persistence of sexual behavior in certain cichlids and in *Platypoecilus maculatus*. Unfortunately, apart from the demonstration that experience affects the development of parental responsiveness in cichlids and gouramis (Noble *et al.*, 1938; Chang and Liley, 1974), the role of experiential factors in the development of reproductive behavior in fish has been virtually ignored.

In spite of considerable interest in sex determination and the use of hormone treatments to alter sex ratios (see Hunter and Donaldson, Chapter 5, this volume), the role of gonadal hormones in the differentiation and development of reproductive behavior have hardly been investigated. It is well established that treatment of adult females with androgens may result in the acquisition of male morphological and behavioral characteristics. Nevertheless, more careful examination reveals that sex reversal of behavior in female sticklebacks, guppies, and *Sarotherodon mossambicus* may be incomplete (Hoar, 1962a; Tavolga, 1949; Billy, 1982), and it suggests that previously established neural mechanisms or genetic determinants can not be completely overridden by hormonal influences. Treatments applied early in development resulted in a more complete functional sex reversal in behavior (as well as in primary and secondary sexual characteristics) in cichlids (Clemens *et al.*, 1966; Billy, 1982) than in poeciliids (Clemens *et al.*, 1966; Lindsay, 1974). Even when sex reversal does not occur, early treatment with androgen may have long lasting behavioral effects and affect the responsiveness to subsequent hormone treatment (Billy, 1982). These findings indicate that it would be of considerable practical and general biological interest to examine in more detail the role of hormones in the differentiation of sexual behavior in fish, and in particular to determine whether there is a process in early development of teleosts comparable to the hormone-dependent differentiation of sexual behavior in mammals (Feder, 1981).

Nonreproductive aggressive behavior appears to be independent of gonadal control. This has prompted a number of researchers to propose that aggressive behavior is regulated directly by pituitary gonadotropin. To date, this hypothesis has not received much experimental support. The fact that the establishment of dominance, maintenance of individual space, or competition for food have been shown in a number of species to occur regardless of age, sex, or season, suggests that it may be both unnecessary and misleading to assume that there must be an endocrine basis common to all forms of aggression.

Other hormones may be involved in certain aspects of reproduction. Paralactin (the teleost homologue of prolactin) has been implicated in parental behavior of certain species. Neurohypophysial hormones induce a "spawning reflex" in some species, but are without effect in others. Although the biological significance of these findings is still not clear, they emphasize the lack of information regarding the physiological mechanisms underlying short-term changes in behavioral responsiveness. The dramatic changes in gonadotropic hormone levels in goldfish and white suckers at the time of spawning provide an intriguing indication that the short-term switching from one activity to another may not be determined solely by appropriate external stimuli. Changes from one phase of the reproductive cycle to another may be accompanied and perhaps governed by endocrine changes superimposed on a tonic state of sexual responsiveness maintained throughout the breeding season by gonadal androgens.

B. Female Reproductive Behavior

As is the case with male vertebrates, reproductive behaviors of female vertebrates are synchronized with specific stages of gonadal development through changes in endocrine and nonendocrine physiology. However, the nature of this behavioral–gonadal synchrony differs fundamentally between the sexes for two important reasons. First, in the male, a general capability for prolonged production and retention of mature and viable sperm results in the potential for "tonic" male sexual competence throughout a breeding season. However, in most externally fertilizing species, sexual behavior of the female is more temporally restricted than that of the male, basically because ovulation occurs only once or several times during a reproductive season and the oocytes must be fertilized soon after ovulation if viability is to be ensured. This distinction in the temporal characteristics of male and female sexual activity is less apparent in species where seasonal spawning is of brief duration (e.g., herring, Stacey and Hourston, 1982) because spawning in both sexes must be highly synchronized. However, when the occurrence of ovulation in an individual female may occur at any time during a breeding season which may last for several weeks or even months (e.g. Salmonids) or when females ovulate more than once during a breeding season (e.g., many belontiids and cichlids, Breder and Rosen, 1966), reproductive activity of males generally extends over considerable periods, but that of females is restricted to a relatively brief period following ovulation. Females of internally fertilizing live-bearing species such as Poeciliids (Liley, 1968, 1972) also exhibit restricted periods of sexual responsiveness, but males are persistently sexually active.

Second, the major difference between the sexes is that in the male, sexual behavior in both externally and internally fertilizing species always culminates in the release of mature gametes. However, in the female sexual behavior is not as rigidly linked to one stage of gamete development but may occur at the time of ovulation in external fertilizers, or prior to ovulation in internal fertilizers.

This temporal lability in the timing of sexual behavior with respect to gamete development appears to have been responsible for major sexual differences in the mechanisms regulating reproductive behavior. As discussed previously (Section II,A), in males, the period of maximum testicular steroidogenesis coincides with the breeding season, and the more or less extensive prespawning behavior is regulated by gonadal hormones. Spawning or mating are of brief duration and at present little is known of endogenous mechanisms specifically concerned with the regulation of these events.

However, in females, particularly among externally fertilizing species, there appears to be no close correlation between maximum steroidogenesis and breeding activity (Section II,B): Maximum levels of circulating estrogen are more closely associated with vitellogenesis than with the breeding season, and reproduction occurs several weeks or months after maximum ovarian growth. In addition, with relatively few exceptions, [e.g. pair formation and nest-site preparation in certain cichlids (Chien and Salmon, 1972; Greenberg et al., 1965)], females perform relatively little in the way of prespawning or premating behvaior. Instead, in most species female reproductive behavior is limited to sexual activities directly involved in oviposition or copulation, and the onset of this behavior appears to be determined by events surrounding ovulation (e.g., the final maturation of the oocytes and their release from the follicles into the ovarian or abdominal cavity). [High levels of androgens have been recorded in females of some species at the time of breeding (Section II,B), but at the present time there is no evidence to suggest that these androgens play a causal role in the onset of female reproductive behavior.]

The possibility of a causal relationship between ovulation (and the resulting presence of ovulated oocytes) and sexual behavior was proposed in 1965, when Yamazaki, as a result of his studies of ovulation and spawning in goldfish, suggested that "ripe eggs in the ovarian lumen stimulate the spawning behavior of females via some pathway." Subsequent investigations (Stacey and Liley, 1974; Stacey, 1976,1981) have confirmed that, in the goldfish at least, events associated with ovulation do play a causal role in the control of female behavior, and it has become apparent that any attempt to understand the physiological control of sexual behavior in female fish should distinguish between causal mechanisms that depend on physiological events preceeding ovulation (referred to here as preovulatory mechanisms), and those dependent on events associated with ovulation (postovulatory mechanisms). In the following account postovulatory mechanisms are considered first because the limited evidence available suggests that such mechanisms may be characteristic of oviparous species, and therefore represent what is assumed to be the ancestral mode of control over reproductive behavior. Preovulatory mechanisms appear to be associated with more specialized reproductive behavior as, for example, in species with internal fertilization or in oviparous species with elaborate prespawning behavior.

1. POSTOVULATORY REGULATION OF FEMALE REPRODUCTIVE BEHAVIOR

Extending Yamazaki's (1965) observations that sexual behavior of female goldfish was terminated if the ovulated oocytes were removed by hand-stripping, Stacey and Liley (1974) demonstrated that sexual behavior in the goldfish could be either restored in ovulated females from which oocytes had been removed, or rapidly induced in nonovulating females, simply by injecting ovulated oocytes through the ovipore and into the ovarian lumen. Similarly, the duration of sexual behavior could be extended considerably if a plug was placed in the ovipore to prevent oviposition (Stacey, 1977). The tendency of nonovulated female goldfish to perform sexual behavior when injected with ovulated oocytes is relatively independent of the state of ovarian maturation. Provided that the ovaries contain oocytes, which at least have begun to accumulate mucosaccharide yolk vesicles (Khoo, 1979), a maturational stage attained many months prior to spawning, injection of ovulated oocytes can induce normal sexual behavior (Stacey, 1977). Therefore, although under natural conditions female sexual behavior in goldfish is observed only in fish which have recently ovulated, the effects of oocyte injection in nonovulated females demonstrate that the physiological changes which synchronize sexual behavior with ovulation are not inextricably linked with normal periovulatory events, but simply result from the presence of ovulated oocytes within the ovarian lumen. As intraovarian injection of inert physical substitutes for ovulated oocytes also can induce female sexual behavior (Stacey, 1977, and unpublished results), it is possible that ovulated oocytes trigger sexual behavior simply by providing an appropriate physical stimulus within the reproductive tract.

Although a close temporal correlation between ovulation and the onset of sexual behavior has been noted in many oviparous teleosts (Liley, 1969, 1980), experiments specifically concerned with the role of ovulated oocytes in female sexual behavior have been conducted only in goldfish. However, it is clear that, in some species, female sexual responsiveness declines even though ovulated oocytes are present in the ovaries. Lam et al. (1978) report that ovulated threespine sticklebacks fail to respond to male courtship if the oocytes become overripe or "berried," a condition which develops about 1 week after ovulation. If ovulated oocytes injected into female goldfish become water hardened, presumably because of water entering the ovipore, they resemble the berried oocytes of the stickleback and often fail to stimulate female sexual behavior (Stacey, 1977). Tan and Liley (1983) found that females of *Puntius gonionotus* cease to respond to sexually active males about 1 hr after the estimated time of ovulation, at which time oocyte viability also decreases. This decrease in response occurred even in females which had not released their ovulated eggs. Other researchers, who did not examine female sexual behavior, have also described in a variety of teleosts, morphological changes in overripened oocytes and a negative correlation between fertilization rate and the period of postovulatory retention of oocytes within the ovarian lumen or body cavity e.g., rainbow trout (Bry, 1981), ayu, *Plecoglossus altivelis* (Hirose *et al.*, 1977), Japanese flounder, *Limanda yokohamae* (Hirose *et al.*, 1979), and *Clarias lazera* (Hogendoorn and Vismans, 1980). It remains to be determined whether overripening and reduced viability of ovulated oocytes are causally related to decreased sexual responsiveness in the postovulatory period, and, if so, what mechanism(s) is involved.

Prostaglandin (PG) apparently mediates the action of ovulated oocytes on sexual behavior of female goldfish. The PG synthesis inhibitor, indomethacin, completely blocks sexual behavior both in ovulated females and in nonovulated females which have been injected with ovulated oocytes; however, PG injection readily overcomes the inhibitory effect of indomethacin (Stacey, 1976, and unpublished results). Of three PGs which have been examined (PGE_1, PGE_2, and $PGF_{2\alpha}$), $PGF_{2\alpha}$ has proved to be the most potent, although all three are effective. Because PG injection also induces apparently normal sexual behavior (but without oocyte release) even in females which have no ovulated oocytes in the reproductive tract, it appears that intraovarian ovulated oocytes trigger female sexual behavior only indirectly, by stimulating PG synthesis.

The sexual behavioral response to PG injection in female goldfish is rapid, brief in duration, and, as with the response to oocyte injection, not dependent on the presence of mature ovaries. Prostaglandin dosages as low as several nanograms per gram of body weight can be effective, and spawning can commence within several minutes of injection, provided the female has access to both a male and aquatic vegetation as spawning substrate. Both the frequency of spawning acts (in which the female enters the vegetation with the male to perform oviposition movements) and the response duration are positively correlated with PG dosage. However, even at dosages which induce high frequencies of spawning acts soon after injection, the response is terminated within several hours (Stacey, 1981).

Because the magnitude of the sexual response decreases rapidly as the time between PG injection and exposure to the spawning situation is increased, but is unaffected by prior PG treatment (Stacey and Goetz, 1982), it is likely that exogenous PG induces a response of brief duration simply because it is quickly metabolized. These findings, together with others discussed later, indicate that the close temporal relationship between ovulation and sexual behavior in female goldfish results from a rapid increase and decrease in PG synthesis in response to the appearance and depletion of intraovarian ovulated oocytes.

Responsiveness to PG is influenced by pituitary–ovary activity, being drastically reduced by hypophysectomy and restored in hypophysectomized fish by injection of salmon gonadotropin, SG-G100 (Stacey, 1976). Whether GtH induces this responsiveness directly, or indirectly by increasing steroidogenesis, is unknown. The responsiveness to ovulated oocytes is restored in intact females with regressed ovaries by injection of estradiol (Stacey and Liley, 1974) and other steroids (Stacey, 1977). Although the ability of steroids to restore or increase responsiveness to PG in intact, regressed females has not been examined, steroid replacement therapy in hypophysectomized fish has been completely ineffective in restoring responsiveness to either PG or oocyte injection (Stacey, 1976, 1977). Therefore, although GtH appears to play a permissive role in the action of PG on goldfish sexual behavior, it is clear that females exhibit similar levels of responsiveness to PG throughout much of the reproductive cycle.

Male goldfish injected with PG perform female sexual behavior, which appears to be indistinguishable from that exhibited by PG-injected females (Stacey, 1981). Males are no less responsive to PG than females, indicating that the brain of the male goldfish is not behaviorally defeminized during development as is the case with many male mammals (Feder, 1981). The PG-induced female behavior in male goldfish does not appreciably interfere with male behavior; male goldfish injected with PG and given simultaneous access to both male and receptive female partners will alternate rapidly between performance of male and female sexual behavior (N. E. Stacey, unpublished results). This male behavioral bisexuality in the goldfish, a gonochoristic species, suggests a physiological basis for the ability of simultaneous hermaphrodite teleosts to switch rapidly between male and female reproductive roles (Fischer, 1980).

A number of preliminary studies indicate that PG-mediated postovulatory sexual behavior may be widespread among externally fertilizing teleosts. Indomethacin blocks nest digging and spawning in ovulated rainbow trout (N. R. Liley, E. S. P. Tan, and J. Cardwell, unpublished results) and the spawning response to milt in ovulated Pacific herring, *Clupea harengus pallasi* (Stacey and Hourston, 1982, and unpublished results). In the cyprinids, *Puntius gonionotus* and *P. tetrazona* (Liley and Tan, 1983; Chuah, 1982), the American flagfish, *Jordanella floridae* (Crawford, 1975), and in a belontiid, *Macropodus opercularis* (Villars and Burdick, 1982), injection of PGF rapidly induces apparently normal sexual behaviors in nonovulated females. Similarly, in the threespine stickleback, indomethacin reduces sexual behaviors of ovulated females, but PG injection partially restores these behaviors both in indomethacin-treated (T. J. Lam, unpublished results) and overripe females (Lam *et al.*, 1978). Finally, PG injection in nonovulated brown acara, *Aequidens portalegrensis*, rapidly induces oviposition behavior (Cole and Stacey, 1982). The case of the acara is of particular interest because females of this and many other cichlid species exhibit a variety of prespawning reproductive behaviors associated with pair formation and preparation of the spawning substrate (Greenberg *et al.*, 1965; Polder, 1971). If prespawning behaviors in the acara are stimulated by steroids, this and other cichlids may provide valuable models for studying the relative contributions of preovulatory and postovulatory mechanisms in the regulation of reproductive behaviors.

The behavioral effects of indomethacin and PG treatments strongly support a physiological role for PG in female sexual behavior of goldfish and several other externally fertilizing teleosts. However, very little is known about where the PG involved in sexual behavior might be synthesized and where it might act. Prostaglandins have been identified in teleost ovaries and have been shown to stimulate follicular rupture (Stacey and Goetz, 1982; Goetz, Chapter 3, this volume). In goldfish, indomethacin blocks the ovulatory action of HCG but does not inhibit final maturation of the oocyte; PG injection, if given near the expected time of follicular rupture, readily restores this process in HCG-treated fish which have also been injected with indomethacin (Stacey and Pandey, 1975). Consistent with these findings, Ogata *et al.* (1979) have shown that ovarian PGF concentration increases near the time of ovulation in HCG-treated loach, *Misgurnus anguillicaudatus*. In addition to these and other studies (review by Stacey and Goetz, 1982) indicating a role for ovarian PG in ovulation, several recent reports demonstrate that PGF increases in the blood in the postovulatory period. Bouffard (1979) found that plasma PGF levels increase at the time of ovulation in goldfish, remain elevated if ovulated oocytes are present in the ovaries, and decrease following oocyte removal; high PGF concentrations in the ovarian fluid bathing the ovulated oocytes suggest that the high plasma levels were of ovarian origin. In brook trout, *Salvelinus fontinalis*, plasma PGF levels increase several hours before spontaneous ovulation and remain elevated for at least 24 hr after ovulation (Cetta and Goetz, 1982).

The simplest interpretation of the increase in periovulatory PG is that elevation of both ovarian and plasma levels results from a single preovulatory mechanism that stimulates synthesis of prostaglandin involved in follicular rupture. However, several observations indicate that different mechanisms cause the preovulatory increase in ovarian PG, which is responsible for ovulation, and the postovulatory rise in plasma PG which may trigger spawning behavior. For example, Bouffard's finding that removal of ovulated oocytes prematurely depresses plasma PGF suggests that intraovarian ovulated oocytes somehow maintain PGF synthesis in the postovulatory period. Furthermore, the fact that intraovarian injection of ovulated oocytes rapidly induces sexual behavior in nonovulated female goldfish indicates that even in females whose ovaries have not been stimulated by preovulatory levels of GtH, intraovarian ovulated oocytes can increase PG synthesis. It is unlikely that oocyte injection induces sexual behavior simply by the introduction of PG synthesized in the ovulated egg donor because indomethacin treatment, which does not reduce responsiveness to exogenous PG (Stacey and Goetz, 1982), is completely effective in blocking the spawning response to oocyte injection (Stacey, 1976).

In goldfish, female sexual behavior induced by PG injection is not affected by removal of the posterior portions of the ovaries, the oviduct, or the area around the ovipore (Stacey and Peter, 1979), indicating that PG does not induce behavior by acting at these peripheral sites. Furthermore, sexual behavior is more effectively stimulated by intracerebroventricular PG injection than by injection given intraperitoneally or intramuscularly (Stacey and Peter, 1979), indicating that PG likely acts at some as yet unknown site within the brain. These behavioral findings alone would be consistent with the hypothesis that PG involved in spawning behavior is synthesized within the brain, perhaps in response to afferent neural activity triggered by the presence of ovulated oocytes. However, in view of the reported postovulatory increases in blood PG, the results of the behavioral studies suggest that female sexual behavior in goldfish is stimulated by ovarian PG which is synthesized during the process of follicular rupture and/or in the presence of ovulated oocytes, released into the peripheral circulation, and acting rapidly within the brain. Because PGs generally function as local hormones, and not as blood-borne hormones in the classical sense, this latter interpretation must be regarded with considerable caution until more definitive experiments have been performed.

The functional significance of a postovulatory control of sexual responsiveness in externally fertilizing species is clear. Such a mechanism ensures that the female is sexually responsive and prepared for oviposition at the time of maximum viability of the ovulated oocytes, and that she remains active only until all mature oocytes have been shed. The time during which mature oocytes remain viable varies considerably among species: less than 1 hr, *Puntius gonionotus* (Liley and Tan , 1983), 12 hr in *Clarias macrocephalus* (Mollah and Tan, 1983), several days in *Catostomus commersoni*

(N. E. Stacey, unpublished results), and several weeks in *Salmo gairdneri* (Escaffre *et al.*, 1977; Bry, 1981). The exact timing of ovulation is probably determined by proximate environmental cues which ensure that oviposition occurs at the season and perhaps the time of day most favorable to the survival of eggs and young. For example, in goldfish which ovulate spontaneously at 20°C in 16 hr light–8 hr dark photoperiod, plasma GtH levels rise dramatically during the latter part of the photophase, peak during the last 4 hr of the scotophase, at which time ovulation occurs, and return to preovulatory levels within several hours of ovulation (Stacey *et al.*, 1979a). Provided that a sexually active male and aquatic vegetation are present, female goldfish which have ovulated begin to spawn at the onset of the photophase. However, if an injection of GtH is administered, females begin to spawn very soon after the induced ovulation, regardless of the time of day at which this occurs. Therefore, it is likely that "wild" goldfish spawn in the early morning because under natural conditions, as in the laboratory, the preovulatory GtH surge is synchronized with photoperiod such that ovulation occurs during the night. Similarly, females of the Japanese medaka, *Oryzias latipes*, which can ovulate and spawn almost every day during the breeding season, exhibit a daily cycle of oocyte maturation which probably is attributable to diel periodicity in release of ovulatory levels of GtH (Iwamatsu, 1978). As in goldfish, the female medaka ovulates during the latter hours of the scotophase and spawns soon after the onset of photophase.

Whether the regulation of ovulation in goldfish and medaka is typical of the majority of oviparous teleosts is not known. However, a great variety of both freshwater and marine species exhibit diel spawning activity (Suzuki and Hioki, 1979; Suzuki *et al.*, 1980; Bruton, 1979; Chien and Salmon, 1972; papers cited in Ferraro, 1980, and Stacey *et al.*, 1979b). Both in species that employ simple broadcast fertilization and in those that exhibit elaborate prespawning behaviors and oviposit on selected substrates, there is much evidence that the precise timing of this diel sexual activity is attributable to rapid changes in the behavior of the female. Although photoperiod may play a direct role in the determination of the time of spawning in these species, it is more likely that, as in goldfish, photoperiod acts only indirectly by determining the time of ovulation (Stacey *et al.*, 1979b) and that spawning then occurs rapidly in response to the presence of ovulated oocytes.

In other species, the synchronization of ovulation with photoperiod may be less precise. For example, in the salmonids, the slow embryonic development and the ability to retain viable oocytes for an extended postovulatory period presumably reduce or eliminate any putative advantage for diel rhythmicity in either ovulation or spawning, and perhaps also explain why spontaneous ovulation occurs readily in the absence of the gravel substrate necessary for successful spawning. Although it is possible that a postovulatory increase in PG plays the same stimulatory role in sexual behavior of both goldfish and salmonids, the precise timing of oviposition in salmonids is evidently determined not by the timing of ovulation, but by other factors, including stimuli from the completed nest site (Tautz and Groot, 1975).

2. PREOVULATORY MECHANISMS

Preovulatory control of sexual behavior has been studied extensively in only one teleost, the ovoviviparous guppy, *Poecilia reticulata* (Crow and Liley, 1979; Liley, 1968, 1972; Liley and Donaldson, 1969; Liley and Wishlow, 1974; Meyer and Liley, 1982). The female guppy undergoes regular cycles of receptivity to male courtship which are correlated with ovarian activity (Liley, 1966): sexual responsiveness is high for several days after parturition, remains low during gestation, and peaks again following the subsequent parturition. Fertilization is intrafollicular; the juveniles are released from the follicles just before parturition. Considerable evidence indicates that both the cycle of receptivity and the associated cycle of pheromone-mediated attractiveness to males are induced by cyclic fluctuations of ovarian estrogen synthesis.

If female guppies are ovariectomized, both sexual behavior and production of a sexual pheromone are reduced. That the behavioral effects of the operation are attributable to removal of estrogen is suggested by the fact that estradiol, estriol, and the synthetic estrogen, diethylstilboestrol, are all able to restore receptivity in ovariectomized, nonreceptive females (Liley, 1972). Earlier studies (Liley, 1968) indicated that GtH or some other pituitary factor might be directly involved in stimulating female sexual behavior. However, the demonstration that SG-G100 restored receptivity of hypophysectomized females only if the ovaries were intact (Liley and Donaldson, 1969) and that estrogen alone restored receptivity in females, which were both hypophysectomized and ovariectomized (Liley, 1972), suggests that GtH regulates receptivity only indirectly, by stimulating ovarian estrogen synthesis.

The postpartum peak in sexual responsiveness coincides with the female's maximum attractiveness to the male. For example, male guppies are more attracted to water in which early postpartum females have been kept than to water used to keep ovariectomized females, or to water used to keep females in the middle of a gestation cycle (Crow and Liley, 1979). Male courtship of ovariectomized females is stimulated by water used to keep hypophysectomized females which have been treated with GtH or estrogen, indicating a role for estrogen in synthesis or release of a sexual pheromone (Meyer and Liley, 1982). Furthermore, water used to keep estrogen-treated, ovariectomized females fails to stimulate male courtship, indicating that estrogen acts via the ovaries to stimulate pheromone production (Meyer and Liley, 1982). The results of these behavioral studies, which demonstrate estrogen-dependent, postpartum peaks of sexual receptivity, and pheromone production, are consistent with the findings that the guppy ovary can synthesize estradiol and that ovarian steroidogenic activity is maximum in the postpartum period (Lambert and van Oordt, 1974). Female *Gambusia* also exhibit a cycle of receptivity which is correlated with the reproductive cycle (Carlson, 1969).

The level of sexual responsiveness in the female guppy is determined not only by ovarian endocrine activity, but also by recent courtship experience. Liley and Wishlow (1974) exposed sexually inexperienced virgin females to various regimens of brief courtship by males that had been gonopodectomized to prevent insemination and pregnancy. Regardless of whether they were intact, or had been ovariectomized for as long as 24 days prior to testing, a large proportion of these virgin females displayed high initial levels of receptivity which declined rapidly after several brief exposures. Ovariectomized virgins did not recover from the decremental effects of courtship experience; however, intact virgins showed transient cyclic increases in receptivity similar to those seen in intact nonvirgins. Therefore, in the naive, virgin, female guppy, initially high levels of receptivity, which evidently are not dependent on ovarian estrogen, probably ensure insemination on the first exposure to a male regardless of the state of oocyte maturation, and rapid habituation of this responsiveness following coitus may serve to reduce the female's exposure to predation (Liley and Wishlow, 1974). If nonvirgin fish display a similar coitus-induced decrease in receptivity, then this effect, in combination with the stimulatory action of ovarian estrogen, would serve to restrict inseminations to the brief period during which mature oocytes are capable of being fertilized.

A preovulatory endocrine control of female reproductive behavior has been demonstrated only in the guppy. However, it is quite likely that analogous mechanisms will be found in ovoviviparous and viviparous species from a number of teleosts groups (Cyprinodontiformes, Belontiformes, Perciformes, Gadiformes: Turner, 1947; Amoroso, 1960; Breder and Rosen, 1966; Hoar, 1969) in which fertilization is intrafollicular and female sexual behavior therefore depends on a preovulatory mechanism. Also, in some oviparous species that utilize internal fertilization, sexual behavior evidently is preovulatory. In the oviparous catfish, *Trachycoristes striatulus*, viable sperm can be held in the oviduct for several months; fertilization apparently occurs during the brief period in which ovulated oocytes are held in the oviduct before oviposition (von Ihering, 1937). In two glandulocaudine species, *Gephyrocharax valencia* (Wohlert, 1934, cited in Breder and Rosen, 1966) and *Corynopoma risii* (Kutaygil, 1959), there is at least strong circumstantial evidence the females are inseminated prior to ovulation.

There have been remarkably few attempts to examine the regulation of female reproductive behavior in teleosts by the use of ovariectomy and steroid-replacement therapy as has been done in the guppy. Ovariectomy caused a loss of sexual responsiveness in the blue gourami, *Trichogaster trichopterus* (Seghers, 1967), and in *Betta splendens* (Noble and Kumpf, 1936); both are externally fertilizing species in which the female's sexual behavior might be expected to be regulated by a postovulatory mechanism. In cichlids, ovariectomy reduced or abolished reproductive behavior (*Sarotherodon macrocephala*, Aronson, 1951; *Hemichromis bimaculatus*, Noble and Kumpf, 1936). However, Noble and Kumpf (1936) briefly reported that, in *Hemichromis*, injection of ovarian extract (source not stated) restored *most* of the female's sexual behavior.

Although ovulation in some cichlids has been reported to occur 1–2 hr prior to the commencement of oviposition (*Sarotherodon macrocephala*, Aronson, 1951; *Aequidens portalegrensis*, Polder, 1971), the precise temporal relationship between ovulation and the various components of the female's complex reproductive behaviors is not clear. In several cichlid species (*Aequidens portalegrensis*, Greenberg *et al.*, 1965; *Sarotherodon macrocephala*, Aronson, 1949; *Pterophyllum scalare*, Chien and Salmon, 1972), nest skimming or nest passing, an incipient oviposition movement in which oocytes are not released, increases dramatically shortly before actual oviposition. Because injection of prostaglandin readily induces skimming (oviposition) behavior in nonovulated *Aequidens portalegrensis* (Cole and Stacey, 1982), it is likely that normal spawning behavior (skimming and oviposition) also are stimulated by prostaglandin. A variety of other female cichlid prespawning behaviors associated with courtship and nest-site preparation precede ovulation by several days and, therefore, are probably regulated by preovulatory mechanisms. Apart from studies on the guppy and the brief report by Noble and Kumpf (1936) on *Hemichromis*, there is no information regarding the nature of these preovulatory mechanisms regulating female reproductive behavior. However, because plasma estrogen and androgen levels increase during vitellogenesis in many teleosts (Section II,B), it is prudent to view prespawning reproductive behaviors in female teleosts as being potentially under the influence of ovarian steroids.

3. FEMALE REPRODUCTIVE BEHAVIOR: DISCUSSION AND CONCLUSIONS

The major finding to emerge from recent studies is that there appear to be at least two very different mechanisms involved in the regulation of female reproductive behavior in teleosts. In oviparous, externally fertilizing species, of which the goldfish is the only species studied in depth, sexual activity, which is limited to spawning behavior, occurs during the postovulatory period and evidently is stimulated by prostaglandin(s) which appears to be synthesized in response to the presence of intraovarian ovulated oocytes. Such a postovulatory mechanism ensures that a female performs sexual behavior soon after ovulation when viability of the eggs is maximal and remains active only until all oocytes have been shed. Gonadal steroids may play a tonic, permissive role, and maintain responsiveness to the stimulus provided by ovulated eggs.

In contrast, it is proposed that in internally fertilizing species, particularly those able to store sperm, the timing of sexual behavior in relation to ovulation may be less critical, and for species in which fertilization is intrafollicular, sexual behavior cannot depend on events associated with ovulation. In the only species investigated in detail, the guppy, gonadal estrogens play a key role in modulating the sexual response of the female.

It should be emphasized that the proposal that there are two basic mechanisms governing reproductive behavior of the female fish rests very heavily on detailed studies of only two species—the goldfish and guppy. Whether this simple distinction between pre- and postovulatory mechanisms will accurately reflect similarities and differences in behavioral regulation when applied to a wider spectrum of teleost reproductive specializations obviously cannot be assessed until a greater variety of species has been examined in detail. However, one can hope that by drawing attention to this distinction, investigators will be encouraged to direct their attention to those species which are likely to provide answers to the most important questions regarding the physiological regulation of reproductive behavior in female teleosts.

In spite of the limited comparative data available, it is interesting and instructive to consider the proposed mechanisms in the broader context of other vertebrate groups. Primitive teleosts undoubtedly possessed the ancestral reproductive patterns from which have evolved the mechanisms regulating female sexual behaviors in extant vertebrates. Although some modern teleosts may retain many elements of their ancestral reproductive functions, it is clear that a long history of independent teleost evolution has resulted in reproductive modifications and specializations in many modern teleost species. Nevertheless, although caution and restraint must be exercised in any discussion of the evolution of mechanisms regulating female vertebrate sexual behavior, comparison of how female sexual behavior is controlled in the various vertebrate classes is of value even if it serves only to stimulate consideration of the processes responsible for this functional diversity.

Postovulatory sexual behavior associated with external fertilization can reasonably be assumed to represent the ancestral mode of vertebrate reproduction. In the few teleost species which have been examined, it appears that postovulatory female sexual behavior is stimulated by PG which is synthesized in response to the presence of ovulated oocytes. Similarly, in externally fertilizing anurans, PG has been noted to have potent stimulatory actions on female sexual behavior. In *Rana pipiens*, nonovulated females emit a release call that inhibits clasping attempts by the male. In ovulated females and in nonovulated females in which water accumulation has been induced by injection of arginine vasotocin (AVT) or ligature of the cloaca, the release call is inhibited, enabling the male to retain his clasp (Diakow and Raimondi, 1981). Although the physiological role of AVT in *Rana* sexual behavior remains to be determined, there is evidence that AVT may be acting by stimulating PG synthesis. Indomethacin inhibits the effect of AVT on release call inhibition; however, injection of PG rapidly inhibits calling in nonovulated, nonreceptive females (Diakow and Nemiroff, 1981). In *Xenopus laevis*, the receptive leg adduction posture demonstrated by receptive females to facilitate clasping by the male, is inhibited by PG synthesis inhibitors; this receptive response is readily induced both in intact, nonovulated and in ovariectomized females by PG injection (D. B. Kelley and R. Bockman, personal communication). Together, these studies of teleosts and anurans suggest that in externally fertilizing vertebrates similar mechanisms may function to activate postovulatory female sexual behavior during the time that ovulated oocytes are ready for release.

The ability of PG to stimulate female sexual behavior is not restricted to externally fertilizing vertebrates. In a variety of internally fertilizing species, PG has been shown to have rapid stimulatory and inhibitory effects on female sexual behaviors which are known to be regulated by periovulatory increases in estrogen. Both in the rat (Hall *et al.*, 1975; Rodriguez-Sierra and Komisaruk, 1977) and the hamster (Buntin and Lisk, 1979), PG injection rapidly stimulates lordosis behavior, but in the guinea pig (Marrone *et al.*, 1979) and the lizard, *Anolis carolinesis* (Tokarz and Crews, 1981), receptivity is rapidly terminated following similar treatment. A possible clue to the normal physiological role of PG in sexual behavior of these species comes from a comparison of the effects of coital stimuli and PG injection on sexual responsiveness. Vaginocervical stimulation and mating rapidly inhibit receptivity in the guinea pig (Marrone *et al.*, 1979) and *Anolis* (Crews, 1973), species in which the effect of PG on sexual behavior also is inhibitory. In contrast, both vaginocervical stimulation (Rodriquez-Sierra *et al.*, 1975) and PG treatment stimulate receptivity in the rat. If this correlation is indicative of an underlying causal relationship between coitus and altered receptivity, then the role of PG in these internally fertilizing species may be comparable to what has been proposed for externally fertilizing teleosts, i.e., that physical stimulation of the reproductive tract rapidly alters sexual activity by stimulating PG synthesis.

Preovulatory and periovulatory female reproductive behaviors in a wide variety of internally fertilizing species (guppy, Liley, 1972; *Anolis*, Crews, 1975; birds, Cheng, 1978; mammals, Morali and Beyer, 1979) evidently are stimulated by increasing blood estrogen levels associated with follicular maturation. Because internal fertilization has arisen independently in teleosts and in other vertebrate classes, it would appear that the mode of reproduction, rather than phylogenetic status, is of primary importance in determining whether estrogen has evolved as a hormonal stimulus for female reproductive behavior. Estrogen has been demonstrated to stimulate synthesis of hepatic vitellogenin in a variety of vertebrates that produce yolky oocytes (teleosts, Ng and Idler, Chapter 8, Volume 9A, this series; amphibia, Wallace and Dumont, 1968; birds, Gruber, *et al.*, 1976). This hormonal role for estrogen in vitellogenesis, which has been observed in species such as the goldfish and *Rana* where female sexual behavior is apparently not regulated by estrogen, might be regarded as a preadaptation allowing estrogen to be incorporated into the regulation of female sexual behavior in species that evolve internal fertilzation. These considerations need not apply only to sexual behaviors: even in externally fertilizing species where sexual behavior is postovulatory, preovulatory reproductive behaviors temporally associated with vitellogenesis (e.g., pair formation and nest building in cichlids) may also be regulated by estrogen.

In summary, we propose that a dichotomy in mechanisms regulating female reproductive behavior in teleosts has arisen as a result of the basic change in the nature of female sexual behavior accompanying the evolution of internal fertilization. In species which retain the presumed ancestral mode of vertebrate reproduction, external fertilization, female sexual behavior necessarily involves oviposition and, therefore, is appropriate only in the postovulatory period. Prostaglandins, which are both rapidly synthesized and rapidly metabolized, evidently serve as a precise endogenous stimulus for sexual behavior in some species, perhaps by increasing and decreasing in the blood in response to the appearance and depletion of ovulated oocytes. With the evolution of internal fertilization and the temporal dissociation of sexual behavior and fertilization, female sexual behavior in teleosts and other vertebrates has become synchronized not with the presence of ovulated oocytes, but rather with that period during which insemination will lead to successful fertilization. This has been achieved by incorporating into the mechanisms regulating sexual behavior two indirect indicators of the state of fertilizable oocytes, the stimulatory action of preovulatory increases in blood estrogen and the inhibitory action of coital experience (guppy, Liley and Wishlow, 1974; *Anolis*, Crews, 1973; mammals, Slimp, 1977). Teleost fishes, in which internal fertilization has arisen in a number of unrelated groups and is associated with both preovulatory and postovulatory sexual behavior, provide valuable opportunities for determining how changes in mode of reproduction have been accompanied by changes in the physiological regulation of female sexual behavior.

VI. BRAIN MECHANISMS OF HORMONE ACTION

A comprehensive understanding of how changes in gonadal function alter reproductive behavior must include an appreciation of hormonal action on the central nervous system. Unfortunately, despite an immense body of research on nonteleost vertebrates dealing with the site (Stumpf and Grant, 1975) and mode (McEwen, 1981) of hormone action on the brain, the neurochemical changes accompanying hormone-stimulated behavior (Crowley and Zemlan, 1981), and the behavioral roles of hormones in early development (Adkins-Regan, 1981), the mechanisms of action of hormones in teleost reproductive behavior are virtually unexplored. Demski and Hornby (1982) reviewed this area; therefore, discussion here is limited to recent reports.

Steroid autoradiography has been employed to identify teleost brain areas which concentrate, and, therefore, are likely to be influenced by, sex steroids. Despite some differences in steroid concentrating sites among the four species which have been examined (green sunfish, *Lepomis cyanellus*, Morell *et al.*, 1975; paradise fish, Davis *et al.*, 1977; goldfish, Kim *et al.*, 1978a; platyfish, *Xiphophorus maculatus*, Kim *et al.*, 1979), all of these studies found steroid uptake in the tuberal hypothalamus, preoptic area, and ventral telencephalon, a pattern consistent with other vertebrate groups (Kim *et al.*, 1978b). Binding in the nucleus lateral tuberus may be related to steroid-feedback regulation of gonadotropin secretion (Billard and Peter, 1977; Crim and Peter, 1978; Peter, 1982). However, demonstration of steroid uptake in the preoptic area and in the supracommisural area of the ventral telencephalon is of particular interest because other experimental approaches have implicated these brain regions in the control of reproductive behaviors.

Whereas partial or total ablation of the telencephalic lobes results in reproductive behavior deficits in a variety of teleosts (see de Bruin, 1980; Demski and Hornby, 1982), in only a few cases have small lesions or electrical stimulation been used to identify discrete telencephalic areas involved in reproductive behavior. Kyle and Peter (1982) have demonstrated that sexual behavior of male goldfish is drastically reduced if small lesions are placed in the area ventralis telencephali pars supracommisuralis (VS) and area ventralis telencephali pars ventralis (PVV), areas known to bind sex steroids in this species; lesions in adjacent telencephalic nuclei and in the preoptic area had no effect. The VS–PVV lesions also inhibit prostaglandin-induced female spawning behavior in males (Kyle *et al.*, 1982a) and females (A. L. Kyle and N. E. Stacey, unpublished results), but did not affect the feeding response to a food odor (Kyle *et al.*, 1982a). In bluegill sunfish (*Lepomis macrochirus*) and in green sunfish, nest building and courtship behavior have been induced by electrical stimulation of the preoptic area (Demski and Knigge, 1971; Demski, 1978; Demski and Hornby, 1982), a sex-steroid-concentrating area implicated in sexual behaviors throughout the vertebrates (Kelley and Pfaff, 1979). Also in *Fundulus heteroclitus*, lesions of the nucleus preopticus inhibited the spawning reflex response to injection of neurohypophyseal hormones (Macey *et al.*, 1974).

Although these studies have identified specific brain areas which may mediate the actions of hormones on reproductive behavior, other recent findings suggest how these and other neural centers may function as an integrated reproductive behavior system. Based primarily on their work with male goldfish, Demski and Hornby (1982) have described a sperm-release (SR) pathway extending caudally from the preoptic area to the rostral spinal cord and apparently stimulating the gonads via sympathetic, cholinergic innervation. The SR pathway also can be activated by electrical stimulation of the medial bundle of the olfactory tract (MOT) (Demski *et al.*, 1982). Demski and Northcutt (1983) have shown that in the goldfish the MOT carried all central projections of the nervus terminalis, which in several teleosts in addition to the goldfish projects at least to the posterior ventral telencephalon and other forebrain areas related to the SR system, and also contains luteinizing hormone releasing hormone (LHRH)-positive cell bodies and fibers (Munz *et al.*, 1981, 1982).

These anatomical and electrophysiological studies are supported by recent behavioral findings. Sexual behavior of male goldfish is severely reduced by section of the MOT; however, section of the lateral olfactory tract (LOT), which does not contain fibers of the nervus terminalis (Demski and Northcutt, 1983) is without effect. In contrast, olfactory tract sections have no effect on prostaglandin-stimulated female sexual behavior in either males or females (N. E. Stacey and A. L. Kyle, unpublished results). These findings raise the question of whether behavioral responses to pheromones are mediated by the olfactory system, or whether in fact a nonolfactory chemosensory system (the nervus terminalis) is involved.

At present, neither the possible behavioral actions of LHRH, nor the physiological significance of the LHRH-containing neuronal network have been examined. However, the fact that LHRH neurons and fibers have been identified in olfactory, optic, and telencephalic areas (Munz *et al.*, 1981, 1982), some of which bind sex steroids (Kim *et al.*, 1978b; Munz *et al.*, 1981), provides strong evidence for an integrated neuronal system regulating reproductive behavior in response to chemosensory, visual, and endocrine stimuli.

REFERENCES

Adkins-Regan, E. (1981). Early organizational effects of hormones: An evolutionary perspective. In "Neuroendocrinology of Reproduction" (N. T. Adler, ed.), pp. 159–228. Plenum, New York.

Ahsan, S. N., and Hoar, W. S. (1963). Some effects of gonadotropic hormones on the three spined stickleback, Gasterosteus aculeatus. Can. J. Zool. 41, 1045–1053.

Algranati, F. D., and Perlmutter, A. (1981) Attraction of zebrafish, Brachydanio rerio, to isolated and partially purified chromatographic fractions. Environ. Biol. Fishes 6, 31–38.

Amoroso, E. C. (1960). Viviparity in fishes. Symp. Zool. Soc. London 1, 153–181.

Amouriq, L. (1964). L'activité et le phénomène social chez Lebistes reticulatus (Poeciliidae: Cyprinodontiformes). C. R. Hebd. Seances Acad. Sci. 259, 2701–2702.

Amouriq, L. (1965). Origine de la substance dynamogène émise par Lebistes reticulatus femelle (Poisson: Poeciliidae, Cyprinodontiformes). C. R. Hebd. Seances Acad. Sci. 260, 2334–2335.

Arai, R. (1967). Androgenic effects of 11-ketotestosterone on some sexual characteristics in the teleost, Oryzias latipes. Annot. Zool. Jpn. 40, 1–5.

Aronson, L. R. (1949). An analysis of reproductive behavior in the mouthbreeding cichlid fish, Tilapia macrocephala (Bleeker). Zoologica (N.Y.) 34, 133–157.

Aronson, L. R. (1951). Factors influencing the spawning frequency in the female cichlid fish, Tilapia macrocephala. Am. Mus. Novit. 1484, 1–26.

Aronson, L. R. (1959). Hormones and reproductive behavior: Some phylogenetic considerations. In "Comparative endocrinology" (A. Gorbman, ed.), pp. 98–120. Wiley, New York.

Aronson, L. R., and Holz-Tucker, M. (1947). Morphological effects of castration and treatment with gonadal hormones on the female cichlid fish, Tilapia macrocephala. Anat. Rec. 99, 572–573.

Aronson, L. R., Scharf, A., and Silverman, H. (1960). Reproductive behavior after gonadectomy in males of the cichlid fish Aequidens latifrons. Anat. Rec. 137, 335.

Atkins, D. L., and Fink, W. L. (1979). Morphology and histochemistry of the caudal gland of Corynopoma riisei Gill. J. Fish Biol. 14, 465–469.

Baenninger, R. (1968). Fighting by Betta splendens: Effects on aggressive displaying by conspecifics. Psychon. Sci. 10, 185–186.

Baggerman, B. (1957). An experimental study on the timing of breeding and migration in the three-spined stickleback (Gasterosteus aculeatus L.), Arch. Neerl. Zool. 16, 159–163.

Baggerman, B. (1966). On the endocrine control of reproductive behaviour in the male three-spined stickleback (Gasterosteus aculeatus L.). Symp. Soc. Exp. Biol. 20, 427–456.

Baggerman, B. (1969). Hormonal control of reproductive and parental behaviour in fishes. In "Perspectives in Endocrinology: Hormones in the Lives of Lower Vertebrates" (E. J. W. Barrington and C. A. Jorgensen, eds.), pp. 321–404. Academic Press, New York and London.

Bailey, R. M., and Harrison, H. M., Jr. (1945). Food habits of the southern channel catfish I. punctatus in the Des Moines River, Iowa. Trans. Am. Fish. Soc. 74, 110–138.

Baldwin, F. M., and Goldin, H. S. (1939). Effects of testosterone propionate on the female viviparous teleost, Xiphophorus helleri Heckel. Proc. Soc. Exp. Biol. Med. 42, 813–819.

Ball, J. N. (1960). Reproduction in female bony fishes. Symp. Zool. Soc. London 1, 105–135.

Balon, E. K. (1975). Reproductive guilds of fishes: A proposal and definition. J. Fish. Res. Board Can. 32, 821–864.

Balon, E. K. (1981). Additions and amendments to the classification of reproductive styles in fishes. Environ. Biol. Fishes 6, 377–390.

Beach, F. A. (1964). Biological bases for reproductive behavior. In "Social Behavior and Organization among Vertebrates" (W. Etkin, ed.), pp. 117–142. Univ. of Chicago Press, Chicago, Illinois.

Billard, R., and Peter, R. E. (1977). Gonadotropin release after implantation of anti-estrogens in the pituitary and hypothalamus of goldfish, *Carassius auratus*. *Gen. Comp. Endocrinol.* **32**, 213–220.

Billard, R., Breton, B., Fostier, A., Jalabert, B., and Weil, C. (1978). Endocrine control of the teleost reproductive cycle and its relation to external factors: Salmonid and cyprinid models. In "Comparative Endocrinology" (P. J. Gaillard and H. H. Boer, eds.), pp. 37–48. Elsevier-North-Holland Biomedical Press, Amsterdam.

Billy, A. J. (1982). The effects of early and late androgen treatments and induced sexual reversals on the behavior of *Sarotherodon mossambicus*. M.Sc. Thesis, University of British Columbia, Vancouver, B.C., Canada.

Blüm, V. (1968). Die Auslösung des Laichreflexes durch Reserpin bei dem südamerikanischen Buntbarsch *Pterophyllum scalare*. *Z. Vergl. Physiol.* **60**, 79–81.

Blüm, V. (1974). Die Rolle des Prolaktins der Cichlidenbrutpflege. *Fortschr. Zool.* **22**, 310–333.

Blüm, V., and Fiedler, K. (1964). Der Einfluss von Prolactin auf das Brutpflegerverhalten von *Symphysodon aequifasciata axelrodi*. L. P. Schultz (Cichlidae, Teleostei). *Naturwissenschaften* **51**, 149.

Blüm, V., and Fiedler, K. (1965). Hormonal control of reproductive behavior in some cichlid fish. *Gen. Comp. Endocrinol.* **5**, 186–196.

Blüm, V., and Fiedler, K. (1974) Prolaktinempfindliche Strukturen in Fischgehirn. *Fortschr. Zool.* **22**, 155–166.

Bommelaer, M. C., Billard, R., and Breton, B. (1981). Changes in plasma gonadotropin after ovariectomy and estradiol supplementation at different stages at the end of the reproductive cycle in the rainbow trout (*Salmo gairdneri* R.). *Reprod. Nutr. Dev.* **21**, 989–997.

Bouffard, R. E. (1979). The role of prostaglandins during sexual maturation, ovulation, and spermiation in the goldfish, *Carassius auratus*. M. Sc. Thesis, University of British Columbia, Vancouver, B.C., Canada.

Braddock, J. C. (1945). Some aspects of the dominance-subordination relationship in the fish *Platypoecilus maculatus*. *Physiol. Zool.* **18**, 176–195.

Breder, C. M., Jr., and Rosen, D. E. (1966). "Modes of Reproduction in Fishes," Natural History Press, New York.

Brett, B. L. H., and Grosse, D. J. (1982). A reproductive pheromone in the Mexican poeciliid fish *Poecilia chica*. *Copeia* pp. 219–223.

Bruton, M. N. (1979). The breeding biology and early development of *Clarias gariepinus* (Pisces: Clariidae) in Lake Sibaya, South Africa, with a review of breeding in species of the subgenus *Clarias (Clarias)*. *Trans. Zool. Soc. London* **35**, 1–45.

Bry, C. (1981). Temporal aspects of macroscopic changes in rainbow trout (*Salmo gairdneri*) oocytes before ovulation and of ova fertility during the post-ovulation period: Effect of treatment with 17α-hydroxy-20β-dihydroprogesterone. *Aquaculture* **24**, 153–160.

Buntin, J. D., and Lisk, R. D. (1979). Prostaglandin E2-induced lordosis in estrogen-primed female hamsters: relationship to progesterone action. *Physiol. Behav.* **23**, 569–575.

Callard, G. V., Petro, Z., and Ryan, K. J. (1978). Conversion of androgen to estrogen and other steroids in the vertebrate brain. *Am. Zool.* **18**, 511–523.

Callard, G. V., Petro, Z., Ryan, K. J., and Claiborne, J. B. (1981). Estrogen synthesis *in vitro* and *in vivo* in the brain of a marine teleost (*Myoxocephalus*). *Gen. Comp. Endocrinol.* **43**, 243–255.

Campbell, C. M., Walsh, J. M., and Idler, D. R. (1976). Steroids in the plasma of the winter flounder (*Pseudopleuronectes americanus* Walbaum). A seasonal study and investigation of steroid involvement in oocyte maturation. *Gen. Comp. Endocrinol.* **29**, 14–20.

Campbell, C. M., Fostier, A., Jalabert, B., and Truscott, B. (1980). Identification and quantification of steroids in the serum of spermiating or ovulating rainbow trout. *J. Endocrinol.* **85**, 371–378.

Carew, B. A. M. (1968). Some effects of methallibure (I.C.I. 33,828) on the stickleback *Gasterosteus aculeatus* L. M. Sc. Thesis, Department of Zoology, University of British Columbia, Vancouver, B.C., Canada.

Carlson, D. R. (1969). Female sexual receptivity in *Gambusia affinis* (Baird and Girard). *Tex. J. Sci.* **21**, 167–173.

Cetta, F., and Goetz, F. W. (1982). Ovarian and plasma prostaglandin E and F levels in brook trout (*Salvelinus fontinalis*) during pituitary-induced ovulation. *Biol. Reprod.* **27**, 1216–1221.

Chang, B. D., and Liley, N. R. (1974). The effect of experience on the development of parental behavior in the blue gourami, *Trichogaster trichopterus*. *Can. J. Zool.* **52**, 1499–1503.

Cheal, M., and Davis, R. E. (1974). Sexual behavior: social and ecological influences in the Anabantoid fish, *Trichogaster trichopterus*. *Behav. Biol.* **10**, 435–445.

Chen, L. C., and Martinich, R. L. (1975). Pheromonal stimulation and metabolite inhibition of ovulation in the zebrafish, *Brachydanio rerio*. *Fish. Bull.* **73**, 889–894.

Cheng, M. F. (1978). Progress and prospects in ring dove research: A personal view. *Adv. Study Behav.* **9**, 97–128.

Chien, A. K. (1973). Reproductive behavior of the angelfish *Pterophyllum scalare* (Pisces: Cichlidae). II. Influence of male stimuli upon the spawning rate of females. *Anim. Behav.* **21**, 457–463.

Chien, A. K., and Salmon, M. (1972). Reproductive behavior of angelfish, *Pterophyllum scalare*. I. A quantitative analysis of spawning and parental behavior. *Forma Funct.* **5**, 45–74.

Chiszar, D., Ashe, V., Seixas, S., and Henderson, D. (1976). Social aggressive behavior after various intervals of social isolation in bluegill sunfish (*Lepomis macrochirus*, Refinesque) in different states of reproductive readiness. *Behav. Biol.* **16**, 475–487.

Chuah, H. P. (1982). The Sexual Behavior of the Tiger Barb, *Barbus Tetrazona* and the Effects of Prostaglandin $F_{2\alpha}$. B.Sc. Thesis, School of Biological Sciences, Universiti Sains Malaysia, Penang, Malaysia.

Chizinsky, W. (1968). Effects of castration upon sexual behavior in male platyfish, *Xiphophorus maculatus*. *Physiol. Zool.* **41**, 466–475.

Clemens, H. P., and Inslee, T. (1968). The production of unisexual broods by *Tilapia mossambica* sex-reversed with methyltestosterone. *Trans. Am. Fish. Soc.* **97**, 18–21.

Clemens, H. P., McDermitt, C., and Inslee, T. (1966). The effects of feeding methyl testosterone to guppies for 60 days after birth. *Copeia* pp. 280–284.

Cole, K. S., and Stacey, N. E. (1982). Prostaglandin and female spawning behavior in a cichlid. *Amer. Zool.* **22**, 924.

Colombo, L., Marconato, A., Belvedere, P. C., and Friso, C. (1980). Endocrinology of teleost reproduction: A testicular steroid pheromone in the black goby, *Gobius jozo* L. *Boll. Zool.* **47**, 355–364.

Colombo, L., Belvedere, P. C., Marconato, A., and Bentivegna, F. (1982). Pheromones in teleost fish. *In* "Proc. Int'l. Symp. Reprod. Physiol. Fish," (C. J. J. Richter and H. J. Th. Goos, eds.), pp. 84–94. Pudoc, Wageningen, Netherlands.

Crapon, de Caprona, M. D. (1980). Olfactory communication in a cichlid fish, *Haplochromis burtoni* Z. *Tierpsychol.* **52**, 113–134.

Crawford, S. (1975). An analysis of the reproductive behaviour of the female American Flagfish, *Jordanella floridae* (Pisces: Cyprinodontidae), including preliminary studies of the histology of the reproductive tract and of the internal factors regulating spawning behavior. B. Sc. Thesis, University of British Columbia, Vancouver, B.C., Canada.

Crews, D. (1973). Coition-induced inhibition of sexual receptivity in female lizards (*Anolis carolinesis*). Physiol. Behav. **11**, 463–468.

Crews, D. (1975). Psychobiology of reptilian reproduction. *Science* **189**, 1059–1065.

Crim, L. W., and Idler, D. R. (1978). Plasma gonadotropin, estradiol and vitellogenin and gonad phosvitin levels in relation to the seasonal reproductive cycles of female brown trout. *Ann. Biol. Anim., Biochim., Biophys.* **18**, 1001–1005.

Crim, L. W., and Peter, R. E. (1978). The influence of testosterone implantation in the brain and pituitary on pituitary gonadotropin levels in Atlantic salmon parr. *Ann. Biol. Anim., Biochim., Biophys.* **18**, 689–694.

Crow, R. T., and Liley, N. R. (1979). A sexual pheromone in the guppy, *Poecilia reticulata* (Peters). *Can. J. Zool.* **57**, 184–188.

Crowley, W. R., and Zemlan, F. P. (1981). The neurochemical control of mating behavior. *In* "Neuroendocrinology of Reproduction: Physiology and Behavior" (N. T. Adler, ed.), pp. 451–484. Plenum, New York.

Davis, R. E., and Kassel, J. (1975). The ontogeny of agonistic behavior and the onset of sexual maturation in the paradise fish *Macropodus opercularis* (Linnaeus). *Behav. Biol.* **14**, 31–39.

Davis, R. E., Mitchell, M., and Dolson, L. (1976). The effect of methallibure on conspecific visual reinforcement and social display frequency in the paradise fish *Macropodus opercularis* (L.). *Physiol. Behav.* **17**, 47–52.

Davis, R. E., Morell, J. I. and Pfaff, D. W. (1977). Autoradiographic localization of sex steroid-concentrating cells in the brain of the teleost fish *Macropodus opercularis* (Osteichthyes: Belontiidae). *Gen. Comp. Endocrinol.* **33**, 496–505.

de Bruin, J. P. C. (1980). Telencephalon and behavior in teleost fish: A neuroethological approach. *In* "Comparative Neurology of the Telencephalon" (S. O. E. Ebbesson, ed.), pp. 175–201. Plenum, New York.

Demski, L. S., Dulka, J. G., and Northcutt, R. G. (1982). Chemosensory control of spawning mechanisms in goldfish. *Soc. Neurosci. Abstr.* **8**, 611.

Demski, L. S., and Hornby, P. J. (1982). Hormonal control of fish reproductive behavior: Brain-gonadal steroid interactions. *Can. J. Fish. Aquat. Sci.* **38**, 36–47.

Demski, L. S., and Knigge, K. M. (1971). The telencephalon and hypothalamus of the bluegill (*Lepomis macrochirus*): Evoked feeding, aggressive and reproductive behavior by representative frontal sections. *J. Comp. Neurol.* **143**, 1–16.

Demski, L. S., and Northcutt, R. G. (1983). The terminal nerve: A new chemosensory system in vertebrates? *Science* **220**, 435–437.

Diakow, C., and Nemiroff, A. (1981). Vasotocin, prostaglandin, and female reproductive behavior in the frog, *Rana pipiens*. *Horm. Behav.* **15**, 86–93.

Diakow, C., and Raimondi, D. (1981). Physiology of *Rana pipiens* reproductive behavior: A proposed mechanism for inhibition of the release call. *Am. Zool.* **21**, 295–304.

Dindo, J. J., and MacGregor, R., III (1981). Annual cycle of serum gonadal steroids and serum lipids in striped mullet. *Trans. Am. Fish. Soc.* **110**, 403–409.

Egami, N. (1959a). Preliminary note on the induction of the spawning reflex and oviposition in *Oryzias latipes* by the administration of neurophypophyseal substances. *Annot. Zool. Jpn.* **32**, 13–17.

Egami, N. (1959b). Effect of testosterone on the sexual characteristics of the gobiid fish, *Pterogobius zonoleucus*. *Annot. Zool. Jpn.* **32**, 123–128.

Egami, N., and Ishii, S. (1962). Hypophyseal control of reproductive functions in teleost fishes. *Gen. Comp. Endocrinol., Suppl.* **1**, 248–253.

Elefthériou, B. E., Norman, R. L., and Summerfelt, R. (1968). Plasma levels of 1,3, 5(10)-estratriene-3,17β-diol and 3,17β-dihydroxy-Δ1,3,5(10)-estratriene-16-one. *Steroids* **11**, 89–92.

Emanuel, M. E., and Dodson, J. J. (1979). Modification of the rheotropic behavior of male rainbow trout (*Salmo gairdneri*) by ovarian fluid. *J. Fish. Res. Board Can.* **36**, 63–68.

Endler, J. A. (1980). Natural selection on color patterns in *Poecilia reticulata*. *Evolution* **34**, 76–91.

Escaffre, A. M., Petit, J., and Billard, R. (1977). Evolution de la quantité d'ovules récoltés et conservation de leur aptitude a être fécondés au cours de la période post ovulataire chez la truite arc en ciel. *Bull. Fr. Piscic.* **265**, 134–142.

Feder, H. H. (1981). Perinatal hormones and their role in the development of sexually dimorphic behaviors. *In* "Neuroendocrinology of Reproduction" (N. T. Adler, ed.), pp. 127–157. Plenum, New York.

Fernald, R. D. (1976). The effect of testosterone on the behavior and coloration of adult male cichlid fish, *Haplochromis burtoni*. *Horm. Res.* **7**, 172–178.

Ferraro, S. P. (1980). Daily time of spawning of 12 fishes in the Peconic Bays, New York. *Fish. Bull.* **78**, 455–464.

Fiedler, K. (1962). Die Wirkung von Prolactin auf das Verhalten des Lippfisches *Crenilabrus ocellatus* (Forskål). *Zool. Jahrb., Abt. Anat. Zool. Physiol. Tiere* **69**, 609–620.

Fiedler, K. (1970). Hormonale Auslösung der Geburtsbewegungen beim Seepferdchen (*Hippocampus*, Syngnathidae, Teleostei). *Z. Tierpsychol.* **27**, 679–686.

Fiedler, K. (1974). Hormonale Kontrolle des Verhaltens bei Fischen. *Fortsch. Zool.* **22**, 269–309.

Fiedler, K., Brüss, R., Christ, H., and Lotz-Zoller, R. (1979). Behavioural effects of peptide hormones in fishes. *In* "Brain and Pituitary Peptides" (W. Wuttke, A. Weindl, K. H. Vogt, and R. R. Dries, eds.), pp. 65–70. Karger, Basel.

Fischer, E. A. (1980). The relationship between mating system and simultaneous hermaphroditism in the coral reef fish, *Hypoplectrus nigricans* (Serranidae). *Anim. Behav.* **28**, 620–633.

Fontaine, M. (1976). Hormones and the control of reproduction in aquaculture. *J. Fish. Res. Board Can.* **33**, 922–939.

Forselius, S. (1957). Studies of anabantid fishes. *Zool. Bidr. Uppsala* **32**, 97–597.

Fostier, A., Weil, C., Terqui, M., Breton, B., and Jalabert, B. (1978). Plasma estradiol-17β and gonadotropin during ovulation in rainbow trout (*Salmo gairdneri* R). *Ann. Biol. Anim., Biochim., Biophys.* **18**, 929–936.

Franck, D., and Hannes, R. P. (1979). Hormones and the social behaviour of the male swordtail (*Xiphophorus helleri*). *Newsl. Int. Assoc. Fish Ethol.* **2**(6), 60.

Gandolfi, G. (1969). A chemical sex attractant in the guppy *Poecilia reticulata* Peters (Pisces: Poeciliidae). *Monit. Zool. Ital.* [N.S.] **3**, 89–98.

Goff, R. (1979). Electrophysiological investigations of the response of male goldfish to the pheromone released by a spawning female. M. Sc. Thesis, University of British Columbia, Vancouver, B.C., Canada.

Golubev, A. V., and Marusov, E. A. (1979). A study of chemoreception in the White Sea threespine stickleback (*Gasterosteus aculeatus*) during the reproductive period. *Biol. Nauki (Alma-Ata)* **9**, 33–35.

Gottfried, H., and van Mullem, P. J. (1967). On the histology of the interstitium and the occurrence of steroids in the stickleback (*Gasterosteus aculeatus* L.) testis. *Acta Endocrinol. (Copenhagen)* **56**, 1–15.

Greenberg, B. (1947). Some relations between territory, social hierarchy, and leadership in the green sunfish (*Lepomis cyanellus*). *Physiol. Zool.* **20**, 267–299.

Greenberg, B., Zijlstra, J. J., and Baerends, G. P. (1965). A quantitative description of the behavior changes during the reproductive cycle of the cichlid fish *Aequidens portalegrensis* (Hensel). *Proc. Ned. Akad. Wet., Ser. C* **68**, 135–149.

Gruber, M., Bos, E. S., and Ab, G. (1976). Hormonal control of vitellogeninsynthesis in avian liver. *Mol. Cell. Endocrinol.* **5**, 41–50.

Hale, E. B. (1956). Effects of forebrain lesions on the aggressive behavior of green sunfish, *Lepomis cyanellus. Physiol. Zool.* **29**, 107–127.

Hall, N. R., Luttge, W. G., and Berry, R. B. (1975). Intracerebral prostaglandin E2: Effects upon sexual behavior, open field activity, and body temperature in ovariectomized rats. *Prostaglandins* **10**, 877–888.

Hannes, R. P., and Franck, D. (1983). The effect of social isolation on the androgen and corticosteroid level in male swordtails, *Xiphophorus helleri*, and cichlid fish, *Haplochromis burtoni*. (In press.)

Hannes, R. P., Franck, D., and Liemann, F. (1983). Rank order fights influencing tissue concentrations of androgens and corticosteroids in male swordtails (*Xiphophorus helleri*). (In press.)

Hara, T. (1967). Electrophysiological studies of the olfactory system of the goldfish, *Carassius auratus* L. III. Effects of sex hormones on olfactory activity. *Comp. Biochem. Physiol.* **22**, 209–225.

Haskins, C. P., Haskins, E., McLaughlin, J. J. A., and Hewitt, R. E. (1961). Polymorphism and population structure in *Lebistes reticulatus*, an ecological study. *In* "Vertebrate Speciation" (W. F. Blair, ed.), pp. 320–395. Univ. of Texas Press, Austin.

Heiligenberg, W., and Kramer, U. (1972). Aggressiveness as a function of external stimulation. *J. Comp. Physiol.* **77**, 332–340.

Heinrich, W. (1967). Untersuchungen zum Sexualverhalten in der Gattung *Tilapia* (Cichlidae, Teleostei) und bei Artbastarden. *Z. Tierpsychol.* **24**, 684–754.

Heller, H. (1972). The effect of neurohypophysial hormones on the reproductive tract of lower vertebrates. *Gen. Comp. Endocrinol., Suppl.* **3**, 703–714.

Henderson, D. L., and Chiszar, D. A. (1977). Analysis of aggressive behaviour in the bluegill sunfish *Lepomis macrochirus* Rafinesque: effects of sex and size. *Anim. Behav.* **25**, 122–130.

Hirose, K., Ishida, R., and Sakai, K. (1977). Induced ovulation of Ayu using human chorionic gonadotropin (HCG), with special reference to changes in several characteristics of eggs retained in the body cavity after ovulation. *Bull. Jpn. Soc. Sci. Fish.* **43**, 409–416.

Hirose, K., Michida, Y., and Donaldson, E. M. (1979). Induced ovulation of Japanese flounder (*Limanda yokohamae*) with human chorionic gonadotropin and salmon gonadotropin, with special references to changes in quality of eggs retained in the ovarian cavity after ovulation. *Bull. Jpn. Soc. Sci. Fish.* **45**, 31–36.

Hishida, T., and Kawamoto, N. (1970). Androgenic and male-inducing effects of 11-ketotestosterone on a teleost, the medaka, *Oryzias latipes. J. Exp. Zool.* **173**, 279–283.

Hoar, W. S. (1962a). Reproductive bheavior of fish. *Gen. Comp. Endocrinol., Suppl.* **1**, 206–216.

Hoar, W. S. (1962b). Hormones and the reproductive behaviour of the male three-spined stickleback (*Gasterosteus aculeatus*) *Anim. Behav.* **10**, 247–266.

Hoar, W. S. (1969). Reproduction. *In* "Fish Physiology" (W. S. Hoar and D. J. Randall, eds.), Vol. 3, pp. 1–72. Academic Press, New York.

Hogendoorn, H., and Vismans, M. M. (1980). Controlled propagation of the African catfish, *Clarias lazera* (C. and V.) II. Artificial reproduction. *Aquaculture* **21**, 39–53.

Honda, H. (1979). Female sex pheromone of the Ayu, *Plecoglossus altivelis*, involved in courtship behaviour. *Bull. Jpn. Soc. Sci. Fish.* **45**, 1375–1380.

Honda, H. (1980a). Female sex pheromone of rainbow trout, *Salmo gairdneri*, involved in courtship behavior. *Bull. Jpn. Soc. Sci. Fish.* **46**, 1109–1112.

Honda, H. (1980b). Female sex pheromone of the loach, *Misgurnus anguillicaudatus*, involved in courtship behavior. *Bull. Jpn. Soc. Sci. Fish.* **46**, 1223–1225.

Hopkins, C. D. (1972). Sex differences in electric signaling in an electric fish. *Science* **176**, 1035–37.

Hunter, J. R., and Hasler, A. D. (1965). Spawning association of the redfin shiner *Notropis umbratilis*, and the green sunfish *Lepomis cyanellus*. *Copeia* pp. 265–281.

Idler, D. R. (1973). "Hormones in the Life of the Atlantic Salmon. International Atlantic Salmon Symposium, 1973." The International Atlantic Salmon Foundation: St. Andrews, New Brunswick, Canada.

Idler, D. R., Bitners, I. I., and Schmidt, P. J. (1961). 11-ketotestosterone: An androgen for sockeye salmon. *Can. J. Biochem. Physiol.* **39**, 1737–1742.

Idler, D. R., Horne, D. A., and Sangalang, G. B. (1971). Identification and quantification of the major androgens in testicular and peripheral plasma of Atlantic salmon (*Salmo salar*) during sexual maturation. *Gen. Comp. Endocrinol.* **16**, 257–267.

Idler, D. R., Hwang, S. J., Crim, L. W., and Reddin, D. (1981). Determination of sexual maturation stages of Atlantic salmon (*Salmo salar*) captured at sea. *Can. J. Fish. Aquat. Sci.* **38**, 405–413.

Ingersoll, D. W., Bronstein, P. M., and Bonventre, J. (1976). Chemical modulation of agonistic display in *Betta splendens*. *J. Comp. Physiol. Psychol.* **90**, 198–202.

Iwamatsu, T. (1978). Studies on oocyte maturation of the medaka, *Oryzias latipes*. VI. Relationship between the circadian cycle of oocyte maturation and activity of the pituitary gland. *J. Exp. Zool.* **206**, 355–364.

Jalabert, B., Moreau, J., Planquette, D., and Billard, R. (1974). Determinisme du sexe chez les hybrides entre *Tilapia macrochir* et *Tilapia nilotica*: Action de la methyltestostérone dans l'alimentation des alévins sur la différenciation sexuelles; proportion du sexes dans les descendance des males "inverses." *Ann. Biol. Anim., Biochim., Biophys.* **14**, 729–739.

Jalabert, B., Goetz, F. W., Breton, B., Fostier, A., and Donaldson, E. M. (1978). Precocious induction of oocyte maturation and ovulation in coho salmon, *Oncorhynchus kisutch*. *J. Fish. Res. Board Can.* **35**, 1423–1429.

Jensen, G. L., and Shelton, W. L. (1979). Effects of estrogens on *Tilapia aurea*: Implications for production of monosex genetic male tilapia. *Aquaculture* **16**, 233–242.

Johns, L. S., and Liley, N. R. (1970). The effects of gonadectomy and testosterone treatment on the reproductive behavior of the male blue gourami, *Trichogaster trichopterus*. *Can. J. Zool.* **48**, 977–987.

Katz, Y., and Eckstein, B. (1974). Changes in steroid concentration in blood of female *Tilapia aurea* (Teleostei, Cichlidae) during initiation of spawning. *Endocrinology* **95**, 963–967.

Keenleyside, M. H. A. (1979). "Diversity and Adaptation in Fish Behaviour," Zoophysiology, Vol. II. Springer-Verlag; Berlin and New York.

Kelley, D. B., and Pfaff, D. W. (1979). Generalizations from comparative studies on neuroanatomical and endocrine mechanisms of sexual behavior. *In* "Biological Determinants of Sexual Behavior" (J. B. Hutchison, ed.), pp. 225–254. Wiley, New York.

Kendle, E. R. (1970). Sexual discrimination by olfaction in the black bullhead, *Ictalurus melas*. *State Nebr. Game Parks Comm. Job Prog. Rep. 1970* pp. 21–31.

Khoo, K. H. (1979). The histochemistry and endocrine control of vitellogenesis in goldfish ovaries. *Can. J. Zool.* **57**, 617–626.

Kim, Y. S., Stumpf, W. E., and Sar, M. (1978a). Topography of estrogen target cells in the forebrain of goldfish, *Carassius auratus*. *J. Comp. Neurol.* **182**, 611–620.

Kim, Y. S., Stumpf, W. E., Sar, M., and Martinez-Vargas, M. C. (1978b). Estrogen and androgen target cells in the brain of fishes, reptiles and birds: Phylogeny and ontogeny. *Am. Zool.* **18**, 425–433.

Kim, Y. S., Stumpf, W. E., and Sar, M. (1979). Topographical distribution of estrogen target cells in the forebrain of platyfish, *Xiphophorus maculatus*, studied by autoradiography. *Brain Res.* **170**, 43–49.

Kramer, B. (1971). Zur hormonalen Steuerung von Verhaltensweisen der Fortpflanzung beim Sonnenbarsch *Lepomis gibbosus*. *Z. Tierpsychol.* **28**, 351–386.

Kramer, B. (1972). Behavioural effects of an antigonadotropin, of sexual hormones, and of Psychopharmaka in the pumpkinseed sunfish, *Lepomis gibbosus* (Centrarchidae). *Experientia* **28,** 1195.

Kramer, B. (1973). Chemische Wirkstoffe in Nestbau, Sexual und Kampfverhalten des Sonnenbarsches *Lepomis gibbosus* (L.) (Centrarchidae, Teleostei). *Z. Tierpsychol.* **32,** 353–373.

Kramer, D. L. (1972). The role of androgens in the parental behavior of the blue gourami, *Trichogaster Trichopterus* (Pisces, Belontiidae). *Anim. Behav.* **20,** 798–807.

Kutaygil, N. (1959). Insemination, sexual differentiation and secondary sex characters in *Stevardia albipinnis* Gill. *Istanbul Univ. Fen Fak. Mecm.*, *Ser. B* **24,** 93–128.

Kyle, A. (1982). Psychoneuroendocrinology of spawning behaviour in the male goldfish. Ph.D. Thesis, University of Alberta, Edmonton.

Kyle, A., and Peter, R. E. (1982). Effects of forebrain lesions on spawning behaviour in the male goldfish. *Physiol. Behav.* **28,** 1103–1109.

Kyle, A. L., Stacey, N. E., and Peter, R. E. (1982a). Ventral telencephalic lesions: Effects on bisexual behavior, activity, and olfaction in the male goldfish. *Behav. Neurol. Biol.* **36,** 229–241.

Kyle, A. L., and Stacey, N. E., and Peter, R. E. (1982b). Sexual stimuli associated with increased gonadotropin and milt levels in goldfish. *In* "Proc. Int'l Symp. Reprod. Physiol. Fish," (C. J. J. Richter and H. J. Th. Goos, eds.), p. 240. Pudoc, Wageningen, Netherlands.

Lam, T. J., Nagahama, Y., Chan, K., and Hoar, W. S. (1978). Overripe eggs and post-ovulatory corpora lutea in the threespine stickleback. *Gasterosteus aculeatus* L., form *trachurus*. *Can. J. Zool.* **56,** 2029–2036.

Lambert, J. G. D., and van Oordt, P. G. W. J. (1974). Ovarian hormones in teleosts. *Fortschr. Zool.* **22,** 340–349.

Laskowski, W. (1954). Einige Verhaltensstudien an *Platypoecilus variatus*. *Biol. Zentralbl.* **73,** 429–438.

Laumen, J., Pern, U., and Blüm, V. (1974). Investigations on the function and hormonal regulation of the anal appendices in *Blennius pavo* (Risso). *J. Exp. Zool.* **190,** 47–56.

Lee, C. T., and Ingersoll, D. W. (1979). Social chemosignals in five Belontiidae (Pisces) species. *J. Comp. Physiol. Psychol.* **93,** 117–1181.

Levy, M., and Aronson, L. R. (1955). Morphological effects of castration and hormone administration in the male cichlid fish *Tilapia macrocephala*. *Anat. Rec.* **122,** 450–451.

Liley, N. R. (1966). Ethological isolating mechanisms in four sympatric species of Poeciliid fishes. *Behaviour, Suppl.* **13,** 1–197.

Liley, N. R. (1968). The endocrine control of reproductive behaviour in the female guppy, *Poecilia reticulata* Peters. *Anim. Behav.* **16,** 318–331.

Liley, N. R. (1969). Hormones and reproductive behavior. *In* "Fish Physiology" (W. S. Hoar and D. J. Randall, eds.), Vol. 3, pp. 73–116. Academic Press, New York.

Liley, N. R. (1972). The effects of estrogens and other steroids on the sexual behavior of the female guppy, *Poecilia reticulata*. *Gen. Comp. Endocrinol.*, *Suppl.* **3,** 542–552.

Liley, N. R. (1980). Patterns of hormonal control in the reproductive behavior of fish, and their relevance to fish management and culture programs. *In* "Fish Behavior and its Use in the Capture and Culture of Fishes" (J. E. Bardach, J. J. Magnuson, R. C. May, and J. M. Reinhart, eds.), ICLARM Conf. Proc. 5, pp. 210–246. Int. Cent. Living Aquat. Resour. Manage., Manila, Philippines.

Liley, N. R., (1982). Chemical communication in fish. *Can. J. Fish. Aquat. Sci.* **39,** 22–35.

Liley, N. R., and Donaldson, E. M. (1969). The effects of salmon pituitary gonadotropin on the ovary and the sexual behavior of the female guppy, *Poecilia reticulata*. *Can. J. Zool.* **47,** 569–573.

Liley, N. R., and Tan, E. S. P. (1983). The induction of spawning behaviour in *Puntius gonionotus* by treatment with prostaglandin $PGF_{2\alpha}$ (In preparation).

Liley, N. R., and Wishlow, W. P. (1974). The interaction of endocrine and experiential factors in the regulation of sexual behaviour in the female guppy *Poecilia reticulata*. *Behaviour* **48**, 185–214.

Lindsay, W. K. (1974). The effects of neonatal androgen treatment on androgen-induced masculinization in the adult female guppy, *Poecilia reticulata*. B.Sc. Thesis, University of British Columbia, Vancouver, B.C., Canada.

Losey, G. S., Jr. (1969). Sexual pheromones in some species of the genus *Hypsoblennius Gill*. *Science* **163**, 181–183.

McBride, J. R., Fagerlund, U. H. M., Smith, M., and Tomlinson, N. (1963). Resumption of feeding by and survival of adult sockeye salmon (*Oncorhynchus nerka*) following advanced gonad development. *J. Fish. Res. Board Can.* **20**, 95–100.

Macey, M. J., Pickford, G. E., and Peter, R. E. (1974). Forebrain localization of the spawning reflex response to exogenous neurohypophysial hormones in the killifish, *Fundulus heteroclitus*. *J. Exp. Zool.* **190**, 269–280.

McEwen, B. S. (1981). Cellular biochemistry of hormone action in brain and pituitary. *In* "Neuroendocrinology of Reproduction" (N. T. Adler, ed.), pp. 485–518. Plenum, New York.

MacGregor, R., III, Dindo, J. J., and Finucane, J. H. (1981). Changes in serum androgens and estrogens during spawning in bluefish, *Pomatomus saltator*, and king mackerel, *Scomberomorus cavalla*. *Can. J. Zool.* **59**, 1749–1754.

Machemer, L. (1971). Synergistische Wirkung von Säuger-Prolaktin und Androgen beim Paradiesfisch, *Macropodus opercularis* L. (Anabantidae). *Z. Tierpsychol.* **28**, 33–53.

Machemer, L., and Fiedler, K. (1965). Zur hormonalen Steuerung des Schaumnestbaues beim Paradiesfisch, *Macropodus opercularis* L. (Anabantidae, Teleostei). *Naturwissenschaften* **52**, 648–649.

MacKenzie, D. S., Scott, A. P., Stacey, N. E., and Kyle, A. L. (1982). Gonadotropin and sex steroid changes during spawning in the white sucker (*Catostomus commersoni*). *Amer. Zool.* **22**, 995.

McPhail, J. D. (1969). Predation and the evolution of a stickleback (*Gasterosteus*). *J. Fish. Res. Board Can.* **26**, 3183–3108.

Mainardi, D., and Rossi, A. C. (1968). Comunicazione chimica in rapporto alla costruzione del nido nel pesce anabantide *Colisa lalia*. *Rend., Ist. Lomb. Accad. Sci. Lett. B* **102**, 23–28.

Marrone, B. L., Rodriguez-Sierra, J. F., and Feder, H. H. (1979). Differential effects of prostaglandins on lordosis behavior in female guinea pigs and rats. *Biol. Reprod.* **20**, 853–861.

Meyer, J. H. (1983). Steroid influences upon the discharge frequencies of weakly electric fish. *J. Comp. Physiol.* (in press).

Meyer, J. H., and Liley, N. R. (1982). Hormonal control of pheromone production in the guppy (*Poecilia reticulata* Peters). *Can. J. Zool.* **60**, 1505–1510.

Meyer, J. H., and Zakon, H. H. (1982). Androgens alter tuning in electroreceptors. *Science* **217**, 635–637.

Molenda, W., and Fiedler, K. (1971). Die Wirkung von Prolaktin auf das Verhalten von Stichlings ♂ ♂ (*Gasterosteus aculeatus* L.). *Z. Tierpsychol.* **28**, 463–474.

Mollah, M. F. A., and Tan, E. S. P. (1983). Viability of catfish (*Clarias macrocephalus* Gunther) eggs fertilised at various times post-ovulation. *J. Fish. Biol.* **22**, 563–566.

Morali, G., and Beyer, C. (1979). Neuroendocrine control of mammalian estrous behavior. *In* "Endocrine Control of Sexual Behavior" (C. Beyer, ed.), pp. 33–75. Raven Press, New York.

Morell, J. I., Kelley, D. B., and Pfaff, D. W. (1975). Sex steroid binding in the brains of vertebrates. Studies with light microscopic autoradiography. *Proc. Int. Symp. Brain-Endocr. Interact.* **2**, 230–256.

Moser, G. H. (1967). Reproduction and development of *Sebastodes paucispinus* and comparison with other rockfishes off Southern California. *Copeia* pp. 773–797.

Munz, H., Stumpf, W. E., and Jennes, L. (1981). LHRH systems in the brain of platyfish. *Brain Res.* **221**, 1–13.

Munz, H., Class, B., Stumpf, W. E., and Jennes, L. (1982). Centrifugal innervation of the retina by luteinizing hormone releasing hormone (LHRH)-immunoreceptive telencephalic neurons in teleostean fishes. *Cell Tissue Res.* **222**, 313–323.

Nelson, K. (1964). Behavior and morphology in the glandulocaudine fishes (Ostariophysi, Characidae). *Univ. Calif., Berkeley, Publ. Zool.* **75**, 55–152.

Newcombe, C., and Hartman, G. (1973). Some chemical signals in the spawning behavior of rainbow trout (*Salmo gairdneri*). *J. Fish. Res. Board Can.* **30**, 995–997.

Noble, G. K., and Borne, R. (1940). The effect of sex hormone on the social hierarchy of *Xiphophorus helleri. Anat. Rec.* **78**, Suppl., 147.

Noble, G. K., and Kumpf, K. F. (1936). The sexual behavior and secondary sex characters of gonadectomized fish. *Anat. Rec.* **67**, Suppl., 113.

Noble, G. K., Kumpf, K. F., and Billings, V. N. (1938). The induction of brooding behavior in the jewel fish. *Endocrinology* **23**, 353–9.

Ogata, H., Normura, T., and Hata, M. (1979). Prostaglandin $F2_\alpha$ changes induced by ovulatory stimuli in the pond loach, *Misgurnus anguillicaudatus. Bull. Jpn. Soc. Sci. Fish.* **45**, 929–931.

Okada, H., Sakai, D. K., and Sugiwaka, K. (1978). Chemical stimulus on the reproductive behavior of the pond smelt. *Sci. Rep. Hokkaido Fish Hatchery* **33**, 89–99.

Oshima, K., and Gorbman, A. (1968). Modification by sex hormones of the spontaneous and evoked bulbar electrical activity in goldfish. *J. Endocrinol.* **40**, 409–420.

Oshima, K., and Gorbman, A. (1969). Effect of estradiol on NaCl-evoked olfactory bulbar potentials in goldfish: Dose-response relationships. *Gen. Comp. Endocrinol.* **13**, 92–97.

Pandey, S. (1969a). Effects of hypophysectomy on the testis and secondary sex characters of the adult guppy *Poecilia reticulata* Peters. *Can. J. Zool.* **47**, 775–781.

Pandey, S. (1969b). Effects of methyl testosterone on the testis and secondary sex characters of the hypophysectomized adult male guppy *Poecilia reticulata* Peters. *Can. J. Zool.* **47**, 783–786.

Pandey, S. (1970). Effects of methallibure on the testes and secondary sex characters of the adult and juvenile guppy, *Poecilia reticulata* Peters. *Biol. Reprod.* **2**, 239–244.

Partridge, B. L., Liley, N. R., and Stacey, N. E. (1976). The role of pheromones in the sexual behaviour of the goldfish. *Anim. Behav.* **24**, 291–299.

Parzefall, J. (1970). Morphologische Untersuchungen an einer Höhlenform von *Mollienesia sphenops* (Pisces, Poeciliidae). *Z. Morphol. Tiere* **68**, 323–342.

Parzefall, J. (1973). Attraction and sexual cycle of poeciliids. *In* "Genetics and Mutagenesis of Fish" (J. H. Schröder, ed.), pp. 357–406. Springer-Verlag, Berlin and New York.

Peter, R. E. (1977). The preoptic nucleus in fishes: A comparative discussion of function - activity relationships. *Am. Zool.* **17**, 775–785.

Peter, R. E. (1982). Neuroendocrine control of reproduction in teleosts. *Can. J. Fish. Aquat. Sci.* **39**, 48–55.

Pickford, G. E., and Strecker, E. L. (1977). The spawning reflex response of the killifish, *Fundulus heteroclitus:* Isotocin is relatively inactive in comparison with arginine vasotocin. *Gen. Comp. Endocrinol.* **32**, 132–137.

Pickford, G. E., Knight, W. R., and Knight, J. N. (1980). Where is the spawning reflex receptor for neurohypophyseal peptides in the killifish, *Fundulus Leteroclitus?* *Rev. Can. Biol.* **39,** 97–105.

Polder, J. J. W. (1971). On gonads and reproductive behavior in the cichlid fish *Aequidens portalegrensis* (Hensel). *Neth. J. Zool.* **21,** 265–365.

Pollack, E. I., Becker, L. R., and Haynes, K. (1978). Sensory control of mating in the blue gourami, *Trichogaster trichopterus* (Pisces, Belontiidae). *Behav. Biol.* **22,** 92–103.

Ramaswami, L. S., and Hasler, A. D. (1955). Hormones and secondary sex characters in the minnow, *Hyborhynchus. Physiol. Zool.* **28,** 62–68.

Reinboth, T., and Rixner, W. (1970). "Verhalten das Kleinen Maulbrüters *Hemihaplochromis multicolor* nach Kastration und Behandlung mit Testosteron," C1019/1970, Inst. Wiss. Film, Göttingen.

Richards, I. S. (1974). Caudal neurosecretory system: possible role in pheromone production. *J. Exp. Zool.* **187,** 405–408.

Richards, I. S. (1976). Chemoreception in the brown bullhead catfish, *Ictalurus nebulosus:* Intraspecific discrimination and territorial behavior and the recognition of four sympatric teleosts. Ph.D. Thesis, New York University, New York.

Robertson, O. H. (1961). Prolongation of the life span of kokanee salmon (*Oncorhynchus nerka kennerlyi*) by castration before beginning of gonad development. *Proc. Natl. Acad. Sci. U.S.A.* **47,** 609–621.

Rodriguez-Sierra, J. F., and Komisaruk, B. R. (1977). Effects of prostaglandin E2 and indomethacin on sexual behavior in the female rat. *Horm. Behav.* **9,** 281–189.

Rodriguez-Sierra, J. F., Crowley, W. R., and Komisaruk, B. R. (1975). Vaginal stimulation in rats induces prolonged lordosis responsiveness and sexual receptivity. *J. Comp. Physiol. Psychol.* **89,** 79–85.

Rossi, A. C. (1969). Chemical signals and nest-building in two species of *Colisa* (Pisces, Anabantidae). *Monit. Zool. Ital.* [N.S.] **3,** 225–237.

Roule, L. (1931). "Les poissons et le monde vivant des eaux," Vol. 4. Delagrave, Paris.

Rouse, E. F., Coppinger, C. J., and Barnes, P. R. (1977). The effect of an androgen inhibitor on behavior and testicular morphology in the stickleback *Gasterosteus aculeatus. Horm. Behav.* **9,** 8–18.

Rubec, P. J. (1979). Effect of pheromones on behavior of ictalurid catfish. Ph.D. Thesis, Texas A & M University, College Station.

Sanchez-Rodriguez, M., Escaffre, A. M., Marlot, S., and Reinaud, P. (1978). The spermiation period in the rainbow trout (*Salmo gairdneri*). Plasma gonadotropin and androgen levels, sperm production and biochemical changes in the seminal fluid. *Ann. Biol. Anim., Biochim., Biophys.* **18,** 943–948.

Sangalang, G. B., and Freeman, H. C. (1974). Effects of sublethal cadmium on maturation and testosterone and 11-ketotestosterone production *in vivo* in brook trout. *Biol. Reprod.* **11,** 429–435.

Sangalang, G. B., and Freeman, H. C. (1977). A new sensitive and precise method for determining testosterone and 11-ketotestosterone in fish plasma by radioimmunoassay. *Gen. Comp. Endocrinol.* **32,** 432–439.

Schmidt, P. J., and Idler, D. R. (1962). Steroid hormones in the plasma of salmon at various states of maturation. *Gen. Comp. Endocrinol.* **2,** 204–214.

Schreck, C. B. (1974). Hormonal treatment and sex manipulation in fishes. *In* "Control of Sex in Fishes" (C. B. Schreck, ed.), p. 84–106. Extension Division, Virginia Polytechnic. Institute and State University, Blacksburg.

Schreck, C. B., and Hopwood, M. L. (1974). Seasonal androgen and estrogen patterns in goldfish, *Carassius auratus. Trans. Am. Fish. Soc.* **103,** 375–392.

Schwanck, E. (1980). The effect of size and hormonal state on the establishment of dominance in young males of *Tilapia mariae* (Pisces: Cichlidae). *Behav. Processes* **5**, 45–53.

Scott, A. P., Bye, V. J., Baynes, S. M., and Springate, J. R. C. (1980a). Seasonal variations in plasma concentrations of 11-ketotestosterone and testosterone in male rainbow trout, *Salmo gairdneri* Richardson. *J. Fish Biol.* **17**, 495–505.

Scott, A. P., Bye, V. J., and Baynes, S. M. (1980b). Seasonal variations in sex steroids of female rainbow trout (*Salmo gairdneri* Richardson). *J. Fish Biol.* **17**, 587–592.

Seghers, B. H. (1967). The role of the ovary in the reproductive behavior of the blue gourami, *Trichogaster trichopterus* (Pallas). B.Sc. Thesis, Department of Zoology, University of British Columbia, Vancouver, B.C., Canada.

Semler, D. E. (1971). Some aspects of adaptation in a polymorphism for breeding color in the threespine stickleback (*Gasterosteus aculeatus*). *J. Zool.* **165**, 291–302.

Shirai, K. (1962). Correlation between the growth of the ovipositor and ovarian condition in the bitterling, *Rhodeus ocellatus. Bull. Fac. Fish., Hokkaido Univ.* **13**, 137–151.

Shirai, K. (1964). Histological study on the ovipositor of the rose bitterling, *Rhodeus ocellatus. Bull. Fac. Fish., Hokkaido Univ.* **14**, 193–197.

Silverman, H. I. (1978). Changes in male courting frequency in pairs of the cichlid fish, *Sarotherodon (Tilapia) mossambicus*, with unlimited or with only visual contact. *Behav. Biol.* **23**, 189–196.

Simpson, T. H., and Wright, R. S. (1977). A radioimmunoassay for 11-oxotestosterone: Its application in the measurement of levels in blood serum of rainbow trout (*Salmo gairdneri*). *Steroids* **29**, 383–398.

Slimp, J. C. (1977). Reduction of genital sensory input and sexual behavior of female guinea pigs. *Physiol. Behav.* **18**, 1027–1031.

Smith, R. J. F. (1969). Control of prespawning behaviour of sunfish (*Lepomis gibbosus* and *L. megalotis*). I. Gonadal androgen. *Anim. Behav.* **17**, 279–285.

Smith, R. J. F. (1970). Control of prespawning behaviour of sunfish (*Lepomis gibbosus* and *Lepomis megalotis*). II. Environmental factors. *Anim. Behav.* **18**, 575–587.

Smith, R. J. F. (1974). Effects of 17α-methyltestosterone on the dorsal pad and tubercles of fathead minnows (*Pimephales promelas*). *Can. J. Zool.* **52**, 1031–1038.

Smith, R. J. F. (1976). Seasonal loss of alarm substance cells in North American cyprinoid fishes and its relation to abrasive spawning behaviour. *Can. J. Zool.* **54**, 1172–1182.

Smith, R. J. F., and Hoar, W. S. (1967). The effects of prolactin and testosterone on the parental behaviour of the male stickleback *Gasterosteus aculeatus. Anim. Behav.* **15**, 342–352.

Stacey, N. E. (1976). Effects of indomethacin and prostaglandins on spawning behavior of female goldfish. *Prostaglandins* **12**, 113–126.

Stacey, N. E. (1977). The regulation of spawning behavior in the female goldfish, *Carassius auratus.* Ph.D. Thesis, University of British Columbia, Vancouver, B. C., Canada.

Stacey, N. E. (1981). Hormonal regulation of female sexual behavior in teleosts. *Am. Zool.* **21**, 305–316.

Stacey, N. E., and Goetz, F. W. (1982). Role of prostaglandins in fish reproduction. *Can. J. Fish. Aquat. Sci.* **39**, 92–98.

Stacey, N. E., and Hourston, A. S. (1982). Spawning and feeding behavior of captive Pacific herring, *Clupea harengus pallasi. Can. J. Fish. Aquat. Sci.* **39**, 489–498.

Stacey, N. E., and Kyle, A. L. (1983). Effects of olfactory tract lesions on sexual and feeding behavior in the goldfish. *Physiol. Behav.* **30**, 621–628.

Stacey, N. E., and Liley, N. R. (1974). Regulation of spawning behavior in the female goldfish. *Nature (London)* **247**, 71–72.

Stacey, N. E., and Pandey, S. (1975). Effects of indomethacin and prostaglandins on ovulation of goldfish. *Prostaglandins* 9, 596–607.

Stacey, N. E., and Peter, R. E. (1979). Central action of prostaglandins in spawning behavior of female goldfish. *Physiol. Behav.* 22, 1191–1196.

Stacey, N. E., Cook, A. F., and Peter, R. E. (1979a). Ovulatory surge of gonadotropin in the goldfish, *Carassius auratus. Gen. Comp. Endocrinol.* 37, 246–249.

Stacey, N. E., Cook, A. E., and Peter, R. E. (1979b). Spontaneous and gonadotropin-induced ovulation in the goldfish, *Carassius auratus* L.: Effects of external factors. *J. Fish Biol.* 15, 349–361.

Stuart-Kregor, P. A. C., Sumpter, J. P., and Dodd, J. M. (1981). The involvement of gonadotropin and sex steroids in the control of reproduction in the parr and adults of Atlantic salmon, *Salmo salar* L. *J. Fish Biol.* 18, 59–72.

Stumpf, W. E., and Grant, L . D., eds. (1975). "Anatomical Neuroendocrinology." Karger, Basel.

Sundararaj, B. I., and Goswami, S. V. (1966). Effects of mammalian hypophysial hormones, placental gonadotrophins, gonadal hormones, and adrenal corticosteroids on ovulation and spawning in hypophysectomized catfish, *Heteropneustes fossilis* (Bloch). *J. Exp. Zool.* 161, 287–296.

Suzuki, K., and Hioki, S. (1979). Spawning behavior, eggs and larvae of the Lutjanid fish, *Lutjanus kasmira*, in an aquarium. *Jpn. J. Ichthyol.* 26, 161–166.

Suzuki, K., Tanaka, Y., and Hioki, S. (1980). Spawning behavior, eggs and larvae of the butterflyfish, *Chaetodon nippon*, in an aquarium. *Jpn. J. Ichthyol.* 26, 334–341.

Takahashi, H. (1975). Masculinization of the gonad of juvenile guppy *Poecilia reticulata* induced by 11-ketotestosterone *Bull. Fac. Fish., Hokkaido Univ.* 26, 11–22.

Tautz, A. F., and Groot, C. (1975). Spawning behavior of chum salmon (*Oncorhynchus keta*) and rainbow trout (*Salmo gairdneri*). *J. Fish. Res. Board Can.* 32, 633–642.

Tavolga, M. C. (1949). Differential effects of estradiol, estradiol benzoate and pregneninolone on *Platypoecilus maculatus. Zoologica (N.Y.)* 34, 215–237.

Tavolga, W. N. (1955). Effects of gonadectomy and hypophysectomy on prespawning behavior in males of the gobiid fish *Bathygobius soporator. Physiol. Zool.* 28, 218–233.

Tavolga, W. N. (1956). Visual, chemical and sound stimuli as cues in the sex discriminatory behavior of the gobiid fish *Bathygobius soporator. Zoologica (N.Y.)* 41, 49–64.

Teeter, J. (1980). Pheromone communication in Sea Lampreys (*Petromyzon marinus*): Implications for Population Management. *Can. J. Fish. Aquat. Sci.* 37, 2123–2132.

Thiessen, D. D., and Sturdivant, S. K. (1977). Female pheromone in the black molly fish (*Mollienesia latipinna*): A possible metabolic correlate. *J. Chem. Ecol.* 3, 207–217.

Timms, A. M., and Kleerekoper, H. (1972). The locomotor response of male *Ictalurus punctatus*, the channel catfish, to a pheromone released by the ripe female of the species. *Trans. Am. Fish. Soc.* 102, 302–310.

Tokarz, R. R., and Crews, D. (1981). Effects of prostaglandins on sexual receptivity in the female lizard, *Anolis carolinesis. Endocrinology* 109, 451–457.

Trewavas, E. (1973). 1. On the cichlid fishes of the genus *Pelmatochromis* with proposal of a new genus for *P. congicus;* on the relationship between *Pelmatochromis* and *Tilapia* and the recognition of *Sarotherodon* as a distinct genus. *Bull. Br. Mus. (Nat. Hist.) Zool.* 25, 3–26.

Turner, C. L. (1947). Viviparity in teleosts. *Sci. Mon.* 75, 508–518.

van Bohemen, C. G., and Lambert, J. G. D. (1981). Estrogen synthesis in relation to estrone, estradiol, and vitellogenin plasma levels during the reproductive cycle of the female rainbow trout, *Salmo gairdneri. Gen. Comp. Endocrinol.* 45, 105–114.

van den Hurk, R. (1977). Arguments for a possible endocrine control of the reproductive behaviour of male zebrafish (Brachydanio rerio). J. Endocrinol. 72, 63.

van den Hurk, R., Hart, L. A.'t, Lambert, J. G. D., and van Oordt, P. G. W. J. (1982). On the regulation of sexual behaviour of male zebrafish, Brachydanio rerio. Gen. Comp. Endocrinol. 46, 403.

Villars, T. A., and Burdick, M. (1982). Rapid decline in the behavioral response of female paradise fish to prostaglandin treatment. Amer. Zool. 22, 948.

Villars, T. A., and Davis, R. E. (1977). Castration and reproductive behavior in the paradise fish Macropodus opercularis, Osteichthyes, Belontiidae. Physiol. Behav. 19, 371–376.

von Ihering, R. (1937). Oviductal fertilization in the South American catfish, Trachycorystes. Copeia, 202–205.

Wai, E. H., and Hoar, W. S. (1963). The secondary sex characters and reproductive behaviour of gonadectomized sticklebacks treated with methyl testosterone. Can. J. Zool. 41, 611–628.

Wallace, R. A., and Dumont, J. N. (1968). The induced synthesis and transport of yolk proteins and their accumulation by the oocyte in Xenopus laevis. J. Cell. Physiol. 72, Suppl. 1, 73–90.

Wapler-Leong, D. C. Y., and Reinboth, R. (1974). The influence of androgenic hormone on the behaviour of Haplochromis burtoni (Cichlidae). Fortschr. Zool. 22, 334–339.

Weiss, C. S., and Coughlin, J. P. (1979). Maintained aggressive behavior in gonadectomized male Siamese fighting fish (Betta splendens). Physiol. Behav. 23, 173–177.

Whitehead, C., Bromage, N. R., and Forster, J. R. M. (1978). Seasonal changes in reproductive function of the rainbow trout (Salmo gairdneri). J. Fish Biol. 12, 601–608.

Wiebe, J. P. (1968). The reproductive cycle of the viviparous seaperch, Cymatogaster aggregata Gibbons. Can. J. Zool. 46, 1221–1234.

Wiley, M. L., and Collette, B. B. (1970). Breeding tubercles and contact organs in fishes: Their occurrence, structure and significance. Bull. Am. Mus. Nat. Hist. 143, 147–153.

Wilhelmi, A. E., Pickford, G. E., and Sawyer, W. H. (1955). Initiation of the spawning reflex response in Fundulus by the administration of fish and mammalian neurohypophyseal preparation and synthetic oxytocin. Endocrinology 57, 243–252.

Wilkens, H. (1972). Über Präadaptationen für das Höhlenleben, untersucht am Laichverhalten ober- und unterirdischer Population des Astyanax mexicanus (Pisces). Zool. Anz. 188, 1–11.

Wingfield, J. C., and Grimm, A. S. (1976). Preliminary identification of plasma steroids in the plaice, Pleuronectes platessa L. Gen. Comp. Endocrinol. 29, 78–83.

Wingfield, J. C., and Grimm, A. S. (1977). Seasonal changes in plasma cortisol, testosterone and oestradiol-17β in the plaice, Pleuronectes platessa L. Gen. Comp. Endocrinol. 31, 1–11.

Wootton, R. J. (1970). Aggression in the early phases of the reproductive cycle of the male three-spined stickleback (Gasterosteus aculeatus). Anim. Behav. 18, 740–746.

Wootton, R. J. (1976). "The Biology of the Sticklebacks." Academic Press, New York.

Yamamoto, T. (1969). Sex differentiation. In "Fish Physiology" (W. S. Hoar and D. J. Randall, eds.), Vol. 3, pp. 117–175. Academic Press, New York.

Yamazaki, F. (1965). Endocrinological studies of the reproduction of the female goldfish, Carassius auratus L., with special reference to the function of the pituitary gland. Mem. Fac. Fish., Hokkaido Univ. 13, 1–64.

Yamazaki, F. (1972). Effects of methyltestosterone on the skin and the gonad of salmonids. Gen. Comp. Endocrinol., Suppl. 3, 741–750.

Yamazaki, F., and Donaldson, E. M. (1969). Involvement of gonadotropin and steroid hormones in the spermiation of the goldfish, *Carassius auratus. Gen. Comp. Endocrinol.* **12,** 491–497.

Yamazaki, F., and Watanabe, K. (1979). The role of sex hormones in sex recognition during spawning behaviour of the goldfish, *Carassius auratus* L. *Proc. Indian Natl. Sci. Acad. Part B*-**45,** 505–511.

Zeiske, E. (1968). Prädispositionen bei *Mollienesia sphenops* (Pisces, Poeciliidae) für einen Übergang zum Leben in subterranen Gewässern. *Z. Vergl. Physiol.* **58,** 190–222.

and back again to summer. At the same time, the more complicated movements of the moon in relation to the earth and the sun produce our lunar and tidal cycles. Then, there is the periodicity of the monsoons in the tropics. Life on earth has evolved in relation to these periodicities, and all animals have the capacity to measure time and use this temporal information advantageously. Reproduction is one function that becomes cyclical in many species because of these periodicities.

Biologists understand that the cycles of reproduction are basic to the survival of the maximum number of young and therefore the success of the species. Baker (1938), in an early and important discussion of the evolution of the breeding season, stressed the difference between proximate factors, which time the development of the reproductive organs and processes in breeding adults, and the ultimate factors, such as abundance of food and favorable growing conditions, which affect the survival of the young. For example, in many salmonid fishes, decreasing or short day lengths coupled with late summer and autumn temperatures serve as proximate factors, triggering gonadal development and spawning; flooding waters, warmer temperatures, and abundant food in the spring and early summer are the ultimate factors influencing speedy seaward migration and rapid growth of the young fry.

Not all breeding cycles are based on proximate factors arising from cyclical environmental changes. Some are based on an endogenous rhythm (biological clock). Such a rhythm is demonstrated by maintaining organisms under constant environmental conditions and recording the presumed endogenous activity for a prolonged period. If the activity persists under these constant conditions, and if it deviates each day by a fixed amount from 24 hr (usually between 22 and 28 hr), it is circadian [circa (about) + dies (day)]; if the persistent rhythm is about 365 days it is circannual. These free-running rhythms gradually drift out of phase with the diurnal or annual cycle and are adjusted by an entrainer, referred to as a zeitgeber. Environmental factors serve as zeitgebers just as they serve as proximate cues or triggers in nonendogenous rhythms.

In this chapter, environmental factors are discussed both as proximate factors, which trigger gonadal development and breeding activities in many fish species, and as zeitgebers, which entrain endogenous rhythm in other species. During the 1970s, several critical reviews and symposia were devoted to this topic and the early literature can be traced through their bibliographies (see de Vlaming, 1972a, 1974; Donaldson, 1975; Htun-Han, 1977; Thorpe, 1978; Scott, 1979; Liley, 1980; Baggerman, 1980). The neuroendocrine mediation of environmental factors has also been extensively reviewed (Peter and Hontela, 1978; Billard and Breton, 1978; Poston, 1978; Peter and Crim, 1979; Peter, 1981; Billard *et al.*, 1981a; Crim, 1982; Peter,

Chapter 3, Volume 9A, this series), and is not considered here beyond passing references.

It has now become apparent that different phases of the reproductive cycle of a particular fish species may require different proximate factors or zeitgebers. Therefore, in this chapter, environmental influences on gametogenesis, spawning, and gonadal regression are considered separately in Sections II, III, and IV, respectively. Because the environmental requirements for reproduction are likely to differ among temperate, subtropical, and tropical species, these three groups of fish are considered separately wherever appropriate. Finally, to illustrate that knowledge gained concerning environmental influences on fish reproduction can be put to practical use, applications (both potential and actual) in aquaculture for broodstock management and induction of spawning are discussed.

II. ENVIRONMENTAL INFLUENCES ON GONADAL DEVELOPMENT (GAMETOGENESIS)

A. Temperate Species

Much research has been conducted using temperate species and the results have been reviewed many times (see reviews cited in Section I). Most of the studies concern freshwater species; only a few marine or estuarine species have been studied (Peter and Crim, 1979). Of the environmental factors, photoperiod and/or temperature are generally recognized as the most important cues in the timing of gametogenesis in temperate species.

1. PHOTOPERIOD

In species which spawn in spring or early summer, gonadal recrudescence is often stimulated by long photoperiods, particularly in combination with warm temperatures (see annotated bibliography by Htun-Han, 1977). This has been demonstrated, for example, in the bridle shiner, *Notropis bifrenatus* (Harrington, 1950, 1957), the threespine stickleback, *G. aculeatus* (Baggerman, 1957, 1972, 1980; Schneider, 1969), the Japanese medaka, *Oryzias latipes* (Yoshioka, 1962, 1963), the sunfish, *Lepomis cyanellus* (Kaya and Hasler, 1972) and *L. megalotis* (Smith, 1970), the mosquitofish, *Gambusia affinis* (Sawara, 1974), the golden shiner, *Notemigonus crysoleucas* (de Vlaming, 1975), and the goldfish, *Carassius auratus* (Kawamura and Otsuka, 1950; Fenwick, 1970; Gillet *et al.*, 1978).

An abrupt increase in photoperiod, which is commonly used in studies such as those previously cited, is not ecologically meaningful because fish are normally exposed to a gradually increasing photoperiod. However, in at least one species, this is actually what happens in nature: the minnow, *Phoxinus phoxinus* (Scott, 1979). In Loch Walton (Scotland), in winter, minnows spend the daylight hours under piles of stones in relative darkness; they emerge only after dark. However, in spring, when the water temperature reaches about 8°C, they also emerge in daylight. Therefore, in effect, the fish are exposed to a sudden increase in photoperiod in spring. In the laboratory, a rapid increase in photoperiod from 8 to 16 hr within 1 week caused a greater stimulation of vitellogenesis than gradually increasing photoperiod (Scott, 1979).

In contrast, in species which spawn in autumn or early winter, gonadal recrudescence is often favored by short or decreasing photoperiods (Htun-Han, 1977). This has been demonstrated in the salmonids (Combs *et al.*, 1959; Henderson, 1963; Shiraishi and Fukuda, 1966; Breton and Billard, 1977; Billard *et al.*, 1981b) and the ayu, *Plecoglossus altivelis* (Shiraishi and Takeda, 1961; Shiraishi, 1965a). In the rainbow trout, *Salmo gairdnerii*, a decreasing photoperiod is much more effective than a constant short photoperiod in stimulating gametogenesis (Breton and Billard, 1977; Billard *et al.*, 1981b), but this was not confirmed in a more recent study (Bromage *et al.*, 1982). Skarphedinsson *et al.* (1982) have even noted that long photoperiods stimulate gonad development in the rainbow trout. Therefore, the situation is not clear.

The action spectrum of photoperiodism appears to be broad (ranging from 388 to 653 nm, i.e., long ultraviolet (UV) to short red) for the stickleback, *G. aculeatus* (McInerney and Evans, 1970). However, in the ayu, *P. altivelis*, only short wavelengths (blue and green) accelerate gonadal maturation; long wavelengths (red and yellow) appear to be inhibitory (Shiraishi, 1965c). Caution should be exercised in the interpretation of these data because rigorous controls were not included in the experiments (de Vlaming, 1974).

Light intensity is often not considered in photoperiod studies, and, in most studies, light intensity is not stated. Shiraishi (1965b) demonstrated that the photoperiod effects in the ayu are dependent on light intensity, being absent or altered if the light intensity is too low or too high. Therefore, light intensity may be an important variable among experiments reported in the literature.

The time-measuring mechanism involved in photoperiodic responses appears to be based on a circadian rhythm of sensitivity to light. This was first demonstrated in the stickleback by Baggerman (1969, 1972) who used "skeleton photoperiods" (i.e., 6 hr of light coupled with an additional 2 hr of light at various times in the ensuing dark period). Maximal response in terms of percentage of fish attaining sexual maturity occurred when the light pulse fell between the hours 14 and 16 of the light cycle. Subsequently, a similar phenomenon has been demonstrated in other species: the Indian catfish, *Heteropneustes fossilis* (see Section II,B), the Japanese medaka, *O. latipes* (Chan, 1976), the honmoroko, *Gnathopogon elongatus caerulescence* (Khiêt, 1975), the bitterling, *Rhodeus ocellatus ocellatus* (Nishi, 1979), and the mosquitofish, *G. affinis affinis* (Nishi, 1981). Baggerman (1980) has provided a good discussion of this phenomenon. However, the stimulatory effect of decreasing photoperiod in the rainbow trout cannot be explained in terms of a shifting photosensitive phase (Billard *et al.*, 1981b). A monthly shift by 1 hr in the nighttime light pulse (1 hr) from the 16th to 10th hr of the light cycle did not stimulate gametogenesis; however a decreasing photoperiod from 16 hr of light to 10 hr during the same experimental period (6 months) was stimulatory.

2. TEMPERATURE

As noted by de Vlaming (1972a, 1974), temperature has often not been considered in photoperiod studies; therefore, it is not clear whether the photoperiod effects reported are temperature dependent. Temperature dependency of photoperiodism has in fact been reported in a number of species. For example, in *Lepomis gibbosus*, a long photoperiod [16 hr light alternating with 8 hr darkness (16L–8D)] induced nest building (sexual maturity) at 25°C but not at 11°–13°C (Smith, 1970). In *O. latipes*, long photoperiods fail to stimulate gonadal recrudescence at temperatures below 10°C (Yoshioka, 1970). Similarly, in *G. affinis affinis* (Sawara, 1974), the brook stickleback, *Culaea inconstans* (Reisman and Cade, 1967), and *N. crysoleucas* (de Vlaming, 1975), long photoperiods stimulate gametogenesis only if combined with warm temperatures. However, in some species, long photoperiods are stimulatory at both warm and cold temperatures: sexual maturation in the stickleback (Baggerman, 1957, 1980; Schneider, 1969), nest building in *L. megalotis* (Smith, 1970), ovarian development in winter goldfish (Gillet *et al.*, 1978), and spermatogenesis in seaperch, *Cymatogaster aggregata* (Wiebe, 1968). Further, in salmonids, decreasing or short photoperiods or an accelerated light cycle (annual light cycle condensed to 9 or 6 months) stimulate gametogenesis regardless of temperature (Henderson, 1963; Breton and Billard, 1977; MacQuarrie *et al.*, 1979). Nevertheless, in such species, temperature still affects the degree of photostimulation (see aforementioned references).

Temperature may even play a dominant role in sexual cycling in some species. For example, in the longjaw goby, *Gillichthys mirabilis*, low temperatures promote sexual maturation regardless of photoperiod, although the effect is enhanced by short photoperiod (de Vlaming, 1972b). Even a brief exposure of 2 hr/day to 27°C will prevent gonadal recrudescence (de Vlaming, 1972c). Temperature was also found to be more important than photoperiod for gametogenesis in the darter, *Etheostoma lepidum* (Hubbs and Strawn, 1957), spermatogenesis in the lake chub, *Couesius plumbeus* (Ahsan, 1966), and oogenesis in *C. aggregata* (Wiebe, 1968).

In the lake chub, low temperatures favor the formation of primary spermatocytes (meiotic phase), but high temperatures promote spermatogonial proliferation (mitosis) and spermiation (Ahsan, 1966). Comparable results were obtained in the killifish, *Fundulus heteroclitus* (Matthews, 1939; Lofts *et al.*, 1968). In contrast, the formation of spermatocytes appears to be stimulated by high temperatures in *Phoxinus laevis* (Bullough, 1939). In the stickleback, spermatogenesis proceeds to completion regardless of temperature and photoperiod (Baggerman, 1980).

Temperature is also important for oogenesis in some species. In the marsh killifish, *F. confluentus*, low temperatures promote the early phases of oocyte growth but high temperatures favor the late phases (Harrington, 1959). A similar situation apparently occurs in the banded sunfish, *Enneacanthus obesus*, although photoperiods are also involved (Harrington, 1956). In contrast, in *C. aggregata*, the formation and primary growth phase of oocytes (probably also the beginning of vitellogenesis, i.e., yolk vesicle formation or endogenous vitellogenesis) are stimulated by warm temperatures and the late phases (yolk granule formation or exogenous vitellogenesis) by low temperatures (Wiebe, 1968). Similarly, high temperatures enhance the early phases of oocyte growth in *P. laevis* (Bullough, 1939). However, in other species, the primary growth phase and early phases of vitellogenesis (endogenous vitellogenesis) occur independently of environmental factors (although the rate may be affected), e.g., *N. crysoleucas* (de Vlaming, 1975) and sticklebacks (Baggerman, 1980).

Temperature may exert its effects by (1) a direct action on gametogenesis (Lofts *et al.*, 1968), (2) an action on pituitary gonadotropin secretion (Breton and Billard, 1977; Peter, 1981), (3) an action on metabolic clearance of hormones (Peter, 1981), (4) an action on the responsiveness of the liver to estrogen in the production of vitellogenins (Yaron *et al.*, 1980), or (5) an action on the responsiveness of the gonad to hormonal stimulation (Jalabert *et al.*, 1977; Bieniarz *et al.*, 1978).

3. SEASONAL VARIATION IN PHOTOTHERMAL EFFECTS

The effects of photoperiod and temperature on fish gonadal development often vary with season. For example, in the stickleback, most fish in autumn are able to attain sexual maturation only when exposed to a long photoperiod (16L–8D); however, as the season progresses from autumn to early spring, an increasing number of fish are able to reach sexual maturity under shorter photoperiods (Baggerman, 1980). Therefore, although fish in November cannot respond to a short photoperiod (8L–16D), some fish in January and all fish in February–March are able to respond to 8L–16D with sexual maturation. The experiments were conducted at a constant temperature of 20°C. It is not clear if a lower temperature would affect the response.

Baggerman (1980) went on to demonstrate that the seasonal variation in responsiveness of sticklebacks to photoperiod is based on a seasonal change in the circadian photosensitivity rhythm. The photosensitive phase occurs progressively earlier in the daily light cycle as the season proceeds from late summer to spring. Therefore, in autumn the photosensitive phase of sticklebacks occurs around the 16th hour of the light cycle, but in spring it moves forward to around the 8th hour. Because during this period the gonads develop from phase 1 (up to completion of spermatogenesis in the male; up to early yolk vesicle stage in the female) to phase 2 (androgen secretion and spermiation in the male; completion of vitellogenesis, oocyte maturation, and ovulation in the female), the seasonal change may reflect a change in photosensitivity of the various gonadal stages. However, even at constant 8L–16D and 20°C beginning in autumn, conditions under which gonadal activity is arrested at phase 1 (Baggerman, 1957; T. J. Lam *et al.*, unpublished; see also Section IV, A), a seasonal change in the daily light-sensitivity rhythm still occurs although at a slower rate than under natural photoperiod–temperature regimes or at constant 8L–16D and 15°C (Baggerman, 1980). Therefore, an endogenous mechanism is indicated. Because the photosensitive phase shift occurs faster at constant 8L–16D and 15°C than at constant 8L–16D and 20°C, an exogenous input (viz. temperature) is also suggested. It is not clear whether gonadal development had in the meantime continued under constant 8L–16D and 15°C. If so, the seasonal variation in responsiveness to photoperiod may still in part be attributable to changing photosensitivity accompanying advancing stages of gametogenesis. Henderson (1963) has in fact demonstrated in the brook trout, *Salvelinus fontinalis*, that the effect of photoperiod depends on the phase of gametogenesis in progress at the start of the experiment.

In goldfish, Fenwick (1970) found that long photoperiods (16L–8D and 24L) at 11°–12°C stimulated gonadal development only during spring. However, Kawamura and Otsuka (1950) reported ovarian stimulation in goldfish by long photoperiods and warm temperature during both winter and spring. In a more recent and detailed study, Gillet *et al.*, (1978) demonstrated that in winter, a long photoperiod (16L–8D) promoted goldfish ovarian development at both 10°C and 20°C, although the effect was greater at 10°C. In autumn, the 16L–8D at 10°C regime also appeared to be stimulatory, but a short photoperiod (8L–16D) at 10°C proved more effective. However, in spring, 8L–16D at 10°C was inhibitory, but 16L–8D at 10°C remained stimulatory.

The discrepancy in the results of the aforementioned three studies may be attributable to differences in the initial gonadal condition of the respective experiments. In Fenwick's experiments, the fish used for the various seasons came from the same stock tank which had been maintained under constant conditions of 8L–16D and 11°–12°C; however, in the experiments of Gillet *et al.*, the fish came from ponds under natural conditions. Unfortunately, although an indication of the initial gonadal condition was given by Gillet *et al.* in terms of gonosomatic index (GSI), no such information was given by Fenwick.

Another point is that only GSI data are given in both these studies. These data do not yield as much information as gonadal histology. In fact, GSI may be misleading if spawning has been missed in the experiment. However, spawning may be detected histologically by the presence of postovulatory follicles. An example of the inadequacy of GSI compared to histological criteria is provided by de Vlaming (1975). In his study of *N. crysoleucas*, a seasonal change in environmental effects on gametogenesis was indicated by the GSI data when in fact no such change occurred based on histology; a long photoperiod–warm temperature regime stimulated gonadal development to the prespawning condition or induced spawning regardless of the season.

A diurnal variation in responsiveness to temperature has also been demonstrated in goldfish (Spieler *et al.*, 1977). Goldfish exposed to a daily increase in temperature from 15° to 24°C (for a 4-hr period) showed the greatest gonadal development when the thermoperiod fell during the last 4 hr of darkness on a 12L–12D photoperiod.

There are a few other studies of seasonal environmental effects on gametogenesis in fish. These include those of Wiebe (1968) in *C. aggregata*, de Vlaming (1972b,c) in *G. mirabilis*, and Sundararaj *et al.* (see Section II,B) in the Indian catfish. More research is needed, but in designing studies, it is important to consider the gonadal phase at the beginning of the experiment in each season, and to include histological data other than GSI.

4. Sexual Difference in Response to Environmental Factors

Although a sexual difference in gonadal response to environmental factors is not evident in many species (e.g., Fenwick, 1970; de Vlaming, 1975), it has been demonstrated in a few species. In *C. aggregata*, males respond mainly to photoperiod, but females respond predominantly to temperature

(Wiebe, 1968; Section II,A,2). In the stickleback, spermatogenesis (including spermiogenesis) appears to be independent of environmental factors (Craig-Bennett, 1931; Baggerman, 1980) but vitellogenesis (yolk granule deposition or exogenous vitellogenesis at least) depends on photoperiod and is influenced by temperature (Baggerman, 1957, 1980; Sections II,A,2 and II,A,3). However, functional sexual maturity in the male (i.e., androgen secretion, spermiation, and reproductive behavior) is similarly dependent on photoperiod and influenced by temperature (Baggerman, 1957, 1980; Sections II,A,2 and II,A,3). In the Gulf croaker, *Bairdeila icistia,* males matured under all laboratory conditions, but females appeared to depend on photoperiod and temperature for sexual maturation (Haydock, 1971). Other examples of sexual differences in environmental response are given by de Vlaming (1972a,b) (see also Sections II,B and II,C).

An interesting example is presented by the brook trout (Pyle, 1969). When brook trouts were maintained under continuous light at 8.3°C during their first reproductive cycle, the males spermiated earlier than when they were kept under either simulated natural photoperiods or continuous darkness at the same temperature. However, the females spawned at about the same time under all three conditions. Therefore, although the first sexual maturation (puberty) in the male is influenced by (although not dependent on) environmental factors, that in the female appears to be totally independent of environmental influence. Henderson (1963) reached the same conclusion that puberty in the brook trout does not depend on environmental factors.

5. POSSIBILITY OF ENDOGENOUS RHYTHM

In the foregoing example of the brook trout, an endogenous rhythm in the timing of puberty is suggested. An endogenous rhythm may also operate in the second reproductive cycle of the brook trout because gonadal recrudescence can occur under continuous light or darkness (at constant 8.3°C) although there are phase deviations from the normal cycle (Poston and Livingston, 1971). Similarly, rainbow trouts (*S. irideus*) kept under continuous light or darkness (for 5 yr) did not exhibit a difference in sexual maturation compared to those kept under natural photoperiods (Bieniarz, 1973). In another species of rainbow trout, *S. gairdneri,* maintenance under constant 12L and 12D and 9°C (flow rate, O_2 content, pH, and feeding rate also kept constant) did not affect the spawning time compared to fish kept under natural photoperiods (other factors being the same) (Whitehead *et al.*, 1978). Therefore, an endogenous rhythm is also indicated in the two species of rainbow trout. Similar findings were also obtained in *F. heteroclitus* (Pang, 1971).

However, in these species, photoperiod and/or temperature have been noted to play an important role in sexual cycling (Sections II,A,1 and II,A,2). Perhaps in such cases, the environmental factors serve as zeitgebers (see Section I) which synchronize the endogenous reproductive cycle with the annual environmental cycle thereby synchronizing the individuals of a population reproductively with one another. As stated by Scott (1979), "although not essential to successful gametogenesis, the environmental cues may well be essential to successful reproduction, because individuals which mature out of phase with the environment—or each other—will be ineffectual."

Baggerman (1957, 1980) suspected the existence of an endogenous reproductive cycle in the stickleback based on her finding that the reproductive cycle persisted with some phase shifts when the fish were kept under constant 16L–8D and 20°C for 420 days. However, the reproductive cycle did not persist under constant 8L–16D and 20°C. Perhaps only phase 1 and the termination of breeding (gonadal regression) are under endogenous control, but phase 2 is under environmental control (Section II,A,1,II,A,2 and II,A,3). This is further discussed in Section IV,A.

That only some phases of the reproductive cycle (rather than the whole cycle) may involve an endogenous rhythm has also been suggested by other studies (Section IV,A). For example, de Vlaming (1975) demonstrated that in N. crysoleucas, although the early phases of gametogenesis may involve an endogenous rhythm, the final phases depend on specific environmental factors.

Other examples of possible involvement of an endogenous rhythm in sexual cycling have been noted in subtropical and tropical species (Sections II,B and C). However, before endogenous rhythmicity can be accepted as a fact, rigorous experimentation is necessary. Several sets of constant conditions, involving not only different constant photoperiods (e.g., continuous light or darkness) but also different constant temperatures, should be used. Other conditions such as water quality (e.g., pH, salinity, O_2 content) and food intake (quantity as well as quality) should be kept constant. These experimental conditions are seldom met in the studies reported. In particular, usually only one constant temperature is studied. This is discussed in Section IV,A.

B. Subtropical or Subtemperate Species

In subtropical or subtemperate regions (regions close to the Tropic of Cancer or Capricorn), seasonal variations in photoperiod and temperature are relatively small. Nevertheless several species have responded to such changes. In the Indian catfish, H. fossilis, both photoperiod and temperature affect gonadal recrudescence, but temperature is apparently the more important factor (Vasal and Sundararaj, 1976; Sundararaj and Vasal, 1976). During the preparatory period of the reproductive cycle (February–April), exposure of Indian catfish to a long photoperiod (14L–10D) for 6 weeks stimulates ovarian recrudescence (vitellogenesis), but the response depends on the temperature, being greater at higher temperatures (>25°C). At 30°C, vitellogenesis is stimulated regardless of photoperiods (e.g., 12L–12D, 14L–10D, continuous light, or continuous darkness). During the postspawning period (September–January), catfish are responsive to 14L–10D at 25°C only after prior exposure to decreasing photoperiod (12L decreasing to 9L). Following 30 days of pretreatment with short photoperiod (9L–15D) at ambient temperatures (23°–20.2°C), exposure of fish to 30°C for 45–60 days induces vitellogenesis regardless of photoperiods (9L–15D or 14L–10D). Vasal and Sundararaj (1975) and Sundararaj and Vasal (1976) have also demonstrated that the photosexual response is based on a circadian rhythm of photosensitivity. Nighttime light pulses (1 hr) administered from 1800 to 0500 hr with or without the primary 6 hr photoperiod result in photosexual stimulation with peaks between 2200 and 0100 hr. This adds to the list of species demonstrating such circadian rhythm of photosensitivity (Section II,A,1).

Maintenance of female *H. fossilis* under continuous darkness or light at 25°C for 34 months did not eliminate the reproductive cycle but only modified it (Sehgal and Sundararaj, 1970a,b; Sundararaj and Sehgal, 1970; Sundararaj and Vasal, 1973, 1976). Therefore, an endogenous component in the control of sexual cycling in *H. fossilis* is suggested.

Long photoperiod also stimulates gonadal development in two other Indian teleosts whose spawning seasons also fall during the summer months when the daylength is slightly longer compared to the winter months. One is the catfish, *Mystus tengara* (Guraya *et al.*, 1976) and the other, a carp, *Cirrhina reba* (Verghese, 1967, 1970, 1975). In *M. tengara*, the response of the ovary to a long photoperiod (14L–10D) increases with the approach of the natural spawning period. In *C. reba*, long photoperiods (14L–10D, 18L–6D, or continuous light; temperature, 19–30°C) accelerate gonadal maturation and a short photoperiod (8L–16D) or total darkness delays it (Verghese, 1970, 1975). Fish under long photoperiods (14L–10D or 18L–6D) also remain sexually mature for 1 month beyond the breeding season (Verghese, 1967). Temperature is apparently less important (Rao *et al.*, 1972).

Similarly, gonadal development in the spangled perch, *Therapon unicolor*, and the East Queensland rainbow fish, *Nematocentris splendida*, two freshwater fishes of northeast Australia, is associated with increasing daylength and temperatures (Beumer, 1979). In *N. splendida*, at all the temperature ranges studied (21°–24°C, 25°–27°C, and 28°–30°C), a long photoperiod (14L–10D) shortened the period to first spawning and increased the number of broods produced. The best results were obtained in the intermediate temperature range of 25°–27°C.

The situation varies from that of subtropical fishes which spawn during the winter months. In the grey mullet, *Mugil cephalus*, which shows peak spawning in January and February in Hawaiian waters (Kuo and Nash, 1975), a short photoperiod of 6L–18D has been shown to induce vitellogenesis after 8 weeks of exposure, but the magnitude of the response depends on the temperature, being greater at lower temperatures (17°C, 21°C) than at higher temperatures (24°–26°C) (Kuo *et al.*, 1974). A constant temperature of 21°C and 6L–18D produced the fastest rate of vitellogenesis. Similarly, a short photoperiod (9L–15D) and low temperature (16°C) stimulated ovarian recrudescence in *Mirogrex terrae-sanctae*, a winter-spawning cyprinid in the Sea of Galilee (Yaron *et al.*, 1980); a high temperature of 27°C appeared to inhibit vitellogenesis.

C. Tropical Species

1. NATIVE SPECIES

In tropical regions (regions near the equator), photoperiod hardly varies throughout the year, although temperature may change slightly in accordance with the wet and dry seasons. Tropical species tend to have an extended spawning period, or even continuous breeding throughout the year, but spawning peaks do occur, which are usually associated with seasonal rainfall and/or floods (Hyder, 1969, 1970; Munro *et al.*, 1973; Lowe-McConnell, 1975; Geisler *et al.*, 1975; Johannes, 1978; Schwassmann, 1978; Kramer, 1978; Nzioka, 1979; Hails and Abdullah, 1982). Little is known of environmental cues for such seasonal peaks in reproductive activity. Factors associated with rainfall or floods are more likely to be related to synchronization of final maturation and spawning rather than gametogenesis (Hyder, 1970; Schwassmann, 1971, 1978; see also Section II). However, gametogenesis may be affected in some species. In *Plectroplites ambiguus*, gonads develop earlier and more uniformly among fish during years when floods and high rivers are common throughout the winter and early spring (Lake, 1967). In a South American gymnotoid, *Eigenmannia virescens*, a weakly electric fish that matures and spawns during the rainy season (Hopkins, 1974), rain simulation in combination with increasing water level and decreasing conductivity induce complete gametogenesis leading to spawning (Kirschbaum, 1975, 1979). Decreasing conductivity alone or rain simulation and rising water level could only induce partial ovarian recrudescence. It is not clear which specific ovarian stage requires all three factors. However, in the male, decreasing conductivity alone or rain simulation combined with rising water level induce complete spermatogenesis although apparently to a lesser extent than when all three factors act together. The conductivity effect is not related to carbonate and total hardness; pH *per se* is not apparently important. Photoperiod and temperature have not been studied.

Photoperiod is often assumed to be unimportant (Schwassmann, 1978). This is indeed the case in the Java medaka, *Oryzias javanicus*, which was collected from the mangrove swamps of Singapore (1° above equator) (S. I. Chong and T. J. Lam, unpublished). The fish spawns daily for brief periods (variable duration, averaging 5–7 days) alternating with quiescent periods (variable duration, averaging 5–7 days) throughout the year. Photoperiods (8L–16D, 16L–8D, continuous light) at ambient temperatures (27±1°C) affect neither this pattern of spawning frequency nor the fecundity of the fish. Salinity also does not produce an effect even though the fish live in an estuarine environment with daily fluctuations of salinity. In contrast, the temperate Japanese medaka, *O. latipes*, which also spawns daily but only during the breeding season (spring–summer), is sexually sensitive to photoperiodic changes (Yoshioka, 1962, 1963, 1970). Interestingly, wild populations of *O. latipes* from the southern warmer latitudes (25°N) of Japan fail to respond sexually to photoperiodic manipulations at 23–25°C (Sawara and Egami, 1977). Populations introduced to Palau Island (5°N) also breed throughout the year as do the *O. javanicus* (Sawara and Egami, 1977).

Similarly, photoperiods (continuous darkness, 4L–20D, 4L–8D/4L–8D, 8L–16D, 12L–12D, and continuous light) at 25°C did not affect the timing of first sexual maturity (puberty) in another freshwater species, *Sarotherodon* (formerly *Tilapia*) *mossambicus*, which is found in Singapore (Poon, 1980). Puberty and spawning can occur in continuous darkness or continuous light implying an endogenous rhythm. As noted previously, the timing of puberty may also be independent of environmental control in some temperate species (Section II,A,4). Whether photoperiod affects postpubertal gonadal recrudescence in *S. mossambicus* has not been investigated.

Based on field observations and data. Hyder (1970) concluded that temperature and light intensity are important cues for seasonal peaks of gonadal activity in *Tilapia (Sarotherodon) leucosticta* living in an equatorial lake of Kenya. However, high light intensities were found to delay sexual maturity in both sexes in another species, *T. zillii*, which was maintained from the fry stage under 12L–12D photoperiod at various light intensities (temperature averages 23°C), ranging from 1.5 μamp (covered tank) to 7.2 μamp (100 W bulb) (Cridland, 1962).

Perhaps temperature is the more important factor in *Tilapia* reproduction. The reproductive rate in *T. mossambica* increased with temperatures up to 28°–31°C (Mironova, 1977). Female *T. aurea* kept at 17°C had regressed ovaries but exposure to 28°C for 2 weeks markedly stimulated ovarian development (Terkatin-Shimony *et al.*, 1980).

In the guppy, (*Poecilia reticulata* = *Lebistes reticulatus*), a tropical ovoviviparous teleost which breeds throughout the year, the effect of photoperiod, although noted, is not clear. While Scrimshaw (1944) found that continuous light shortened the interval between successive broods with the appearance of superfetation, Dildine (1936) did not obtain an effect of continuous light or darkness even after 100 days. Bong (1972) found that the photosexual response differed between the wild guppy (collected from monsoon drains in Singapore) and a cultured variety, the Tuxedo guppy. In the wild guppy, photoperiods (continuous darkness, 8L–16D, 16L–8D, and continuous light) did not affect ovarian development, but continuous light and 16L–8D appeared to inhibit spermatogenesis as compared to total darkness. In contrast, in the Tuxedo guppy, continuous light and 16L–8D stimulated ovarian development as compared to 8L–16D and continuous darkness. This difference probably reflects an effect of acclimatization to different ecological factors present in the wild and under culture conditions.

Bong (1972) has also demonstrated that gametogenesis in the wild guppy is enhanced by high light intensities (100 and 180 foot-candles) as compared to low intensity (20 foot-candles). The Tuxedo guppy was not studied.

In another study (Seah and Lam, 1973a,b), differences were obtained in gonadal response to temperature and salinity between the wild guppy and another cultured variety, the Cobra guppy. In the wild guppy, spermatogenesis was not affected by temperature (26.5° and 29.0°C) and salinity (fresh water, 33.3% seawater, and seawater), but ovarian development in fresh water was significantly greater at 26.5°C than at 29.0°C (Seah and Lam, 1973b). This temperature effect in the female wild guppy disappeared when the fish were kept in 33.3% seawater suggesting that the effect may be attributable to enhanced osmoregulatory expenditure of energy thereby reducing energy resources for ovarian development. However, in the Cobra guppy, neither temperature (26.5°C, 29.0°C, and 32.0°C) nor salinity (fresh water, 33% seawater, and 50% seawater) affected gonadal development, although 50% seawater did have an initial transient inhibitory effect and 33.0°C appeared to prevent or inhibit gestation (Seah and Lam, 1973a). Guppies kept in France show optimal spermatogenesis at 25°C (Billard, 1968).

Two points have emerged from the aforementioned studies. First, acclimatization and/or genetic selection may alter the gonadal response of a species to enrivonmental factors. This is also observed in *O. latipes* as previously mentioned (Sawara and Egami, 1977), and in the following three other cases. In *M. cephalus* and *M. capito*, reproduction cannot occur in fresh water (Abraham *et al.*, 1966; Eckstein, 1975), but it proceeds normally when the fish have been maintained in fresh water from the fry stage (Eckstein and Eylath, 1970). In *E. virescens,* wild fish sometimes reach sexual maturity in the laboratory, but rarely or never spawn; however, fish born and raised in captivity spawn regularly when mature (Kirschbaum, 1979). Second, males and females may respond differently to environmental factors. This has already been noted in some temperate species (Section II,A,4).

In another cultured tropical freshwater aquarium fish, the neon tetra (*Paracheirodon innesi*), studies by Tay (1983) demonstrated that temperature, water quality (pH and conductivity), and light intensity are important factors that influence gonadal development. Gonadal development was enhanced when the fish were maintained at 25°C (compared to 20°C or 30°C), low pH and conductivity, and low light intensity. Like another South American fish mentioned earlier, *E. virescens* (Kirschbaum, 1979), *P. innesi* is truly halophobic or alkaliphobic; even a low salt content of 16.6% seawater (5 ‰) inhibited gonadal development. This was attributed to the increased presence of calcium ions which exerted a marked inhibitory effect. Surprisingly, the fish actually thrive and breed better in acidic deionized water. These findings may have ecological significance because the neon tetra originates from the blackwaters of the Amazonia where (1) the water is extremely soft (extremely low calcium concentration) and acidic (pH 4.0–4.8); (2) the water temperature may drop 4°C during the rainy season from the normal range of 28°–30°C; and (3) forest cover reduces light penetration (Geisler *et al.,* 1975).

A number of tropical species show a well defined seasonal reproductive cycle typical of temperate species (Lam, 1974; Payne, 1975; Johannes, 1978; Beumer, 1979; Kuo and Nash, 1979; Kumagai, 1981). The rabbitfish, *Siganus canaliculatus* (Lam, 1974; Soh, 1976), and the milkfish *Chanos chanos* (Kuo and Nash, 1979; Kumagai, 1981), are two such examples. Regulating environmental factors have not been identified although some suggestions have been put forth (Soh, 1976; Kumagai, 1981). Laboratory studies demonstrated that a long photoperiod of 18L–6D retarded gonadal maturation in *S. canaliculatus* compared to the natural photoperiod of 12L–12D (Lam and Soh, 1975). Temperature has not been studied, but may be of importance because the fish probably migrate to deeper waters, where temperatures are lower, prior to returning to the coast to spawn. This may also be the case with the milkfish. However, laboratory studies did not reveal a difference in gonadal development between 3-year-old immature milkfish kept at 28°–32°C and at 23°–26°C for 6 weeks (Lacanilao *et al.,* 1982). Further, immature milkfish (2–4 years old) held in a large floating net-cage (10 m diameter × 3 m depth) matured and spawned spontaneously after about 18 months, although wild "spent" milkfish, similarly maintained, failed to re-mature (Lacanilao and Marte, 1980). Nevertheless, the particular area where the net–cages were located (shallow, clean, calm, and clear sea) had an annual temperature range of 25°–31°C. Milkfish have also matured sexually in large concrete tanks (8.25 or 12 m diameter) in Taiwan, and the common feature is a temperature range of 21.4–30.7°C (Liao and Chen, 1979; Tseng and Hsiao, 1979).

Another common feature is that fish were fed a high-protein diet. The role of nutrition in fish gonadal development has received little attention. However, nutrition may be an important environmental factor in terms of seasonal changes in abundance and quality of food. It is well-known that plankton undergoes seasonal changes in abundance and species composition even in the tropics (Chua, 1970a,b). This would have profound effects down the food chain particularly in the tropics where the waters are generally of low productivity. In this regard, it is interesting to note that adult milkfish feed on a single species of macrozooplankton (e.g., *Lucifer* sp., *Acetes* sp., *Stolepholus* sp.) at a time (Kumagai, 1981). This implies that the fish are feeding on a large plankton mass; this author has personally observed numerous anchovies (*Stolepholus* sp.) in the stomach of a sexually mature milkfish captured from the wild.

Salinity is not apparently important for milkfish gametogenesis at least within the ranges of 7–12 ‰ (Nash and Kuo, 1976; Kuo *et al.*, 1979), 13.7–29.8 ‰ (Liao and Chen, 1979; Tseng and Hsiao, 1979), and 28–35 ‰ (Lacanilao and Marte, 1980). However, vitellogenesis is inhibited in fresh water (Kumagai, 1981). Light intensity or related factors may be important (Kumagai, 1981).

2. INTRODUCED SPECIES

A few temperate species have been introduced to the tropics; these include the goldfish, *C. auratus*, the Chinese carps, and the common carp, *Cyprinus carpio*. It is of interest to determine if acclimatization and/or genetic selection have produced a change in the gonadal response of these species to environmental factors. Unfortunately little experimental work has been conducted on these species in the tropics, although much has been done on them in temperate regions. As noted previously, in temperate goldfish, gametogenesis is affected by both photoperiod and temperature, but the photosexual response is dependent on the seasons (Section II,A,3). When goldfish were kept at 30°C, gonads regressed even though gonadotropin secretion was enhanced, but after 4 months of acclimation, gonadal recrudescence was restored (Gillet and Billard, 1977; Gillet *et al.*, 1978). Therefore, acclimation apparently produced a change in the gonadal response of goldfish to high temperature. This may also be the case with tropical goldfish which can breed throughout the year in waters of high temperatures (26°–31°C).

It is not known whether photoperiod affects goldfish in the tropics as it does in temperate regions. One study demonstrated that total darkness causes gonadal regression (as measured by depression in gonadal [^{3}H] thymidine incorporation) in tropical goldfish (Yadav and Ooi, 1977). A similar conclusion was reported for temperate goldfish (Ogneff, 1911).

Another factor which has been shown to affect gametogenesis in temperate goldfish is dissolved oxygen level; a low level causes gonadal regression (Gillet *et al.*, 1981). This is likely to occur also in tropical goldfish.

In carps, temperature appears to be the most important factor controlling sexual cycling in temperate regions (Billard *et al.*, 1978). Chinese carps and the common carp (*C. carpio*) attain sexual maturity earlier in the warm south than the north in both China and Europe (Chung *et al.*, 1980; Bakos *et al.*, 1975; Kausch, 1975). Gupta (1975) maintained *C. carpio* at 23°C and found that 25% of the females commenced spawning at 15 months compared to 4 years under natural temperature regimes. In the tropics, sexually mature carps can be obtained throughout the year (Kausch, 1975, also personal observations). This is assumed to be an effect of sustained high temperatures (Kausch, 1975). At 20°–24°C, *C. carpio* can undergo continuous gametogenesis (Kossman, 1975). The common carp appears to be fairly independent of photoperiod, provided that temperature is optimal (Meske *et al.*, 1968).

D. Role of Social Factors

There are occasional reports which suggest that social factors may influence gametogenesis in fish. (Here all factors associated with the social environment are considered to be social factors whether they are chemical, e.g., pheromones, visual, auditory, or tactile). In the platyfish, *Xiphophorus variatus*, adult males inhibit the maturation of juveniles but not growth; the inhibition is overcome when the juveniles reach a certain size (Borowsky, 1973, 1978). Therefore, the relationship of more adult males in the population and fewer maturing males obtains; also, the larger the average juvenile, the greater the number of males maturing. In the guppy, *P. reticulata*, high population density retards ovarian development (Dahlgren, 1979).

There is also evidence of social facilitation of gonadal development. Marshall (1972) reported that recordings of sounds produced by male *S. mossambicus* hastened spawning of isolated females by about 10 days suggesting acceleration of ovarian recrudescence and–or ovulation or oviposition. Visual stimuli are also important because isolated females attain first spawning 10 days later if deprived of visual contact with a conspecific of the same age in an adjacent aquarium (Silverman, 1978b). Although the results may mean a delay in ovulation or oviposition rather than ovarian development, the latter was favored because visually isolated females also showed greater growth, which is suggestive of lesser energy drain for ovarian development (Silverman, 1978b). However, in another study involving several spawning cycles and fish subjected to various levels of sensory contact, Silverman (1978a) concluded that visual stimuli affected ovulation more than it affected vitellogenesis, but nonvisual stimuli (specific stimuli involved not known) hastened both vitellogenesis and ovulation.

In the threespine stickleback (*G. aculeatus*), males in winter condition came into breeding more readily if kept one on each side of a glass partition than did solitary males (Van den Assem, 1967). Reisman (1968) also noted that the presence of a conspecific (particularly a female) stimulated the development of androgen-dependent secondary sexual characters in male sticklebacks. All the foregoing studies suggest social influence on gametogenesis or steroidogenesis, but exactly what stage is affected, or whether pheromones are involved in some of them, remains to be investigated.

III. ENVIRONMENTAL INFLUENCES ON SPAWNING

Considered under spawning are several physiological processes: oocyte maturation (germinal vesicle breakdown), ovulation and oviposition in the female, and spermiation and sperm release in the male. Although these processes can and do occur separately, they are often not considered separately as far as environmental influences are concerned. In fact, most of the information available is based on field observations of spawning activity in relation to environmental factors. However, these stages are expected to require precise environmental cues for synchronization (Scott, 1979). Failure at these stages (particularly ovulation) is often reported for captive fish in aquaculture.

A. Temperate Species

1. TEMPERATURE

In goldfish (*C. auratus*), ovulation is influenced by water temperature (Stacey *et al.*, 1979a,b). In cold water ($12 \pm 1°C$), vitellogenesis can proceed to the tertiary yolk–granule stage (Yamazaki, 1965, also personal observation) at a faster rate than in warm water as mentioned earlier (Gillet *et al.*, 1978), but ovulation will not occur unless vegetation is present (Pandey and

Hoar, 1972; Pandey *et al.*, 1973, 1977; Stacey and Pandey, 1975; Lam *et al.*, 1975, 1976; Peter *et al.*, 1978; Stacey *et al.*, 1979b). However, when the water temperature is raised to about 20°C or higher, or the fish are transferred from cold to warm water, ovulation will occur in sexually mature goldfish within a few days even in the absence of vegetation (Yamamoto *et al.*, 1966; Yamazaki, 1965; Yamamoto and Yamazaki, 1967; Pandey *et al.*, 1977; Stacey *et al.*, 1979a,b). Similarly, in an Australian freshwater fish, *P. ambiguus*, ovulation does not occur below 23.6°C, although spermiation can occur (Lake, 1967). In the gulf croaker, *B. icistia*, oocyte hydration and ovulation did not occur below 17°C (Haydock, 1971). The temperature requirement for spawning is even more exact in another Australian freshwater fish, *Maccullochella macquariensis;* below 20°C no spawning (probably ovulation) occurs, but above 20°C atresia sets in (Lake, 1967).

Warm temperatures have also been suggested to stimulate final maturation stages of spermatogenesis in the killifish, *F. heteroclitus* (Matthews, 1939; Pickford *et al.*, 1972), and of oogenesis in the marsh killifish, *F. confluentus* (Harrington, 1959) and the sunfish, *E. obesus* (Harrington, 1956). In *N. crysoleucas*, final ooctye maturation, ovulation, and spermiation occurred only in fish exposed to a long photoperiod–high temperature regime; neither long photoperiod nor high temperature alone was effective (de Vlaming, 1975).

In Oita Ecological Aquarium (Japan), many marine species can be induced to spawn by an increase in the water temprature (H. Nakajima, personal communication). A rise in temperature is also implicated in the spawning of the dace, *Leuciscus leuciscus* (Mills, 1980). Several Australian freshwater species (other than those already mentioned) spawn at or above specific temperatures (23°–24°C) (Lake, 1967; Lake and Midgley, 1970).

In the channel catfish, *Ictalurus punctatus*, spawning occurs in the spring when the water temperature is around 21°–24°C (Huet, 1975). In the common carp, *C. carpio*, gametogenesis may be completed by October, but spawning does not occur until the following spring or summer (Billard *et al.*, 1978) when water temperatures rise above 17°C, the minimum spawning temperature (Shikhshabekov, 1974). In the tench (*Tinca tinca*), spawning never occurs below 20°C (Breton *et al.*, 1980b) as was also observed in the stickleback by Baggerman (1969). Similarly, the spawning of the pike (*Esox lucius*) requires warm temperatures (Billard and Breton, 1978).

On the contrary, autumn or winter breeders spawn at relatively low temperatures (Hokanson *et al.*, 1973). In the rainbow trout, low temperatures are important to ovulation, otherwise the ova survive only a short time (Billard and Breton, 1977). However, Peterson (1972) believed that changes in barometric pressure may be more important. Also, in the brook trout, spermiation and spawning may be affected by photoperiodic manipulation

(Pyle, 1969; Poston and Livingston, 1971). In the sea bass, *Dicentrarchus labrax*, spawning occurs in winter in the Mediterranean area, in spring in Brittany, and during early summer in Ireland, when the temperature reaches 10°–12°C (Billard and Breton, 1978). The perch, *Perca fluviatilis*, spawns at around 11.5°C (Lake, 1967). The winter flounder, *Pseudopleuronectes americanus*, may ovulate at temperatures as low as 6°C but not lower (Smigielski, 1975).

2. Spawning–Nesting Substrates

Aquatic vegetation enhances the ovulatory response of goldfish to warm temperatures (it can even induce ovulation in cold water) (Stacey *et al.*, 1979b). Nesting substrate (e.g., hollow logs) is necessary for the spawning of *M. macquariensis* (Lake, 1967). Whether vegetation or other spawning–nesting substrates play a similar role in other teleosts is not known, but it is a common practice to introduce such substrates to spawning ponds of cultured fishes such as giant gouramy (*Osphronemus gouramy*), catfish (e.g., channel catfish, *I. punctatus*), and *C. carpio* (Huet, 1975; Suseno and Djajadiredja, 1981).

3. Other Factors

Several other factors have been reported to influence spawning.

(1). *Water current*. The minnow, *P. phoxinus*, will not spawn in still water (Scott, 1979).

(2). *Oxygen*. Low dissolved oxygen levels reduce or prevent spawning in the fathead minnow, *Pimphales promelas* (Brungs, 1971) and the black crappie, *Pomoxis nigromaculatus* (Carlson and Herman, 1978). However, a high oxygen concentration enhances or triggers ovulation in the carp, *C. carpio* (Horvath and Peteri, as quoted by Billard and Breton, 1982).

(3). *pH*. Spawning of some species may be inhibited in acidic waters (Beamish, 1976).

(4). *Salinity*. In the sea bass, *D. labrax*, oocyte maturation and ovulation will not occur in fresh water (Stequert, 1972), although spermiation may occur in salinities as low as 1–2 ‰ (Roblin, 1980).

(5). *Barometric pressure*. In rainbow trout, spawning activity appeared to coincide with an increase or decrease in barometric pressure (but not with high or low pressure as such) (Peterson, 1972).

(6). *Rainfall, flood, lunar cycle, and social factors*. These are discussed together with tropical species in Section III, B and C.

4. Environmental Factors and Spermiation

There is a paucity of experimental data on environmental influences on spermiation or sperm release. Available evidence suggests that spermiation is less dependent on environmental modulation than are oocyte maturation and ovulation. As mentioned earlier, in *P. ambiguus*, spermiation can occur below 23.6°C, but ovulation cannot (Lake, 1967); in *D. labrax*, spermiation can occur in low salinities (Roblin, 1980), but ovulation cannot (Stequert, 1972). In cyprinids, spermiation may occur almost all year round, but ovulation normally can only occur in the warm season (Billard *et al.*, 1978). However, in a few species, spermiation is affected by environmental factors. In the lake chub, *C. plumbeus*, high temperatures promote spermiation (Ahsan, 1966). In contrast, in rainbow trout, spermiation occurs at low temperatures under decreasing photoperiod (Breton and Billard, 1977). In the stickleback, androgen secretion and perhaps spermiation are controlled by photoperiod and influenced by temperature (Baggerman, 1980; T. J. Lam, unpublished). In other species, social factors may be more important (see Section III,C).

B. Subtropical and Tropical Species

In tropical and subtropical species, peak spawning activity is often associated with rainfall, floods, or the lunar cycle (de Vlaming, 1974; Lowe-Mc-Connell, 1975; Schwassmann, 1971, 1978, 1980; Gibson, 1978; Billard and Breton, 1978; Liley, 1980). Some temperate species living in lower latitudes may also spawn during floods, but only when the temperature is appropriate (John, 1963; Lake, 1967; Mackay, 1973).

1. RAINFALL AND FLOODS

Species that have been reported to spawn in relation to rainfall and/or floods include African catfish, *Clarias qariepinus* (Bruton, 1979), Indian catfish, *H. fossilis* (Sundararaj and Vasal, 1976), Indian major carps (Sinha *et al.*, 1974), barbs, *Puntius* spp. (Lake, 1967; Huet, 1975), sparid, *Pagrus ehrenbergii* (Stepkina, 1973), *Scleropages formosus* (Scott and Fuller, 1976), characids, *Bryconamericus emperador* and *Piabucina panamensis* (Kramer, 1978), and *T. unicolor* (Beumer, 1979).

Some of the species are incapable of spawning in the absence of rainfall or floods (Lowe-McConnell, 1975; Bruton, 1979; Khanna, 1958; Sinha *et al.*, 1974), and spawning can be induced by flood simulation or a rise in pond water level or refill of a sun-dried pond (Sinha *et al.*, 1974; Bruton, 1979).

It is not clear which of the terminal reproductive events (oocyte maturation, ovulation, and/or oviposition) is triggered or enhanced by rainfall or flood, or whether spermiation and/or sperm release is involved. In the Indian major carps, ovulation (possibly also oocyte maturation) is implicated because the fish can spawn without flood or flood simulation if ovulation is first induced by hypophysation (Chaudhuri, 1976). In *P. ambiguus* (a temperate species) yolky oocytes fail to mature and ovulate but become atretic if a flood fails to occur (Mackay, 1973); when floods are of a minor nature, incomplete ovulation is common (Lake, 1967).

Further, it is not clear what specific factor or factors associated with rainfall or floods are involved in spawning stimulation . Lake (1967) suggested a factor (possibly an oil, petrichor) from the dried soil when water comes into contact with it. Sinha *et al.* (1974) and Bruton (1979) suggested numerous related factors, among them lowering of water temperature, petrichor from newly wetted soil, dilution of electrolytes, e.g., chlorides (decrease in conductivity), increase in oxygen content, and a change of pH. No single factor has been identified; perhaps a consortium of factors is involved. This area deserves more serious attention, not only because of its academic interest, but also because the information obtained will have practical application in the induction of spawning in some cultured species (Section V,B).

As noted earlier, factors associated with rainfall may be involved in gonadal recrudescence in some species. In *E. virescens*, mature fish continue to spawn in the absence of three crucial factors for gonadal recrudescence (viz., rain simulation, rising water level, and decreasing conductivity) when the fish were held at a constant water level and a constant low conductivity (Kirschbaum, 1979).

2. LUNAR CYCLE

Many tropical or subtropical marine fishes exhibit lunar or semilunar spawning periodicity (see reviews by Johannes, 1978; Gibson, 1978; Schwassmann, 1971, 1980). These include rabbitfishes, siganid species (Lam, 1974; Popper *et al.*, 1976; Hasse *et al.*, 1977), milkfish, *C. chanos* (Kumagai, 1981), threadfin, *Polydactylus sexfilis* (May *et al.*, 1979), and anemonefish, *Amphiprion melanopus* (Ross, 1978). Some temperate species also show this phenomenon. Johannes (1978) listed six species out of a total of 51 known to have such spawning rhythms. The best known examples are the California grunions, *Leuresthes tenuis* and *L. sardina* (Walker, 1949, 1952; Thomson and Muench, 1976). Other examples include the mummichog or killifish, *F. heteroclitus* (Taylor *et al.*, 1979; Taylor and DiMichele, 1980), the puffer, *Fugu niphobles* (Nozaki *et al.*, 1976), the New Zealand fish, *Galaxias attenuatus* (Hefford, 1931), the gadoid, *Enchelyopus cimbrius* (Battle, 1930), and the Atlantic silverside, *Menidia menidia* (Middaugh, 1981).

Most of the fish spawn on or around the new or full moon in synchrony with the spring tides (Johannes, 1978). Timing of spawning to coincide with ebbing spring tides may have the adaptive value of maximizing tidal transport of eggs offshore (Johannes, 1978). However, the milkfish (*C. chanos*)

spawns during the first- and last-quarter moon (neap tides) with corresponding peak appearance of fry (estimated at age 3 weeks) during the new and full
moon (Kumagai, 1981). This quarter moon periodicity has also been demonstrated in the anemone fish, *A. melanopus* (Ross, 1978). It may serve the
opposite strategy of minimizing the offshore flushing of eggs and ensuring
that the larvae remain near the coast.

Because the lunar influence occurs only during the spawning season
when vitellogenesis or spermatogenesis has already been initiated, it probably concerns the terminal events of the reproductive cycle. Taylor and Di-
Michele (1980) studied the ovarian changes during the lunar spawning cycle
of *F. heteroclitus*. They found marked cyclical changes only in ovarian hydration and the occurrence of mature oocytes; vitellogenic oocytes were present
throughout the lunar cycle with much less dramatic changes. The findings
suggest lunar involvement in final oocyte maturation and ovulation more
than in vitellogenesis. Whether spermiation is similarly affected is not
known.

What specific factor(s) determines the lunar or semilunar spawning periodicity are unknown. In the Gulf of California grunion (*L. sardina*), the
semilunar spawning runs appear to be a response to tide height rather than
to moon phase as such (Thomson and Muench, 1976), but tides are not
apparently important for the California grunion (*L. tenuis*) (Gibson, 1978).
Also, in two species of rabbitfish, *S. canaliculatus* and *S. guttatus*, the lunar
spawning rhythm persists in outdoor tanks where the water level is constant
(personal observations; J. V. Juario, personal communication). Availability of
insect food, which follows a lunar periodicity, has been suggested as a possible cue for *Mormyrus kannume* (Scott, 1979). Endogenous rhythmicity has
also been suggested (Gibson, 1978), but experimental evidence is lacking.
Even so, the endogenous rhythm may still need some lunar or related
factor(s) for synchronization or entrainment. Recent evidence of a lunar
rhythm of thyroxine surge in coho salmon, *Onchorhynchus kisutch* (Grau *et
al.*, 1981), raises the possibility of a similar phenomenon for gonadotropin.

3. TEMPERATURE AND OTHER FACTORS

In the neon tetra, *P. innesi*, abrupt transfer of gravid fish from 25° to 20°C
induced ovulation, but neither transfer from 25°C to 30°C or from 25° to
25°C induced ovulation (Tay, 1983). Whether this applies to other tropical
species awaits investigation.

In *S. canaliculatus*, abrupt transfer of gravid fish from 91.4 cm of water in
a circular tank to 17.8–22.9 cm of water in a flat, rectangular tank induced
spawning behavior and oviposition (McVey, 1972). In another species, *S.
rivulatus*, spawning was induced by abrupt water change (Popper *et al.*,

1973). Aquarists often relate similar experience with tropical aquarium fishes, but no written reports are available. Specific environmental factors involved have not been determined.

C. Role of Social Factors

Factors associated with the presence of the opposite sex and/or courtship behavior may be important in synchronizing spawning (spermiation, ovulation and/or oviposition) in some teleosts (see reviews by Aronson, 1965; Solomon, 1977; Liley, 1980, 1982, Chapter 1, this volume). In goldfish, sexually active males were reported to induce ovulation (Yamazaki, 1965; Yamamoto et al., 1966), although this has not been confirmed experimentally (Stacey et al., 1979b). Conversely, contact with a pair of spawning goldfish enhanced gonadotropin secretion and milt production (spermiation?) (Kyle et al., 1979, 1982), which was apparently not mediated by visual or chemical cues (Kyle et al., 1982). Oviposition is apparently in turn triggered by the attracted male "pushing" against the ovulated female at the water surface (Partridge et al., 1976). In rainbow trout, the presence of females in the next pond upstream stimulated milt production (Kausch, 1975). In the Pacific herring (*Clupea harengus pallaci*), oviposition was triggered by the presence of milt, apparently in response to a pheromone (Stacey and Hourston, 1982).

Pheromone release by males has also been implicated in the stimulation of ovulation and/or oviposition in two tropical species, zebrafish, *Brachydanio rerio* (Chen and Martinich, 1975) and angelfish, *Pterophyllum scalare* (Chien, 1973).

Visual stimuli may also induce spawning (ovulation and/or oviposition?) in some teleosts (Aronson, 1951; Polder, 1971; Chien, 1973; Silverman, 1978a). For example, in *Aequidens portalegrensis*, even the image of herself in a mirror induced an isolated female to spawn (Polder, 1971).

Finally, factors associated with crowding have been shown to retard or inhibit spawning in several species (Swingle, 1957; Whiteside and Richan, 1969; Yu and Perlmutter, 1970; Chew, 1972; FitzGerald and Keenleyside, 1978).

Although fragmentary, the aforementioned examples serve to emphasize the importance of social intervention in the synchronization of spawning activity in fishes. It should be examined in more species.

D. Circadian Spawning Rhythm

Many species spawn at a specific time or period of the day. Goldfish in warm water (21°C) always ovulate during the latter part of the dark phase despite alterations of photoperiods (Stacey *et al.*, 1979a,b). This diurnal periodicity of ovulation is disrupted in cold water (12°C) but appears to occur also in goldfish in the tropics (26°–31°C) (personal observations). In the medaka (*O. latipes*), oviposition normally occurs within 1 hr after the onset of light, and shifts in daily photoperiod will induce corresponding shifts in the time of oviposition (Egami, 1954; Takano *et al.*, 1973). Germinal vesicle breakdown and ovulation occur in the later part of the dark phase as in goldfish, about 6 hr (Yamauchi and Yamamoto, 1973) and 2–3 hr (Takano *et al.*, 1973; Yamauchi and Yamamoto, 1973) before oviposition, respectively.

There are many other examples from both temperate and tropical species. In most cases, spawning occurs in the daytime: zebrafish, soon after the onset of light (LeGault, 1958; Hisaoka and Firlit, 1962; Eaton and Farley, 1974); rainbow cichlid (*Herotilapia multispinosa*), hour 6 of the light cycle (Brown and Marshall, 1978); two species of anabantids (*Trichopsis vittatus* and *T. pumilus*), last 3 hr of the light cycle (Marshall 1967); anemone fish (*A. melanopus*), 2–3 hr after sunrise in Guam (Ross, 1978). Other species spawn at dusk or night: golden perch (*P. ambiguus*), 3–5 hr after sunset (Lake, 1967); South American characoid (*Prochilodus scrofa*), at dusk or night (Lowe-McConnell, 1975); milkfish (*C. chanos*), around midnight (F. L. Lacanilao and C. L. Marte, personal communication). However, other species have a more variable spawning time, but some form of periodicity still exists. For example, in *Hemichromis bimaculatus*, spawning may occur at any time between 0930 and 1600 hr (independent of light intensity), but never at night (Nobel and Curtis, 1939). In the angelfish, *P. scalare*, spawning may occur at all times of the day, but shows a peak in the last 2 hr of light (Chien and Salmon, 1972). Presumably all these spawning strategies have adaptive significance for the survival of the spawn in the respective ecological niches.

In at least four of the aforementioned species (medaka, rainbow cichlid, and the two anabantids), it has been noted that constant light disrupts the periodicity and reduces the spawning frequency (Takano *et al.*, 1973; Brown and Marshall, 1978; Marshall, 1967). This suggests that the spawning rhythm is synchronized by the onset of light or darkness.

IV. ENVIRONMENTAL INFLUENCES ON GONADAL REGRESSION

Postspawning gonadal regression is another phase of the reproductive cycle which may be synchronized by environmental factors. This has received relatively little attention, and temperate, subtropical, and tropical species are considered together in this section.

A. Endogenous Rhythm

Endogenous timing of postspawning gonadal regression has been suggested for several teleosts: the bridle shiner, *N. bifrenatus* (Harrington, 1957), the stickleback, *G. aculeatus* (Baggerman, 1957, 1980), the green sunfish, *L. cyanellus* (Kaya, 1973), the Indian catfish, *H. fossilis* (Sehgal and Sundararaj, 1970a,b; Sundararaj and Sehgal, 1970; Sundararaj and Vasal, 1973, 1976), another Indian catfish, *M. tengara* (Guraya *et al.*, 1976), the minnow, *P. phoxinus* (Scott, 1979), and the tench, *T. tinca* (Breton *et al.*, 1980a,b). The evidence is based mostly on the observation that continuation of environmental conditions conducive to gonadal recrudescence (at other times) or constant photoperiod and temperature (e.g., continuous light or darkness) cannot prevent gonadal regression. However, the evidence may not be rigorous enough. As noted by de Vlaming (1972a), the possible role of temperature has often not been considered. For example, in the tench, spermatogenesis ceases in midsummer even though environmental conditions and gonadotropin levels still seem favorable (Breton *et al.*, 1980a). However, this may be attributable to the high summer temperatures because low temperatures are necessary for initiating a new spermatogenic wave (Breton *et al.*, 1980a). In other words, the testis once spent remains regressed because there will not be any spermatogenic recruitment until the temperature has dropped to a sufficiently low level. Therefore, gonadal regression in this case may be timed by temperature rather than an endogenous mechanism.

Even if an endogenous mechanism does exist, it may still be influenced by environmental factors. In the Indian catfish, *H. fossilis*, although postspawning gonadal regression cannot be prevented by environmental manipulations, it can be accelerated by low temperatures and short photoperiods and can be delayed by warm temperatures (30°C) (Sundararaj and Vasal, 1976). Pretreatment with a decreasing or short photoperiod restores the gonadal response to long photoperiods and warm temperatures.

In the stickleback, under constant 16L–8D and 20°C, breeding is followed by gonadal regression, but occurs again in due course (Baggerman, 1957, 1980). However, under constant 8L–16D and 20°C, breeding is not only terminated earlier, but is also prevented from recurring (Baggerman, 1957, 1980; T. J. Lam *et al.*, unpublished).

In the green sunfish, *L. cyanellus*, long photoperiod in combination with either high or low temperatures cannot prevent postspawning gonadal regression. Gonadal regression is nevertheless more rapid at 24°C than at a lower temperature (Kaya, 1973).

In all these cases and others (Guraya *et al.*, 1976; Yoshioka, 1966), there appears to be a "refractory period" right after the spawning season in which

fish are unresponsive to environmental conditions favorable to gametogenesis at other times. The basis for this refractoriness is unknown. Baggerman
(1972, 1980) suggested an endogenous mechanism that induces an increase
in the photoreactivity threshold. There is evidence in goldfish which suggests that ovarian regression is related to the disappearance of a significant
daily cycle of serum gonadotropin levels (Peter, 1981). Gonadal regression
can occur in spite of a relatively high blood level of gonadotropin if a daily
cycle is absent. The reason for this is not clear and it is not clear what causes
the daily gonadotropin cycle to disappear. It is possible that sustained
gonadotropin secretion (absence of daily cycle) causes constant stimulation of
gonadotropin receptors in the ovary leading to their inactivation.

B. Temperature and Photoperiod

A thorough investigation by de Vlaming (1972b,c) demonstrated that in
the longjaw goby, *G. mirabilis*, gonadal regression is timed by high summer
temperatures. During the spawning period, high temperatures (24°–32°C)
cause gonadal regression regardless of photoperiod. Only brief daily exposures to high temperatures are sufficient to initiate gonadal regression; the
actual thermoperiod needed varies inversely with the temperature (6 hr/day
for 27°C; 8 hr/day for 24°C). During the regression and postspawning periods, high temperatures prevent gonadal recrudescence in response to photoperiodic manipulations, and during the preparatory period, high temperatures again cause gonadal regression regardless of photoperiod. It is interesting to note that a longer period of exposure to a given high temperature is required to cause gonadal regression than to inhibit recrudescence.

High temperatures also induce or accelerate gonadal regression or inhibit gonadal recrudescence in goldfish, *C. auratus* (Gillet *et al.*, 1978), in *F.
heteroclitus* (Burger 1940), in *C. plumbeus* (Ahsan, 1966), in *M. terrae-
sanctae* (Yaron *et al.*, 1980), and in the grey mullet, *M. cephalus* (Kuo *et al.*,
1974; Kuo and Nash, 1975). However, it should be noted that in *F. heteroclitus* and in *C. plumbeus*, high temperatures do not inhibit, but rather
accelerate, spermatogonial mitosis, although later stages are inhibited (Lofts
et al., 1968; Ahsan, 1966). Similarly, in *M. terrae-sanctae*, high temperatures inhibit vitellogenesis, but do not inhibit oogonial mitosis (Yaron *et al.*,
1980). The mitotic effect may be a direct thermal effect (Lofts *et al.*, 1968).

In contrast, low temperatures (<20°C) cause gonadal regression in some
warm-water species, e.g., *T. aurea* (Terkatin-Shimony *et al.*, 1980) and *P.
innesi* (Tay, 1983).

In the golden shiner, *N. crysoleucas*, warm temperatures in combination
with short photoperiod cause gonadal regression (de Vlaming, 1975; de

1. HEAVY METALS

a. Cadmium. Chronic exposure to sublethal levels of cadmium inhibits spermatogenesis in the brook trout, *S. fontinalis* (Sangalang and O'Halloran, 1972, 1973; Sangalang and Freeman, 1974), the guppy, *P. reticulata* (C. K. Quek and T. J. Lam, unpublished), and the Java medaka, *O. javanicus* (Y. H. Lim and T. J. Lam, unpublished). Androgen synthesis is also inhibited in the brook trout. Cadmium also induces atresia of oocytes in the Java medaka (Y. H. Lim and T. J. Lam, unpublished). However, gametogenesis in the blue-gill, *Lepomis macrochirus*, is less sensitive to cadmium (Eaton, 1974).

b. Mercury. Methylmercury induces gonadal regression and inhibits ovulation and oviposition in the Japanese medaka (*O. latipes*) possibly by blocking the release of gonadotropin (Chan, 1977). Effective chronic doses include 4.3, 10.7, and 21.5 ppb, the effect being dose dependent. Similarly, phenylmercuric acetate inhibits spawning in the zebrafish, *B. rerio* (Kihlstrom *et al.*, 1971). Another mercury compound, mercuric chloride, reduces enzymic activity in the ovary of the snakehead, *Channa punctatus* (Sastry and Agrawal, 1979a).

c. Other Heavy Metals. Chronic exposure to sublethal levels of lead, zinc, or copper inhibits fish reproduction. Lead nitrate (3.8 ppm) reduces ovarian enzymic activity in *C. punctatus* after 30 days of treatment (Sastry and Agrawal, 1979b). Zinc (5 ppm for 9 days) delays spawning and reduces the number of eggs spawned in zebrafish (Speranza *et al.*, 1977). Zinc (173 ppb for 70 days) also retards sexual maturation in the female guppy (Pierson, 1981). Copper inhibits spawning in the bluegill at 162 ppb for 22 months (Benoit, 1975) and reduces the number of eggs produced per female in fathead minnows, *P. promelas*, at 37 ppb for 3 months (Pickering *et al.*, 1977).

Reproductive effects of heavy metals in combination have received little attention (Spehar *et al.*, 1978). Because the various heavy metals are often present in the same polluted environment at the same time, studies of their interactive effects on gonadal activity would be more meaningful.

2. PESTICIDES

a. Chlorinated Hydrocarbons. These compounds, also known as organochlorines, are notoriously toxic to life processes, and gametogenesis is not spared. Endrin (10 ppb for 1 year) caused atresia of oocytes in the cutthroat trout, *Salmo clarki*, without apparently affecting the testis (Eller, 1971). Endrin (0.6 ppb "Hexadrin" for 96 hr or 4 weeks) also suppressed gonadotropin secretion and gonadal ^{32}P uptake in both male and female Indian catfish, *H. fossilis* (Singh and Singh, 1980a,b,d,); gonadal steroid biosynthesis may also be reduced (Singh and Singh, 1980c,d). Similar findings have been obtained for aldrin (17 ppb "Aldrin" for 4 weeks) in female Indian catfish (Singh and Singh, 1981). Also, DDT retarded ovarian development when fed to brook trout (S. fontinalis) at high sublethal levels (Macek, 1968).

Polychlorinated biphenyls (PCB's) administered orally or intraperitoneally retarded or inhibited steroid synthesis and caused gonadal regression in Atlantic cod, Gadus morhua (Freeman and Sangalang, 1977; Freeman et al., 1978, 1980), brook trout, S. fontinalis (Freeman and Idler, 1975), rainbow trout, S. gairdneri, and carp, C. carpio (Sivarajah et al., 1978a,b). Reproduction is also adversely affected in other species (Nebeker et al., 1974; Bengtsson, 1980). Interestingly, both DDT and PCB's appear to have a stimulatory effect on gametogenesis and steroidogenesis at low dosages (Macek, 1968; Freeman and Sangalang, 1977; Freeman et al., 1978).

Another organochlorine, γ-benzenehexachloride (BHC) (5 ppb for 4 weeks), causes atresia of oocytes and inhibits luteinizing hormone (LH)-induced in vitro ovulation in the Japanese medaka, O. latipes (Hirose, 1975).

b. Organophosphate and Other Nonchlorinated Pesticides. Although generally less hazardous to life, these compounds are not without any harmful effect on fish gametogenesis. Malathion (9 ppm "Cython" for 96 hr or 4 weeks) and parathion (630 ppb "Paramar M50" for 4 weeks) produced an effect on gonadotropin secretion, gonadal ^{32}P uptake, and steriodogenesis similar to that of endrin and aldrin mentioned previously (Singh and Singh, 1980a–d, 1981). Parathion also inhibited spermatogenesis in the guppy (Billard and Kinkelin, 1970). Other nonchlorinated pesticides such as fenitrothion, carbaryl, lebaycid, and diazinon all inhibit oogenesis to a greater or lesser extent in various species (Saxena and Garg, 1978; Carlson, 1971; Kling, 1981; Allison, 1977; Goodman et al., 1979); steroidogenesis may also be affected (Kapur et al., 1978).

3. OIL

Chronic exposure to crude oil causes partial testicular regression in the cunner, Tautogolabrus adspersus (Payne et al., 1978). However, there have also been reports of a lack of effect of oil on fish reproduction (Kiceniuk et al., 1980; Hodgins et al., 1977).

4. pH

Acidification of surface waters is one of the most serious problems of environmental pollution in North America (Fromm, 1980). In flagfish, Jordanella floridae, vitellogenesis was impaired in acidic waters (pH<6).

Spawning failure caused by acid stress has also been reported (Beamish, 1976). These effects may be related to an upset in calcium metabolism and to faulty deposition of yolk proteins in developing oocytes (Fromm, 1980).

High pH also poses a problem. Chronic exposure to an alkaline environment upsets guppy reproduction (Rustamova, 1979). At 75 ppm sodium hydroxide and above, no guppy reached sexual maturity.

5. RADIATION

Radiation contamination may pose a problem in areas with nuclear power installations. Exposure of fish to radiation retards or inhibits gametogenesis, or induces ovo–testis formation, or even destroys germ cells; the effect depends on the dosages, the stage at which the radiation is applied, and the duration of treatment (Kobayashi and Mogami, 1958; Yoshimura *et al.*, 1969; Egami and Hyodo-Taguchi, 1969; Hyodo-Taguchi and Egami, 1971; Bonham and Donaldson, 1972; Egami and Hama-Furukawa, 1981; and literature reviewed therein). Complete sterility may result in some cases (Egami and Hama-Fukukawa, 1981).

V. APPLICATIONS IN AQUACULTURE

Knowledge of the environmental control of reproduction in cultured species can be put to practical use in at least four ways. First, environmental factors involved in gonadal recrudescence and regression may be manipulated to accelerate, maintain, or delay gametogenesis in brookstock so that spawning may be scheduled to yield fry whenever needed. Second, environmental factors involved in spawning may be manipulated to induce or inhibit spawning for the same purpose as stated. Third, environmental factors involved in gonadal regression may be manipulated to inhibit gametogenesis in growing fish so that somatic growth may be encouraged in the absence of energy drain for reproduction. Fourth, environmental factors involved in gonadal development and spawning may be manipulated to advance spawning time so that generation time may be reduced and genetic improvement of stocks accelerated. In the discussion that follows, only examples of applications (both potential and actual) in broodstock management and induction of spawning are given.

A. Broodstock Management

There have been many attempts to induce gonadal maturation in captive broodstock through hormonal manipulation. This approach has encountered many difficulties and has met with little success (Lam, 1982; Lacanilao *et al.*, 1982). The alternative approach of environmental manipulation is more promising (Shehadeh, 1970; Pullin and Kuo, 1981; Billard, 1981; Billard and Breton, 1982).

Environmental manipulation has successfully accelerated or prolonged spawning in several cultured species. In the turbot (*Scophthalmus maximus*) and the sole (*Solea solea*), long photoperiods (18L–6D or 20L–4D) in combination with a cold temperature (12°C) accelerated gonadal recrudescence resulting in early spawning (Htun-Han and Bye, cited by Htun-Han, 1977). In the dab, *Limanda limanda*, a photoperiod regime of 4L–8D/4L–8D at 11 ± 1°C promoted ovarian recrudescence so that spawning occurred 4–5 months ahead of season (Htun-Han, 1975). This has found commercial application in fry production (Bye and Htun-Han, 1978, 1979; Jones *et al.*, 1979).

In salmonids, a decreasing photoperiod or an accelerated seasonal light cycle (annual light cycle condensed to 9 or 6 months) induces full gametogenesis and spawning several months ahead of season (Hazard and Eddy, 1951; Corson, 1955; Combs *et al.*, 1959; Nomura, 1962; Henderson, 1963; Shehadeh, 1970; Kunesh *et al.*, 1974; Breton and Billard, 1977; Billard and Breton, 1977; Whitehead *et al.*, 1978; MacQuarrie *et al.*, 1978, 1979; see also Htun-Han, 1977, for other references). This has found application in salmon farming (Nomura, 1962; Billard, 1981; Billard and Breton, 1982).

Similar photoperiodic manipulations accelerate sexual maturation (by several months) in the ayu, *P. altivelis* (Nomura, cited by Kuronoma, 1968; Fushiki, 1979) and the sea bass, *D. labrax* (Barnabe, cited by Billard, 1981; Billard and Breton, 1982). Compressing the annual photothermal cycle into 10 months each year for 3 successive years has also accelerated the time of spawning in the sea bass and the turbot, and in another marine species, the gilthead sea bream, *Sparus aurata* (Girin and Devauchelle, 1978).

Exposure of Indian catfish, *H. fossilis*, to a long photoperiod (14L–10D) at 25°C during the nonrefractory period induces in the same female four batches (one per month) of yolky oocytes which can be spawned with injections of ovine luteinizing hormone. This gives four spawnings per fish where only one is possible in nature (Sundararaj and Vasal, 1976). Sundararaj and Vasal (1976) believed that six spawnings are attainable if a higher temperature (30°C) is used.

Maintenance of common carp, *C. carpio*, in recirculating water at 23°C causes some females to commence spawning at age 15 months and the males to commence spermiating at age 6 months, compared to age 4 years and age 3 years, respectively, under normal temperature regimes (Shehadeh, 1970; Gupta, 1975). When broodcarps are kept at 20°C, spermiation is continuous and females become sexually mature several times a year (Kossman, 1975).

Keeping fish in heated water only a few degrees centigrade above the ambient temperature (under natural photoperiod) has also been shown to accelerate gametogenesis in the tench, *T. tinca* (Breton *et al.*, 1980a,b), and the male pink salmon, *Oncorhynchus gorbuscha* (MacKinnon and Donaldson, 1976).

Furthermore, as noted in Section II,B, photothermal manipulations markedly enhance the reproductive capacity of the Indian minor carp (*C. reba*) and the grey mullet (*M. cephalus*). Although these and most of the aforementioned findings have not been applied to commercial operations, the potential seems obvious.

Another approach to broodstock management is to delay gonadal maturation so as to stagger fry availability outside the normal season. In salmonids, a prolonged or continuous photoperiod or a delayed rate of change in photoperiod delays sexual maturation by several weeks or months (Allison, 1951; Hazard and Eddy, 1951; Combs *et al.*, 1959; Henderson, 1963; Shiraishi and Fukuda, 1966; Billard *et al.*, 1978; MacQuarrie *et al.*, 1978, 1979; Lundqvist, 1980). A prolonged photoperiod of 18L–6D also retards gonadal maturation in the rabbitfish, *S. canaliculatus* (Lam and Soh, 1975).

One problem that may arise is the inability of some species to respond to photoperiodic manipulations during their first reproductive cycle (Henderson, 1963; Pyle, 1969; Poon, 1980; Billard and Breton, 1982). Therefore, environmental induction of precocious puberty may not be possible in such fishes.

B. Induction of Spawning

It is a common practice among aquaculturists to use hormones to induce spawning in gravid fish (Harvey and Hoar, 1979; Lam, 1982). The environmental approach has not been fully explored, although this would circumvent problems associated with hormonal administration (Liley, 1980; Lam, 1982). Environmental factors conducive to spawning (Section IV), once known, can be manipulated to trigger natural spawning in gravid fish. For example, goldfish can be induced to spawn by the simple procedure of transferring a gravid female from 13–14° to 20°C in the presence of water plants and a sexually mature male (Yamamoto *et al.*, 1966).

In India, the major carps are routinely induced to spawn in special spawning ponds called bundhs where flooding is simulated (Sinha *et al.*, 1974). There are two types of bundhs, perennial bundhs (wet bundhs) and seasonal bundhs (dry bundhs). A perennial bundh is usually located in the slope of a vast catchment area with proper embankments. The bundh has a large shallow area adjoining a deeper pond area which retains water throughout the year and where an adequate stock of breeders is maintained. During heavy showers the bundh is flooded and spawning occurs.

A seasonal bundh is a shallow pond located along the runway of a catchment area. During heavy showers, rain water flows from the catchment area and accumulates in the pond. Breeders are then introduced and breeding occurs soon after, especially when the bundh as well as the catchment area is flooded.

In Indonesia, several cultured freshwater species, such as gouramies (*Trichogaster pectoralis, Osphronemus gouramy, Helostoma temmincki*), common carp (*C. carpio*), Java carp (*Puntius javanicus*), mata merah or "red eye" (*P. orphroides*), nilem (*Osteochilus hasselti*), and catfish (*Clarias batrachus*), are bred in specially prepared spawning ponds. The ponds are dried in the sun for several days until the bottom is completely dry and "cracking"; then clean fresh water is introduced followed by gravid fish (Huet, 1975; Suseno and Djajadiredja, 1981). In some cases (common carp, giant gouramy, *O. gouramy,* and catfish), nesting or spawning substrates are also introduced; in nilem, running water is necessary. A similar method can be used to induce spawning in catfishes of the genus *Clarias* in Africa (Christensen, 1981). In this case, the pond is dried for 7 days or more in strong sunshine and then refilled 1 week before the new moon with an application of 50 liters dry cattle manure, 25 liters dry soil, and 25 ml phosphoric acid per 100 m^2. This is followed by the introduction of two or three gravid fish of each sex.

Some *Clarias* species may also be induced to spawn by simply raising the pond water level (Bruton, 1979). In the United States, the channel catfish (*I. punctatus*) is bred in special ponds provided with nesting substrates in the form of recepticles placed at the bottom (Huet, 1975).

Such simple low-technology methods of induced fish breeding based on environmental manipulations are particularly suitable for developing countries (Christensen, 1981); they should be more extensively explored.

VI. CONCLUSIONS

Environmental influences are different for gonadal recrudescence (gametogenesis), spawning, and gonadal regression, respectively. For gonadal recrudescence, photoperiod and temperature are the important proximate factors in temperate and subtropical (subtemperate) species, but the relative importance of these two factors, their respective effects, and their interactions are dependent on the species, season, gametogenic stage, and sexes. In general, long or increasing photoperiod and/or high or rising temperature are stimulatory to gametogenesis in those species which spawn in the spring and summer but the reverse holds for autumn or winter spawners. However, puberty may be independent of environmental control in some species.

There is good evidence of a circadian rhythm of photosensitivity in several species. A seasonal shift in the photosensitive phase of the daily cycle (demonstrated in sticklebacks) may be the basis for seasonal variation in gonadal responsiveness to photoperiod shown in a number of species.

Much less is known about tropical species, although many species do show seasonal peaks in reproductive activity or even a well defined seasonal reproductive cycle. Photoperiod, which hardly varies in the tropics, is less important than temperature. Other factors such as conductivity, light intensity, food supply, oxygen level, acoustic stimulation of rain, and rising water level may be important. The environmental control, if present, is much more complex in tropical species than in temperate fishes.

Temperature is more important than photoperiod in the control of spawning. There is often a minimum temperature requirement for spawning, and abrupt changes of temperature (e.g., sudden warming in temperate species) often precipitate spawning. In tropical or subtropical species, spawning is often associated with flooding or the lunar (tidal) cycle, but the specific factors involved are not known. Other factors affecting spawning in both temperate and tropical or subtropical species include the presence of nesting or spawning substrates such as green vegetation, and social factors such as pheromones. There is good evidence of a circadian spawning rhythm in many species: spawning occurs at specific time or period of the daily light–dark cycle.

In some species, gonadal regression follows spawning regardless of environmental conditions. In such species, gonadal regression may be timed by an endogenous mechanism, and is accompanied by a refractory period during which environmental manipulations cannot stimulate gonadal recrudescence. In some other species, gonadal regression is timed by high temperatures and/or short or decreasing photoperiods. Little is known of environmental control of gonadal regression (if any) in tropical species. Increased conductivity may be involved in some fishes of the Amazonia. Other factors known to cause gonadal regression in fish include stress, reduced food availability, and pollutants (e.g., heavy metals, pesticides, oil, pH, and radiation).

There is some evidence (albeit not sufficiently rigorous) of involvement of a circannual rhythm in sexual cycling in some species. Even so, environmental factors may still function as zeitgebers synchronizing or entraining the rhythm.

Two other points emerge from the data. First, acclimatization to tropical or culture conditions and/or genetic selection may alter gonadal response to environmental factors. Second, both gametogenesis and spawning may be subject to social inhibition or facilitation. In other words, social factors may have a "primer" function on gametogenesis as well as a "releaser" function on spawning.

Finally, this area of research is of great relevance to aquaculture because the findings can be applied, *inter alia*, to stimulate or delay gonadal maturation in broodstock and to induce spawning in breeders.

ACKNOWLEDGMENTS

The assistance of Dr. W. S. Hoar in writing most of the introduction and in reading the draft manuscript is gratefully acknowledged. Thanks is extended to Mrs. Jessie Mui who typed the manuscript and Ms G. L. Loy who assisted in listing the references.

REFERENCES

Abraham, M., Blanc, N., and Yoshouv, A. (1966). Oogenesis in five species of grey mullets (Teleostei, Mugilidae) from natural and landlocked habitats. *Isr. J. Zool.* **15**, 155–172.

Ahsan, S. N. (1966). Effects of temperature and light on the cyclical changes in the spermatogenetic activity of the lake chub, *Couesius plumbeus*. *Can. J. Zool.* **44**, 161–171.

Allison, D. T. (1977). Use of exposure units for estimating aquatic toxicity of organophosphate pesticides. *Ecol. Res. Series, U.S. Environ. Protect. Agency.* EPA-600/3-77-077.

Allison, L. N. (1951). Delay of spawning in eastern brook trout by means of artificially prolonged light intervals. *Prog. Fish-Cult.* **13**, 111–116.

Aronson, L. R. (1965). Environmental stimuli altering the physiological condition of the individual among lower vertebrates. *In* "Sex and Behaviour" (F. A. Beach, ed.), pp. 290–318. Wiley, New York.

Bagenal, T. B. (1969). The relationship between food supply and fecundity in brown trout *Salmo trutta* L. *J. Fish Biol.* **1**, 167–182.

Baggerman, B. (1957). An experimental study of the timing of breeding and migration in the three-spined stickleback (*Gasterosteus aculeatus* L.). *Arch. Neerl. Zool.* **12**, 105–318.

Baggerman, B. (1969). Influence of photoperiod and temperature on the timing of the breeding season in the stickleback, *Gasterosteus aculeatus*. *Gen. Comp. Endocrinol.* **13**, 491.

Baggerman, B. (1972). Photoperiodic responses in the stickleback and their control by a daily rhythm of photosensitivity. *Gen. Comp. Endocrinol., Suppl.* **3**, 466–476.

Baggerman, B. (1980). Photoperiodic and endogenous control of the annual reproductive cycle in teleost fishes. *In* "Environmental Physiology of Fishes" (M. A. Ali, ed.), pp. 533–567. Plenum, New York.

Baker, J. R. (1938). The evolution of breeding seasons. *In* "Evolution" (G. R. de Beer, ed.), pp. 161–177. Oxford Univ. Press, London and New York.

Bakos, J., Horvath, L., Jaczo, I., and Tamas, G. (1975). Breeding habits of the major cultivated fishes of EIFAC region and problems of sexual maturation in captivity. *EIFAC Tech. Pap.* **25**, 25–42.

Battle, H. I. (1930). Spawning periodicity and embryonic death rate of *Enchelyopus cimbrius* (L.) in Passamaquoddy Bay. *Contrib. Can. Biol. Fish.* [N.S.] **5**, 363–380.

Beamish, R. J. (1976). Acidification of lakes in Canada by acid precipitation and the resulting effects on fishes. *Water, Air, Soil Pollut.* **6**, 501–514.

Bengtsson, B. E. (1980). Long-term effects of PCB (clophen A50) on growth, reproduction and swimming performance in the minnow, *Phoxinus phoxinus*. *Water Res.* **14**, 681–687.

Benoit, D. A. (1975). Chronic effects of copper on survival, growth and reproduction of the bluegill (*Lepomis macrochirus*). *Trans. Am. Fish. Soc.* **104**, 353–358.

Beumer, J. P. (1979). Reproductive cycles of two Australian freshwater fishes: the spangled perch, *Therapon unicolor* Gunther, 1859 and the East Queensland rainbowfish, *Nematocentris splendida* Peters, 1866. *J. Fish Biol.* **15**, 111–134.

Bieniarz, K. 1973. Effect of light and darkness on incubation of eggs, length, weight and sexual maturity of sea trout (*Salmo trutta fario* L.) and rainbow trout (*Salmo irideus* Gibbons). *Aquaculture* **2**, 299–315.

Bieniarz, K., Epler, P., Breton, B., and Thuy, L. N. (1978). The annual reproductive cycle in adult carp in Poland: Ovarian state and serum gonadotropin level. *Ann. Biol. Anim., Biochim., Biophys.* **18**, 917–921.

Billard, R. (1968). Influence de la température sur la durée et l'efficacité de la spermatogénèse du guppy *Poecilia reticulata*. *C.R. Hebd. Seances Sci., Ser. D* **265**, 2287–2290.

Billard, R. (1981). The control of fish reproduction in aquaculture. *World Conf. Aquacult., 1981* (in press).

Billard, R., and Breton, B. (1977). Sensibilité à la température des différentes étapes de la reproduction chez la truite arc-en-ciel. *Cah. Lab. Montereau* **5**, 5–24.

Billard, R., and Breton, B. (1978). Rhythms of reproduction in teleost fish. In "Rhythmic Activity of Fishes" (J. E. Thorpe, ed.), pp. 31–53. Academic Press, New York.

Billard, R., and Breton, B. (1983). Control of reproduction and fish farming. *Proc. Int. Symp. Comp. Endocrinol., 9th, 1981* (in press).

Billard, R., and Kinkelin, P. (1970). Sterilization of the testicles of guppies by means of non-lethal doses of parathion. *Ann. Hydrobiol.* **1**, 91–99.

Billard, R., Breton, B., Fostier, A., Jalabert, B., and Weil, C. (1978). Endocrine control of the teleost reproductive cycle and its relation to external factors: Salmonid and cyprinid models. In "Comparative Endocrinology" (P. J. Gaillard and H. H. Boer, eds.), pp. 37–48. Elsevier/North-Holland Biomedical Press, Amsterdam.

Billard, R., Bry, C., and Gillet, C. (1981a). Stress, environment and reproduction in teleost fish. In "Stress and Fish" (A. D. Pickering, ed.), pp. 185–208. Academic Press, New York.

Billard, R., Reinaud, P., and Le Brenn, P. (1981b). Effects of changes of photoperiod on gametogenesis in the rainbow trout (*Salmo gairdneri*). *Reprod. Nutr. Dev.* **21**, 1009–1014.

Bong, K. P. (1972). Effect of light on growth and reproduction in *Poecilia reticulata* (Peters). B.Sc. Hons. Thesis, Dept. of Zoology, National University of Singapore, Singapore.

Bonham, K., and Donaldson, L. R. (1972). Sex ratios and retardation of gonadal development in chronically gamma-irradiated Chinook salmon smolts. *Trans. Am. Fish. Soc.* **101**, 428–434.

Borowsky, R. L. (1973). Social control of adult size in males of *Xiphophorus variatus*. *Nature (London)* **245**, 332–335.

Borowsky, R. L. (1978). Social inhibition of maturation in natural populations of *Xiphophorus variatus* (Pisces: Poeciliidae). *Science* **201**, 933–935.

Breton, B., and Billard, R. (1977). Effects of photoperiod and temperature on plasma gonadotropin and spermatogenesis in the rainbow trout *Salmo gairdneri* Richardson. *Ann. Biol. Anim., Biochim., Biophys.* **17**, 331–340.

Breton, B., Horoszewicz, L., Billard, R., and Bieniarz, K. (1980a). Temperature and reproduction in tench: Effect of a rise in the annual temperature regime on gonadotropin level, gametogenesis and spawning. I. The male. *Reprod. Nutr. Dev.* **20**, 105–118.

Breton, B., Horoszewicz, L., Bieniarz, K., and Epler, P. (1980b). Temperature and reproduction in tench: Effect of a rise in the annual temperature regime on gonadotropin level, gametogenesis and spawning. II. The female. *Reprod. Nutr. Dev.* **20**, 1011–1024.

Bromage, N. R., Whitehead, C., Elliott, J. A. K., and Matty, A. J. (1982). Investigations into the importance of day length on the photoperiodic control of reproduction in the female rainbow trout. *Abstr. Pap. Posters, Int. Symp. Reprod. Physiol. Fish, 1982*, p. 7.

Brown, D. H., and Marshall, J. A. (1978). Reproductive behaviour of the rainbow cichlid *Herotilapia multispinosa* (Pisces, Cichlidae). *Behaviour*, **67**, 3–4.

Brungs, W. A. (1971). Chronic effects of low dissolved oxygen concentrations on the fathead minnow (*Pimephales promelas*). *J. Fish. Res. Board Can.* **20**, 1119–1123.

Bruton, M. N. (1979). The breeding biology and early development of *Clarias gariepinus* (Pisces: Clariidae) in Lake Sibaya, South Africa, with a review of breeding in species of the subgenus *Clarias* (Clarias). *Trans. Zool. Soc. London* **35**, 1–45.

Bullough, W. S. (1939). A study of the reproductive cycle of the minnow in relation to the environment. *Proc. Zool. Soc. London* **109**, 79–102.

Burger, J. W. (1940). Some further experiments on the relation to the external environment to the spermatogenetic cycle of *Fundulus heteroclitus. Bull. Mt. Desert Isl. Biol. Lab.* **42**, 20–21.

Bye, V., and Htun-Han, M. (1978). Out of season spawning. Day-length control-key to flexible production. *Fish Farmer* **1**, 10–12.

Bye, V., and Htun-Han, M. (1979). New developments with cod and turbot. Light and temperature is key to controlled spawning. *Fish Farmer* **2**, 27–28.

Carlson, A. R. (1971). Effects of long term exposure to carbaryl (SEVIN) on survival, growth and reproduction of the fathead minnow (*Pimpephales promelas*). *J. Fish. Res. Board Can.* **29**, 583–587.

Carlson, A. R., and Herman, L. J. (1978). Effects of long-term reduction and diel fluctuation in dissolved oxygen on spawning of black crappie, *Pomoxis nigromaculatus. Trans. Am. Fish. Soc.* **107**, 742–746.

Chan, K. K. S. (1976). A photosensitive daily rhythm in the female medaka, *Oryzias latipes. Can. J. Zool.* **54**, 852–856.

Chan, K. K. S. (1977). Chronic effects of methylmercury on the reproduction of the teleost fish, *Oryzias latipes.* Ph.D. Thesis, Dept. of Zoology, University of British Columbia, Vancouver, B.C., Canada.

Chaudhuri, H. (1976). Uses of hormones in induced spawning of carps. *J. Fish. Res. Board Can.* **33**, 940–947.

Chen, L. C., and Martinich, R. L. (1975). Pheromonal stimulation and metabolite inhibition of ovulation in the zebrafish, *Brachydanio rerio. Fish. Bull.* **73**, 889–894.

Chew, R. L. (1972). The failure of largemouth bass, *Micropterus salmoides floridanus* (Le Sneur), to spawn in eutrophic, over-crowded environments. *Proc. Conf. South Assoc. Game Fish Commn.* **26**, 1–28.

Chien, A. K. (1973). Reproductive behaviour of the angelfish *Pterophyllum scalare* (Pisces: Chichilidae). II. Influence of male stimuli upon the spawning rate of females. *Anim. Behav.* **21**, 457–463.

Chien, A. K., and Salmon, M. (1972). Reproductive behaviour of the angelfish, *Pterophyllum scalare.* I. A quantitative analysis of spawning and parental behaviour. *Forma Funct.* **5**, 45–74.

Christensen, M. S. (1981). A note on the breeding and growth rates of the catfish *Clarias mossambicus* in Kenya. *Aquaculture* **25**, 285–288.

Chua, T. E. (1970a). A preliminary study on the plankton of the Ponggol Estuary. *Hydrobiologia* **35**, 254–272.

Chua, T. E. (1970b). Notes on the abundance of the diatom *Hemidiscus hardmanianus* (Grev.-Mann) in the Singapore Straits. *Hydrobiologia* **36**, 61–64.

Chung, L., Lee, Y. K., Chang, S. T., Liu, C. C., and Chen, F. C. (1980). The biology and artificial propagation of farm fishes. *IDRC-MR* **15**, 7–242.

Combs, B. D., Burrows, R. E., and Bigej, R. G. (1959). The effect of controlled light on maturation of adult blueback salmon. *Prog. Fish-Cult.* **21**, 63–69.

Corson, B. W. (1955). Four years' progress in the use of artificially controlled light to induce early spawning of brook trout. *Prog. Fish-Cult.* **17**, 99–102.

Craig-Bennett, A. (1931). The reproductive cycle of the three-spined stickleback *Gasterosteus aculeatus* Linn. *Philos. Trans. R. Soc. London* **219**, 197–279.

Cridland, C. C. (1962). Laboratory experiments on the growth of *Tilapia* sp. The effect of light and temperature on the growth of *T. zilli* in aquaria. *Hydrobiologia* **20**, 155–166.

Crim, L. W. (1982). Evnironmental modulation of annual and daily rhythms associated with reproduction in teleost fishes. *Can. J. Fish. Aquat. Sci.* **39**, 17–21.

Dahlgren, B. T. (1979). The effects of population density on fecundity and fertility in the guppy, *Poecilia reticulata* (Peters). *J. Fish Biol.* **15**, 71–91.

De Montalambert, G., Jalabert, B., and Bry, C. (1978). Precocious induction of maturation and ovulation in northern pike (*Esox lucius*). *Ann. Biol. Anim., Biochim. Biophys.* **18**, 969–975.

de Vlaming, V. L. (1971). The effects of food deprivation and salinity changes on reproductive function in the estuarine gobiid fish, *Gillichthys mirabilis*. *Biol. Bull. (Woods Hole, Mass.)* **141**, 458–471.

de Vlaming, V. L. (1972a). Environmental control of teleost reproductive cycles: A brief review. *J. Fish Biol.* **4**, 131–140.

de Vlaming, V. L. (1972b). The effects of temperature and photoperiod on reproductive cycling in the estuarine gobiid fish, *Gillichthys mirabilis*. *Fish. Bull.* **70**, 1137–1152.

de Vlaming, V. L. (1972c). The effects of diurnal thermoperiod treatments on reproductive function in the estuarine gobiid fish *Gillichthys mirabilis* Cooper *J. Exp. Mar. Biol. Ecol.* **9**, 155–163.

de Vlaming, V. L. (1974). Environmental and endocrine control of teleost reproduction. *In* "Control of Sex in Fishes" (C. B. Schreck, ed.), pp. 13–83. Virginia Polytechnic Institute and State University, Blacksburg.

de Valming, V. L. (1975). Effects of photoperiod and temperature on gonadal activity in the cyprinid teleost, *Notemigonus crysoleucas*. *Biol. Bull. (Woods Hole, Mass.)* **148**, 402–415.

de Vlaming, V. L., and Paquette, G. (1977). Photoperiod and temperature effects on gonadal regression in the golden shiner, *Notemigonus crysoleucas*. *Copeia* No. 4, pp. 793–796.

Dildine, G. C. (1936). The effect of light and temperature on the gonads of *Lebistes*. *Anat. Rec.* **67**, 61.

Donaldson, E. M. (1975). Physiological and physico-chemical factors associated with maturation and spawning. *EIFAC Tech. Pap.* **25**, 53–71.

Donaldson, E. M. (1981). The pituitary-interrenal axis as an indicator of stress in fish. *In* "Stress and Fish" (A. D. Pickering, ed.), pp. 11–47. Academic Press, New York.

Donaldson, E. M., and Scherer, E. (1982). Methods to test and assess effects of chemicals on reproduction in fish. *Proc. Workshop Methods Assess. Effects Chem. Reprod. Funct. Biota, 1981* (in press).

Eaton, J. G. (1974). Chronic cadmium toxicity to the bluegill (*Lepomis macrochirus* Rafinesque). *Trans. Am. Fish. Soc.* **103**, 729–735.

Eaton, R. C., and Farley, R. D. (1974). Spawning cycle and egg production of zebrafish, *Brachydanio rerio*, in the laboratory. *Copeia* No. 1, pp. 195–204.

Eckstein, B. (1975). Possible reasons for the infertility of grey mullets confined to fresh water. *Aquaculture* **5**, 9–17.

Eckstein, B., and Eylath, U. (1970). The occurrence and biosynthesis *in vitro* of 11-ketotestosterone in ovarian tissue of the mullet, *Mugil capito*, derived from two biotypes. *Gen. Comp. Endocrinol.* **14**, 396–403.

Egami, N. (1954). Effect of artificial photoperiodicity on time of oviposition in the fish, *Oryzias latipes*. *Annot. Zool. Jpn.* **27**, 57–62.

Egami, N., and Hama-Furukawa, A. (1981). Response to continuous Y-irradiation of germ cells in embryos and fry of the fish, *Oryzias latipes*. *Int. J. Radiat. Biol.* **40**, 563–568.

Egami, N., and Hyodo-Taguchi, Y. (1969). Hermaphroditic gonads produced in *Oryzias latipes* by x-radiation during embryonic stages. *Copeia* No. 1, pp. 195–196.

Eller, L. L. (1971). Histopathologic lesions in cutthroat trout (*Salmo clarki*) exposed chronically to the insecticide Endrin. *Am. J. Pathol.* **64**, 321–326.

Fenwick, J. C. (1970). The pineal organ: photoperiod and reproductive cycles in the goldfish, *Carassius auratus* L. *J. Endocrinol.* **46**, 101–111.

FitzGerald, G. J., and Keenleyside, M. H. A. (1978). The effects of numerical density of adult fish on reproduction and parental behaviour in the convict cichlid fish *Cichlasoma nigrofasciatum* (Günther). *Can. J. Zool.* **56**, 1367–1371.

Freeman, H. C., and Idler, D. R. (1975). The effect of polychorinated biphenyl on steriodogenesis and reproduction in the brook trout (*Salvelinus fontinalis*). *Can. J. Biochem.* **53**, 666–670.

Freeman, H. C., and Sangalang, G. (1977). The effects of polychorinated biphenyl (Arochlor 1252) contaminated diet on steroidogenesis and reproduction in the Atlantic cod (*Gadus morhua*). *ICES CM* **1977/E 67**, 1–7 (mimeo).

Freeman, H. C., Sangalang, G., and Flemming, B. (1978). The effects of a polychlorinated biphenyl (PCB) diet on Atlantic cod (*Gadus morhua*). *ICES CM* **1978/E 18**, 1–7 (Mimeo).

Freeman, H. C., Uthe, J. F., and Sangalang, G. (1980). The use of steroid hormone metabolism studies in assessing the sublethal effects of marine pollution. *Rapp. P.-V. Reun., Cons. Int. Explor. Mer* **179**, 16–22.

Fromm, P. O. (1980). A review of some physiological and toxicological responses of freshwater fish to acid stress. *Environ. Biol. Fishes* **5**, 79–93.

Fushiki, S. (1979). Studies on the effect of long photoperiod in spring season on gonadal maturation of ayu-fish *Plecoglossus altivelis* T. et S. *Res. Rep. Shigo Prefect. Fish. Exp. Stn. (Shigasuichi Kempo)* **31**, 1–56.

Geisler, R., Knoppel, H. A., and Sioli, H. (1975). The ecology of freshwater fishes in Amazonia present status and future tasks for research. *Anim. Res. Dev.* **1**, 102–119.

Gibson, R. N. (1978). Lunar and tital rhythms in fish. *In* "Rhythmic Activity of Fishes" (J. E. Thorpe, ed.), pp. 201–213. Academic Press, New York.

Gillet, C., and Billard, R. (1977). Stimulation of gonadotropin secretion in goldfish by elevation of rearing temperature. *Ann. Biol. Anim., Biochim., Biophys.* **17**, 673–678.

Gillet, C., Breton, B., and Billard, R. (1978). Seasonal effects of exposure to temperature and photoperiod regimes on gonad growth and plasma gonadotropin in goldfish (*Carassius auratus*). *Ann. Biol. Anim., Biochim., Biophys.* **18**, 1045–1049.

Gillet, C., Billard, R., and Breton, B. (1981). La reproduction du poisson rouge *Carassius auratus* élevé à 30°C. Effet de la photopériode, de l' alimentation et de l' oxygénation. *Cah. Lab. Montereau* **11**, 49–56.

Girin, M., and Devauchelle, N. (1978). Shift in the reproduction period of salt water fish using shortened photoperiod and temperature cycles (in french). *Ann. Biol. Anim., Biochim., Biophys.* **18**, 1059–1065.

Goodman, L. R., Hansen, D. J., Coppage, D. L., Moor, J. C., and Mathews, E. (1979). Diazinon: Chronic toxicity to, and brain acetyl cholinesterase inhibition in the sheepshead minnow. *Cyprinodon variegatus. Trans. Am. Fish. Soc.* **108**, 479–488.

Grau, E. G., Dickhoff, W. W., Nishioka, R. S., Bern, H. A., and Folmar, L. C. (1981). Lunar phasing of the thyroxine surge preparatory to seaward migration of salmonid fish. *Science* **211**, 607–609.

Gupta, S. (1975). The development of carp gonads in warm water aquaria. *J. Fish Biol.* **7**, 775–782.

Guraya, S. S., Saxena, P. K., and Gill, M. (1976). Effect of long photoperiod on the maturation of ovary of the catfish, *Mystus tengara* (Ham.). *Acta Morphol. Neerl.-Scand.* **14**, 331–338.

Hails, A. J., and Abdullah, Z. (1982). Reproductive biology of the tropical fish *Trichogaster pectoralis* (Regan). *J. Fish. Biol.* **21**, 157–170.

Hanyu, I., Asahina, K., and Shimizu, A. (1982). The roles of light and temperature in the reproductive cycles of three bitterling species: *Rhodeus ocellatus*, *Acheilognathus tabira* and *Pseudoperilampus typus*. *Abstr. Pap. Posters, Int. Symp. Reprod. Physiol. Fish, 1982* p. 28.

Harrington, R. W., Jr. (1950). Preseasonal breeding by the bridled shiner, *Notropis bifrenatus*, induced under light-temperature control. *Copeia* No. 4, pp. 304–311.

Harrington, R. W., Jr. (1956). An experiment on the effects of contrasting daily photoperiods on gametogenesis and reproduction in the centrarchid fish *Enneacanthus obesus* (Girard). *J. Exp. Zool.* **131**, 203–224.

Harrington, R. W., Jr. (1957). Sexual photoperiodicity of the cyprinid fish, *Notropis bifrenatus* (Cope), in relation to the phases of its annual reproductive cycle. *J. Exp. Zool.* **135**, 529–555.

Harrington, R. W., Jr. (1959). Effects of four combinations of temperature and daylength on the ovogenetic cycle of a low-latitude fish, *Fundulus confluentus* Goode and Bean. *Zoologica* (*N.Y.*) **44**, 149–168.

Harvey, B. J., and Hoar, S. W. (1979). "The Theory and Practice of Induced Breeding in Fish," TS 21e. IDRC, Ottawa.

Hasse, J. J., Madraisau, B. B., and McVey, J. P. (1977). Some aspects of the life history of *Siganus canaliculatus* (Park) (Pisces: Siganidae) in Palau. *Micronesica* **13**, 297–312.

Haydock, I. (1971). Gonadal maturation and hormone-induced spawning of the Gulf croaker, *Bairdiella icistia*. *Fish Bull.* **69**, 157–180.

Hazard, T. P., and Eddy, R. E. (1951). Modification of the sexual cycle in brook trout (*Salvelinus fontinalis*) by control of light. *Trans. Am. Fish. Soc.* **80**, 158–162.

Hefford, A. E. (1931). "Report on Fisheries for the Year Ended 31st March, 1941," pp. 1–20. New Zealand Marine Dept.

Henderson, N. E. (1963). Influence of light and temperature on the reproductive cycle of the eastern brook trout, *Salvelinus fontinalis* (Mitchell). *J. Fish. Res. Board Can.* **20**, 859–897.

Hester, F. J. (1964). Effects of food supply on fecundity in the female guppy *Lebistes reticulatus* (Peters). *J. Fish. Res. Board Can.* **21**, 757–764.

Hirose, K. (1975). Reproduction in medaka, *Oryzias latipes*, exposed to sublethal concentrations of γ-benzenehexachloride (BHC). *Bull. Tokai. Reg. Fish. Res. Lab.* **81**, 139–149.

Hisaoka, K. K., and Firlit, C. F. (1962). Ovarian cycle and egg production in the zebrafish, *Brachydario rerio*. *Copeia* No. 4, pp. 788–792.

Hodgins, H. O., Gronlund, W. D., Mighell, J. L., Hawkes, J. W., and Robisch, P. A. (1977). Effect of crude oil on trout reproduction. *In* "Fate and Effects of Petroleum Hydrocarbons in Marine Organisms and Ecosystems" (D. A. Wolfe, ed.), pp. 143–144. Pergamon, Oxford.

Hokanson, K. E. F., McCormick, J. H., Jones, B. R., and Tucker, J. H. (1973). Thermal requirement for maturation, spawning and embryo survival of the brook trout *Salvelinus fontinalis*. *J. Fish. Res. Bd. Can.* **30**, 975–984.

Hopkins, C. D. (1974). Patterns of electrical communication: Functions in the social behaviour of *Eigenmannia virescens*. *Behaviour* **50**, 270–305.

Htun-Han, M. (1975). The effects of photoperiod on maturation in the dab, *Limanda limanda* (L.). Ph.D. Thesis, University of East Anglia, U.K.

Htun-Han, M. (1977). The effects of photoperiod on reproduction in fishes: An annotated bibliography. *Libr. Inf. Leafl. Minist. Agric. Fish. Food*, (*G.B.*) No. 6, pp. 1–30.

Hubbs, C., and Strawn, K. (1957). The effects of light and temperature on the fecundity of the greenthroat darter, *Ethestoma lupidum*. *Ecology* 38, 596–602.

Huet, M. (1975). "Textbook of Fish Culture: Breeding and Cultivation of Fish." Fishing News (Book) Ltd., West Byfleet, England.

Hyder, M. (1969). Gonadal development and reproductive activity of the cichlid fish *Tilapia leucosticta* (Trewavas) in an equitorial lake. *Nature, (London)* 224, 1112.

Hyder, M. (1970). Gonadal and reproductive patterns in *Tilapia leucosticta* (Teleostei: Cichlidae) in an equatorial lake, (Lake Naivasha Kenya). *J. Zool.* 162, 179–195.

Hyodo-Taguchi, Y., and Egami, N. (1971). Notes on X-ray effects on the testis of the goby, *Chasmichthys glosus*. *Annot. Zool. Jpn.* 44, 19–22.

Jalabert, B., Breton, B., Brzuska, E., Fostier, A., and Wieniawski, J. (1977). A new tool for induced spawning: The use of 17α-hydroxy-20 β-dihydroprogesterone to spawn carp at low temperature. *Aquaculture* 10, 353–364.

Johannes, R. E. (1978). Reproductive strategies of coastal marine fishes in the tropics. *Environ. Biol. Fishes* 3, 65–84.

John, K. R. (1963). The effect of torrential rains on the reproductive cycle of *Rhinichthys osculus* in the Chiricahua mountains, Arizona. *Copeia* No. 2, pp. 286–291.

Jones, A., Prickett, R. A., and Douglas, M. T. (1979). Recent developments in rearing marine flatfish larvae, particularly turbot (*Scophthalmus maximus* L.), on a pilot commercial scale. *Symp. Early Life Hist. Fish, 1979* ICES/ELH Symp./RA8.

Kapur, K., Kamaldeep, K., and Toor, H. S. (1978). The effect of fenitrothion on reproduction of a teleost fish. *Cyprinus carpio communis* Linn: A biochemical study. *Bull. Environ. Contam. Toxicol.* 20, 438–442.

Kausch, H. (1975). Breeding habits of the major cultivated fishes of EIFAC region and problems of sexual maturation in captivity. *EIFAC Tech. Pap.* 25, 43–52.

Kawamura, T., and Otsuka, S. (1950). On acceleration of the ovulation in the goldfish. *Jpn. J. Ichthyol.* 1, 157–165.

Kaya, C. M. (1973). Effects of temperature and photoperiod on seasonal regression of gonads of green sunfish, *Lepomis cyanellus*. *Copeia* No. 2, pp. 369–373.

Kaya, C. M., and Hasler, A. D. (1972). Photoperiods and temperature effects on the gonads of green sunfish, *Lepomis cyanellus* (Rafinesque), during the quiescent winter phase of its annual sexual cycle. *Trans. Am. Fish. Soc.* 101, 270–275.

Khanna, D. V. (1958). Observations on the spawning of the major carps at a fish farm in the Punjab. *Indian J. Fish.* 5, 282–290.

Khiêt, L. V. (1975). Physiological studies on influences of environmental factors upon maturation of fishes. Thesis Dr. Agric., Tokyo University, Tokyo.

Kiceniuk, J. W., Fletcher, G. L., and Misra, R. (1980). Physiological and morphological changes in a cold torpid marine fish upon acute exposure to petroleum. *Bull. Environ. Contam. Toxicol.* 24, 313–319.

Kihlstrom, J. E., Landberg, C., and Halth, L. (1971). Number of eggs and young produced by zebrafishes (*Brachydanio rerio*, Ham.-Buch.) spawning in water containing small amounts of phenylmercuric acetate. *Environ. Res.* 4, 355–359.

Kirschbaum, F. (1975). Environmental factors control the periodical reproduction of tropical electric fish. *Experientia* 31, 1159–1160.

Kirschbaum, F. (1979). Reproduction of the weakly electric fish *Eigenmannia virescens*. (Rhamphichtyidae, Teleostei) in captivity. *Behav. Ecol. Sociobiol.* 4, 331–355.

Kling, D. (1981). Total atresia of the ovaries of *Tilapia leucosticta* (Cichlidae) after intoxication with the insecticide lebaycig. *Experientia* 37, 73–74.

Kobayashi, S., and Mogami, M. (1958). Effects of x-irradiation upon rainbow trout (*Salmo irideus*). III. Ovary growth in the stages of fry and fingerling. *Bull. Fac. Fish., Hokkaido Univ.* 9, 89–94.

Kossman, H. (1975). Reproduction experiments on carp (*Cyprinus carpio*). *EIFAC Tech. Pap.* **25**, 122–126.

Kramer, D. L. (1978). Reproductive seasonality in the fishes of a tropical stream. *Ecology* **59**, 976–985.

Kumagai, S. (1981). "Ecology of milkfish with emphasis on Reproductive Periodicity," Terminal Report. SEAFDEC, Philippines.

Kunesh, W. H., Freeman, W. J., Hoehm, M., and Nordin, N. G. (1974). Altering the spawning cycle of rainbow trout by control of artificial light. *Prog. Fish.-Cult.* **36**, 225–226.

Kuo, C. M., and Nash, C. E. (1975). Recent progress on the control of ovarian development and induced spawning of the grey mullet (*Mugil cephalus* L.). *Aquaculture* **5**, 19–29.

Kuo, C. M., and Nash, C. E. (1979). Annual reproductive cycle of milkfish, *Chanos chanos* Forskal, in Hawaiian waters. *Aquaculture* **16**, 247–252.

Kuo, C. M., Nash, C. E., and Shehadeh, Z. H. (1974). The effects of temperature and photoperiod on ovarian development in captive grey mullet (*Mugil cephalus* L.). *Aquaculture* **3**, 25–43.

Kuo, C. M., Nash, C. E., and Watanabe, W. D. (1979). Induced breeding experiments with milkfish, *Chanos chanos* (Forskal), in Hawaii. *Aquaculture* **18**, 95–105.

Kuronama, K. (1968). New systems and new fishes for culture in the Far East. *FAO Fish. Rep.* **44**, 123–142.

Kuznetsov, V. A., and Khalitov, N. Kh. (1978). Alterations in the fecundity and egg quality of the roach, *Rutilus rutilus*, in connection with different feeding conditions. *J. Ichthyol.* **18**, 63–70.

Kyle, A. L., Stacey, N. E., Billard, R., and Peter, R. E. (1979). Sexual stimuli rapidly increase milt and gonadotropin levels in goldfish. *Am. Zool.* **19**, Abstr. 19.

Kyle, A. L., Stacey, N. E., and Peter, R. E. (1982). Sexual stimuli associated with increased gonadotropin and milt levels in goldfish. *Abstr. Pap. Posters*, Int. Symp. Reprod. Physiol. Fish, 1982. p. 38.

Lacanilao, F. L., and Marte, C. L. (1980). Sexual maturation of milkfish in floating cages. *Asian Aquacult.* **3**, 4–6.

Lacanilao, F. L., Marte, C. L., and Lam, T. J. (1983). Problems associated with hormonal induction of gonad development in milkfish (*Chanos chanos* Forskal). *Proc. Int. Symp. Comp. Endocrinol., 9th, 1981* (in press).

Lake, J. S. (1967). Rearing experiments with five species of Australian freshwater fishes. I. Inducement of spawning. *Aust. J. Mar. Freshwater Res.* **18**, 137–153.

Lake, J. S., and Midgley, S. H. (1970). Australian Osteoglossidae (Teleostei). *Aust. J. Sci.* **32**, 442.

Lam, T. J. (1974). Siganids: Their biology and mariculture potential. *Aquaculture* **3**, 325–354.

Lam, T. J. (1982). Applications of endocrinology to fish culture. *Can. J. Fish. Aquat. Sci.* **39**, 111–137.

Lam, T. J., and Soh, C. L. (1975). Effect of photoperiod on gonadal maturation in the rabbitfish, *Siganus canaliculatus*. *Aquaculture* **5**, 407–410.

Lam, T. J., Pandey, S., and Hoar, W. S. (1975). Induction of ovulation in goldfish by synthetic luteinizing hormone-releasing hormone (LH-RH). *Can. J. Zool.* **53**, 1189–1192.

Lam, T. J., Pandey, S., Nagahama, Y., and Hoar, W. S. (1976). Effect of synthetic luteinizing hormone-releasing hormone (LH-RH) on ovulation and pituitary cytology of the goldfish *Carassius auratus*. *Can. J. Zool.* **54**, 816–824.

LeGault, R. (1958). A technique for controlling the time of daily spawning and collecting of eggs of the zebrafish, *Brachydanio rerio* (Hamilton-Buchanan). *Copeia* No. 4, pp. 328–330.

Liao, I. C., and Chen, T. (1979). Report on the induced maturation and ovulation of milkfish (*Chanos chanos*) reared in tanks. *Proc. Annu. Meet.—World Maricult. Soc.* **10**.

Liley, N. R. (1980). Patterns of hormonal control in the reproductive behavior of fish, and their relevance to fish management and culture programs. *In* "Fish Behavior and its Use in the Capture and Culture of Fishes" (J. E. Bardach, J. J. Magnuson, R. C. May, and J. M. Reinhart, eds.), ICLARM Conf. Proc. 5, pp. 210–246. Int. Cent. Living Aquat. Resour. Manage., Manila, Philippines.

Liley, N. R. (1982). Chemical communication in fish. *Can. J. Fish. Aquat. Sci.* **39,** 22–35.

Lofts, B., Pickford, G. E., and Atz, J. W. (1968). The effects of low temperature and cortisol, on testicular regression in the hypophysectomized cyrpinodont fish, *Fundulus heteroclitus. Biol. Bull. (Woods Hole, Mass.)* **134,** 74–86.

Lowe-McConnell, R. H. (1975). "Fish Communities in Tropical Freshwaters: Their Distribution, Ecology, and Evolution." Longmans, Green, New York.

Lundqvist, H. (1980). Influence of photoperiod on growth in Baltic salmon parr (*Salmo salar* L.) with special reference to effect of precocious sexual maturation. *Can. J. Zool.* **58,** 940–944.

Macek, K. J. (1968). Reproduction in brook trout (*Salvelinus fontinalis*) fed sublethal concentrations of DTT. *J. Fish. Res. Board Can.* **25,** 1787–1796.

McInerney, J. E., and Evans, D. O. (1970). Action spectrum of the photoperiod mechanism controlling sexual maturation in the threespine stickleback, *Gasterosteus aculeatus. J. Fish. Res. Board Can.* **27,** 749–763.

Mackay, N. J. (1973). Histological changes in the ovaries of the golden perch, *Plectroplites ambiguus,* associated with the reproductive cycle. *Aust. J. Mar. Freshwater Res.* **24,** 95–101.

MacKinnon, C. N., and Donaldson, E. M. (1976). Environmentally induced precocious sexual development in the male pink salmon (*Oncorhynchus gorbuscha*). *J. Fish. Res. Board Can.* **33,** 2602–2605.

McQuarrie, D. W., Markert, J. R., and Vanstone, W. E. (1978). Photoperiod induced off-season spawning of coho salmon (*Oncorhynchus kisutch(. Ann. Biol. Anim., Biochim., Biophys.* **18,** 1051–1058.

MacQuarrie, D. W., Vanstone, W. E., and Markert, J. R. (1979). Photoperiod induced off-season spawning of pink salmon (*Oncorhynchus gorbuscha*). *Aquaculture* **18,** 289–302.

McVey, J. P. (1972). Observations on the early-stage formation of rabbitfish, *Siganus fuscescens* (should be *S. canaliculatus*) at Palau Mariculture Demonstration Centre. *South Pac. Fish. Newsl., 1st, 1972* Vol. 6, pp. 11–12.

Marshall, J. A. (1967). Effect of artificial photoperiodicity on the time of spawning of *Trichopsis vittatus* and *T. pumilus* (Pisces, Belontiidae). *Anim. Behav.* **15,** 510–513.

Marshall, J. A. (1972). Influence of male sound production on oviposition in female *Tilapia mossambica* (Pisces, Cichlidae). *Bull. Ecol. Soc. Am.* **53,** 29.

Matthews, S. A. (1939). The effects of light and temperature on the male sexual cycle in *Fundulus. Biol. Bull. Woods Hole, Mass.)* **70,** 92–95.

May, R. C., Akiyama, G. S., and Santerre, M. T. (1979). Lunar spawning of the threadfin, *Polydactylus sexfilis,* in Hawaii, *Fish. Bull.* **76,** 900–904.

Meske, C., Lühr, B., and Szablewski, W. (1968). Hypophysierung von Aquarienkarpfen und künstliche Laicherbrütung als Methode zur Züchtung neuer Karpfenrassen. *Theor. Appl. Genet.* **38,** 47–51.

Middaugh, D. P. (1981). Reproductive ecology and spawning periodicity of the Atlantic silverside, *Menidia menidia* (Pisces: Atherinidae). *Copeia* No. 4, pp. 766–776.

Mills, C. A. (1980). Spawning and rearing eggs of the dace *Leuciscus leuciscus* (L.). *Fish. Manage.* **11,** 67–72.

Mironova, N. V. (1977). Energy expenditure on egg production in young *Tilapia mossambica* and the influence of maintenance conditions on their reproductive intensity. *J. Ichthyol* **17,** 627–633.

Munro, J. L., Gaut, V. C., Thompson, R., and Reeson, P. H. (1973). The spawning seasons of Caribbean reef fishes. *J. Fish. Biol.* **5**, 69–84.

Nash, C. E., and Kuo, C. M. (1976). Preliminary capture, husbandry and induced breeding results with the milkfish, *Chanos chanos* (Forskal). *Proc. Int. Milkfish Workshop, 1976* pp. 139–159.

Nebeker, A. V., Puglisi, F. A., and Defoe, D. L. (1974). Effect of polychlorinated biphenyl compounds on survival and reproduction of the fathead minnow and flagfish. *Trans. Am. Fish. Soc.* **103**, 562–568.

Nishi, K. (1979). A daily rhythm in the photosensitive development of the ovary in the bitter-ling, *Rhodeus ocellatus ocellatus. Bull. Fac. Fish., Hokkaido Univ.* **30**, 109–115.

Nishi, K. (1981). Circadian rhythm in the photosensitive development of the ovary in the mosquitofish, *Gambusia affinis affinis* (Baird et Girard). *Bull. Fac. Fish., Hokkaido Univ.* **32**, 211–220.

Nobel, G. K., and Curtis, B. (1939). The social behavior of the jewel fish, *Hemichromis bimaculatus. Bull. Am. Mus. Nat. Hist.* **76**, 1–46.

Nomura, M. (1962). Studies on reproduction of rainbow trout *Salmo gairdnerii* with special reference to egg taking. III. Acceleration of spawning by control of light. *Bull. Jpn. Soc. Sci. Fish.* **28**, 1070–1076.

Nozaki, M., Tsutsumi, T., Kobayashi, H., Takei, Y., Ichikawa, T., Tsuneki, K., Miyagawa, K., Uemura, H., and Tatsumi, V., (1976). Spawning habit of the puffer, *Fugu niphobles* (Jordan et Snyder) I. *Zool. Mag.* **85**, 156–168.

Nzioka, R. M. (1979). Observations on the spawning seasons of East African reef fishes. *J. Fish Biol.* **14**, 329–342.

Ogneff, J. (1911). Ueber die Aendarungen in den organen der goldfische nach dreijahrigem verbeiben in finsternis. *Anat. Anz.* **40**, 81–87.

Pandey, S., and Hoar, W. S. (1972). Induction of ovulation in goldfish by clomiphene citrate. *Can. J. Zool.* **50**, 1679–1680.

Pandey, S., Stacey, N. E., and Hoar, W. S. (1973). Mode of action of clomiphene citrate in inducing ovulation in goldfish. *Can. J. Zool.* **51**, 1315–1316.

Pandey, S., Lam, T. J., Nagahama, Y., and Hoar, W. S. (1977). Effects of dexamethasone metopirone on ovulation in the goldfish, *Carassius auratus. Can. J. Zool.* **55**, 1342–1350.

Pang, P. K. T. (1971). The effects of complete darkness and vitamin C supplement on the killifish, *Fundulus heteroclitus*, adapted to sea water. I. Calcium metabolism and gonadal maturation. *J. Exp. Zool.* **178**, 15–22.

Partridge, B. L., Liley, N. R., and Stacey, N. E. (1976). The role of pheromones in the sexual behaviour of the goldfish. *Anim. Behav.* **24**, 291–299.

Payne, A. I. (1975). The reproductive cycle, condition and feeding in *Barbus liberiensis*, a tropical stream-dwelling cyrpinid. *J. Zool.* **176**, 247–269.

Payne, J. F., Kiceniuk, J. W., Squires, W. R., and Fletcher, G. L. (1978). Pathological changes in a marine fish after a 6-month exposure to petroleum *J. Fish. Res. Board Can.* **35**, 665–667.

Peter, R. E. (1981). Gonadotropin secretion during reproductive cycles in teleosts: Influences of environmental factors. *Gen. Comp. Endocrinol.* **45**, 294–305.

Peter, R. E., and Crim, L. W. (1979). Reproductive endocrinology of fishes: Gonadal cycles and gonadotropin in teleost. *Annu. Rev. Physiol.* **41**, 323–335.

Peter, R. E., and Hontela, A. (1978). Annual gonadal cycles in teleosts: Environmental factors and gonadotropin levels in blood. *In* "Environmental Endocrinology" (I. Assenmacher and D. S. Farners, eds.), pp. 20–25. Springer-Verlag, Berlin and New York.

Peter, R. E., Hontela, A., Cook, A. F., and Paulencu, C. R. (1978). Daily cycles in serum cortisol levels in the goldfish: Effects of photoperiod, temperature, and sexual condition. *Can. J. Zool.* **56**, 2443–2448.

Peterson, D. A. (1972). Barometric pressure and its effect on spawning activities of rainbow trout. *Prog. Fish-Cult.* **34,** 110–112.

Pickering, Q. H., Brungs, W., and Gast, M. (1977). Effects of exposure time and copper concentration on reproduction of the fathead minnow (*Pimephales promalas*). *Water Res.* **11,** 1079–1083.

Pickford, G. E., Lofts, B., Bara, G., and A-z, J. W. (1972). Testis stimulation in hypophysectomized male killifish, *Fundulus heteroclitus,* treated with mammalian growth hormone and/or luteinizing hormone. *Biol. Reprod.* **7,** 370–386.

Pierson, K. B. (1981). Effects of chronic zinc exposure on the growth, sexual maturity, reproduction and bioaccumulation of the guppy, *Poecilia reticulata. Can. J. Fish. Aquat. Sci.* **38,** 23–31.

Polder, J. J. W. (1971). On gonads and reproductive behaviour in the cichlid fish *Aequidens portalgrensis* (Hensel). *Neth. J. Zool.* **21,** 265–365.

Poon, K. H. (1980). Some ecological factors governing growth and maturation of the tropical freshwater fish *Sarotherodon mossambicus* (Peters): Effects of photoperiod and stocking conditions. M.Sc. Thesis, Dept. of Zoology, National University of Singapore, Singapore.

Popper, D., Gordin, H., and Kissil, G. W. (1973). Fertilization and hatching of rabbitfish *Siganus rivulatus. Aquaculture* **2,** 37–44.

Popper, D., May, R. C., and Lichatowich, T. (1976). An experiment in rearing larval *Siganus vermiculatus* (Valenciennes) and some observations on its spawning cycle. *Aquaculture* **7,** 281–290.

Poston, H. A. (1978). Neuroendocrine mediation of photoperiod and other environmental influences on physiological responses in slamonids: A review. *Tech. Pap. U.S. Fish Wildl. Serv.* **96.**

Poston, H. A., and Livingston, D. L. (1971). The effect of continuous darkness and continuous light on the functional maturity of brook trout during their second reproductive cycle. *Fish. Res. Bull., N.Y. Conserv. Dep.* **33,** 25–29.

Pullin, R. S. V., and Kuo, C. M. (1981). Developments in the breeding of cultured fishes. *In* "Advances in Food Producing Systems for Arid and Semiarid Lands" (J. T. Manassah and E. J. Briskey, eds.), Part B, pp. 899–978. Academic Press, New York.

Pyle, E. A. (1969). The effect of constant light or constant darkness on the growth and sexual maturity of brook trout. *Fish. Res. Bull., N.Y. Conserv. Dep.* **31,** 13–19.

Rao, N. G. S., Ray, P., and Gopinathan, K. (1972). Observations on the spawning of *Cirrhina reba* (Hamilton) in the Cauvery and Bhavani rivers. *J. Inland Fish. Soc. India* **4,** 69–73.

Rasquin, P., and Hafter, E. (1951). Effects of ACTH and cortisone on the pituitary, thyroid and gonads of the teleost *Astyanax mexicanus. Anat. Rec.* **111,** 41–42.

Rasquin, P., and Hafter-Atz, E. (1952). Effects of ACTH and cortisone on the pituitary, thyroid and gonads of the teleost *Astyanax mexicanus. Zoologica* (*N.Y.*) **37,** 77–87.

Reisman, H. M. (1968). Effects of social stimuli on the secondary sex characters of male three-spined sticklebacks, *Gasterosteus aculeatus. Copeia* No. 4, pp. 816–826.

Reisman, H. M., and Cade, T. J. (1967). Physiological and behavioral aspects of reproduction in the brook stickleback, *Culea inconstants Am. Midl. Nat.* **77,** 257–295.

Roblin, C. (1980). Etude comparée de la biologie du développement (gonadogénèse, croissance, nutrition) du loup (*Dicentrarchus labrax*) en milieu natural et en élevage contrôlé. Acad. de Montpellier Thèse 3ème cycle, Univ. Sci. Technol., Languedoc.

Ross, R. M. (1978). Reproductive behaviour of the anemonefish *Amphiprion melanopus* on Guam. *Copeia* No. 1, pp. 103–107.

Rustamova, Sh. A. (1979). The chronic effect of alkali on the growth development and fecundity of the guppy. *Hydrobiol. J.* **13,** 83–85.

Sangalang, G. B., and Freeman, H. C. (1974). Effects of sublethal cadmium on maturation and testosterone and 11-ketotestosterone production *in vivo* in brook trout. *Biol. Reprod.* 11, 429–435.

Sangalang, G. B., and O'Halloran, M. J. (1972). Cadmium induced testicular injury and alterations of androgen synthesis in brook trout. *Nature (London)* 240, 470–471.

Sangalang, G. B., and O Halloran, M. J. (1973). Adverse effects of cadmium on brook trout testis and on *in vitro* testicular androgen synthesis. *Biol. Reprod.* 9, 394–403.

Sastry, K. V., and Agrawal, M. K. (1979a). Mercuric chloride induced enzymbiological changes in kidney and ovary of a teleost fish, *Channa punctatus*. *Bull. Environ. Contam. Toxicol.* 22, 38–43.

Sastry, K. V., and Agrawal, M. K. (1979b). Effects of lead nitrate on the activities of a few enzymes in the kidney and ovary of *Heteropneustes fossilis* (actually *Channa punctatus*). *Bull. Environ. Contam. Toxicol.* 22, 55–59.

Sawara, Y. (1974). Reproduction of the mosquitofish (*Gambusia affinis affinis*), a freshwater fish introduced into Japan. *Jpn. J. Ecol.* 24, 140–146.

Sawara, Y., and Egami, N. (1977). Note on the differences in the response of the gonad to the photoperiod among population of *Oryzias latipes* collected in different localities. *Annot. Zool. Jpn.* 50, 147–150.

Saxena, P. K., and Garg, M. (1978). Effect of insecticidal pollution on ovarian recrudescence in the freshwater teleost *Channa punctatus* (Bl.). *Indian J. Exp. Biol.* 16, 689–691.

Schneider, L. (1969). Experimentelle Untersuchungen über den Einfluss von Tageslänge und Temperatur auf die Gonadenreifung beim Dreistachligen Stichling (*Gasterosteus aculeatus*). *Oecologia* 3, 249–265.

Schwassmann, H. O. (1971). Biological rhythms. *In* "Fish Physiology" (W. S. Hoar and D. J. Randall, eds.), Vol. 6, pp. 371–428. Academic Press, New York.

Schwassmann, H. O. (1978). Times of annual spawning and reproductive strategies in Amazonian fishes. *In* "Rhythmic Activity of Fishes" (J. E. Thorpe, ed.), pp. 187–200. Academic Press, New York.

Schwassmann, H. O. (1980). Biological rhythms: Their adaptive significance. *In* "Environmental Physiology of Fishes" (M. A. Ali, ed.), pp. 613–630. Plenum, New York.

Scott, D. B. C. (1964). Reproduction in female *Phoxinus* Ph.D. Thesis, University of Glasgow, Scotland.

Scott, D. B. C. (1979). Environmental timing and the control of reproduction in teleost fish. *Symp. Zool. Soc. London* 44, 105–132.

Scott, D. B. C., and Fuller, J. D. (1976). The reproductive biology of *Scleropages formosus* (Müller & Schlegel) (Osteoglossomorpha, Osteoglossidae) in Malaya, and the morphology of its pituitary gland. *J. Fish Biol.* 8, 45–53.

Scott, D. P. (1962). Effect of food quantity on fecundity of rainbow trout, *Salmo gairdneri*. *J. Fish. Res. Board Can.* 19, 715–730.

Scrimshaw, N. S. (1944). Superfetation in Poeciliid fishes. *Copeia* No. 3, pp. 180–183.

Seah, K. P., and Lam, T. J. (1973a). Effect of salinity and temperature on growth and reproduction in the guppy, *Poecilia reticulata* (Peters) I. Cobra guppy. *J. Singapore Natl. Acad. Sci.* 3(1), 29–35.

Seah, K. P., and Lam, T. J. (1973b). Effect of salinity and temperature on growth and reproduction in the guppy, *Poecilia reticulata* (Peters). II. Wild guppy. *J. Singapore Natl. Acad. Sci.* 3(1), 36–42.

Sehgal, A., and Sundararaj, B. I. (1970a). Effects of various photoperiodic regimens on the ovary of the catfish, *Heteropneustes fossilis* (Bloch) during the spawning and postspawning periods. *Biol. Reprod.* 2, 425–434.

Sehgal, A., and Sundararaj, B. I. (1970b). Effects of blinding and/or total darkness on ovarian recrudescence or aggression during the appropriate periods of the reproductive cycle of the catfish, *Heteropneustes fossilis* (Bloch). *J. Interdiscip. Cycle Res.* **1**, 147–159.

Shehadeh, Z. H. (1970). Controlled breeding of culturable species of fish—a review of progress and current problems. *Proc., Indo-Pac. Fish. Counc.* **14**, 2–33.

Shehadeh, Z. H. (1975). Induced breeding techniques—a review of progress and problems. *EIFAC Tech. Pap.* **25**, 72–89.

Shikhshabekov, M. M. (1974). Features of the sexual cycles of some semi-diadromous fishes in the lower reaches of the terek. *J. Ichthyol.* **14**, 79–87.

Shiraishi, Y. (1965a). The influence of photoperiodicity on the maturation of Ayu-fish, *Plecoglossus altivelis*. II. Relation between the maturation and the daylength. *Bull. Freshwater Fish. Res. Lab.* **15**, 59–68.

Shiraishi, Y. (1965b). The influence of photoperiodicity on the maturation of Ayu-fish, *Plecoglossus altivelis*. III. The limit of the intensity of light. *Bull. Freshwater Fish. Res. Lab.* **15**, 69–76.

Shiraishi, Y. (1965c). The influence of photoperiodicity on the maturation of Ayu-fish, *Plecoglossus altivelis*. IV. The effect of wavelength and of conversion of darkness and light on the maturation. *Bull. Freshwater Fish. Res. Lab.* **15**, 77–84.

Shiraishi, Y., and Fukuda, Y. (1966). The relation between the daylength and the maturation in four species of salmonid fish. *Bull. Freshwater Fish. Res. Lab.* **16**, 103–111.

Shiraishi, Y., and Takeda, T. (1961). The influence of photoperiodicity on the maturation of the Ayu-fish, *Plecoglossus altivelis*. *Bull. Freshwater Fish. Res. Lab.* **11**, 69–81.

Silverman, H. I. (1978a). Effects of different levels of sensory contact upon reproductive activity of adult male and female *Sarotherodon (Tilapia) mossambicus* (Peters); Pisces: Cichlidae. *Anim. Behav.* **26**, 1081–1090.

Silverman, H. I. (1978b). The effects of visual social stimulation upon age at first spawning in the mouth-brooding Cichlid fish *Sarotherodon (Tilapia) mossambicus* (Peters). *Anim. Behav.* **26**, 1120–1125.

Singh, H., and Singh, T. P. (1980a). Effect of two pesticides on ovarian ^{32}p uptake and gonadotropin concentration during different phases of annual reproductive cycle in the freshwater catfish, *Heteropneustes fossilis* (Bloch). *Environ. Res.* **22**, 190–200.

Singh, H., and Singh, T. P. (1980b). Effects of two pesticides on testicular ^{32}p uptake, gonadotrophic potency, lipid and cholesterol content of testis, liver and blood serum during spawning phase in *Heteropneustes fossilis* (Bloch). *Endokrinologie* **76**, 288–296.

Singh, H., and Singh, T. P. (1980c). Effect of two pesticides on total lipid and cholesterol contents of ovary, liver and blood serum during different phases of the annual reproductive cycle in the freshwater teleost *Heteropneustes fossilis* (Bloch). *Environ. Pollut.* **23**, 9–17.

Singh, H., and Singh, T. P. (1980d). Short-term effect of two pesticides on the survival, ovarian ^{32}p uptake and gonadotrophic potency in a freshwater catfish, *Heteropneustes fossilis* (Bloch). *J. Endocrinol.* **85**, 193–199.

Singh, H., and Singh, T. P. (1981). Effect of parathion and aldrin on survival, ovarian ^{32}p uptake and gonadotrophic potency in freshwater catfish, *Heteropneustes fossilis* (Bloch). *Endokrinologie* **77**, 173–178.

Sinha, V. R. P., Jhingran, V. G., and Ganapati, S. V. (1974). A view on spawning of the Indian major carps. *Arch. Hydrobiol.* **73**, 518–536.

Sivarajah, K., Franklin, C. S., and Williams, W. P. (1978a). The effects of polychlorinated biphenyls on plasma steroid levels and hepatic microsomal enzymes in fish. *J. Fish Biol.* **13**, 401–409.

Sivarahaj, K., Franklin, C. S., and Williams, W. P. (1978b). Some histopathological effects of Arochlor 1254 on the liver and gonads of rainbow trout, *Salmo gairdneri* and carp. *Cyppinus carpio. J. Fish Biol.* **13**, 411–414.

Skarphedinsson, O., Scott, A. P., and Bye, V. J. (1982). Long photoperiods stimulate gonad development in rainbow trout. *Abstr. Pap. Posters, Int. Symp. Reprod. Physiol. Fish, 1982*, p. 62.

Smigielski, A. S. (1975). Hormonal-induced ovulation of the winter flounder, *Pseudopleuronectes americanus. Fish. Bull.* **73**, 431–438.

Smith, R. J. F. (1970). Control of prespawning behavior of sunfish (*Lepomis gibbosus* and *Lepomis megalotis*). II. Environmental factors. *Anim. Behav.* **18**, 575–587.

Soh, C. L. (1976). Some aspects of the biology of *Siganus canaliculatus* (Park) 1797. Ph.D. Thesis, Dept. of Zoology, National University of Singapore, Singapore.

Solomon, D. J. (1977). A review of chemical communication in freshwater fish. *J. Fish Biol.* **11**, 369–376.

Spehar, R. L., Leonard, E. N., and DeFoe, D. L. (1978). Chronic effects of cadmium and zinc mixtures on flagfish (*Jordanella floridae*). *Trans. Am. Fish. Soc.* **107**, 354–360.

Speranza, A. W., Seeley, R. J., Seeley, V. A., and Perlmutter, A. (1977). The effect of sublethal concentrations of zinc on reproduction in the zebrafish *Brachydanio rerio* Hamilton Buchanan. *Environ. Pollut.* **12**, 217–222.

Spieler, R. E., Noeske, T. A., de Vlaming, V., and Meier, A. H. (1977). Effects of thermocycles on body weight gain and gonadal growth in the goldfish, *Carassius auratus. Trans. Am. Fish. Soc.* **106**, 440–444.

Stacey, N. E., and Hourston, A. S. (1982). Spawning and feeding behavior of captive Pacific herring, *Clupea harengus pallasi. Can. J. Fish. Aquat. Sci.* **39**, 489–498.

Stacey, N. E., and Pandey, S. (1975). Effects of indomethacin and prostaglandins on ovulation of goldfish. *Prostaglandins* **9**, 597–607.

Stacey, N. E., Cook, A. F., and Peter, R. E. (1979a). Ovulatory surge of gonadotropin in the goldfish, *Carassius auratus. Gen. Comp. Endocrinol.* **37**, 246–249.

Stacey, N. E., Cook, A. F., and Peter, R. E. (1979b). Spontaneous and gonadotropin-induced ovulation in the goldfish, *Carassius auratus* L.: Effects of external factors. *J. Fish Biol.* **15**, 349–361.

Stepkina, M. V. (1973). Some biological characteristics of *Pagrus ehrenbergii* Val. *J. Ichthyol.* **13**, 641–649.

Stequert, B. (1972). Contribution à l'étude de la biologie du bar (*Dicentrarchus labrax* L.) des réservoirs à poissons de la région d'Arcachon. Thèse 3ème cycle, Université de Bordeaux, France.

Sundararaj, B. I., and Sehgal, A. (1970). Short and long-term effects of imposition of total darkness on the annual ovarian cycle of the catfish, *Heteropneustes fossilis* (Bloch). *J. Interdiscip. Cycle Res.* **1**, 291–301.

Sundararaj, B. I., and Vasal, S. (1973). Photoperiodic regulation of reproductive cycle in the catfish *Heteropneustes fossilis* (Bloch). *Int. Congr. Ser.—Excerpta Med.* **273**, 180–184.

Sundararaj, B. I., and Vasal, S. (1976). Photoperiod and temperature control in the regulation of reproduction in the female catfish *Heteropneustes fossilis. J. Fish. Res. Board Can.* **33**, 959–973.

Suseno, D., and Djajadiredja, R. R. (1981). A review on the status of freshwater finfish induced breeding in Indonesia. *Induced Fish Breed. Workshop, 1980* IDRC-178e.

Swingle, H. S. (1957). A repressive factor controlling reproduction in fishes. *Proc. Pac. Sci. Congr., 8th, 1953* Vol. 3a, pp. 865–871.

Takano, K., Kasuga, S., and Sato, S. (1973). Daily reproductive cycle of the medaka, *Oryzias latipes* under artificial photoperiod. *Bull. Fac. Fish., Hokkaido Univ.* **24**, 91–99.

Tay, L. L. (1983). Ph.D. Thesis (in preparation).

Taylor, M. H., and DiMichelle, L. (1980). Ovarian changes during the lunar spawning cycle of *Fundulus heteroclitus. Copeia* No. 1, pp. 118–125.

Taylor, M. H., Leach, G. J., DiMichelle, L. D., Levitan, W. M., and Jacob, W. F. (1979). Lunar spawning cycle in the mumichog, *Fundulus heteroclitus* (Pisces: Cyprinodontidae). *Copeia* pp. 291–297.

Terkatin-Shimony, A., Ilan, Z., Yaron, Z., and Johnson, D. W. (1980). Relationship between temperature, ovarian recrudescence, and plasma cortisol level in *Tilapia aurea* (Cichlidae, Teleostei). *Gen. Comp. Endocrinol.* **40**, 143–148.

Thomson, D. A., and Muench, K. A. (1976). Influence of tides and waves on the spawning behavior of the Gulf of California grunion, *Leuresthes sardina* (Jenkins and Evermann). *Bull. South. Calif. Acad. Sci.* **75**, 198–203.

Thorpe, J. E. (1978). "Rhythmic Activity of Fishes." Academic Press, New York.

Tseng, L. C., and Hsiao, S. M. (1979). First successful case of artificial propagation of pond-reared milkfish. *China Fish.* **320**, 9–10.

Tyler, A. V., and Dunn, R. S. (1976). Ration, growth, measures of somatic and organ condition in relation to meal frequency in winter flounder, *Pseudopleuronectes americanus*, with hypotheses regarding population homeostasis. *J. Fish. Res. Board Can.* **33**, 63–75.

van den Assem, J. (1967). Territory in the three-spined stickleback *Gasterosteus aculeatus* L., an experimental study in intraspecific competition. *Behaviour, Suppl.* **16**, 1–164.

Vasal, S., and Sundararaj, B. I. (1975). Responses of the regressed ovary of the catfish *Heteropneustes fossilis* (Bloch) to interrupted night photoperiods. *Chronobiologia* **2**, 224–239.

Vasal, S., and Sundararaj, B. I. (1976). Response of the ovary in the catfish, *Heteropneustes fossilis* (Bloch), to various combinations of photoperiod and temperature. *J. Exp. Zool.* **197**, 247–264.

Verghese, P. U. (1967). Prolongation of spawning season in the carp*Cirrhina reba* (Ham.) by artificial light treatment. *Curr. Sci.* **36**, 465–467.

Verghese, P. U. (1970). Preliminary experiments on the modification of the reproductive cycle of an Indian carp *Cirrhina reba* (Ham.) by control of light and temperature. *Proc., Indo-Pac. Fish. Counc.* **13**, 171–184.

Verghese, P. U. (1975). Internal rhythm of sexual cycle in a carp *Cirrhina reba* (Ham.) under artificial conditions of darkness. *J. Inland Fish. Soc. India* **7**, 182–188.

Walker, B. W. (1949). Periodicity of spawning in the grunion, *Leuresthes tenuis*. Ph.D. Thesis, University of California, Los Angeles.

Walker, B. W. (1952). A guide to the grunion. *Calif. Fish Game* **38**, 409–420.

Whitehead, C., Bromage, N. R., Forster, J. R. M., and Matty, A. J. (1978). The effects of alterations in photoperiod on ovarian development and spawning time in the rainbow trout. (*Salmo gairdneri*). *Ann. Biol. Anim., Biochim., Biophys.* **18**, 1035–1043.

Whiteside, B. G., and Richan, F. J. (1969). Regressive factors controlling reproduction in goldfish. *Proq. Fish-Cult.* **31**, 165.

Wiebe, J. P. (1968). The effects of temperature and daylength on the reproductive physiology of the viviparous sea perch, *Cymatogaster aggregata* Gibbons. *Can. J. Zool.* **46**, 1207–1219.

Wootton, R. J. (1973). Fecundity of the three-spined stickleback, *Gasterosteus aculeatus* (L.). *J. Fish Biol.* **5**, 683–688.

Wootton, R. J. (1977). Effect on food limitation during the breeding season on the size, body components and egg production of female sticklebacks (*Gasterosteus aculeatus*). *J. Anim. Ecol.* **46**, 823–834.

Yadav, M., and Ooi, H. S. D. (1977). Effect of photoperiod on the incorporation of ^{3}H-thymidine into the gonads of *Carassius auratus*. *J. Fish Biol.* **11**, 409–416.

Yamamoto, K., and Yamazaki, F. (1967). Hormonal control of spermiation and ovulation in goldfish. *Gunma Symp. Endocrinol.* **4**, 131–145.

Yamamoto, K., Nagahama, Y., and Yamazaki, F. (1966). A method to induce artificial spawning of goldfish all through the year. *Bull. Jpn. Soc. Sci. Fish.* **32**, 977.

Yamauchi, K., and Yamamoto, K. (1973). *In vitro* maturation of the oocytes in the medaka, *Oryzias latipes*. *Annot. Zool. Jpn.* **46**, 144–153.

Yamazaki, F. (1965). Endocrinological studies on the reproduction of the female goldfish *Carassius auratus* with special reference to the function of the pituitary gland. *Mem. Fac. Fish., Hokkaido Univ.* **13**, 1–64.

Yaron, Z., Cocos, M., and Salzer, H. (1980). Effects of temperature and photoperiod on ovarian recrudescence in the cyprinid fish *Mirogrex terrae-sanctae*. *J. Fish Biol.* **16**, 371–382.

Yoshimura, N., Etoh, H., Egami, N., Asami, K., and Yamada, T. (1969). Note on the effects of β-rays from ^{90}Sr − ^{90}Y on spermatogenesis in the teleost, *Oryzias latipes*. *Annot. Zool. Jpn.* **42**, 75–79.

Yoshioka, H. (1962). On the effects of environmental factors upon reproduction of fishes. I. The effect of daylength on the reproduction of the Japanese killifish, *Oryzias latipes*. *Bull. Fac. Fish., Hokkaido Univ.* **13**, 123–136.

Yoshioka, H. (1963). On the effects of environmental factors upon the reproduction of fishes. II. Effects of short and long daylengths on *Oryzias latipes* during the spawning season. *Bull. Fac. Fish., Hokkaido Univ.* **14**, 137–171.

Yoshioka, H. (1966). On the effects of environmental factors upon the reproduction of fishes. III. The occurrence and regulation of refractory period in the photoperiodic response of medaka, *Oryzias latipes*. *J. Hokkaido Univ. Educ.* **17**, 23–33.

Yoshioka, H. (1970). On the effects of envionmental factors upon the reproduction of fishes. IV. Effects of long photoperiod on the development of ovaries of adult medaka, *Oryzias latipes*, at low temperatures. *J. Hokkaido Univ. Educ.* **21**, 14–20.

Yu, M. L., and Perlmutter, A. (1970). Growth inhibiting factors in the zebrafish (*Brachydanio rerio*) and the blue gourami (*Trichogaster trichopterus*). *Growth* **34**, 153–175.

HORMONAL CONTROL OF OOCYTE FINAL MATURATION AND OVULATION IN FISHES

FREDERICK W. GOETZ

Department of Biology
University of Notre Dame
Notre Dame, Indiana

I. INTRODUCTION

The ease with which lower vertebrate oocytes can be maintained in *in vitro* incubations, and, in general, the large number of oocytes contained in an individual female, have facilitated investigations of the hormonal regulation of final maturation and ovulation in these organisms. As a result, a great deal is known about the hormonal regulation of these events. Much of the information gained from *in vitro* investigations (see Sections II,B and D) is now being applied to *in vivo* studies or being correlated with naturally occurring reproductive events. Therefore, the importance of *in vitro* investi-

FISH PHYSIOLOGY, VOL. IXB

gations is now being realized. Several reviews have been published regarding specific aspects of final maturation in teleosts or concerning research conducted on individual species of fish (Jalabert, 1976; Hirose, 1976; Sundararaj and Goswami, 1971a, 1974a, 1977; Wallace and Selman, 1978; Van Ree *et al.*, 1977; Goetz, 1979; Epler, 1981d).

II. FINAL MATURATION

A. Characteristics

In fish, several distinct stages of development occur during the complete maturation of the oocyte. Oocytes undergo previtellogenic and vitellogenic growth phases during which the oocytes enlarge, the follicular layers develop, and yolk accumulates. Following the vitellogenic phase, or beginning in the latter portion of this phase, several further maturational processes occur prior to ovulation. In this discussion, these processes are collectively referred to as final maturation, although it is obvious that there are certain differences in these processes when individual species are considered. In general, final maturation includes changes in the nucleus and chromosomes and in the cytoplasm and its inclusions.

Following the vitellogenic phase of the oocyte, or in some species during this phase, the nucleus or germinal vesicle (GV) migrates from a central or slightly eccentric position to the oocyte periphery (Fig. 1). In several species, it has been demonstrated that the migration is directed at the micropyle (Yamazaki, 1965; Goetz, 1976). Following the migration, the germinal vesicle disperses; this is termed germinal vesicle breakdown (GVBD). The area in which this occurs may remain rather inconspicuous as in some salmonids, or become quite distinct as in yellow perch (*Perca flavescens*) in which an elevated cap is formed (Goetz, 1976). The time and rate at which the GV migrates to a peripheral position varies between species, although some confusion has been introduced concerning the natural migratory process of the GV as a result of the hormonal manipulation of oocytes in *in vitro* incubations. In some species, such as members of the Salmonidae, the GV migrates from the center of the oocyte before vitellogenesis is complete and requires a rather long period for complete migration (Sakun, 1966). An extreme case may be *Lebistes reticulatus* in which the GV is already eccentric even prior to the growth phase of the oocyte (Raven, 1961). In contrast, in other species, GV migration does not occur until vitellogenesis is complete (Yamazaki, 1965; Yamamoto, 1956a; Malservisi and Magnin, 1968). In the temperate zone, vitellogenesis in yellow perch occurs in the fall and

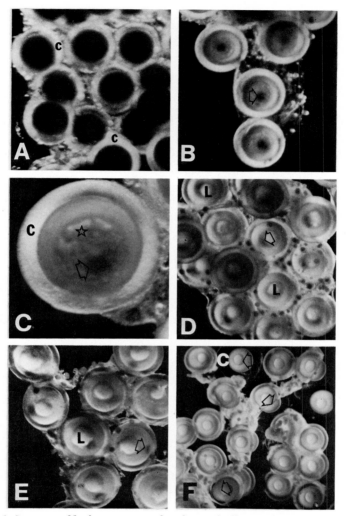

Fig. 1. Sequence of final maturation and ovulation of yellow perch (*Perca flavescens*) oocytes during *in vitro* incubation with 17α,20β-DHP. All oocytes, except those in A and F, were treated with a "clearing" fixative (Goetz and Bergman, 1978a) prior to observation. (A) Oocytes prior to incubation. Transparent chorion (C) surrounds opaque oocyte (×10). (B) Same oocytes as shown in A after treatment in a "clearing" fixative. Small lipid droplets surround a slightly eccentric germinal vesicle (GV) (arrows) (×10). (C) Enlarged view of oocytes shown in B. Arrow indicates GV, star indicates beginning of lipid droplet coalescence and C designates chorion (×28). (D) Oocytes in which lipid droplet (L) coalescence is complete. The GV (arrow) is on the surface of the coalesced lipid droplet approximately one-half the distance from the oocyte center (×10). (E) Oocytes in which the GV (arrow) has migrated to the surface of the oocyte and is beginning to spread laterally. The lipid droplet (L) is also beginning to move toward the oocyte surface (×10). (F) Oocytes having undergone GVBD and beginning ovulation. Arrows indicate ovulated or ovulating oocytes. Note the nearly transparent, expanded chorion layer (C) around ovulated oocytes. Development of an elevated animal region and formation of a perivitelline space occur following exposure to water and prior to fertilization (see Goetz, 1979) (×6.5).

early winter (Malservisi and Magnin, 1968; T. Kingsley, unpublished results). By February the oocytes complete vitellogenesis, although the nucleus remains located in the oocyte center. From February until the time of spawning (April–June), the GV moves closer to the periphery as the lipid droplets coalesce. It appears that the GV actually lies on the surface of the coalescing lipid droplets, and, as a result, is pushed toward the periphery as coalescence proceeds. Finally, prior to ovulation, the GV migrates a short distance from the lipid droplet to the surface of the oocyte and breaks down. In nature, the entire process of migration and GVBD in yellow perch may take several months; however, *in vitro*, under appropriate hormonal stimulation, it can be induced within 96 hr (Goetz and Theofan, 1979; Fig. 1).

Developing oocytes of many fish species have distinct lampbrush chromosomes (Braekevelt and McMillan, 1967; Moser, 1967; K. Yamamoto, 1956b; Malservisi and Magnin, 1968; T. S. Yamamoto, 1963), and, just prior to GVBD, descriptions of the chromosomes in various stages of prophase have been given (Sundararaj and Goswami, 1977; Van Ree *et al.*, 1977; Iwamatsu *et al.*, 1976). In vertebrates, it is generally assumed that GVBD marks the resumption of meiosis; however, very few investigations on fish have specifically addressed this topic. In the Japanese medaka (*Oryzias latipes*), it has been demonstrated *in vitro* and *in vivo* that the transition from prophase I to prophase II requires 1–4 hr and occurs following GVBD (Iwamatsu *et al.*, 1976; Iwamatsu, 1965). By 5 hr post-GVBD, *in vitro*, the chromosomes are in metaphase II. Similarly, in the pond loach (*Misgurnus angillicaudatus*) the second meiotic metaphase was observed in oocytes approximately 6–7 hr following GVBD *in vitro* (Iwamatsu and Katoh, 1978). Chromosomes in carp (*Cyprinus carpio*) oocytes that have completed final maturation *in vitro* have also been observed in metaphase II (Sakun and Gureeva-Preobrazhenskaya, 1975); in several other fish species, chromosomes in various phases of the first meiotic division have been reported (Van Ree *et al.*, 1977; Yamamoto, 1956b; Sakun, 1966).

In addition to the nuclear events, there are also several obvious cytoplasmic processes that occur during GV migration and breakdown. These include the coalescence of lipid droplets and yolk globules, and an overall increase in oocyte translucency. Before GV migration and breakdown, oocytes are generally opaque, making the centrally or eccentrically located GV of some species difficult to observe without the use of a "clearing" solution (Fig. 1). During GV migration and breakdown, oocytes become more translucent. This phenomenon occurs in most fish species studied and has been reported for yellow perch, brook trout (*Salvelinus fontinalis*), rainbow trout (*Salmo gairdneri*), catfish, killifish (*Fundulus heteroclitus*), striped bass (*Morone saxatilis*), goldfish (*Carassius auratus*), zebrafish (*Brachydanio rerio*), rockfish (*Sebastodes* spp.), and pond loach (Goetz, 1976; Jalabert *et al.*, 1972; Goswami and Sundararaj, 1971a; Stevens, 1966; Yamazaki, 1965; Wallace and Selman, 1978; Moser, 1967; Hitz and DeLacy, 1965; Iwamatsu and Katoh, 1978). Changes in oocyte translucency during final maturation are probably related to the fusion of yolk globules that occurs in many species at this time (Wallace and Selman, 1981). However, in some species, complete fusing of yolk globules does not occur at final maturation (Yamamoto, 1958; Shackley and King, 1977). During final maturation there may also be a coalescence of lipid droplets within the oocyte. The degree of coalescence varies among species of different orders of fish and may be related to the quantity and distribution in the oocyte. Interestingly, it appears that the degree of lipid droplet coalescence follows a phylogenetic pattern. For instance, lipid coalescence in oocytes of higher teleosts such as yellow perch, walleye (*Stizostedion vitreum*), striped bass, and paradise fish (*Macropodus opercularis*) results in the formation of one major droplet (Goetz, 1976, 1979; Stevens, 1966; DeNeff, 1980). In contrast, in lower teleosts, such as brook trout and rainbow trout, there is relatively less lipid coalescence resulting in ovulated oocytes that still contain a large number of lipid droplets (Goetz, 1976; Jalabert *et al.*, 1973). Further, it appears that in some intermediate teleosts such as sticklebacks and killifish, the degree of coalescence is also intermediate (Wallace and Selman, 1978, 1979; Yamamoto, 1963). A major exception to this trend appears in the European eel (*Anguilla anguilla*) and Japanese eel (*Anguilla japonica*) in which one to several large lipid droplets are present in oocytes following GVBD (Epler and Bieniarz, 1978; Yamauchi and Yamamoto, 1974).

Finally, it has also been demonstrated that the oocytes of killifish, several species of stickleback, and the medaka hydrate significantly during final maturation, resulting in an enlarged oocyte following GVBD (Wallace and Selman, 1978, 1979; Hirose, 1976).

For many species, the transition from vitellogenesis to final maturation is indistinct. If final maturation is delineated as the period when oocytes are responsive (will undergo GVBD) to various hormonal manipulations *in vitro* (see Sections II,B and D), then the onset of this period varies from species to species and is related to the position of the GV within the oocyte and the size of the oocyte (Goetz and Theofan, 1979; Wallace and Selman, 1980). For example, in killifish, oocytes that have reached a certain size in the ovary are capable of responding to hormone treatment *in vitro*, but smaller ones are not (Wallace and Selman, 1980). Obviously, the ability of the oocyte to respond to various hormonal treatments depends on the competence of the oocyte and follicle. This, in turn, depends on the hormones present at a particular time. In the killifish, competency appears to be related to the presence of gonadotropin and this may be related to food availability (Wallace and Selman, 1980). The competence of the oocyte and follicle to respond to hormone manipulation may be acquired slowly in some species or more rapidly in others, depending on the length of the gonadal cycle.

B. Effects of Steroids

Several early investigations of fish demonstrated that certain steroids, applied *in vivo*, were capable of inducing oocyte final maturation and ovulation (Kirshenblat, 1959; Sundararaj and Goswami, 1966). In an attempt to more precisely study the effects of steroids on final maturation and to completely remove extraovarian factors, *in vitro* incubation systems were employed. Since the initial use of these *in vitro* systems in fish, many other investigations have been completed on the *in vitro* effects of steroids in various species. Table I lists these investigations, indicating the incubation conditions employed and the steroids tested. Where possible, the relative potencies of the steroids are indicated using steroid concentrations, types and numbers of steroids tested, and relative results as overall criteria. The ratings are obviously somewhat subjective and strict comparisons between investigations are limited because in many studies insufficient numbers, types, or concentrations of steroids were tested.

As is obvious from the table, many different incubation conditions have been employed, and various types of media have been used successfully. The effects of various media or the components of the incubation medium on the ability to induce final maturation or ovulation *in vitro* have been investigated in several species (Iwamatsu, 1973; Goswami and Sundararaj, 1971b; Van Ree *et al.*, 1977).

Figure 2 is a compilation of some of the data from Table I, and is intended to give a clearer view of the effective steroid types in several fish species. Only investigations in which a wide range of steroids at a sufficient number of concentrations were used. The relative potencies of major steroid types were determined by the criteria used for Table I. As Fig. 2 indicates, estrogens are generally not effective in inducing final maturation in fish oocytes. The one major exception to this is the significant stimulatory effect of estrone, but not estradiol or estriol, on GVBD in zebrafish oocytes (Van Ree *et al.*, 1977). Androgens are the second least effective steroid type, followed by the 11-oxygenated corticosteroids. However, in several fish species certain 11-oxygenated corticosteroids are quite effective. For example, Goswami and Sundararaj (1971b, 1974) reported that 21-deoxycortisol and cortisol were quite effective in inducing final maturation in catfish oocytes. Cortisone was less effective, and corticosterone was nearly ineffective. Hirose (1972a) reported that cortisol was the most effective steroid in Japanese medaka when applied 10 or 4 hr prior to natural ovulation. When

Table I

Investigations on the *in Vitro* Effects of Steroids on Final Maturation in Fish Oocytes

Species	Incubation conditions			Steroids tested[a]	References
	Medium	Temp (°C)	Length (hr)		
Acipenser stellatus	Ringer (+ 0.1% egg albumin)	?	?	5. Progesterone	Dettlaff and Skoblina (1969)
Anguilla anguilla L.	Cortland (+ 0.3% NaCl)	18–20	24	4. Estriol, estrone, estradiol, testosterone, progesterone, corticosterone, DOC, DOCA, cortisone[b]	Epler and Bieniarz (1978)
Brachydanio rerio	Tyrode Solution (Ca^{2+} and Mg^{2+}-free)	26	0.3–24	1. DOCA, estrone 2. 17α-OH-DHP, 20β-OH-DHP, 17α,20β-DHP, testosterone, cortisol 3. epipregnanolone 4. progesterone, estradiol, 11-ketotestosterone	Van Ree *et al.* (1977)
Carassius auratus	Developed—"CBSS" (Jalabert *et al.*, 1973; Jalabert, 1976)	25	24	1. 17α,20β-DHP 2. 11-DC, DOC, progesterone	Jalabert (1976)
Carassius auratus	"Slightly Modified" from CBSS (Jalabert *et al.*, 1973)	20	24	1. 17α,20β-DHP 2. 17α-OH-DHP 3. Progesterone, 17α,20β-dihydroxy-5β-pregnane-3-one, 5β-pregnane-3β,17α,20β-triol 4. 3α-hydroxy-5β-pregnan-20-one, 17α-hydroxy-5β-pregnane-3,20-dione, 3α,17α-dihydroxy-5β-pregnan-20-one, 5β-pregnane-3α,17α,20β-triol	Nagahama *et al.* (1983)

(continued)

Table I *Continued*

Species	Incubation conditions			Steroids tested[a]	References
	Medium	Temp (°C)	Length (hr)		
Carp	Cortland + 10% calf serum	18–20	24	1. 17α,20β-DHP, DOC 2. Testosterone 4. Estradiol	Epler (1981b)
Cyprinus carpio	Cortland	18–19	48	5. Progesterone	Sakun and Gureeva-Preobrazhenskaya (1975)
Esox lucius	Developed—"TBSS" (Jalabert et al., 1973; Jalabert, 1976)	12	40–60	1. 17α,20β-DHP 2. 20β-OH-DHP, DOC, progesterone, 20α-OH-DHP, 17α-OH-DHP, pregnenolone 3. Cortisone, cortisol, corticosterone	Jalabert (1976); Jalabert and Breton (1974)
Fundulus heteroclitus	Developed—"FO" Wallace and Selman (1978)	14–15	45–50	1. DOC, progesterone 2. Cortisol	Wallace and Selman (1978)
Gasterosteus aculeatus	"FO" Wallace and Selman (1978)	16	60	5. DOC	Wallace and Selman (1979)
Heteropneustes fossilis	Wolf and Quimby (modified, see Goswami and Sundararaj, 1971b)	25	15	1. 11-DC, DOC, DOCA, 21-DC 2. Cortisol, cortisol acetate, progesterone, 17α-OH prog, 11β-OH prog, 11α,17α prog 3. DHEA, androstenedione, cortisone, cortisone acetate, corticosterone, corticosterone acetate, pregnenolone, 17α-OH preg,	Goswami and Sundararaj (1971b, 1974)

124

Species	Medium	Temp.	Days	Steroids	Reference
Hypseleotris galii	Hank's	22	24	11-ketotestosterone, 19-nortestosterone 4. 17α-Estradiol or 17β, pregnanediol-20α or 20β, testosterone, estrone, estriol, 1α-OH corticosterone	Mackey (1975)
Misgurnus anguillicaudatus	Earl's Medium 199	25	10	4. Progesterone, estrone, methyltestosterone[b] 4. Estrone, estradiol 5. Cholesterol, corticosterone, cortisone, DOC, 11-DC, progesterone, testosterone, androstanolone, 16-dehydro-DHP, 5α-pregnandiol, 17α-OH-DHP	Iwamatsu and Katoh (1978)
Misgurnus fossilis	Ringer's	?	26–32	5. Cortisone acetate	Kirshenblat (1959)
Oncorhynchus keta	Balanced salt solution	15–20	20–24	5. Progesterone	Osanai *et al.* (1973)
Oncorhynchus kisutch	Cortland	15.6	?	1. 17α,20β-DHP, 20β-OH-DHP 2. Progesterone, cortisone, cortisol	Sower (1980)
Oncorhynchus rhodurus	Balanced salt solution (Nagahama *et al.*, 1980) or "Trout Ringer" (Kagawa *et al.*, 1982)	15	72–96	1. 17α,20β-DHP, 17α,20β-dihydroxy-5β-pregnan-3-one, 17α-OH prog 2. Progesterone, 17α-hydroxy-5β-pregnane-3,20-dione, 5β-pregnane-3α,17α,20β-triol, 5β-pregnane-3β,17α,20β-triol, testosterone 4. Cortisol, estradiol-17β, 3α-hydroxy-5β-pregnan-20-one,	Nagahama *et al.* (1980, 1983)

(*continued*)

Table I *Continued*

Species	Incubation conditions			Steroids tested[a]	References
	Medium	Temp (°C)	Length (hr)		
Oryzias latipes	Medium 199	24	13–24	3α,17α-dihydroxy-5β-pregnan-20-one 1. Cortisol 2. Progesterone, DOC, 17β-estradiol, testosterone	Hirose (1972a)
Oryzias latipes	Earl's Medium 199	26	22–23	1. Pregnanolone, 17α-OH prog, 17α-OH preg, progesterone, 11β-OH-DHP, DOC, aldosterone, testosterone, 20α- and 20β-OH-DHP 17α,20β-DHP[c] 2. Cortisol, corticosterone 3. Dihydrocholesterol, 11α-OH-DHP, 11-keto DHP 4. Cholesterol, androstenedione, estrone, estriol, estradiol	Iwamatsu (1974, 1978a, 1980)
Perca flavescens	Cortland	12–15	36–48	1. 17α,20β-DHP, 20β-OH-DHP 2. 11-DC, DOC, progesterone, 17α-OH-DHP, 20α-OH-DHP, 17α,20α-DHP 3. 21-DC, cortisol, cortisone, corticosterone, testosterone 4. *17β-Estradiol, 11β-OH testosterone	Goetz and Bergman (1978a); Goetz and Theofan (1979); Theofan (1981); *F. W. Goetz, unpublished data

Species	Medium			Steroids	References
Plecoglossus altivelis	Developed (Nagahama *et al.*, 1983)	18	42	1. 17α,20β-DHP, progesterone, 17α-OH-DHP 2. 5β-pregnane-3β,17α,20β-triol 3. 5β-pregnane-3α,17α,20β-triol 4. 17α-hydroxy-5β-pregnane-3,20-dione, 3α,17α-dihydroxy-5β-pregnan-20-one	Nagahama *et al.* (1983)
Salmo gairdneri	"TBSS" (Jalabert *et al.*, 1973; Jalabert, 1976)	10 or 15	72–120	1. 17α,20β-DHP, 20β-OH-DHP 2. Progesterone, 17α-OH-DHP, 20α-OH-DHP 3. Testosterone, DOCA 4. 17β-Estradiol, estrone, cortisone, cortisol	Fostier *et al.* (1973); Jalabert *et al.* (1972, 1973); Jalabert (1976)
Salmo gairdneri	Balanced salt solution (Nagahama *et al.*, 1980) or "Trout Ringer" (Kagawa *et al.*, 1982)	15	72–96	1. 17α,20β-DHP, 17α,20β-dihydroxy-5β-pregnan-3-one 2. Progesterone, DOC, 5β-pregnane-3α,17α,20β-triol, 17α-hydroxy-5β-pregnan-3,20-dione 4. Cortisol, testosterone, 17β-estradiol, 3α,17α-dihydroxy-5β-pregnan-20-one, 3α-hydroxy-5β-pregnan-20-one	Nagahama *et al.* (1980); Nagahama *et al.* (1983)
Salmo irideus	Cortland	13–14	48	5. Progesterone	Sakun and Gureeva-Preobrazhenskaya (1975)

(continued)

127

Table I *Continued*

| Species | Incubation conditions | | | | Steroids tested[a] | References |
	Medium	Temp (°C)	Length (hr)			
Salvelinus fontinalis	Cortland	15	45–48		1. 17α,20β-DHP, 20β-OH-DHP 2. DOC, 17α-OH-DHP, progesterone, 11-DC 3. Pregnenolone, cortisone, cortisol, 21-DC, corticosterone, DHEA, testosterone, DHEA, and androstenedione 4. 17β-Estradiol	Goetz and Bergman (1978a); Duffey and Goetz (1980)

[a]Relative potencies: 1. very active, 2. moderately active, 3. slightly active, 4. inactive, 5. active, induced maturation but other steroids were not tested or dose response curves were not determined for comparisons. Abbreviations used are: DOC, deoxycorticosterone; DOCA, deoxycorticosterone acetate; 11-DC, 11-deoxycortisol; 21-DC, 21-deoxycortisol; 17α-OH-DHP, 17α-hydroxyprogesterone; 17α-OH preg, 17α-hydroxypregnenolone; 20α-OH-DHP, 20α-dihydroprogesterone; 20β-OH-DHP, 20β-dihydroprogesterone; 17α,20β-DHP, 17α,20β-dihydroxy-4-pregnen-3-one; 17α,20α-DHP, 17α,20α-dihydroxy-4-pregnen-3-one; 11β-OH-DHP, 11β-hydroxyprogesterone; 11α,17α-DHP, 11α,17α-dihydroxyprogesterone; DHP, progesterone; 1α-OH corticosterone, 1α-hydroxycorticosterone; DHEA, dehydroepiandrosterone.

[b]Only assayed for ovulation.

[c]Many more steroids were tested; however, only selected steroids are listed.

	PROGESTOGENS	11-DEOXYCORTICOIDS	11-OXYCORTICOIDS	ANDROGENS	ESTROGENS
RAINBOW TROUT	++++	+	O	O	O
BROOK TROUT	++++	+++	+	+	O
PIKE	++++	+++	+	−	−
GOLDFISH	++++	+++	+	−	O
YELLOW PERCH	++++	+++	+	+	O
MEDAKA (1)	++++	+++	++	++	O
MEDAKA (2)	++	++	++++	+	+
ZEBRAFISH	++	++++	++	++	++
CATFISH	++	++++	+++	+	O

Fig. 2. Composite of steroid effects on final maturation in oocytes of several species listed in Table I. A "++++" indicates the most effective steroid type; other designations (+++, ++, +) represent effectiveness in relation to the most potent type. No response is indicated by "O" and "—" indicates that the steroid type was not tested. (Data are from the following: rainbow trout, Jalabert et al., 1972, 1973, Jalabert, 1976, Fostier, et al., 1973; brook trout, Goetz and Bergman, 1978a; Duffey and Goetz, 1980; pike, Jalabert, 1976, Jalabert and Breton, 1974; goldfish, Jalabert et al., 1973, Jalabert, 1976; yellow perch, Goetz and Bergman, 1978a, Goetz and Theofan, 1979, Theofan, 1981; medaka (1), Iwamatsu, 1974, 1978a, 1980; medaka (2) Hirose, 1972a; zebrafish, Van ree et al., 1977; catfish, Goswami and Sundararaj, 1971b, 1974.)

applied 17 hr prior to ovulation, progesterone was as potent as cortisol although neither steroid was very effective at this time. In addition, exposure to cortisol for 2 hr produced the same results as with continuous exposures of 17 hr (Hirose, 1972b). In other fish species, such as zebrafish, brook trout, yellow perch, pike (Esox lucius), and goldfish, 11-oxycorticosteroids are either moderately effective or stimulatory only at much higher concentrations (Van Ree et al., 1977; Goetz and Bergman, 1978a; Jalabert et al., 1973; Jalabert, 1976).

As is evident from Fig. 2, for the majority of fish species studied the most effective steroids are either 11-deoxycorticosteroids or progestogens. Progesterone has been demonstrated to be effective in a number of fish species (see Table I), although its stimulatory effect might be a result, to some extent, of conversion to more active metabolites rather than direct stimulation. In brook trout, rainbow trout, yellow perch, goldfish, pike, ayu (Plecoglossus altivelis), and amago salmon (Oncorhynchus rhodurus) several progestogens can induce GVBD, but the most effective is 17α,20β-dihydroxy-4-pregnen-3-one (17α,20β-DHP) (Duffey and Goetz, 1980; Goetz and Theofan, 1979; Jalabert, 1976; Fostier et al., 1973; Nagahama et al., 1983). In brook trout, rainbow trout, yellow perch, and pike, 20β-dihydroprogesterone is the second most potent steroid, and 17α-hydroxyprogesterone is usually only as effective as progesterone. In rainbow trout and yellow perch, exposure to $17\alpha,20\beta$-DHP for as short a period as 1 min can result in 100% GVBD (Jalabert, 1976; F. W. Goetz, unpublished results).

Iwamatsu (1974, 1978a, 1980) also reported the *in vitro* final maturational effects of steroids in oocytes from Japanese medaka. Many steroids were tested with either continuous or short exposures. Effective C-19 steroids had in common a ketol or α-hydroxyl group at the 3 position and a 17β-hydroxyl group; effective C-21 steroids had in common a ketol or β-hydroxyl group at the 3 position, and a ketol or α-hydroxyl group at the 20 position, with Δ^4 or Δ^5-unsaturation or 5α-saturation (Iwamatsu, 1978a). From Table I it can be seen that Iwamatsu (1978a, 1980) found several progestational steroids, including progesterone, 17α-hydroxyprogesterone, 20β-dihydroprogesterone and $17\alpha,20\beta$-DHP to be very effective. In contrast, cortisol, cortisone, and corticosterone were less potent. These results are in apparent contradiction with those of Hirose (1972a) who found cortisol to be more potent than progesterone. Unfortunately, the results of the two investigators cannot be strictly compared because Iwamatsu (1974) recorded GVBD and ovulation separately, but Hirose (1972a) apparently combined data on GVBD and ovulation. In fact, Iwamatsu (1974) did not observe any ovulation (although GVBD could still be induced) in steroid-treated incubates as long as oocytes were removed well in advance (14–15 hr) of the natural time of GVBD. If oocytes were removed later and then treated with progesterone, significant ovulation was observed. However, even some controls removed 13–14 hr prior to natural GVBD underwent GVBD, and if removed 10 hr before natural GVBD some controls also ovulated. Therefore, in Iwamatsu's experiments it appeared that the inertia for undergoing final maturation was attained approximately 14 hr before natural GVBD and for ovulation approximately 10 hr. Iwamatsu (1978a) suggested that oocytes in Hirose's investigation had been affected by the natural release of gonadotropin prior to removal and, therefore, would ovulate following *in vitro* incubation with steroids. Although this may explain the discrepancy in the results on ovulation, this still does not explain the differences in the relative results from cortisol and progesterone between investigations.

In several species, 11-deoxygenated corticosteroids are quite effective in inducing final maturation. For example, deoxycorticosterone acetate was the most effective steroid in the zebrafish (Van Ree *et al.*, 1977), and in the catfish, 11-deoxycortisol and deoxycorticosterone were among the most potent in inducing final maturation (Goswami and Sundararaj, 1971b, 1974). In brook trout, yellow perch, goldfish, and pike, 11-deoxycorticosteroids are also quite effective, but not as potent as $17\alpha,20\beta$-DHP (Goetz and Theofan, 1979; Duffey and Goetz, 1980; Jalabert, 1976; Jalabert *et al.*, 1973).

Although it is difficult to generalize, it appears that in several species there is a certain degree of specificity for 11-deoxygenated, C-21 steroids (Goetz and Bergman, 1978a). This specificity is apparently greatest in brook trout, yellow perch, goldfish, and pike. In catfish, specificity around the 11 position may not be as significant because both 11-deoxygenated and 11-oxygenated corticosteroids induce maturation. Apparently, 17α,20β-DHP is the most potent stimulator of final maturation in every species in which it has been tested, with the exception of the zebrafish. In the zebrafish, 17α,20β-DHP is not as potent as deoxycorticosterone acetate (Van Ree *et al.*, 1977). However, it should be noted that in investigations of zebrafish the effects of steroids on final maturation would be of an enhancing nature because control oocytes spontaneously mature when incubated *in vitro*. Whether the effects of various steroids in this capacity would be the same as in the initiation of final maturation is unknown.

Several investigators have demonstrated that combinations of certain steroids produce a greater maturational response than of the individual steroids alone. Sundararaj and Goswami (1971b) were the first to note that cortisol "synergized" with deoxycorticosterone in inducing GVBD in catfish oocytes. Synergism was obtained only at a specific ratio of cortisol to deoxycorticosterone. Later, Jalabert (1975) reported that cortisol and cortisone increased the efficiency of 17α,20β-DHP-induced final maturation of rainbow trout oocytes. At the levels tested, neither cortisol or cortisone alone induced GVBD in rainbow trout (Jalabert *et al.*, 1972). Theofan (1981) also observed that ineffective 11-deoxycortisol concentrations enhanced the response observed with 17α,20β-DHP in yellow perch oocytes, and deoxycorticosterone enhanced the maturational response observed with 17α,20β-DHP in amago salmon oocytes (Young *et al.*, 1982a).

Synergism or facilitation by 11-oxygenated or 11-deoxygenated corticosteroids may have an important role in the overall stimulation of natural final maturation because it has been noted that in some species plasma levels of corticosteroids increase at the time of reproduction (Schmidt and Idler, 1962; Cook *et al.*, 1980; Wingfield and Grimm, 1977). Presumably, these steroids are of interrenal origin. Further, investigators have demonstrated that ovaries of several teleosts can produce 11-deoxygenated corticosteroids (Colombo *et al.*, 1973; Colombo and Colombo, 1977; Theofan and Goetz, 1983; Tesone and Charreau, 1980); therefore, facilitation of a maturational steroid(s) could originate from an ovarian or extraovarian source.

C. Steroid Mechanism of Action

Similar to fish, amphibian oocytes also undergo GVBD when stimulated *in vitro* with certain steroids (for review, see Masui and Clarke, 1979). Further research on the mechanism of steroid action in amphibians has demonstrated that the maturational steroid most likely acts at the level of the oocyte membrane, possibly inhibiting adenylate cyclase and thereby decreasing cyclic AMP levels (Finidori-Lepicard *et al.*, 1981; Mulner *et al.*, 1979). In amphibians, a decrease in cyclic AMP appears to be necessary for the stimulation of GVBD by steroids (Morrill *et al.*, 1977). Although similar studies have not been conducted in fish oocytes, several studies have indicated that denuded (follicle-free) oocytes are still responsive to steroid stimulation, indicating that the follicular layers are not necessary for steroids to induce final maturation (Dettlaff and Skoblina, 1969; Jalabert *et al.*, 1973; Iwamatsu, 1980). In addition, it appears that only those steroids that are extremely effective in inducing final maturation in intact oocytes can effectively induce GVBD in denuded oocytes. Therefore, in rainbow trout, only 20β-dihydroprogesterone and 17α,20β-DHP induce GVBD in denuded oocytes, but all progestogens can induce GVBD in intact oocytes (Jalabert *et al.*, 1973; Jalabert, 1976). In medaka oocytes, more steroids induced GVBD in denuded oocytes; however, only 20β-dihydroprogesterone, 17α,20β-DHP, and pregnenolone were effective at the lowest concentrations tested (Iwamatsu, 1980). In that experiment, pregnenolone, 20β-dihydroprogesterone, and 17α,20β-DHP were also the only effective steroids at similar concentrations on intact oocytes. In contrast, Hirose (1972c) reported that cortisol-induced final maturation of medaka oocytes was inhibited by the removal of the follicle. However, the inhibition may have been a result of the method used to remove the follicles (Iwamatsu, 1980).

From studies using translational and transcriptional inhibitors, it has been clearly demonstrated that protein synthesis is necessary for steroid-induced full maturation. Dettlaff and Skoblina (1969) were the first to report that puromycin, a translational inhibitor, blocked progesterone-induced maturation in sturgeon oocytes, but actinomycin D, a transcriptional inhibitor, did not. Later, Goswami and Sundararaj (1973) found that neither actinomycin D nor mitomycin C (transcriptional inhibitor) could inhibit deoxycorticosterone-induced final maturation in catfish oocytes, but puromycin and cycloheximide (translational inhibitor) did. Subsequent experiments on rainbow trout and yellow perch have yielded similar results (Jalabert, 1976; Theofan and Goetz, 1981). In addition, in rainbow trout, the phase during which cycloheximide can inhibit final maturation extends for approximately 14 hr. From the combined results, it appears that transcription of RNA is not a requisite for steroid-induced final maturation, although translation is.

Various studies using amphibian oocytes have demonstrated that an intermediate in the stimulation of GVBD by steroids is a "maturational promoting factor" (MPF) (Masui and Clarke, 1979). The MPF is probably a phosphoprotein, present in the cytoplasm of the oocyte. Its initial production or activation requires protein synthesis; however, once MPF is present its production appears to be autocatalytically amplified. The stimulation of GVBD by MPF in amphibian oocytes cannot be inhibited by cycloheximide. The presence of MPF in sturgeon oocytes has been demonstrated (Masui and Clarke, 1979); however, similar work has yet to be completed on other species.

D. Effects of Gonadotropins or Pituitary Preparations

The efficacy of various pituitary and gonadotropin preparations in the *in vivo* induction of final maturation and ovulation in fish is well known (Chaudhuri, 1976). As a result, it is logical to assume that these preparations might have an effect when incubated *in vitro* with oocytes. As with steroids, many studies have been completed on the *in vitro* effects of pituitary and gonadotropin preparations. These investigations are listed in Table II, and the types

Table II

Investigations on the *in Vitro* Effects of Pituitary and Gonadotropin Preparations on Final Maturation in Fish Oocytes

Species	Preparations tested[a]	References
Acipenser stellatus	5. Sturgeon pituitaries	Dettlaff and Skoblina (1969) Dettlaff and Davydova (1979)
Anguilla anguilla L.	2. Carp pituitaries + HCG[b] 4. Carp pituitaries	Epler and Bieniarz (1978)
Brachydanio rerio	2. Zebrafish pituitaries 4. o-LH, o-FSH, o-LtH, ACTH$_{1-24}$, SG-G100, c-GTH	van Ree *et al.* (1977)
Carassius auratus	1. Carp pituitary 2. c-GTH 3. SG-G100, HCG 4. o-LH, o-FSH	Jalabert (1976); Jalabert *et al.* (1973)
Carp	2. Carp pituitary homogenate 3. Pituitaries of species from Cyprinidae, Esocidae, Salmonidae, Percidae 4. o-LH, o-FSH, HCG	Epler *et al.* (1979); Epler (1981a)
Cyprinus carpio	4. Carp pituitaries	Sakun and Gureeva-Preobrazhenskaya (1975)
Esox lucius	1. SG-G100 4. o-LH, o-FSH, HCG 5. Trout, carp, pike gonadotropic extracts	Jalabert (1976); Jalabert and Breton (1974)
Fundulus heteroclitus	1. o-LH, HCG, killifish pituitaries 2. o-FSH	Wallace and Selman (1980)

(*continued*)

Table II *Continued*

Species	Preparations tested[a]	References
Heteropneustes fossilis	3. o-LH, SG-G100 4. o-FSH, b-TSH, o-GH, o-PRL, ACTH, catfish pituitaries	Goswami and Sundararaj (1971b); Goswami *et al.* (1974); Sundararaj Goswami (1974b); Sundararaj *et al.* (1972)
Hypseleotris galii	1. Goldfish pituitaries[b] 4. HCG, PMS	Mackay (1975)
Misgurnus anguillicaudatus	2. HCG 3. PMS	Iwamatsu and Katoh (1978)
Misgurnus fossilis	4. Loach pituitaries, o-LH	Kirshenblat (1959)
Oncorhynchus keta	5. Salmon pituitaries	Osanai *et al.* (1973)
Oncorhynchus kisutch	5. SG-G100	Sower (1980)
Oncorhynchus rhodurus	4. PMS 5. SG-G100	Nagahama *et al.* (1980)
Oryzias latipes	1. SG-G100[c] 2. b-LH, o-FSH, HCG, PMS	Hirose (1971); Hirose and Donaldson (1972)
Oryzias latipes	1. PMS, LH 2. FSH, TSH 4. ACTH, HCG	Iwamatsu (1978b)
Perca flavescens	3. o-LH, PMS, o-FSH, SG-G100, carp pituitaries 4. HCG	Goetz and Bergman (1978b)
Salmo gairdneri	1. SG-G100, c-GTH, t-GTH 2. o-LH 4. o-FSH, HCG	Jalabert *et al.* (1972, 1974)
Salmo gairdneri	4. HCG, o-LH 5. SG-G100	Nagahama *et al.* (1980)
Salmo irideus	5. Salmon pituitaries	Sakun and Gureeva-Preobrazhenskaya (1975)
Salvelinus fontinalis	1. Carp pituitaries 2. o-LH 3. HCG 4. PMS, o-FSH	Goetz (1976)
Stizostedion vitreum	1. SG-G100, carp pituitaries, o-LH 2. HCG, o-FSH, PMS	Goetz and Bergman (1978b)

[a]Relative potencies are indicated as follows: 1. very active, 2. moderately active, 3. slightly active, 4 inactive, 5. active. Maturation was induced, but other preparations were not tested or dose response curves were not determined for relative comparisons. Abbreviations are as follows: LH, luteinizing hormone; FSH, follicle stimulating hormone; TSH, thyroid stimulating hormone; ACTH, adrenocorticotropic hormone; PRL, prolactin; LtH, lactogenic hormone; HCG, human chorionic gonadotropin; PMS, pregnant mare's serum gonadotropin; o, ovine; b, bovine; c-GTH, carp gonadotropin; t-GTH, trout gonadotropin; SG-G100, salmon gonadotropin.

[b]Assayed only for ovulation.

[c]Assay criteria for ovulation and GVBD were combined.

of preparations tested and the relative potencies are indicated. As a result of the nonuniformity in purity, types, and concentrations of preparations, it is very difficult to make comparisons between fish concerning biological activity. However, from these studies it is apparent that gonadotropins or pituitary preparations are effective in inducing *in vitro* final maturation in oocytes of many species, and that, generally, piscine preparations are the most effective. It is difficult to generalize about the effectiveness of mammalian gonadotropic preparations; however, in those fish in which mammalian preparations induce GVBD, luteinizing hormone (LH) is usually more effective than follicle stimulating hormone (FSH). This is apparently true for medaka, rainbow trout, brook trout, walleye, and killifish (Wallace and Selman, 1980; Iwamatsu, 1978b; Jalabert *et al.*, 1972,1974; Hirose, 1971; Goetz, 1976; Goetz and Bergman, 1978b). In fact, the effectiveness of FSH may be a result of the contamination of the preparations by LH. Although human chorionic gonadotropin (HCG) has been used extensively for the *in vivo* induction of final maturation and ovulation, it is surprisingly less active *in vitro* in many fish.

Pituitary or gonadotropin preparations are very effective in inducing *in vitro* final maturation in a number of species; however, it is obvious from the data in Table II that in some species these preparations are ineffective. For instance, in catfish neither LH or salmon gonadotropin could induce greater than 15% maturation over a wide range of concentrations (Goswami and Sundararaj 1971b; Goswami *et al.*, 1974; Sundararaj and Goswami, 1974b; Sundararaj *et al.*, 1972). Homologous pituitary preparations were also ineffective, as were other mammalian pituitary preparations such as FSH, adrenocorticotropic hormone, growth hormone, or prolactin (Goswami and Sundararaj, 1971b). Similarly, in yellow perch, mammalian gonadotropins and piscine pituitary preparations, including perch pituitaries, cannot induce significant maturation *in vitro* (Goetz and Bergman, 1978b; F. W. Goetz, unpublished results). Only a homologous pituitary preparation can enhance final maturation in zebrafish oocytes (Van Ree *et al.*, 1977).

The ineffectiveness of gonadotropic preparations *in vitro* in oocytes of some fish species could result from a variety of factors. *In vivo*, gonadotropins reach target cells by means of capillary beds in the follicle. However, in *in vitro* incubations gonadotropins must traverse the surface epithelium of the follicle, and this may be relatively impermeable to gonadotropins in certain species or under certain incubation conditions. Incubation conditions might not be suitable or optimal for gonadotropin stimulation. For instance, Jalabert (1976) has shown that the pH of the incubation medium has a substantial influence on the *in vitro* effect of pituitary or gonadotropin preparations in rainbow trout and goldfish oocytes. In both species the median effective dose (50% effective dose) of a gonadotropin or pituitary preparation was reduced by increasing the pH of the incubation medium from 7.1 to 8.1 in trout and 7.5 to 8.1 in goldfish (Jalabert, 1976). In addition, the media used for most *in vitro* incubations are relatively simple and may lack certain essential nutrients or factors. The length of the incubation time is also important, if, as has been hypothesized (see Section II,E), gonadotropins must stimulate the production of maturational steroids in the ovary. In *Fundulus heteroclitus*, the incubation time with gonadotropins had to be doubled, from the time needed with steroids, to observe an optimal effect (Wallace and Selman, 1980). The relative effectiveness of gonadotropin preparations *in vitro* may be influenced by the lack of other hormones acting directly or indirectly on the ovary. For instance, in pike, rainbow trout, goldfish, and killifish, cortisol at levels that alone are nonstimulatory or partially stimulatory, can greatly enhance the *in vitro* effects of pituitary or gonadotropin preparations when applied together (Jalabert, 1975, 1976; Wallace and Selman, 1980). Deoxycorticosterone and testosterone also increase the effectiveness of salmon gonadotropin in stimulating GVBD in amago salmon (Young *et al.*, 1982a). In intact sturgeons subjected to a sharp decrease in holding temperatures or prolonged captivity ("reserved"), pituitary preparations are ineffective in inducing final maturation *in vivo* or in *in vitro* (Dettlaff and Davydova, 1979). However, they are quite effective at higher temperatures. If cooled or reserved females are given an injection of triiodothyronine (T_3), pituitary preparations will then induce complete final maturation either *in vivo* or in oocytes removed and incubated *in vitro* (Dettlaff and Davydova, 1979). The effect of T_3 is apparently indirect or requires other hormones because cooled oocytes incubated with T_3 and a pituitary preparation do not undergo final maturation. In addition, the T_3 effect is apparently involved with gonadotropin stimulation because progesterone still induces GVBD *in vitro* in oocytes from cooled or reserved females.

Finally, it is possible that in some species gonadotropins do not directly stimulate the ovaries in the pathway leading to the induction of final maturation (see Section II,E).

E. Gonadotropin Mechanism of Action

From various experimental evidence, it has been hypothesized that, in some fish species, pituitary gonadotropins induce final maturation by stimulating the synthesis of a maturational steroid(s) in the follicle (Fig. 3) (Jalabert, 1976; Hirose, 1976; Iwamatsu, 1978b; Sundararaj and Goswami, 1977). In turn, the steroid(s) induces final maturation as previously described (see Section II,C). Theoretically, this hypothesis explains the mecha-

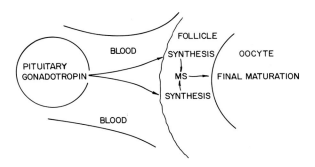

Fig. 3. Pituitary–ovarian relay for the stimulation of final maturation (MS, maturational steroid).

nism of action of pituitary preparations or gonadotropins in any fish species in which these preparations are able to induce significant maturation *in vitro*. Therefore, this would include *Acipenser stellatus, Anguilla anguilla,** *Brachydanio rerio, Carassius auratus, Esox lucius, Fundulus heteroclitus, Hypseleotris galii,** *Misgurnus anguillicaudatus, Oncorhynchus kisutch, Oncorhynchus rhodurus, Oncorhynchus keta, Oryzias latipes, Salmo gairdneri, Salvelinus fontinalis,* and *Stizostedion vitreum* (see Table II).

Results of various investigations have substantiated the hypothesis that gonadotropins act first at the follicle and that steroid synthesis is an intermediate for the induction of final maturation. First, although several researchers have reported that steroids are still capable of stimulating final maturation in denuded oocytes, it has been demonstrated in rainbow trout, medaka, and sturgeon that denuded oocytes do not undergo GVBD in the presence of pituitary or gonadotropin preparations (Jalabert, 1976; Iwamatsu, 1980; Hirose, 1972c; Dettlaff and Skoblina, 1969). Further, Iwamatsu (1980) observed that in oocytes in which the thecal layers, but not the granulosa layers were removed, *in vitro* maturation could still be induced by gonadotropins. Even in totally defolliculated oocytes (thecal and granulosa layers gone), maturation was induced by gonadotropin when cocultured with granulosa cells from pre- or postovulatory follicles. These results indicate that some process, presumably steroid synthesis occurring in the granulosa layer, is responsible for maturation. The observation that 3β-hydroxysteroid dehydrogenase activity is only found in the granulosa layer in medaka is consistent with the hypothesis that gonadotropins stimulate the synthesis and/or release of a maturational steroid(s) from this cell layer (Iwamatsu, 1980).

*It is assumed that ovulated oocytes had also undergone final maturation.

In several species, it has been demonstrated that *in vitro* gonadotropin or pituitary-stimulated final maturation requires a longer time than steroid-induced maturation (Jalabert, 1976; Wallace and Selman, 1980). This suggests that an additional step must intervene in gonadotropin-induced final maturation in comparison to steroid induction. Further, although *in vitro* steroid-induced final maturation is unaffected by transcriptional inhibitors, gonadotropin or pituitary-induced maturation is blocked in both rainbow trout and sturgeon by actinomycin D and mitomycin C (Jalabert, 1976; Dettlaff and Skoblina, 1969). This indicates an additional step, presumably involving steroidogenesis, in the gonadotropin-induction of final maturation.

That the follicle is producing maturational steroid under gonadotropin stimulation has been well substantiated. In amago salmon (*Oncorhynchus rhodurus*), denuded oocytes completed final maturation when incubation in media that had previously been exposed to preovulatory follicles that had been stimulated with salmon gonadotropin (Kanatani and Nagahama, 1980). In medaka, gonadotropin-induced final maturation was blocked by metopirone, an inhibitor of steroid synthesis (Hirose, 1973). This was interpreted as indicating that 11β-hydroxylase activity was necessary for gonadotropin-induced maturation because one effect of metopirone is to inhibit this enzyme. However, the most convincing evidence for the stimulation of maturational steroid synthesis by gonadotropins has only recently been published. In amago salmon, cyanoketone (inhibitor of 3β-hydroxysteroid dehydrogenase) inhibited salmon gonadotropin (SG-G100) and pregnenolone-stimulated final maturation *in vitro* but did not block maturation induced by Δ^4-steroids such as progesterone or $17\alpha,20\beta$-DHP (Young *et al.*, 1982a).

Although a direct pituitary–ovarian relay may exist in some fish, in at least one species, the Indian catfish, a pituitary–interrenal–ovarian relay has been proposed (Fig. 4) (Sundararaj and Goswami, 1977). In this system, pituitary gonadotropins stimulate the interrenals to produce a maturational steroid(s) that would, in turn, induce final maturation. The basic evidence in support of this hypothesis has been thoroughly reviewed by Sundararaj and Goswami (1977). Briefly, gonadotropins and homologous pituitary preparations cannot induce significant maturation in *in vitro* incubations with catfish oocytes (Goswami and Sundararaj, 1971b; Sundararaj *et al.*, 1972). Gonadotropins induce final maturation and ovulation when injected into hypophysectomized catfish or if incubated *in vitro* with cocultures of oocytes and interrenal tissue (Sundararaj and Goswami, 1966, 1969, 1974b; Goswami *et al.*, 1974). Finally, LH stimulates cortisol and deoxycorticosterone synthesis in catfish interrenals (Sundararaj and Goswami, 1969), and it stimulates an increase in plasma cortisol following injection into gravid or regressed and intact or ovariectomized catfish (Truscott *et al.*, 1978). It does not appear that catfish ovaries produce 21-hydroxylated steroids (Ungar *et al.*, 1977),

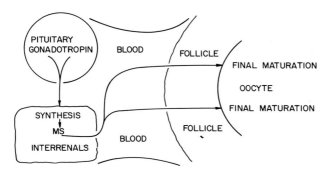

Fig. 4. Pituitary–interrenal–ovarian relay for the stimulation of final maturation (MS, maturational steroid).

and, therefore, would not be the source of maturational steroid if corticosteroids are the natural mediators. However, investigations (Ungar *et al.*, 1977) on the steroidogenic capacity of the catfish ovary using radiolabeled incorporation, employed very short incorporation times and the possible production of other steroids that might have had maturation-inducing abilities was not explored.

Although oocytes of several other species listed in Table II appear relatively unresponsive to gonadotropin or pituitary stimulation *in vitro*, it should not be automatically assumed that a pituitary–interrenal–ovarian relay for the stimulation of final maturation exists in them as well. For example, although gonadotropins and pituitary preparations are relatively ineffective in *in vitro* incubations with yellow perch oocytes, the same preparations are very effective in walleye (Goetz and Bergman, 1978b), a species in the same family (Percidae) as yellow perch. It is unlikely that in two species so closely related, that two different pathways for the stimulation of final maturation would exist. Further, yellow perch ovaries are capable of synthesizing a number of steroids that are extremely effective in inducing final maturation *in vitro* (Theofan and Goetz, 1983).

As a result of investigations of the facilitation of steroid and pituitary-induced final maturation by corticosteroids, it has been hypothesized that the interrenals may indirectly influence oocyte final maturation (Fig. 5) (Jalabert, 1976). Stress, arising from various factors during the time of reproduction, could result in an increased output of 11-oxygenated corticosteroids by the interrenals and, therefore, increased plasma titers as observed in some species (Schmidt and Idler, 1962; Wingfield and Grimm, 1977). In turn, these corticosteroids could enhance the effectiveness of a maturational steroid such as $17\alpha,20\beta$-DHP, or gonadotropin stimulation (see Sections II,B and D). Alternatively, because 11-deoxygenated corticosteroids

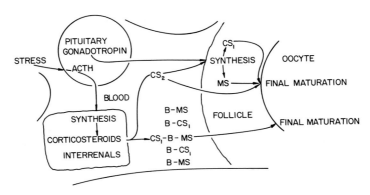

Fig. 5. Pituitary–ovarian relay and corticosteroidal facilitation (MS, maturational steroid; CS_1, 11-deoxygenated corticosteroids; CS_2, 11-oxygenated corticosteroids; B, plasma binding protein; ACTH, adrenocorticoptropic hormone).

such as 11-deoxycortisol and deoxycorticosterone are apparently bound to the same plasma proteins as are progestational steroids (Fostier and Breton, 1975), these steroids if produced in higher quantities by the interrenals could displace bound maturational steroid so that it would be free to act on the oocyte. Results from studies of rainbow trout indicate that only unbound maturational steroid is effective in inducing GVBD (Fostier and Breton, 1975). In part or together, these mechanisms could represent the indirect influence of the interrenals on final maturation in those species employing a direct pituitary–ovarian relay. In addition, it should be noted that aside from the interrenals, ovarian production of 11-deoxygenated corticosteroids (Colombo *et al.*, 1973, 1978; Colombo and Colombo, 1977; Theofan and Goetz, 1983; Tesone and Charreau, 1980) may also play a role in enhancing final maturation.

F. Production of Maturational Steroids

Results of the *in vitro* studies mentioned earlier (see Section II,B), indicate for various species which steroid(s) would be the most potent stimulators of final maturation and, therefore, candidates for the natural mediators of oocyte maturation. However, fewer studies have been conducted to determine if these effective maturational steroids are present in a given species, and if they are produced at the time of final maturation.

Progestational steroids, effective in the induction of final maturation in trout and salmon, have been identified in the plasma of several salmonid species at the time of natural reproduction. For example, $17\alpha,20\beta$-DHP,

17α-hydroxyprogesterone, and progesterone were identified in the plasma of sockeye salmon (*Onchorhynchus nerka*) females just prior to and following spawning (Schmidt and Idler, 1962; Idler *et al.*, 1959). Various corticosteroids, including cortisol and cortisone were also identified. Campbell *et al.* (1980) also observed high levels of 17α-hydroxyprogesterone and 17α,20β-DHP in sera from rainbow trout undergoing oocyte final maturation. Lower concentrations of 11-deoxycorticosteroids were also identified. Plasma of sexually mature winter flounder (*Pseudopleuronectes americanus*) also contained 17α-hydroxyprogesterone and 17α,20β-DHP (Campbell *et al.*, 1976). Progestational steroids have been suggested as being important in the induction of oocyte final maturation in this species (Campbell *et al.*, 1976). In the Indian catfish, a species in which 11-oxygenated and 11-deoxygenated corticosteroids are potent stimulators of *in vitro* final maturation, plasma concentrations of cortisol, but not 11-deoxygenated corticosteroids, increased following LH stimulation of gravid females (Truscott *et al.*, 1978). The presence of 17α,20β-DHP was detected in the plasma, but concentrations were always low and did not vary with LH treatment.

Studies of the steroidogenic capacity of ovaries from many fish species have been conducted (see Chapter 7, Volume 9A, this series); however, relatively few have been conducted on fish for which relevant biological data have been collected on the *in vitro* effects of steroids. Therefore, it is difficult to assess the biological importance of many of these studies in relation to the stimulation of final maturation. Fortunately, for some species included in Table I, information is available concerning the steroidogenic capacity of the ovary. Ovaries of gonadotropin-stimulated catfish, incubated with [^{14}C]pregnenolone, produced 3α-hydroxy-5β-pregnan-20-one and 5β-pregnan-3α,20α-diol; however, interrenal tissue produced primarily corticosterone and cortisol (Ungar *et al.*, 1977). The finding that catfish ovaries were unable to produce corticosteroids that are the most effective in the *in vitro* induction of final maturation, lends support to the hypothesis that in this species the maturational steroid(s) comes from the interrenals (see Section II,E). However, because basal corticosteroid production by interrenals was not determined, it is still uncertain whether gonadotropins can increase the production of corticosteroids from the interrenals. Similarly, in incubations with zebrafish ovaries and [^{14}C]progesterone, no corticosteroids were observed (Lambert and van Oordt, 1975), although deoxycorticosterone acetate is the most effective steroid in this species *in vitro*. Therefore, if deoxycorticosterone is the naturally occurring final maturational steroid in zebrafish, it may come from an extraovarian source. However, the observation that a zebrafish pituitary homogenate enhances final maturation in zebrafish oocytes *in vitro* (Van Ree *et al.*, 1977) indicates that an extraovarian source of maturational steroid may not be necessary.

Investigations of the steroidogenic capacity of brook trout and yellow perch ovaries indicate that both species can produce a number of steroids active in the *in vitro* induction of final maturation (Theofan and Goetz, 1983). Brook trout ovarian tissue incubated with [^{14}C]progesterone produced 5β-pregnanedione, 17α-hydroxyprogesterone, and deoxycorticosterone, which all stimulate final maturation *in vitro* in this species (Goetz and Bergman, 1978a; Duffey and Goetz, 1980; Theofan and Goetz, 1983). In addition, a metabolite was isolated that was very active when bioassayed with brook trout oocytes. This metabolite comigrated with the standards, 17α,20β-DHP and 11-deoxycortisol in bidimensional thin-layer chromatography, but was not characterized further. Yellow perch ovarian tissue converted [^{14}C]progesterone to 5α-pregnanedione, 17α-hydroxyprogesterone, 20α-dihydroprogesterone, 17α-hydroxy-20α-dihydroprogesterone (identified on basis of chromatographic mobility), and a metabolite that comigrated with 11-deoxycortisol and 17α,20β-DHP (Theofan and Goetz, 1983). This latter metabolite was the most active in inducing final maturation when bioassayed in perch oocytes *in vitro*.

From cell-free homogenates of ayu ovaries, Suzuki *et al.* (1981a) demonstrated (by recrystallization and derivatization) the production of 17α,20β-DHP and several other metabolites including 17α-hydroxyprogesterone, 3α, 17α-dihydroxy-5β-pregnan-20-one, 17α,20β-dihydroxy-5β-pregnan-3-one, 5β-pregnane-3β,17α-20β-triol, and 5β-pregnane-3β,17α,20β-triol. The production of the various metabolites depended on the substrate used and whether or not the ovaries had come from gonadotropin-stimulated fish. The 17α,20β-DHP was only observed in ovarian homogenates from fish stimulated by salmon gonadotropin; however, it was probably produced by ovaries of untreated fish and metabolized (Suzuki *et al.*, 1981a). The overall results of the investigation indicated that gonadotropin stimulation increased both the 3β- and 20β-hydroxysteroid dehydrogenases (Suzuki *et al.*, 1981a). Although various 17,20-hydroxylated steroid metabolites were produced by gonadotropin-stimulated ayu ovaries, it has subsequently been noted (Nagahama *et al.*, 1983) that 17α,20β-DHP is still the most potent steroid in inducing final maturation of ayu oocytes (Table I). The ovarian production of 17α,20β-DHP from labeled 17α-hydroxyprogesterone has also been confirmed in amago salmon (Suzuki *et al.*, 1981b). In *in vitro* incubations with amago salmon oocytes, the 17α,20β-DHP levels (measured by radioimmunoassay) in the incubation medium were significantly elevated in the presence of salmon gonadotropin (Young *et al.*, 1983a).

Finally, in rainbow trout, in which 17α,20β-DHP is the most effective steroid in *in vitro* incubations, no mention was made of the isolation of 17α,20β-DHP from incubations with [^{14}C]pregnenolone, although 17α-hydroxyprogesterone, progesterone, and several androgens were observed

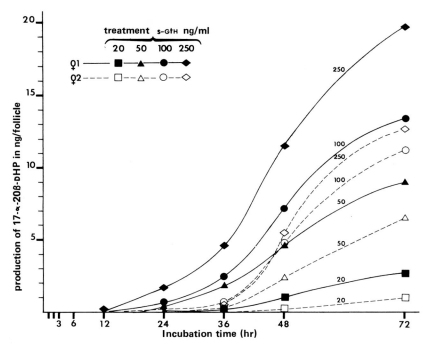

Fig. 6. Kinetics of 17α,20β-DHP release in the medium in incubations of trout oocytes, in the presence of increasing gonadotropin concentrations after 72 hr of incubation (levels in control incubates, without gonadotropin, for female 1 and 2 were 0.09 and 0.13 ng, respectively) (from Fostier *et al.*, 1981a).

(Lambert and van Bohemen, 1978). In contrast, Fostier *et al.* (1981a) reported high levels of 17α,20β-DHP (measured by radioimmunoassay) in media from incubations of rainbow trout ovaries stimulated with salmon gonadotropin (Fig. 6). The levels of 17α,20β-DHP increased significantly in a dose-dependent manner from 24 to 72 hr of incubation in gonadotropin incubates, although only very low levels were measured in unstimulated incubates. The failure to observe 17α,20β-DHP in previous incorporation experiments may be a result of the reproductive stage of the ovary (Fostier *et al.*, 1981a), length of incorporation, or absence of gonadotropin stimulation (Fostier *et al.*, 1981a).

Using radioimmunoassays, the circulating levels of 17α,20β-DHP have been followed in several salmonids during the reproductive season (Table III). A large and rapid increase in plasma 17α,20β-DHP was observed in individual rainbow trout around the time of ovulation (Fostier *et al.*, 1981b). The rise of 17α,20β-DHP apparently began at the same time that plasma

TABLE III

Circulating Levels of 17α,20β-Dihydroxy-4-pregnen-3-one (17α,20β-DHP) during
Reproduction in Salmonids

Species	Condition[a]	Concentration 17α,20β-DHP (ng/ml)	
		Prefinal maturation[b]	Ovulatory[c]
Rainbow trout	N	1–2	272–520[d]
	N	16	354–416[e]
	N	<10	~230–270[f]
	I	<10	~270–780[e]
Atlantic salmon	N	<10	~120–280[f]
	I	<10	~30–150[f]
Coho salmon	I	<10	~130–220[f]
Amago salmon	N	~1	50–70[g]

[a]Abbreviations are: N, natural and I, gonadotropin or pituitary-induced.

[b]Levels were recorded in fish with fully developed oocytes but prior to final maturation or ovulation. For several of the studies cited this period was estimated.

[c]Levels were measured in fish at or near the time of ovulation.

[d]From Fostier et al., 1981b.

[e]From Scott et al., 1982.

[f]From Wright and Hunt, 1982.

[g]From Young et al., 1983b.

gonadotropin levels increased, and peaked about the same time that the first peak in gonadotropin was recorded. In rainbow trout, the circulating levels of 17α,20β-DHP reached an extremely high value (300–500 ng/ml) around the time of ovulation (Fostier et al., 1981b). However, extrapolation from earlier data indicates that if GVBD and ovulation are separated by approximately 2–3 days (Jalabert, 1978), the levels of 17α,20β-DHP during GVBD may not be as high. Certainly, the increase in 17α,20β-DHP coincided with GV maturation. High levels of 17α,20β-DHP may be necessary for the induction of ovulation (Jalabert, 1978) or function in a feedback capacity (Fostier et al., 1981b; Jalabert et al., 1980). Several other investigations have also observed comparable increases in the circulating levels of 17α,20β-DHP in rainbow trout during natural or pituitary-induced final maturation and ovulation (Table III). Circulating levels of 17α,20β-DHP have also been reported for Atlantic salmon (Salmo salar), coho salmon, and amago salmon (Table III). As with rainbow trout, the same basic trend has been observed. Prior to final maturation and ovulation 17α,20β-DHP levels are low and rise to very high levels at ovulation or shortly thereafter. The only major difference in results has been the lower absolute levels of 17α,20β-DHP recorded in amago salmon at ovulation compared to those for trout. This may

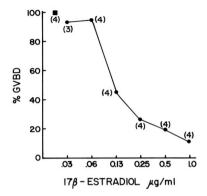

Fig. 7. *In vitro* effects of 17β-estradiol on GVBD in brook trout oocytes stimulated with 0.05 mg/ml alcohol-extracted carp pituitary preparation. Numbers in parenthesis indicate the amount of donors tested. Pituitary preparation alone is indicated by (■). No maturation occurred in controls (from Theofan, 1981).

result from an intrinsically lower capacity to synthesize 17α,20β-DHP in amago follicles (Young *et al.*, 1983b).

Although estrogenic steroids do not generally appear to have a stimulatory role in final maturation (see Section II,B), they may still have an indirect role in the timing of oocyte final maturation. Antiestrogens stimulated ovulation in goldfish (Pandey and Hoar, 1972) and loach (Ueda and Takahashi, 1976) and 17β-estradiol decreased the effects of gonadotropin and pituitary preparations on *in vitro* final maturation in incubations with rainbow trout (Jalabert, 1975) and brook trout (Theofan, 1981) (Fig. 7) oocytes. *In vitro*, estradiol (as well as testosterone and several other androgens) inhibits hydrocortisone acetate-induced final maturation of catfish oocytes (Sundararaj *et al.*, 1978), and, therefore, may directly inhibit the effect of a maturational steroid. However, in rainbow trout (Jalabert, 1975) and brook trout (F. W. Goetz, unpublished results), 17β-estradiol had no effect on steroid-induced final maturation. Alternatively, from studies on amphibian ovaries it has been suggested that estrogens inhibit the conversion of pregnenolone to progesterone (Spiegel *et al.*, 1978), and, in this manner, may block gonadotropin-induced final maturation (Schuetz, 1972; Spiegel *et al.*, 1978). Although conversion of [^{14}C]-pregnenolone to steroid metabolites in incubations with brook trout ovaries was relatively low with or without gonadotropin stimulation, Theofan (1981) did observe a decrease in the conversion of pregnenolone to δ^4-3-ketosteroids and an increase in the conversion to δ^5-3β-hydroxysteroid metabolites in gonadotropin-stimulated incubates containing 17β-estradiol. In addition to an effect on the conversion of pregnenolone to progesterone, estrogens may have a more dramatic effect earlier in the steroidogenic pathway. Possibly, estrogens may have a role in regulat-

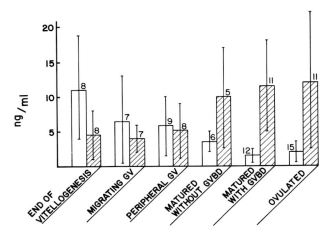

Fig. 8. Plasma 17β-estradiol (clear bar) and gonadotropin (hatched bar) in female trout at various stages of maturity (mean ± SD) (from Fostier *et al.*, 1978).

ing the time of natural final maturation by affecting the gonadotropin stimulation of ovarian steroidogenesis.

Both estradiol and gonadotropin levels, determined in rainbow trout from vitellogenesis through ovulation, indicate that there is a progressive and opposing shift in the levels of the two hormones throughout this period (Fostier *et al.*, 1978) (Figure 8). Estradiol is found in relatively high concentrations at the end of vitellogenesis and progressively decreases during final maturation and ovulation. In opposing fashion, gonadotropin is relatively low during vitellogenesis and increases steadily toward the time of ovulation (and see Section II,G). A similar decrease in estrogen levels prior to ovulation has also been observed in coho salmon, whitespotted char (*Salvelinus leucomaenis*), and in plaice (*Pleuronectes platessa*) (Jalabert *et al.*, 1978a; Kagawa *et al.*, 1981; Wingfield and Grimm, 1977). Decreasing levels of estrogen may trigger the increase in gonadotropin observed during GVBD and ovulation (Fostier *et al.*, 1978, 1981b). Further, the decrease in estrogen may remove any inhibition on steroidogenesis at the ovarian level.

In summary, from those studies conducted, it does appear that several of the fish species listed in Table I produce maturational steroids that have been determined to be effective in *in vitro* assays. In particular, the maturational steroid, 17α,20β-DHP, has been firmly identified as an ovarian metabolite in ayu and amago salmon (Suzuki *et al.*, 1981a,b). Its production by ovaries of other species, in which it is the most potent maturational steroid, has yet to be absolutely confirmed, although indirect evidence indicates its production (Fostier *et al.*, 1981a; Theofan and Goetz, 1983). Further, from the combined results of several investigations on rainbow trout, very predictable changes in the circulating levels of 17β-estradiol, gonadotropins, and 17α,20β-DHP are observed during final maturation and ovulation. Specifically, as estradiol decreases, gonadotropin levels increase, followed closely by a rapid rise in 17α,20β-DHP around the time of ovulation.

G. Gonadotropins during Final Maturation and Ovulation

From various investigations it is clear that circulating gonadotropin levels increase during final maturation and ovulation in several species (Fostier *et al.*, 1978; Stacey *et al.*, 1979; Crim *et al.*, 1973, 1975; Young *et al.*, 1983b) (Figs. 8 and 9). In goldfish, a surge of serum gonadotropin was observed that

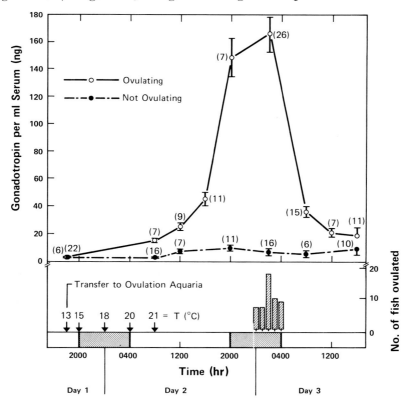

Fig. 9. Times of ovulation and serum gonadotropin (GtH:mean ± SE) profiles of sexually mature female goldfish kept on a 16L–8D photoperiod and warmed from 13° ± 1° to 21° ± 1°C. The number of fish ovulating each hour is shown in the lower graph. The GtH levels measured in serum from ovulatory fish sampled at the time of ovulation (0000–0400 hr on day 3) were pooled and expressed as a mean value at 0200 hr on day 3 (from Stacey *et al.*, 1979).

was coincident with ovulation (Stacey *et al.*, 1979) (Fig. 9). Goldfish that did not ovulate also did not exhibit this surge. It has also been shown that the gonadotropin surge is preceded slightly by a sharp increase in circulating cortisol (Cook *et al.*, 1980). Possibly the increased cortisol levels could facilitate the action of gonadotropin on ovulation.

In contrast, in rainbow trout, gonadotropins begin to increase approximately 10 days prior to ovulation and reach an initial peak of 10–20 ng/ml just prior to or at ovulation (Fostier *et al.*, 1981b). Following this peak, gonadotropin levels decline slightly, probably as a result of feedback by 17α-20β-DHP (Fostier *et al.*, 1981b; Jalabert *et al.*, 1980), and then increase even further to approximately 20–30 ng/ml following ovulation. In addition, the postovulatory levels in circulating gonadotropin are significantly higher in females subject to egg retention than in spent females (Jalabert and Breton, 1980).

III. OVULATION

A. Characteristics

If ovulation is defined as the actual expulsion of the oocyte from the follicle, then there are several very critical preparatory processes preceding ovulation (Jalabert and Szöllösi, 1975; Goetz *et al.*, 1982). Following final maturation, or during this time, the many microvillar connections between the oocyte and the follicular layer become detached (follicular separation). Following follicular separation, a very distinct hole is formed in the follicle through which the oocyte leaves (rupture). Release of the oocyte is apparently not a passive process in which the follicle simply degrades, leaving the oocyte free from the follicle; rather, the descriptions of ovulation and the effects of various chemical agents on this process (See Sections III,B and C) indicate that expulsion of the oocyte occurs by active contraction of the follicle, pushing the oocyte out (expulsion). The fact that the oocyte is constricted as it is expelled, may indicate that a very specific area of the follicle is weakened at ovulation and involved in hole formation.

B. Follicular Separation and Rupture

Very few studies have specifically investigated the mechanism by which the follicle separates from the oocyte and ruptures. In mammals, proteolytic enzymes may be involved in disrupting oocyte–follicle connections and in weakening the follicle at ovulation (Gilula *et al.*, 1978; Espey, 1974, 1980).

Specifically, the protease, plasmin, may play a role in degrading the follicular wall for rupture to occur (for review, see Strickland and Beers, 1979). Presumably, enzymes are also involved in these processes in fish ovulation. Using a modified gelatin–silver film procedure, Oshiro and Hibiya (1975) observed a substantial amount of presumed protease activity surrounding loach (*Misgurnus anguillicaudatus*) oocytes that had completed GVBD, just prior to ovulation. This activity was not present in oocytes far from ovulation. The enzyme activity could be involved in disrupting the follicle–oocyte connections. Jalabert and Szöllösi (1975) reported that *in vitro*, a trypsin inhibitor was partially inhibitory to $PGF_{2\alpha}$-induced ovulation of rainbow trout oocytes. It was hypothesized that the inhibitor blocked the breakdown of the follicle or in some cases the disruption of the follicle from the oocyte. However, the results were not totally conclusive, because the trypsin inhibitor actually stimulated ovulation by itself in oocytes from several females. Using a transcriptional inhibitor, actinomycin D, Theofan and Goetz (1981) observed two major phases of RNA synthesis involved in ovulation of yellow perch oocytes induced *in vitro* by $17\alpha,20\beta$-DHP. Actinomycin D completely inhibited ovulation when introduced early (0–6 hr) in the incubation, but later (14–22 hr) it was only partially inhibitory. Possibly, the early phase of RNA synthesis is involved in the production of enzymes necessary for follicle–oocyte disruption and/or in the formation of a hole in the follicle (Theofan and Goetz, 1981).

The hormonal regulation of follicular separation is unclear. Oocytes of very few fish undergo ovulation following steroid stimulation *in vitro* (Theofan and Goetz, 1981). There could be several reasons for this. First, steroids may not be the only hormonal agents involved in preparing the follicle for ovulation, and, therefore, other hormones may be necessary. Although contradictory to earlier reports (Hirose, 1972a), Iwamatsu (1974, 1978a) reported that *in vitro*, steroids would induce final maturation but not ovulation of medaka oocytes. In contrast, ovulation could be induced *in vitro* with pregnant mare's serum and *in vivo* in hypophysectomized females with several gonadotropins (Iwamatsu, 1978b). These results, taken together with the observation that progesterone can induce final maturation, but not ovulation when injected into hypophysectomized fish, suggested that steroids may interact with gonadotropins to induce ovulation of medaka oocytes; however, they are not the only mediators (Iwamatsu, 1978a).

Alternatively, the functional integrity of the follicular envelopes may be altered to such an extent in routine *in vitro* incubations that all of the processes normally stimulated by the steroid *in vivo* are not completed (Jalabert, 1978). In support of this, rainbow trout oocytes incubated under long-term sterile conditions with $17\alpha,20\beta$-DHP did ovulate following prostaglandin treatment (Jalabert, 1978), although under routine *in vitro* incubations they do not (Jalabert, 1976). Therefore, follicular separation and hole formation had to occur under the sterile incubation conditions. Unlike most fish, yellow perch oocytes ovulate predictably following *in vitro* steroid stimulation (Goetz and Bergman, 1978a; Goetz and Theofan, 1979). In this species all of the necessary preparatory steps to ovulation are apparently stimulated easily by the maturational steroid *in vitro*, or occur coincidentally as artifacts of the incubation.

Finally, no investigations of fish have explored the possibility that the agents responsible for rupture are distinct from those involved in follicular separation, and, therefore, could possibly be under separate hormonal control. In several species, during ovulation the oocyte is apparently forced out of a hole in the follicular envelope smaller than its diameter, and, therefore, it is quite possible that only a discrete section of the follicle is actually weakened and involved in hole formation. In mammalian ovaries it has been hypothesized that the plasminogen system is involved in weakening the follicular wall during ovulation (Strickland and Beers, 1979). The production of plasminogen activator by granulosa cells can be stimulated by gonadotropins. In turn, the activator could increase the conversion of plasminogen to plasmin, a protease that might act to weaken the follicle wall. Although no studies along these lines have been completed with fish, quite possibly an analogous system exists in the follicle of fish oocytes to stimulate the degradation of a discrete portion of the follicle.

C. Oocyte Expulsion

How the oocyte is actually expelled from the follicle is not entirely clear; however, several investigations have indicated that contraction of cells, possibly like those of smooth muscle, may be involved. Ultrastructural studies of trout, goldfish, and medaka follicles, have found microfilaments present in the thecal cells prior to ovulation (Pendergrass and Schroeder, 1976; Szöllösi and Jalabert, 1974; Szöllösi *et al.*, 1978; Nagahama *et al.*, 1976). These particular thecal cells are similar to other types of smooth muscle cells (Szöllösi and Jalabert, 1974), although certain differences apparently exist (Pendergrass and Schroeder, 1976). In medaka, the microfilaments are sparce in thecal cells far from ovulation, but increase in number and organization as the time of ovulation approaches (Pendergrass and Schroeder, 1976). This period of thecal development undoubtedly requires hormonal stimulation. Whether or not the development is induced by maturational steroids alone or in combination with other hormones such as gonadotropin is unclear. However, the observation (Jalabert, 1978) that rainbow trout oocytes, incubated under sterile conditions with 17α,20β-DHP, were capable of ovulat-

ing following prostaglandin treatment may indicate that gonadotropins are
not necessary.

That contraction of some type of muscular or microfilamentous system is
necessary for ovulation has been indirectly supported by several investigations. Cytochalasin B totally inhibited *in vitro* ovulation of medaka oocytes
and partially inhibited *in vitro* ovulation of trout oocytes (Schroeder and
Pendergrass, 1976; Jalabert and Szöllösi, 1975). One effect of cytochalasin B
is the inhibition of actin polymerization (Weihing, 1978). Therefore, if actin
is involved in a contractile system in the follicle then cytochalasin might have
its greatest effect during the period of thecal morphogenesis. This could
explain the slightly different results obtained with cytochalasin B in trout
and medaka. The degree of thecal development prior to cytochalasin treatment might determine the strength of inhibition by this compound. Agents
such as La^{3+} and Mn^{2+} that block Ca^{2+} influx, necessary for contraction of
muscle, have also been reported to inhibit ovulation of trout oocytes *in vitro*
(Jalabert and Szöllösi, 1975). Finally, in brook trout oocytes that have undergone GVBD and follicular separation *in vivo*, dibutyryl cyclic 3',5'-ade-
nosine monophosphate (DBcAMP) and phosphodiesterase inhibitors such as
theophylline, SQ20,006 and 3-isobutyl-1-methyl-xanthine are potent inhibitors of ovulation *in vitro* in incubates with or without prostaglandins (Goetz
et al., 1982) (Fig. 10). Cyclic AMP inhibits contraction of other types of

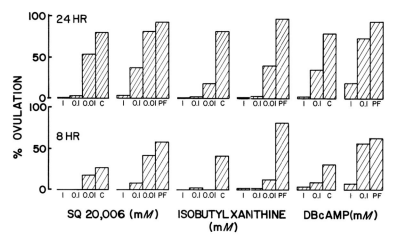

Fig. 10. Effects of SQ20,006, 3-isobutyl-1-methyl-xanthine, and dibutyryl cyclic 3',5'-ade-
nosine monophosphate on *in vitro* spontaneous and $PGF_{2\alpha}$-enhanced ovulation at 8 and 24 hr of
incubation (PF, 2.0 μg/ml $PGF_{2\alpha}$; C, spontaneous control ovulation) (from Goetz *et al.*, 1982).

smooth muscle by interfering with the phosphorylation of myosin, a step necessary for contraction (Adelstein and Klee, 1980). Therefore, the effects observed on brook trout oocytes with DBcAMP and phosphodiesterase inhibitors, may further indicate the necessity for muscular contraction in ovulation.

From ultrastructural studies, it has been proposed that expulsion of lamprey (*Petromyzon marinus*) oocytes is accomplished by a change in the shape of specific follicular cells from a cuboidal to columnar configuration (Yorke and McMillan, 1980). The change in shape constricts the follicle at one end and pushes the oocyte out of the opposite end of the follicle (Fig. 11). The hormonal control of this has not been investigated. Although the structural details of follicle and oocyte observed in lampreys at ovulation may not be exactly comparable to that in teleosts, the mechanism of follicular constriction by changes in cell shape might be the same.

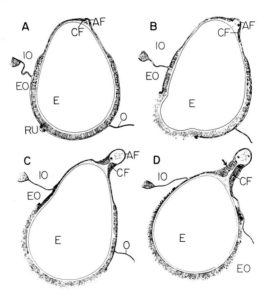

Fig. 11. Lamprey ovulation (from Yorke and McMillan, 1980). A. Follicle just after the rupture of the follicular and thecal layers. Follicular cells surrounding the apical follicular cells (AF) are becoming columnar (CF) (E, egg; EO, extraovarian space; IO, intraovarian space; O, ovarian wall; RU, site of rupture; ×65). B. A follicle during the early stage of egg emergence. There is a continued dramatic increase in the height of the columnar follicular cells (CF) (×65). C. Midstage of egg emergence. The differentiation of the columnar follicular cells creates a constriction near the apical end of the follicle, which pinches off the apical follicular cells (×65). D. Late stage in the emergence of the egg into the extraovarian space. Progressive differentiation of adjacent follicular cells to form columnar follicular cells results in an elongation of the constricted portion of the follicle (arrow), pushing the egg ahead (×65).

The hormonal control of oocyte expulsion has been the object of several investigations. In mammals and fish, prostaglandins are believed to be involved in ovulation, possibly by stimulating follicular contraction (for review, see Espey, 1980; Stacey and Goetz, 1982). Ovulation in goldfish, induced *in vivo* by HCG and elevated holding temperatures, was inhibited by indomethacin (prostaglandin synthesis inhibitor) when applied 6 hr following HCG injections (Stacey and Pandey, 1975). Prostaglandin E_2 (PGE_2), E_1 (PGE_1), and $F_{2\alpha}$ ($PGF_{2\alpha}$) all induced ovulation in indomethacin-blocked females and other studies with PGE_2 indicated that the time of injection was critical. Further studies have also demonstrated that goldfish ovulation can be inhibited by indomethacin if administered during the spontaneous preovulatory surge, or following preoptic lesions that stimulate gonadotropin release (Stacey and Goetz, 1982). The combined results from goldfish studies suggest that one effect of indomethacin is at the ovarian level.

Besides affecting the ovary, prostaglandins might also have an effect on the hypothalamic–hypophysial system. In mammals, there is ample evidence that prostaglandins stimulate gonadotropin release and it appears that the primary site of action is on the hypothalamus (Behrman, 1979). There is some evidence for a hypothalamic–hypophysial site of action for prostaglandins in fish. In catfish (*Heteropneustes fossilis*), Singh and Singh (1976) reported that PGE_1 and $PGF_{2\alpha}$ induced ovulation in intact females but not in hypophysectomized fish. In addition, an increase in the gonadotropin content (measured by bioassay) of the pituitary and serum was observed in females receiving PGE_1 or $PGF_{2\alpha}$. In a study of carp (*Cyprinus carpio*), ovulation, which was induced by holding conditions, was dramatically inhibited by indomethacin injection (Kapur and Toor, 1979). Because clomiphene citrate, a compound that stimulates gonadotropin release in carp (Breton *et al.*, 1975), was able to stimulate ovulation in indomethacin-blocked carp, it is possible that indomethacin blocks an endogenous increase in gonadotropins necessary for ovulation. These investigations suggest that the effect of prostaglandins on the hypothalamic–hypophysial system, if functional, would be stimulatory. However, injection of PGE_2 and $PGF_{2\alpha}$ (2.0 μg) into the third ventricle of goldfish resulted in a decrease in serum gonadotropin 30 min postinjection (Peter and Billard, 1976).

That prostaglandins can exert a direct effect at the ovarian level has been more clearly shown by several *in vitro* investigations. Jalabert and Szöllösi (1975) reported that $PGF_{2\alpha}$ could induce ovulation *in vitro* of rainbow trout oocytes that had previously undergone GVBD and follicular separation *in vivo;* however, PGE_2 was ineffective at the concentrations tested. Goldfish oocytes, from females induced to mature *in vivo* with HCG, underwent ovulation within 15–30 min in the presence of prostaglandins (Kagawa and Nagahama, 1981). Although PGE_1, PGE_2, $PGF_{1\alpha}$, and $PGF_{2\alpha}$ all induced ovulation, $PGF_{2\alpha}$ was the most potent. Brook trout oocytes that have undergone GVBD and follicular separation *in vivo*, normally ovulate without hormonal stimulation within 24 hr following removal to an *in vitro* incubation system (Fig. 10) (Goetz and Smith, 1980; Goetz *et al.*, 1982). Prostaglandin $F_{2\alpha}$ enhanced this spontaneous ovulation (Fig. 10), but PGE_1 decreased it (Goetz *et al.*, 1982). The inhibitory effect of PGE_1 might be a result of increased cyclic AMP levels, because E prostaglandins stimulate cyclic nucleotides in various systems (Kuehl *et al.*, 1976). As indicated earlier, cAMP and phosphodiesterase inhibitors would be inhibitory to brook trout ovulation (Fig. 10).

In vitro, yellow perch oocytes ovulated predictably within 30–35 hr at 15°C in incubations with $17\alpha,20\beta$-DHP (Geotz and Theofan, 1979). When perch oocytes were incubated with $17\alpha,20\beta$-DHP and indomethacin, ovulation was totally inhibited by 33 hr although some ovulation was observed by 48 hr. Addition of PGE_2, PGE_1, and $PGF_{2\alpha}$-induced significant ovulation in a dose-dependent manner in indomethacin-blocked incubates (Goetz and Theofan, 1979); PGE_2 was the most potent. Although inhibition of prostaglandin synthesis by indomethacin was undoubtedly responsible for blocking ovulation, it should not be automatically assumed that $17\alpha,20\beta$-DHP stimulated prostaglandin synthesis; however, $17\alpha,20\beta$-DHP may have modified it. Control ovaries also produce prostaglandins in *in vitro* incubations (Stacey and Goetz, 1982; Goetz, 1980), but this may be an artifact of the incubation conditions. However, regardless of the source of stimulation, prostaglandins are capable of inducing ovulation in yellow perch oocytes. Prostaglandins have also been reported to stimulate *in vitro* ovulation of pike and eel oocytes (Jalabert, 1976; Epler and Bieniarz, 1978; Epler, 1981c).

Catecholamines can also stimulate ovulation of oocytes *in vitro*. Jalabert (1976) reported that epinephrine stimulated *in vitro* ovulation of rainbow trout oocytes. The stimulatory effect could be blocked by the α-adrenergic receptor blockers, dibenzyline, dibenamine, and phentolamine, but not by chemical blockers of the β-adrenergic receptors. The results suggest that epinephrine stimulates ovulation through interaction with α-adrenergic receptors; however, the mechanism following this activation was still unclear. Epinephrine also stimulates *in vitro* ovulation of yellow perch, brook trout and carp oocytes (Stacey and Goetz, 1982; F. W. Goetz, unpublished results; Epler, 1981c). The epinephrine effect on yellow perch ovulation may be a result of increased prostaglandin synthesis because levels of PGE and PGF, measured by radioimmunoassay, are higher in incubates containing epinephrine than in controls (Stacey and Goetz, 1982). The physiological significance of catecholamines in fish ovulation is still unclear. If involved, catecholamines could theoretically come from sympathetic nerve endings in the ovary or from chromaffin tissue. Several investigations have been completed on the nervous innervation of mammalian ovaries (Stefenson *et al.*, 1981; Morimoto *et al.* 1981); however, similar studies on fish are lacking. For rainbow trout, it has been suggested that the sympathetic nervous system (SNS) may represent the relay between final maturation and ovulation (Jalabert, 1976). Therefore, the SNS would receive a signal at the end of final maturation, and, in turn, stimulate the release of epinephrine in the ovary. Epinephrine could stimulate oocyte expulsion directly or indirectly through prostaglandins. Other than the observations that epinephrine can stimulate *in vitro* ovulation and prostaglandin synthesis, this hypothesis has yet to be tested.

D. Prostaglandin Production at Ovulation

From investigations on the effects of prostaglandins in fish, it does appear that certain prostaglandins have a stimulatory effect on ovulation. The question remains as to whether or not the effect is physiologically significant. In this regard, Bouffard (1979) has shown that the plasma levels of PGFs (measured by RIA) increase in female goldfish, induced to ovulate with increased holding temperatures and HCG. The mean PGF levels increased (although not significantly) at the time of ovulation (about 980 pg/ml versus 317 pg/ml at preinjection) and rose at least fourfold by 12 hr following ovulation (about 4100 pg/ml). In contrast, in several fish that did not ovulate, PGF levels remained as low as preinjection values. Although controls were not run on gravid females, levels of PGF in nongravid, saline-injected fish remained low over a similar period of sampling. Further, PGF levels in ovarian fluid from ovulated fish were even higher than in the plasma. In contrast, plasma PGE levels (measured by RIA) were comparatively high in preinjected, gravid females (about 2600 pg/ml), decreased by the second period of sampling, and were approximately 1600 pg/ml by the time of ovulation (Bouffard, 1979). Following ovulation, PGE levels declined further to approximately 1100 pg/ml. The PGE levels measured in nongravid, saline-injected fish did not change over a similar sampling period; however, these results cannot be adequately compared to those of gravid fish because the values of PGE were always quite low (about 100 pg/ml) and, therefore, an effect from serial sampling would not be observed unless it was stimulatory. That repeated sampling may have been responsible for the decrease in PGE levels at ovulation is further supported by the observation that spontaneously ovulating fish had PGE levels of approximately 7200 pg/ml at ovulation (first sample), as opposed to the 1600 pg/ml observed in fish injected with HCG and sampled several times earlier (see previous discussion). The PGE levels in the spontaneously ovulating fish also dropped to approximately 2100 pg/ml by the second sample 24 hr later. Therefore, the validity of the PGE levels observed in females throughout ovulation is questionable. Because PGF levels did not increase significantly until after ovulation, their role in the ovulatory process in goldfish is still unclear. However, many factors would be involved in evaluating these data. Extrapolating from *in vitro* data on goldfish ovulation, it appears that $PGF_{2\alpha}$ levels below 0.1 μg/ml might be nonstimulatory (Kagawa and Nagahama, 1981), although this cannot be conclusively ascertained. If so, the levels of PGF recorded in goldfish at ovulation or even after ovulation appear to be low. However, oocytes treated *in vitro* might not develop the sensitivity that would occur *in vivo* or be at exactly the proper receptive stage when removed for incubation. Alternatively, small increases in plasma prostaglandin levels might reflect larger synthetic rates within the ovarian tissue, especially if any of the prostaglandins are being metabolized.

Plasma and ovarian PGE and PGF levels have also been measured (by RIA) in brook trout induced to ovulate with a single injection of an alcohol-extracted carp pituitary preparation (Cetta and Goetz, 1983). Rather than using repetitive sampling that might affect plasma prostaglandin levels, fish were sampled once at various stages before, during, and after ovulation. Gravid, uninjected female brook trout had low plasma PGF levels and very high ovarian PGE levels (Fig. 12). In contrast, females that had ovulated approximately 80–100% of their oocytes or were assayed 24 hr postovulation, had significantly elevated plasma and ovarian PGF levels. Plasma PGE levels did not change significantly in any group of fish; however, there was a significant decrease in ovarian PGE levels in all groups compared to the levels in gravid controls. Again, in investigations of brook trout, PGF levels in the plasma or ovarian tissue did not increase significantly until after substantial ovulation (80–100%) had occurred (Fig. 12). Therefore, even though $PGF_{2\alpha}$ was stimulatory to brook trout ovulation *in vitro* (Goetz *et al.*, 1982), it is not clear what the natural role of F prostaglandins is at ovulation. It is possible that F prostaglandins could be responsible for expulsion without being substantially elevated. Alterations in the sensitivity of the follicle to prostaglandins or in the levels of other prostaglandins might enable relatively low levels of PGF to be stimulatory to ovulation. For example, ovarian PGE levels in brook trout progressively decreased from just prior to ovulation until 24 hr postovulation (Fig. 12). Because PGE_1 was inhibitory to brook trout ovulation *in vitro* (Goetz *et al.*, 1982), it is conceivable that the progressive decrease in E prostaglandins might actually determine the time at which ovulation occurs. Therefore, the absolute values of individual prostaglandins may not be as important as their relative concentrations.

Levels of PGE in the ovarian tissue of the pond loach (*Misgurnus anguillicaudatus*) have been measured at the time of ovulation (Ogata *et al.*,

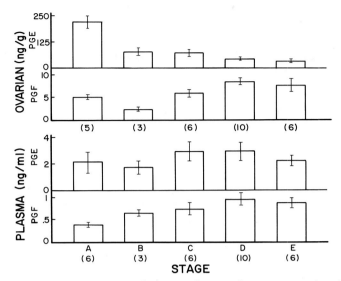

Fig. 12. Concentrations of PGF and PGE in plasma and ovarian tissue from brook trout during pituitary-induced ovulation: stage A, gravid controls; B, preovulatory (GVBD and follicular separation); C, < 50% ovulated; D, 80–100% ovulated; E, 24 hr postovulatory. Numbers in parentheses are sample size. Significant differences as follows: Ovarian PGE, stage A significant from B, C, D, E ($P < 0.05$); ovarian PGF, stage D significant from A, B, C ($P < 0.05$) and stage E significant from B ($P < 0.05$); plasma PGE, none; plasma PGF, stage A significant from D, E ($P < 0.05$) (from Cetta and Goetz, 1983).

1979). Ovarian PGF levels increased significantly by 24 hr following HCG injection, at which time all of the fish had ovulated. Indomethacin blocked the increase in PGF levels induced by HCG treatment.

Considering the various prostaglandin precursors that might be available in fish (Stacey and Goetz, 1982; Mai *et al.*, 1981), it is possible that the RIAs used to date are not measuring relevant prostaglandins. RIAs used to measure fish prostaglandins have employed antibodies directed against one or two series prostaglandins; however, it is possible that three and four series prostaglandins may also be produced (Mai *et al.*, 1981) and might be important. Further, other prostaglandin-related compounds such as prostacyclin, thromboxanes, and leukotrienes have not been investigated and could possibly be involved.

Finally, it should be noted that all investigations conducted to date on the prostaglandin levels in fish at ovulation have used either HCG or pituitary-induced fish (Bouffard, 1979; Ogata *et al.*, 1979; Cetta and Goetz, 1983). Whether similar results will be obtained with naturally ovulating fish is unknown.

In both goldfish and brook trout, PGF levels measured in the ovarian or coelomic fluid of ovulated fish were actually higher than in the plasma of ovulated females (Bouffard, 1979; Cetta and Goetz, 1983). Considering the elevation of PGF synthesis in ovarian tissue in fully ovulated brook trout it is possible that these prostaglandins come from the ovaries.

IV. SYNCHRONY IN THE SEQUENCE AND CONTROL OF FINAL MATURATION AND OVULATION

Because most investigations have focused on the hormonal regulation of either ovulation or final maturation, it is understandable that relatively less is known about the synchrony in the sequence or control of the two processes.

Because oocytes of most fish do not ovulate *in vitro* following final maturation, the relationship between the timing and hormonal control of each process is difficult to study. However, in yellow perch, in which ovulation and final maturation both occur *in vitro*, the synchrony between these events has been studied. Oocytes collected from perch early in the reproductive season, containing a centrally located germinal vesicle (GV), respond to $17\alpha,20\beta$-DHP stimulation *in vitro*. However, at this time, higher concentrations (>2.0 ng/ml) of $17\alpha,20\beta$-DHP generally induce ovulation prior to GVBD and the oocytes die (Goetz and Theofan, 1979). In contrast, oocytes taken from females closer to the natural spawning period, and in which the GV has begun some migration, undergo GVBD prior to ovulation under similar steroid stimulation (Goetz and Theofan, 1979). The difference in results is probably attributable to the sensitivity of the oocyte and how quickly final maturation can be induced in relation to ovulation. Therefore, at higher $17\alpha,20\beta$-DHP concentrations, *in vitro* final maturation and ovulation of yellow perch oocytes appear to be unsynchronized events. If oocytes are more mature (as indicated by GV position) prior to incubation, then $17\alpha,20\beta$-DHP will induce complete GVBD prior to ovulation. If the oocytes are very early, ovulation occurs before GVBD and the oocytes die. However, at very low $17\alpha,20\beta$-DHP concentrations (≤ 0.48 ng/ml) GV migration and GVBD can be stimulated before ovulation even in oocytes with a centrally located GV. At these concentrations, ovulation does eventually follow GVBD provided long enough incubation times are used (Goetz and Theofan, 1979). It is possible that at lower steroid concentrations some form of synchrony exists between final maturation and ovulation that is not present at higher concentrations.

That a proper sequence of hormones is necessary for normal final maturation and ovulation, has been demonstrated in several *in vivo* experiments with steroids. Injection of a maturational steroid in some fish species does result in normal ovulation (Bakiber and Ibrahim, 1979; Sundararaj and Goswami, 1966); however, injection of 17α,20β-DHP induced GVBD, but either failed or was only partially successful in stimulating ovulation in coho salmon, pike, and trout (in which the GV was premigratory) (Jalabert *et al.*, 1978a,b; De Montalembert *et al.*, 1978). Similarly, cortisone acetate induced GVBD but not ovulation in loach (*Misgurnus fossilis*) (Kirshenblat, 1959). In carp held at low temperatures, injection of 17α,20β-DHP alone did not induce final maturation or ovulation *in vivo* (Jalabert *et al.*, 1977). The inability of 17α,20β-DHP to induce ovulation and/or maturation under these conditions appears to be related to the sensitivity of the ovary, indicated indirectly by the position of the GV. Therefore, in every case in which 17α,20β-DHP was unable to induce maturation or ovulation *in vivo*, the oocytes initially contained GVs that were nonperipheral (Jalabert *et al.*, 1977, 1978a,b; De Montalembert *et al.*, 1978). In fact, in rainbow trout containing oocytes with peripheral or subperipheral GVs, 17α,20β-DHP could induce GVBD and ovulation (Jalabert *et al.*, 1976). Further, in carp, coho, pike, and trout (with nonperipheral GV) a low priming dose of gonadotropin given several days prior to an injection of 17α,20β-DHP, significantly increased the maturational and/or ovulatory response (Jalabert *et al.*, 1977, 1978a,b; De Montalembert *et al.*, 1978). In carp, the effect of priming was to begin GV migration to a peripheral position (Jalabert *et al.*, 1977).

It has been suggested that the inability of 17α,20β-DHP to induce ovulation *in vivo* in trout oocytes in which the GV is nonperipheral, may be attributable to the absence of an ovulation "mediator" necessary for oocyte expulsion (Jalabert *et al.*, 1978b). In support of this, mature oocytes (undergone GVBD) taken from female trout that would not ovulate following 17α,20β-DHP treatment, did ovulate when incubated *in vitro* with PGF$_{2\alpha}$ (Jalabert *et al.*, 1978b). This indicated that under the *in vivo* conditions the mechanism for expulsion was present, but the stimulus was not. The production of such a "mediator" is apparently related to the circulating gonadotropin levels because priming with gonadotropin increases the ovulatory response observed with 17α,20β-DHP. The nature of this "mediator" is unknown.

V. CONCLUSIONS

Results of *in vitro* investigations have conclusively shown that certain steroids are potent stimulators of oocyte final maturation in fish. Although phylogenetic differences exist in the stimulation of final maturation by steroids, in at least eight different species representing several teleost orders, the most effective steroid yet tested for the induction of final maturation is $17\alpha,20\beta$-DHP. $17\alpha,20\beta$-dihydroxy-4-pregnen-3-one has been identified as an ovarian metabolite in ayu and amago salmon. That this progestational steroid is the primary natural mediator of oocyte final maturation in rainbow trout and amago salmon, has been further substantiated by the observations that in these species (1) $17\alpha,20\beta$-DHP levels are significantly increased in *in vitro* incubations stimulated with gonadotropins, and (2) circulating levels of $17\alpha,20\beta$-DHP increase substantially during final maturation and ovulation.

In several species, including the catfish and the zebrafish, it appears that 11-deoxygenated or 11-oxygenated corticosteroids are the most potent maturational steroids.

Although it is known that translation of RNA is necessary for steroid-induced final maturation, very little is known about the exact mechanism of steroid action in the induction of final maturation in fish, and this is an interesting area for future research.

From *in vitro* investigations, gonadotropins appear to stimulate final maturation, in most fish directly at the ovarian level, although in at least one species, the Indian catfish, the stimulation appears to be through the interrenals. That gonadotropins stimulate final maturation by stimulating follicular steroidogenesis has been confirmed by several methods. Further, increases in circulating gonadotropin levels correlate well with final maturation and ovulation in several species.

In species with a direct pituitary–ovarian relay for the stimulation of final maturation, the interrenals may indirectly influence this stimulation through the production of 11-oxygenated or 11-deoxygenated corticosteroids. These corticosteroids might enhance the effects of gonadotropins and/or maturational steroids, or alter the relative quantities of free and bound circulating maturational steroid. Estrogens may also influence the overall timing of final maturation by affecting the release of gonadotropins from the pituitary or the stimulation of steroidogenesis by gonadotropins.

The combined results of studies of ovulation of fish oocytes, although informative, still leave many questions concerning the hormonal regulation for investigation. Taking into account the preparatory steps to oocyte expulsion, the entire process of ovulation is extremely complex. Some of the processes leading to ovulation, such as follicular separation, might be stimulated by maturational steroids; however, the actual process of oocyte expulsion is probably under separate control.

Oocyte expulsion may be a result of the contraction of specific cells in the follicle. Because gonadotropin-induced (endogenous and exogenous) ovulation can be inhibited by indomethacin, and because prostaglandins can stimulate ovulation *in vivo* and *in vitro*, it appears that prostaglandins may be the natural mediators of oocyte expulsion. However, the exact prostaglandins responsible are not known and phylogenetic differences may exist. Following ovulation, levels of PGF increase in the plasma in goldfish and brook trout and in the ovarian tissue of brook trout and loach. Epinephrine stimulates *in vitro* ovulation of oocytes from several species and may have a physiological role in directly or indirectly stimulating oocyte expulsion.

ACKNOWLEDGMENTS

Thanks is extended to A. Fostier, Y. Nagahama, and G. Young for supplying copies of manuscripts in press, and R. Wallace for reviewing an initial draft of this chapter.

REFERENCES

Adelstein, R. S., and Klee, C. B. (1980). Smooth muscle myosin light chain kinase. *In* "Calcium and Cell Function" (W. Y. Cheung, ed.), Vol. 1, pp. 167–182. Academic Press, New York.

Babiker, M. M., and Ibrahim, H. (1979). Studies on the biology of reproduction in the cichlid *Tilapia nilotica* (L.): Effects of steroid and tropic hormones on ovulation and ovarian hydration. *J. Fish Biol.* **15**, 21–30.

Behrman, H. R. (1979). Prostaglandins in hypothalamo-pituitary and ovarian function. *Annu. Rev. Physiol.* **41**, 685–700.

Bouffard, R. E. (1979). The role of prostaglandins during sexual maturation, ovulation and spermiation in the goldfish, *Carassius auratus*. M.Sc. Thesis, University of British Columbia, Vancouver, B.C., Canada.

Braekevelt, C. R., and McMillan, D. B. (1967). Cyclic changes in the ovary of the brook stickleback *Eucalia inconstans* (Kirtland). *J. Morphol.* **123**, 373–396.

Breton, B., Jalabert, B., Fostier, A., and Reinaud, P. (1975). Induction de décharges gonadotropes hypophysaires chez la carpe (*Cyprinus carpio* L.) á l'aide du citrate de cisclomiphene. *Gen. Comp. Endocrinol.* **25**, 400–404.

Campbell, C. M., Walsh, J. M., and Idler, D. R. (1976). Steroids in the plasma of the winter flounder (*Pseudopleuronectes americanus* Walbaum). A seasonal study and investigation of steroid involvement in oocyte maturation. *Gen. Comp. Endocrinol.* **29**, 14–20.

Campbell, C. M., Fostier, A., Jalabert, B., and Truscott, B. (1980). Identification and quantification of steroids in the serum of rainbow trout during spermiation and oocyte maturation. *J. Endocrinol.* **85**, 371–378.

Cetta, F., and Goetz, F. W. (1982). Ovarian and plasma prostaglandin E and F levels in brook trout (*Salvelinus fontinalis*) during ovulation. *Biol. Reprod.* **27**, 1216–1221.

Chaudhuri, H. (1976). Use of hormones in induced spawning of carps. *J. Fish. Res. Board Can.* **33**, 940–947.

Colombo, L., and Colombo, B. P. (1977). Gonadal steroidogenesis in teleost fishes. *Invest. Pesq.* **41**, 147–164.

Colombo, L., Bern, H. A., Pieprzyk, J., and Johnson, D. W. (1973). Biosynthesis of 11-deoxycorticosteroids by teleost ovaries and discussion of their possible role in oocyte maturation and ovulation. *Gen. Comp. Endocrinol.* **21**, 168–178.

Colombo, L., Colombo, B. P., and Arcarese, G. (1978). Emergence of ovarian 11-deoxycorticosteroid biosynthesis at ovulation time in the sea bass, *Dicentrarecus labrax* L. *Ann. Biol. Anim., Biochim., Biophys.* **18**, 937–941.

Cook, A. F., Stacey, N. E., and Peter, R. E. (1980). Periovulatory changes in serum cortisol levels in the goldfish, *Carassius auratus*. *Gen. Comp. Endocrinol.* **40**, 507–510.

Crim, L. W., Meyer, R. K., and Donaldson, E. M. (1973). Radioimmunoassay estimates of plasma gonadotropin levels in the spawning pink salmon. *Gen. Comp. Endocrinol.* **21**, 69–76.

Crim, L. W., Watts, E. G., and Evans, D. M. (1975). The plasma gonadotropin profile during sexual maturation in a variety of salmonid fishes. *Gen. Comp. Endocrinol.* **27**, 62–70.

De Montalembert, G., Jalabert, B., and Bry, C. (1978). Precocious induction of maturation and ovulation in northern pike (*Esox lucius*). *Ann. Biol. Anim., Biochim., Biophys.* **18**, 969–975.

DeNeff, S. J. (1980). The social stimulation of reproductive maturation in the female paradise fish, *Macropodus opercularis*. M.Sc. Thesis, University of Notre Dame, Notre Dame, Indiana.

Dettlaff, T. A., and Davydova, S. I. (1979). Differential sensitivity of cells of follicular epithelium and oocytes in the stellate sturgeon to unfavorable conditions and correlating influence of triiodothyronine. *Gen. Comp. Endocrinol.* **39**, 236–243.

Dettlaff, T. A., and Skoblina, M. N. (1969). The role of germinal vesicle in the process of oocyte maturation in *Anura* and *Acipenseridae*. *Ann. Embryol. Morphog., Suppl.* **1**, 133–151.

Duffey, R. J., and Goetz, F. W. (1980). The *in vitro* effects of 17α-hydroxy-20β-dihydroprogesterone on germinal vesicle breakdown in brook trout (*Salvelinus fontinalis*) oocytes. *Gen. Comp. Endocrinol.* **41**, 563–565.

Epler, P. (1981a). Effect of steroid and gonadotropic hormones on the maturation of carp ovaries. Part II. Effect of fish and mammalian gonadotropins on the maturation of carp oocytes *in vitro*. *Pol. Arch. Hydrobiol.* **28**, 95–102.

Epler, P. (1981b). Effect of steroid and gonadotropic hormones on the maturation of carp ovaries. Part III. Effect of steroid hormones on the carp oocyte maturation *in vitro*. *Pol. Arch. Hydrobiol.* **28**, 103–110.

Epler, P. (1981c). Effects of steroid and gonadotropic hormones on the maturation of carp ovaries. Part V. Ovulation, fertilization, and embryonal development of carp oocytes *in vitro*. *Pol. Arch. Hydrobiol.* **28**, 119–125.

Epler, P. (1981d). Effect of steroid and gonadotropic hormones on the maturation of carp ovaries. Part VI. A supposed mechanism of carp oocyte maturation and ovulation. *Pol. Arch. Hydrobiol.* **28**, 127–134.

Epler, P., and Bieniarz, K. (1978). *In vitro* ovulation of European eel (*Anguilla anguilla* L.) oocytes following *in vivo* stimulation of sexual maturation. *Ann. Biol. Anim., Biochim., Biophys.* **18**, 991–995.

Epler, P., Marosz, E., and Bieniarz, K. (1979). Effect of teleost pituitary gonadotropins on the *in vitro* maturation of carp oocytes. *Aquaculture* **18**, 379–382.

Espey, L. L. (1974). Ovarian proteolytic enzymes and ovulation. *Biol. Reprod.* **10**, 216–235.

Espey, L. L. (1980). Ovulation as an inflammatory reaction—A hypothesis. *Biol. Reprod.* **22**, 73–106.

Finidori-Lepicard, J., Schorderet-Slatkine, S., Hanoune, J., and Baulieu, E.-E. (1981). Progesterone inhibits membrane-bound adenylate cyclase in *Xenopus laevis* oocytes. *Nature (London)* **292**, 255–257.

Fostier, A., and Breton, B. (1975). Binding of steroids by plasma of a teleost: The rainbow trout, *Salmo gairdnerii*. *J. Steroid Biochem.* **6**, 345–351.

Fostier, A., Jalabert, B., and Terqui, M. (1973). Action prédominate d'un dérivé hydroxylé de la progestérone sur la maturation *in vitro* des oocytes de la Truite arc-en-ciel, *Salmo gairdnerii*. *C.R. Hebd. Seances Acad. Sci.* **277**, 421–424.

Fostier, A., Weil, C., Terqui, M., Breton, B., and Jalabert, B. (1978). Plasma estradiol-17β and gonadotropin during ovulation in rainbow trout (*Salmo gairdneri* R.). *Ann. Biol. Anim., Biochim., Biophys.* **18**, 929–936.

Fostier, A., Jalabert, B., Campbell, C., Terqui, M., and Breton, B. (1981a). Cinétique de libération *in vitro* de 17α hydroxy-20β dihydroprogesterone par des follicles de Truite arc-en-ciel, *Salmo gairdnerii*. *C.R. Hebd. Seances Acad. Sci.* **292**, 777–780.

Fostier, A., Breton, B., Jalabert, B., and Marcuzzi, O. (1981b). Evolution des niveaux plasmatiques de la gonadotropine glycoprotéique et de la 17αhydroxy-20β dihydroprogesterone au cours de la maturation et de l'ovulation chez la Truite arc-en-ciel, *Salmo gairdnerii*. *C.R. Hebd. Seances Acad. Sci.* **293**, 817–820.

Gilula, N. B., Epstein, M. L., and Beers, W. H. (1978). Cell-to-cell communication and ovulation. A study of the cumulus-oocyte complex. *J. Cell Biol.* **78**, 58–75.

Goetz, F. W. (1976). The *in vitro* induction of final maturation and ovulation in brook trout (*Salvelinus fontinalis*) and yellow perch (*Perca flavescens*) ova and the hormonal control of final maturation in ova of other fish species. Ph.D. Thesis, University of Wyoming, Laramie.

Goetz, F. W. (1979). The role of steroids in the control of oocyte final maturation and ovulation in yellow perch (*Perca flavescens*). *Proc. Indian Natl. Sci. Acad., Part B* **45**, 497–504.

Goetz, F. W. (1980). Prostaglandin levels during steroid-induced *in vitro* final maturation of brook trout (*Salvelinus fontinalis*) oocytes. *Am. Zool.* **20**, 730.

Goetz, F. W., and Bergman, H. L. (1978a). The effects of steroids on final maturation and ovulation of oocytes from brook trout (*Salvelinus fontinalis*) and yellow perch (*Perca flavescens*). *Biol. Reprod.* **18**, 293–298.

Goetz, F. W., and Bergman, H. L. (1978b). The *in vitro* effects of mammalian and piscine gonadotropin and pituitary preparations on final maturation in yellow perch (*Perca flavescens*) and walleye (*Stizostedion vitreum*). *Can. J. Zool.* **56**, 348–350.

Goetz, F. W., and Smith, D. G. (1980). *In vitro* effects of theophylline and dibutyryl adenosine 3′5′-cyclic monophosphoric acid on spontaneous and prostaglandin $PGF_{2\alpha}$-induced ovulation of brook trout (*Salvelinus fontinalis*) oocytes. *Biol. Reprod.* **22**, Suppl. 1, 114A.

Goetz, F. W., and Theofan, G. (1979). *In vitro* stimulation of germinal vesicle breakdown and ovulation of yellow perch (*Perca flavescens*) oocytes. Effects of 17α-hydroxy-20β-dihydroprogesterone and prostaglandins. *Gen. Comp. Endocrinol.* **37**, 273–285.

Goetz, F. W., Smith, D. C., and Krickl, S. P. (1982). The effects of prostaglandins, phosphodiesterase inhibitors and cyclic AMP on ovulation of brook trout (*Salvelinus fontinalis*) oocytes. *Gen. Comp. Endocrinol.* **48**, 154–160.

Goswami, S. V., and Sundararaj, B. I. (1971a). Temporal effects of ovine luteinizing hormone and desoxycorticosterone acetate on maturation and ovulation of oocytes of the catfish, *Heteropneustes fossilis* (Bloch): An *in vivo* and *in vitro* study. *J. Exp. Zool.* **178**, 457–465.

Goswami, S. V., and Sundararaj, B. I. (1971b). *In vitro* maturation and ovulation of oocytes of the catfish, *Heteropneustes fossilis* (Bloch): Effects of mammalian hypophyseal hormones, catfish pituitary homogenate, steroid precursors and metabolites and gonadal and adrenocortical steroids. *J. Exp. Zool.* **178**, 467–477.

Goswami, S. V., and Sundararaj, B. I. (1973). Effect of actinomycin D, mitomycin C, puromycin and cycloheximide on desoxycorticosterone-induced *in vitro* maturation in oocytes of the catfish, *Heteropenustes fossilis* (Bloch). *J. Exp. Zool.* **185**, 327–332.

Goswami, S. V., and Sundararaj, B. I. (1974). Effects of C_{18}, C_{19} and C_{21} steroids on *in vitro* maturation of oocytes of the catfish, *Heteropneustes fossilis* (Bloch). *Gen. Comp. Endocrinol.* **23**, 282–285.

Goswami, S. V., Sundararaj, B. I., and Donaldson, E. M. (1974). *In vitro* maturation response of oocytes of the catfish *Heteropneustes fossilis* (Bloch) to salmon gonadotropin in ovary-head kidney co-culture. *Can. J. Zool.* **52**, 745–748.

Hirose, K. (1971). Biological study on ovulation *in vitro* of fish. I. Effects of pituitary and chorionic gonadotropins on ovulation *in vitro* of medaka, *Oryzias latipes*. *Bull. Jpn. Soc. Sci. Fish.* **37**, 585–591.

Hirose, K. (1972a). Biological study on ovulation *in vitro* of fish. IV. Induction of *in vitro* ovulation in *Oryzias latipes* oocyte using steroids. *Bull. Jpn. Soc. Sci. Fish.* **38**, 457–461.

Hirose, K. (1972b). Effectiveness of duration of exposure to hydrocortisone and HCG on ovulation *in vitro* in *Oryzias latipes*. *Bull. Jpn. Soc. Sci. Fish.* **38**, 869.

Hirose, K. (1972c). Biological study on ovulation *in vitro* of fish. V. Induction of *in vitro* ovulation in *Oryzias latipes* oocytes removed from their follicular tissues. *Bull. Jpn. Soc. Sci. Fish.* **38**, 1091–1096.

Hirose, K. (1973). Biological study on ovulation *in vitro* of fish. VI. Effects of metopirone (SU-4885) on salmon gonadotropin and cortisol-induced *in vitro* ovulation in *Oryzias latipes*. *Bull. Jpn. Soc. Sci. Fish.* **39**, 765–769.

Hirose, K. (1976). Endocrine control of ovulation in medaka (*Oryzias latipes*) and ayu (*Plecoglossus altivelis*). *J. Fish. Res. Board Can.* **33**, 989–994.

Hirose, K., and Donaldson, E. M. (1972). Biological study on ovulation *in vitro* of fish. III. The induction of *in vitro* ovulation of *Oryzias latipes* oocytes using salmon pituitary gonadotropin. *Bull. Jpn. Soc. Sci. Fish.* **38**, 97–100.

Hirose, K. (1973). Biological study on ovulation *in vitro* of fish. VI. Effects of metopirone (SU-4885) on salmon gonadotropin and cortisol-induced *in vitro* ovulation in *Oryzias latipes*. *Bull. Jpn. Soc. Sci. Fish* **39**, 765–769.

Hitz, C. R., and DeLacy, A. C. (1965). Clearing of yolk in eggs of the rockfishes; *Sebastodes caurinus* and *S. auriculatus*. *Trans. Am. Fish. Soc.* **94**, 194–195.

Idler, D. R., Ronald, A. P., and Schmidt, P. J. (1959). Biochemical studies on sockeye salmon during spawning migration. VII. Steroid hormones in plasma. *Can. J. Biochem. Physiol.* **37**, 1227–1238.

Iwamatsu, T. (1965). On fertilizability of pre-ovulation eggs in the medaka, *Oryzias latipes*. *Embryologia* **8**, 327–336.

Iwamatsu, T. (1973). Studies on oocyte maturation in the medaka, *Oryzias latipes*. Improvement of culture medium for oocytes *in vitro*. *Jpn. J. Ichthyol.* **20**, 218–224.

Iwamatsu, T. (1974). Studies on oocyte maturation of the medaka, *Oryzias latipes*. II. Effects of several steroids and calcium ions and the role of follicle cells on *in vitro* maturation. *Annot. Zool. Jpn.* **47**, 30–42.

Iwamatsu, T. (1978a). Studies on oocyte maturation of the medaka. *Oryzias latipes*. V. On the structure of steroids that induce maturation *in vitro*. *J. Exp. Zool.* **204**, 401–408.

Iwamatsu, T. (1978b). Studies on oocyte maturation of the medaka, *Oryzias latipes*. VI. Relationship between the circadian cycle of oocyte maturation and activity of the pituitary gland. *J. Exp. Zool.* **206**, 355–364.

Iwamatsu, T. (1980). Studies on oocyte maturation of the medaka, *Oryzias latipes*. VIII. Role of follicular constituents in gonadotropin- and steroid-induced maturation of oocytes *in vitro*. *J. Exp. Zool.* **211**, 231–239.

Iwamatsu, T., and Katoh, T. (1978). Maturation *in vitro* of oocytes of the loach, *Misgurnus anguillicaudatus* by steroid hormones and gonadotropins. *Annot. Zool. Jpn.* **51**, 79–89.

Iwamatsu, T., Ohta, T., Nakayama, N., and Shoji, H. (1976). Studies of oocyte maturation of the medaka, *Oryzias latipes*. *Annot. Zool. Jpn.* **49**, 28–37.

Jalabert, B. (1975). Modulation par différents stéroides non maturants de l'efficacité de la 17α-hydroxy-20β-dihydroprogestérone ou d'un extrait gonadotrope sur la maturation intra-folliculaire *in vitro* des ovocytes de la Truite arc-en-ciel *Salmo gairdnerii*. *C.R. Hebd. Seances Acad. Sci.* **281**, 811–814.

Jalabert, B. (1976). *In vitro* oocyte maturation and ovulation in rainbow trout (*Salmo gairdneri*), northern pike (*Esox lucius*) and goldfish (*Carassius auratus*). *J. Fish. Res. Board Can.* **33**, 974–988.

Jalabert, B. (1978). Production of fertilizable oocytes from follicles of rainbow trout (*Salmo gairdnerii*) following *in vitro* maturation and ovulation. *Ann. Biol. Anim., Biochim., Biophys.* **18**, 461–470.

Jalabert, B., and Breton, B. (1974). *In vitro* maturation of pike (*Esox lucius*) oocytes. *Gen. Comp. Endocrinol.* **22**, 391 (abstr.).

Jalabert, B., and Breton, B. (1980). Evolution de la gonadotropine plasmatique t-GtH après l'ovulation chez la Truite arc-en-ciel (*Salmo gairdneri* R.) et influence de la rétention des ovules. *C. R. Hebd. Seances Acad. Sci.* **290**, 799–801.

Jalabert, B., and Szöllösi, D. (1975). *In vitro* ovulation of trout oocytes: Effects of prostaglandins on smooth muscle-like cells of the theca. *Prostaglandins* **9**, 765–778.

Jalabert, B., Breton, B., and Bry, C. (1972). Maturation et ovulation *in vitro* des ovocytes de la Truite Arc-en-ciel *Salmo gairdnerii*. *C. R. Hebd. Seances Acad. Sci.* **275**, 1139–1142.

Jalabert, B., Bry, C. Szöllösi, D., and Fostier, A. (1973). Etude comparée de l'action des hormones hypophysaires et stéroides sur la maturation *in vitro* des ovocytes de la truite et du carassin (poissons Téléostéens). *Ann. Biol. Anim., Biochim., Biophys.* **13**, hors-ser., 59–72.

Jalabert, B., Breton, B., and Billard, R. (1974). Dosage biologique des hormones gonadotropes de poissons par le test de maturation *in vitro* des ovocytes de Truite. *Ann. Biol. Anim., Biochim., Biophys.* **14**, 217–228.

Jalabert, B., Bry, C., Breton, B., and Campbell, C. (1976). Action de la 17α-hydroxy-20β dihydroprogestérone et de la progestérone sur la maturation et l'ovulation *in vivo* et sur le niveau d'hormone gonadotrope plasmatique t-GtH chez la Truite Arc-en-ciel *Salmo gairdneri*. *C. R. Seances Hebd. Acad. Sci.* **283**, 1205–1208.

Jalabert, B., Breton, B., Brzuska, E., Fostier, A., and Wieniawski, J. (1977). A new tool for induced spawning: The use of 17α-hydroxy-20βdihydroprogesterone to spawn carp at low temperature. *Aquaculture* **10**, 353–364.

Jalabert, B., Goetz, F. W., Breton, B., Fostier, A., and Donaldson, E. M. (1978a). Precocious induction of oocyte maturation and ovulation in coho salmon, *Oncorhynchus kisutch. J. Fish. Res. Board Can.* **35**, 1423–1429.

Jalabert, B., Breton, B., and Fostier, A. (1978b). Precocious inducation of oocyte maturation and ovulation in rainbow trout (*Salmo gairdneri*): Problems when using 17α-hydroxy-20β-dihydroprogesterone. *Ann. Biol. Anim., Biochim., Biophys.* **18**, 977–984.

Jalabert, B., Breton, B., and Bry, C. (1980). Evolution de la gonadotropine plasmatique t-GtH apres synchronisation des ovulations par injection de 17α-hydroxy-20β dihydroprogestérone chez la Truite arc-en-ciel (*Salmo gairdneri* R.). *C. R. Hebd. Seances Acad. Sci.* **290**, 1431–1434.

Kagawa, H., and Nagahama, Y. (1981). *In vitro* effects of prostaglandins on ovulation in goldfish *Carassius auratus. Bull. Jpn. Soc. Sci. Fish.* **47**, 1119–1121.

Kagawa, H., Takano, K., and Nagahama, Y. (1981). Corelation of plasma estradiol-17β and progesterone levels with ultrastructure and histochemistry of ovarian follicles in the white-spotted char, *Salvelinus leucomaenis. Cell Tissue Res.* **218**, 315–329.

Kagawa, H., Young, G., and Nagahama, Y. (1982). Estradiol-17β production in isolated amago salmon (*Oncorhynchus rhodurus*) ovarian follicles and its stimulation by gonadotropins. *Gen. Comp. Endocrinol.* **47**, 361–365.

Kanatani, H., and Nagahama, Y. (1980). Mediators of oocyte maturation. *Biomed. Res.* **1**, 273–291.

Kapur, K., and Toor, H. S. (1979). The effect of clomiphene citrate on ovulation and spawning in indomethacin treated carp, *Cyprinus carpio. J. Fish Biol.* **14**, 59–66.

Kirshenblat, Y. D. (1959). The action of cortisone on the ovaries of the loach. *Bull. Eskp. Biol. Med.* **47**, 503–507.

Kuehl, F. A., Jr., Cirillo, B. J., and Oien, H. G. (1976). Prostaglandin-cyclic nucleotide interactions in mammalian tissues. *In* "Prostaglandins: Chemical and Biochemical Aspects" (S. M. M. Karim, ed.), pp. 191–225. Univ. Park Press, Baltimore, Maryland.

Lambert, J. G. D., and van Bohemen, C. G. (1978). Steroidogenesis in the ovary of the rainbow trout, *Salmo gairdneri*. *Satellite Symp. Int. Congr. Horm. Steroids, 5th, 1978*, Abstracts, p. 4.

Lambert, J. G. D., and van Oordt, P. G. W. (1975). Steroid transformations *in vitro* by the ovary of the zebrafish, *Brachydanio rerio*. *J. Endocrinol.* **64**, 73p (abstr.).

Mackay, N. J. (1975). The reproductive cycle of the firetail gudgeon, *Hypseleotris galii* IV. Hormonal control of ovulation. *Aus. J. Zool.* **23**, 43–47.

Mai, J., Goswami, S. K., Bruckner, G., and Kinsella, J. E. (1981). A new prostaglandin, C22-$PGF_{2\alpha}$, synthesized from docosahexaenoic acid (C22:6n3) by trout gill. *Prostaglandins* **21**, 691–698.

Malservisi, A., and Magnin, E. (1968). Changments cycliques annuels se produisant dans les ovaries de *Perca fluviatilis flavescens (Mitchill) de la région de Montréal*. *Nat. Can.* **95**, 929–945.

Masui, Y., and Clarke, H. J. (1979). Oocyte maturation. *Int. Rev. Cytol.* **57**, 185–282.

Morimoto, K., Okamura, H., Kanzaki, H., Okuda, Y., Takenaka, A., and Nishimura, T. (1981). The adrenergic nerve supply to bovine ovarian follicles. *Int. J. Fertil.* **26**, 14–19.

Morrill, G. A., Schatz, F., Kostellow, A. B., and Poupko, J. M. (1977). Changes in cyclic AMP levels in the amphibian ovarian follicle following progesterone induction of meiotic maturation. *Differentiation* **8**, 97–104.

Moser, H. G. (1967). Seasonal histological changes in the gonads of *Sebastodes paucispinis* Ayres, an ovoviviparous teleost (Family Scorpaenidae). *J. Morphol.* **123**, 329–354.

Mulner, O., Huchon, D., Thibier, C., and Ozon, R. (1979). Cyclic AMP synthesis in *Xenopus laevis* oocytes, inhibition by progesterone. *Biochim. Biophys. Acta* **582**, 179–184.

Nagahama, Y., Chan, K., and Hoar, W. S. (1976). Histochemistry and ultrastructure of pre- and post-ovulatory follicles in the ovary of the goldfish, *Carassius auratus*. *Can. J. Zool.* **54**, 1128–1139.

Nagahama, Y., Kagawa, H., and Tashiro, F. (1980). The *in vitro* effects of various gonadotropins and steroid hormones on oocyte maturation in amago salmon *Oncorhynchus rhodurus* and rainbow trout *Salmo gairdneri*. *Bull. Jpn. Soc. Sci. Fish.* **46**, 1097–1102.

Nagahama, Y., Hirose, K., Young, G., Aduchi, S., Suzuki, K., and Tamaoki, B. (1983). Relative *in vitro* effectiveness of 17α,20β-dihydroxy-4-pregnen-3-one and other pregnene derivatives on germinal vesicle breakdown in oocytes of four species of teleosts, ayu (*Plecoglossus altivelis*), amago salmon (*Oncorhynchus rhodurus*), rainbow trout (*Salmo gairdneri*) and goldfish (*Carassius auratus*). *Gen. Comp. Endocrinol.* (in press).

Ogata, H., Nomura, T., and Hata, M. (1979). Prostaglandin $F_{2\alpha}$ changes induced by ovulatory stimuli in the pond loach, *Misgurnus anguillicaudatus*. *Bull. Jpn. Soc. Sci. Fish.* **45**, 929–931.

Osanai, K., Hirai, S., and Sato, R. (1973). *In vitro* induction of oocyte maturation in the chum-salmon. *Zool. Mag.* **82**, 68–71.

Oshiro, T., and Hibiya, T. (1975). Presence of ovulation-inducing enzymes in the ovarian follicles of loach. *Bull. Jpn. Soc. Sci. Fish.* **41**, 115.

Pandey, S., and Hoar, W. S. (1972). Induction of ovulation in goldfish by clomiphene citrate. *Can. J. Zool.* **50**, 1679–1680.

Pendergrass, P., and Schroeder, P. (1976). The ultrastructure of the thecal cell of the teleost, *Oryzias latipes*, during ovulation *in vitro*. *J. Reprod. Fertil.* **47**, 229–233.

Peter, R. E., and Billard, R. (1976). Effect of third ventricle injection of prostaglandins on gonadotropin secretion in goldfish, *Carassius auratus*. *Gen. Comp. Endocrinol.* **30**, 451–456.

Raven, C. P. (1961). "Oogenesis: The Storage of Developmental Information." p. 164. Pergamon, Oxford.

Sakun, O. F. (1966). Transition to meiotic division of oocytes with incomplete vitellogenesis under the influence of hormonal stimulation in the Atlantic salmon *Salmo salar* L. *Dokl. Akad. Nauk SSSR* **169**, 241–244.

Sakun, O. F., and Gureeva-Preobrazhenskaya, E. (1975). A study of the maturation of the oocytes of teleostei *in vitro*, incubation methods and ways of analyzing the results. *Vestn. Leningr. Univ., Biol.* **15**, 15–22.

Schackley, S. E., and King, P. E. (1977). Oögenesis in a marine teleost, *Blennius pholis* L. *Cell Tissue Res.* **181**, 105–128.

Schmidt, P. J., and Idler, D. R. (1962). Steroid hormones in the plasma of salmon at various stages of maturation. *Gen. Comp. Endocrinol.* **2**, 204–214.

Schroeder, P. C., and Pendergrass, P. (1976). The inhibition of in vitro ovulation from follicles of the teleost, *Oryzias latipes*, by cytochalasin B. *J. Reprod. Fertil.* **48**, 327–330.

Schuetz, A. W. (1972). Estrogens and ovarian follicular functions in *Rana pipiens*. *Gen. Comp. Endocrinol.* **18**, 32–36.

Scott, A. P., Sheldrick, E. L., and Flint, A. P. F. (1982). Measurement of 17α,20β-dihydroxy-4-pregnen-3-one in plasma of trout (*Salmo gairdneri* Richardson): Seasonal changes and response to salmon pituitary extract. *Gen. Comp. Endocrinol.* **46**, 444–451.

Singh, A. K., and Singh, T. P. (1976). Effect of clomid, sexovid and prostaglandins on induction of ovulation and gonadotropin secretion in a freshwater catfish, *Heteropneustes fossilis* (Bloch). *Endokrinologie* **68**, 129–136.

Sower, S. A. (1980). Sexual maturation of coho salmon (*Oncorhynchus kisutch*): induced ovulation, *in vitro* induction of final maturation and ovulation, and serum hormone and ion levels of salmon in seawater and fresh water. Ph.D. Thesis, Oregon State University, Corvallis.

Spiegel, J., Jones, E., and Snyder, B. W. (1978). Estradiol-17β interference with meiotic maturation in *Rana pipiens* ovarian follicles: Evidence for the inhibition of 3β-hydroxysteroid dehydrogenase. *J. Exp. Zool.* **204**, 187–192.

Stacey, N. E., and Goetz, F. W. (1982). Role of prostaglandins in fish reproduction. *Can. J. Fish. Aquat. Sci.* **39**, 92–98.

Stacey, N. E., and Pandey, S. (1975). Effects of indomethacin and prostaglandins on ovulation of goldfish. *Prostaglandins* **9**, 597–607.

Stacey, N. E., Cook, A. F., and Peter, R. E. (1979). Ovulatory surge of gonadotropin in the goldfish, *Carassius auratus*. *Gen. Comp. Endocrinol.* **37**, 246–249.

Stefenson, A., Owman, C., Sjöberg, N.-O., Sporrong, B., and Walles, B. (1981). Comparative study of the autonomic innervation of the mammalian ovary, with particular regard to the follicular system. *Cell Tissue Res.* **215**, 47–62.

Stevens, R. E. (1966). Hormone-induced spawning of striped bass for reservoir stocking. *Prog. Fish-Cult.* **28**, 19–28.

Strickland, S., and Beers, W. H. (1979). Studies on the enzymatic basis and hormonal control of ovulation. *In* "Ovarian Follicular Development and Function" (A. R. Midgley and W. A. Sadler, eds.), pp. 143–153. Raven Press, New York.

Sundararaj, B. I., and Goswami, S. V. (1966). Effects of mammalian hypophysial hormones, placental gonadotropins, gonadal hormones, and adrenal corticosteroids on ovulation and spawning in hypophysectomized catfish, *Heteropneustes fossilis* (Bloch). *J. Exp. Zool.* **161**, 287–296.

Sundararaj, B. I., and Goswami, S. (1969). Role of interrenal in luteinizing hormone-induced ovulation and spawning in the catfish *Heteropneustes fossilis* (Bloch). *Gen. Comp. Endocrinol., Suppl.* **2**, 374–384.

Sundararaj, B. I., and Goswami, S. V. (1971a). *In vivo* and *in vitro* induction of maturation and ovulation in oocytes of the catfish, *Heteropneustes fossilis* (Bloch) with protein and steroid hormones. *Int. Congr. Ser.—Excerpta Med.* **219**, 966–975.

Sundararaj, B. I., and Goswami, S. V. (1971b). Effects of desoxycorticosterone and hydrocortisone singly and in various combinations on *in vitro* maturation of oocytes of the catfish, *Heteropneustes fossilis* (Bloch). *Gen. Comp. Endocrinol.* **17**, 570–573.

Sundararaj, B. I., and Goswami, S. V. (1974a). Comparative aspects of oocyte maturation in submammalian vertebrates: A review. *Proc. Natl. Inst. Sci. India, Part B* **39**, 286–295.

Sundararaj, B. I., and Goswami, S. V. (1974b). Effects of ovine luteinizing hormone and procine adrenocorticotropin on maturation of oocytes of the catfish, *Heteropneustes fossilis* (Bloch), in ovary-interrenal coculture. *Gen. Comp. Endocrinol.* **23**, 276–281.

Sundararaj, B. I., and Goswami, S. V. (1977). Hormonal regulation of *in vivo* and *in vitro* oocyte maturation in the catfish, *Heteropneustes* fossilis (Bloch). *Gen. Comp. Endocrinol.* **32**, 17–28.

Sundararaj, B. I., Goswami, S. V., and Donaldson, E. M. (1972). Effect of salmon gonadotropin on *in vitro* maturation of oocytes of a catfish, *Heteropneustes fossilis*. *J. Fish. Res. Board Can.* **29**, 435–437.

Sundararaj, B. I., Goswami, S. V., and Virenderjeet (1978). Some aspects of ovum maturation in catfish. *Abstr. Int. Cong. Hormo. Steroids, 5th, 1978* p. 95.

Suzuki, K., Tamaoki, B., and Hirose, K. (1981a). *In vitro* metabolism of 4-pregnenes in ovaries of a freshwater teleost, the ayu (*Plecoglossus altivelis*): Production of $17\alpha,20\beta$-dihydroxy-4-pregnen-3-one and its 5β-reduced metabolites, and activation of 3β- and 20β-hydroxysteroid dehydrogenases by treatment with a fish gonadotropin. *Gen. Comp. Endocrinol.* **45**, 473–481.

Suzuki, K., Nagahama, Y., and Tamaoki, B. (1981b). *In vitro* synthesis of an inducer for germinal vesicle breakdown of fish oocytes, $17\alpha,20\beta$-dihydroxy-4-pregnen-3-one by ovarian tissue preparation of amago salmon (*Oncorhynchus rhodurus*). *Gen. Comp. Endocrinol.* **45**, 533–535.

Szöllösi, D., and Jalabert, B. (1974). La thèque de follicule ovarien de la Truite. *J. Micros. (Paris)* **20**, 92 (abstr.).

Szöllösi, D., Jalabert, B., and Breton, B. (1978). Postovulatory changes in the theca folliculi of the trout. *Ann. Biol. Anim., Biochim., Biophys.* **18**, 883–891.

Tesone, M., and Charreau, E. H. (1980). Steroid biosynthesis in the gonads of the teleost fish, *Jenynsia lineata*. *Comp. Biochem. Physiol. B* **65B**, 631–637.

Theofan, G. (1981). The *in vitro* synthesis of final maturational steroids by ovaries of brook trout (*Salvelinus fontinalis*) and yellow perch (*Perca flavescens*). Ph.D. Thesis, University of Notre Dame, Notre Dame, Indiana.

Theofan, G., and Goetz, F. W. (1981). The *in vitro* effects of transcriptional and translational protein synthesis inhibitors on final maturation and ovulation of yellow perch (*Perca flavescens*) oocytes. *Comp. Biochem. Physiol. A* **69A**, 557–561.

Theofan, G., and Goetz, F. W. (1983). The *in vitro* synthesis of final maturational steroids by ovaries of brook trout (*Salvelinus fontinalis*) and yellow perch (*Perca flavescens*). *Gen. Comp. Endocrinol.* **51**, 84–95.

Truscott, B., Idler, D. R., and Sundararaj, B. I., and Goswami, S. V. (1978). Effects of gonadotropins and adrenocorticotropin on plasmatic steroids of the catfish, *Heteropneustes fossilis* (Bloch). *Gen. Comp. Endocrinol.* **34**, 149–157.

Ueda, H., and Takahashi, H. (1976). Acceleration of ovulation in the loach, *Misgurnus anguillicaudatus* by the treatment with clomiphene citrate. *Bull. Fac. Fish., Hokkaido Univ.* **27**, 1–5.

Ungar, F., Gunville, R., Sundararaj, B. I., and Goswami, S. V. (1977). Formation of 3α-hydroxy-5β-pregnan-20-one in the ovaries of catfish, *Heteropneustes fossilis* (Bloch). *Gen. Comp. Endocrinol.* **31**, 53–59

Van Ree, G. E., Lok, D., and Bosman, G. (1977). *In vitro* induction of nuclear breakdown in oocytes of the zebrafish *Brachydanio rerio* (Ham. Buch.). Effects of the composition of the medium and of protein and steroid hormones. *Proc. K. Ned. Akad. Wet., Ser. C* **80**, 353–371.

Wallace, R. A., and Selman, K. (1978). Oogenesis in *Fundulus heteroclitus* I. Preliminary observations on oocyte maturation *in vivo* and *in vitro*. *Dev. Biol.* **62**, 354–369.

Wallace, R. A., and Selman, K. (1979). Physiological aspects of oogenesis in two species of sticklebacks, *Gasterosteus aculeatus* L. and *Apeltes quadracus* (Mitchill). *J. Fish Biol.* **14**, 551–564.

Wallace, R. A., and Selman, K. (1980). Oogenesis in *Fundulus heteroclitus* II. The transition from vitellogenesis into maturation. *Gen. Comp. Endocrinol.* **42**, 345–354.

Wallace, R. A., and Selman, K. (1981). Cellular and dynamic aspects of oocyte growth in teleosts. *Am. Zool.* **21**, 325–343.

Weihing, R. R. (1978). *In vitro* interactions of cytochalasins with contractile proteins. *In* "Cytochalasins—Biochemical and Cell Biological Aspects" (S. W. Tanenbaum, ed.), pp. 431–444. Elsevier/North-Holland Biomedical Press, New York.

Wingfield, J. C., and Grimm, A. S. (1977). Seasonal changes in plasma cortisol, testosterone and oestradiol-17β in the plaice, *Pleuronectes platessa* L. *Gen. Comp. Endocrinol.* **31**, 1–11.

Wright, R. S., and Hunt, S. M. V. (1982). A radioimmunoassay for 17α,20β-dihydroxy-4-pregnen-3-one: Its use in measuring changes in serum levels at ovulation in Atlantic salmon (*Salmo salar*), coho salmon (*Oncorhynchus kisutch*), and rainbow trout (*Salmo gairdneri*). *Gen. Comp. Endocrinol.* **47**, 475–482.

Yamamoto, K. (1956a). Studies on the formation of fish eggs. I. Annual cycle in the development of ovarian eggs in the flounder, *Liopsetta obscura. J. Fac. Sci., Hokkaido Univ. Ser. 6* **12**, 362–373.

Yamamoto, K. (1956b). Studies on the formation of fish eggs. II. Changes in the nucleus of the oocyte of *Liopsetta obscura*, with special reference to the activity of the nucleolus. *J. Fac. Sci., Hokkaido Univ., Ser. 6* **12**, 375–390.

Yamamoto, K. (1958). Studies on the formation of fish eggs. XII. On the non-massed yolk in the egg of the herring, *Clupea pallasii. Bull. Fac. Fish., Hokkaido Univ.* **8**, 270–277.

Yamamoto, T. S. (1963). Eggs and ovaries of the stickleback, *Pungitius tymensis*, with a note on the formation of a jelly-like substance surrounding the egg. *J. Fac. Sci., Hokkaido Univ., Ser. 6* **15**, 190–199.

Yamauchi, K., and Yamamoto, K. (1974). *In vitro* maturation of Japanese eel eggs and early development of the egg. *Bull. Jpn. Soc. Sci. Fish.* **40**, 153–157.

Yamazaki, F. (1965). Endocrinological studies on the reproduction of the female goldfish, *Carassius auratus* L., with special reference to the function of the pituitary gland. *Mem. Fac. Fish., Hokkaido Univ.* **13**, 1–64.

Yorke, M. A., and McMillan, D. B. (1980). Structural aspects of ovulation in the lamprey, *Petromyzon marinus. Biol. Reprod.* **22**, 897–912.

Young, G., Kagawa, H., and Nagahama, Y. (1983a). Oocyte maturation in the amago salmon (*Oncorhynchus rhodurus*): *In vitro* effects of salmon gonadotropin, steroids and cyanoketone, an inhibitor of 3β-hydroxy-δ⁵-steroid dehydrogenase. *J. Exp. Zool.* (in press).

Young, G., Crim, L. W., Kagawa, H., Kambegawa, A., and Nagahama, Y. (1983b). Plasma 17α,20β-dihydroxy-4-pregnen-3-one levels during sexual maturation of amago salmon (*Oncorhynchus rhodurus*): Correlation with plasma gonadotropin and *in vitro* production by ovarian follicles. *Gen. Comp. Endocrinol.* (in press).

4

SEX CONTROL AND SEX REVERSAL IN FISH UNDER NATURAL CONDITIONS

S. T. H. CHAN AND W. S. B. YEUNG

Department of Zoology
University of Hong Kong
Hong Kong

I. INTRODUCTION

Sex control in fishes is invariably related to reproduction and sex differentiation. These two subjects were discussed in Volume III of this series and

FISH PHYSIOLOGY, VOL. IXB

one should refer to these earlier reviews (Hoar, 1969; Yamamoto, 1969) for a broader background of the subject.

With regard to the expression of sexuality, Pisces, with over 20,000 species in highly varied ecological habitats, is undoubtedly the most diversified group among vertebrates. Although a vast majority of species are gonochorists, hermaphrodites, including sex reversed individuals, are not uncommon. Among the various piscean groups, the superorder Teleostei is unique in having functional sex reversal occurring as a natural phenomenon. In many species of some families, hermaphroditism and sex reversal constitute the normal mode of reproduction. The studies of sex control and of natural sex-reversal mechanism(s) in fishes are not only of academic importance to the understanding of sex determination and sex differentiation, but also of economic value in aquaculture and fish farming.

The genetic and physiological duality of the reproductive cells forms the fundamental basis of sexuality in living organisms (Witschi and Opitz, 1963). In vertebrates, sexual dimorphism is further exhibited by the gonadal sex, body sex, and behavioral sex. The evolution of complex sexual systems is undoubtedly the result of natural selection, and the emergence of certain sexual traits is often affected by environmental factors and serves to enhance reproductive success of the species. In nature, the control of sex is primarily genetic, and the expression of the sex gene(s) first manifests itself in the primary sex organs. The differentiated gonads, notably the testes, in turn secrete steroids (Jost, 1965) and nonsteroidal substances, such as the anti-Müllerian hormone (AMH) (Picard et al., 1978), that control the differentiation of the body sex and the behavioral sex. Although the primary determinants of sex reside in the genome of the individual, the transformation from genotypic sex to phenotypic sex is accomplished only by biochemical processes, which are susceptible to environmental influences. Therefore, sex control in fishes, as in most other organisms, is governed by both genetic (intrinsic) and environmental (extrinsic) factors. These factors, and their related events, are analyzed in this discussion. On the subject of sex control and sex reversal in fish, comparison and reference is invariably made to both gonochoristic and hermaphroditic species, although the latter deserve more emphasis. It should also be stressed at this juncture that natural sex reversal should not be confused with experimental sex reversal. The former refers to sex reversal that occurs as a spontaneous process under natural conditions in a hermaphroditic species, and natural sex reversals in teleosts constitute various patterns of hermaphroditism which function as successfully as gonochorism. The latter usually involves artificial manipulations of the embryonic sex differentiation of a normally gonochoristic species, resulting in a phenotypic sex disharmonious with the genotypic sex. This category of experimental sex reversal is, strictly speaking, not within the scope of the present discussion. However, discussions of experimental sex reversals of gonochoristic fish cannot be completely omitted and are briefly considered. In fact, it is mainly through information derived from experimental manipulations of sex differentiation and of sex reversal of both gonochoristic and hermaphroditic animals that researchers in this field are able to broaden our knowledge and to gain an insight into the mechanism(s) of sex control in nature.

Although a detailed account of the terminology of hermaphroditism and sex reversal has been given by Atz (1964), a few points require amplification. Sex reversal, the transformation of an individual from one sex to another, is defined by Atz (1964) as change "from the possession of recognizable ovarian tissue to that of testicular tissue, or vice versa." The broad meaning of this term, even after the addition of self-explanatory modifier such as "functional" sex reversal, does not distinguish (artificial) sex reversal of gonochorists with a functional phenotypic sex opposite to their genotypic sex from (natural) sex reversal of consecutive hermaphrodites which function as both sexes in temporal sequence during their life cycle. Reinboth (1970) proposed the abandonment of the term "sex reversal" in his discussion of "ambosexual" (intersexual) fishes arguing that this term "may imply the meaning of reverting to a primary sex condition"; he prefers the more neutral term "sex inversion," which Atz (1964) defined as "the acquisition by an individual belonging to one sex of characteristics similar to those of the opposite sex, but not including recognizable gonadal tissue of that sex." Therefore, the same term is used quite differently by these two researchers. Chan (1970), Harrington (1975), and Chan et al. (1975, 1977) introduced the term "sex succession" to define the temporal nature of sex reversal occurring in consecutive hermaphrodites. The term sex succession is undoubtedly more appropriate than sex reversal in those hermaphroditic species where heterosexual areas and tissues coexist before sexual transformation; the change of sex in these instances is, in fact, an ontogenetic event whereby the preexisting male and female tissue undergo a set pattern of successive sex maturation. Nevertheless, in the present chapter, the term sex reversal is used because it is still unclear whether in some species natural sex reversal, in fact, represents a successive expression of the male and female sex of a genotype that is primarily consecutive hermaphroditic, or a phenotypic reversal of what is originally a gonochoristic (female) genotype, or some other spontaneous biological phenomenon yet to be discovered. The terms *experimental* sex reversal, *natural* sex reversal, *functional*, and *nonfunctional* are used when appropriate. The use of the terms "*hermaphroditism*" and "intersexuality," with no restrictive distinction between them, are also preferred to "*ambosexuality*" or "*amphisexuality*", because the former two are by far terms of general usage in the field of reproductive physiology.

II. SEX PATTERNS IN FISHES

A. Gonochorism, Hermaphroditism, and Sex Reversal

Fishes exhibit various patterns of sex ontogeny by which the gonads become functional ovaries or testes. Apart from some teleosts, most fishes are gonochorists, and hermaphroditism or sex reversal is rare in the elasmobranchs, the ganoids, and the lungfishes. Even in the gonochoristic teleosts, there exist marked variations in the mode of sex differentiation and sex determination. Yamamoto (1969) described "differentiated" and "undifferentiated" species of gonochoristic fishes according to the stability of sexuality and the occurrence of spontaneous intersexes and sex reversal. It is generally assumed that the diverse sex pattern in the teleosts is related to the lack of cortical and medullary organization in the embryonic gonad, which is well established in the elasmobranchs, amphibians, and amniotes. Further, the sexually indifferent period in some teleost species is relatively prolonged and sex differentiation is often a protracted process. In the European eel, *Anguilla anguilla*, the indifferent period may last for several years. Juvenile intersexuality has been found both in male and female individuals in the yellow eel phase; although oogonial cysts and solitary oocytes are found together with spermatozoa mounted in interstitial cell cords, the animal cannot reproduce (Bertin, 1958; Kuhlmann, 1975). The final sexual differentiation to gonochoristic forms involves the maturation of either the male or the female elements accompanied by the degeneration of the opposite sex components. Gonochorists with juvenile and rudimentary hermaphroditism have also been reported in a number of fishes, e.g., *Salmo gairdneri* (Mřsić, 1923), *Macropodus concolor* and *M. opercularis* (Schwier, 1939), and *Brachydanio rerio* (Takahashi, 1977). In these species, those individuals that are destined to become males apparently pass through a nonfunctional female phase and then an intersexual phase during their juvenile period.

A great variety of functional hermaphroditism, both normal and abnormal, has been found in various taxonomic groups of teleosts. Atz (1964) provided the most complete review of the subject. According to Atz (1964), normal hermaphroditism refers to those that exist "in a uniform way at some time during the ontogeny of all or many members of a species" and accordingly all other forms of hermaphroditism would be abnormal. Abnormal hermaphroditism does not always imply sterility, and some reported abnormal hermaphrodites are in fact functional. A few of the female offsprings obtained from "virgin birth" of the normally gonochoristic guppy *Poecilia reticulata* produced fatherless broods (Spurway, 1957), and testicular tissue was found in the ovaries of these "females" suggesting that they were self-fertilizing intersexes. Spermatozoa have been reported in the "ovotestis" of an hermaphroditic salmon (Fontaine and Vibert, 1950, quoted by Atz, 1964) and a perch (Jellyman, 1976). The eggs and sperms obtained from the hermaphroditic salmon could be artificially fertilized and developed into young fish.

Normal hermaphroditism, with or without sex reversal, is known to occur in a number of teleost orders. Protandrous hermaphroditism has been reported in Ostariophysi, Polynemidae, Scorpaeniformes (Platycephalidae), and probably in Stomiatiformes. Synchronous (simultaneous) hermaphroditism occurs in Aulopiformes and Antheriniformes, and protogynous hermaphroditism is found in Synbranchiformes. However, the most diverse display of functional hermaphroditism is found in the Perciformes. In this order, all types of sexuality known to occur in vertebrates can be found among the various subgroups. In Serranidae, all members of the genus *Serranus* and the related *Hypoplectrus* are synchronous hermaphrodites (Smith, 1975). Members of the *Epinephelus* (Serranidae), Labridae, Maenidae, and Scaridae are protogynous. Among the sparids, some members are protogynous hermaphrodites (e.g., *Taius tumifrons*, *Spondyliosoma cantharus*), others protandrous (e.g., *Sparus auratus*, *S. longispinis*, *Rhabdosargus sarba*), and still others exhibit all grades of rudimentary hermaphroditism (e.g., *Boops boops*, *Diplodus vulgaris*) ending with true gonochorists (e.g., *Dentex dentex*), in which only the embryonic development of the gonads suggests their link to an hermaphroditic origin. The recent surge of interest in coral fishes reveals the extensive occurrence of protandry and protogyny in Scaridae and Labridae (Robertson, 1972; Shen and Liu, 1976; Fricke and Fricke, 1977; Moyer and Nakazono, 1978a,b). It is apparently only a matter of time before more and more naturally occurring functional hermaphrodites will be discovered among teleosts. More detailed reviews on the occurrence of hermaphroditism in fishes can be found elsewhere (Atz, 1964; Reinboth, 1970; Smith, 1975).

B. Structural Organization of the Germ and Somatic Elements of the Gonads

The structural pattern of the gonads of hermaphroditic fishes varies according to the taxonomic group and the type of hermaphroditism found in the particular species. In general, the anatomy of the gonads is well described in Sparidae, Serranidae, Synbranchiformes, and a number of other groups.

In the sparids (e.g., *Rhabdosargus sarba*), a protandrous hermaphrodite, the gonads are paired ovotestes; each with the ovarian lobe in the middorsal region of the abdominal cavity and the testicular lobe as a band along the ventrolateral wall in the more posterior region of the gonad. As in other sparids and maenids, the two heterosexual regions are separated by connective tissue (see Fig. 1). The testicular lobes, composed of male germ cells and connective tissue, are a solid structure with lacunae along the ovarian wall and between the two sexual regions; these lacunae serve as "sperm ducts" (Fig. 1). In the ovarian lobes, the germinal tissues form ovarian lamellae projecting from the gonadal wall into the ovarian cavity into which ova are discharged (Fig. 2). In the female phase, the ovarian lobe expands with a large number of maturing oocytes and the testicular lobe regresses into a small remnant of connective tissue (Fig. 2). In the serranids (e.g., *Serranus, Chelidoperca, Rypticus*) and the pomacentrids (e.g., *Amphiprion*), there is no connective tissue between the testicular and ovarian tissues, both being in the same cavity of the gonad (Smith, 1975); the organization of the male and female gonoducts are similar to the patterns in the sparids (Atz, 1964). In *Epinephelus*, there is no distinct boundary segregating the male-and female tissues into heterosexual zones. The male and female germ cells are apparently mixed within the same gonadal lumen (Fig. 3). In other hermaphrodites, such as *Gonostoma* (Stomiatiformes) and *Monopterus* (Synbranchiformes), the gonadal structures are so dissimilar to the sparid or serranid pattern and that they can be considered unique. Thus in *Gonostoma*, a serial alternation of male and female zones is found in the cordlike germinal ridge (Kawaguchi and Marumo, 1967); in *Monopterus*, the testicular tissues arise from groups of cells along the inner edges of the two germinal cords which form two loops, the gonadal lamellae, in the central cavity of the gonad (Chan and Phillips, 1967).

1. GERM ELEMENTS

The coexistence of both sex tissues in the gonad of hermaphroditic species apparently provides a structural basis for sex reversal in fishes by sequential maturation of the male and female germinal elements (Chan, 1970, 1977), although the cause of and mechanism by which this coexistence of male and female gonia result remain to be elucidated (see Chan and O, 1981). In the protandrous sparids, the functional ovaries and the female germ cells derive from preexisting female zones which are formed together with the testicular zones very early during gonadal ontogeny (Zohar and Abraham, 1978, see also Fig. 1). Oocytes are present in the functional male phase, although most of them are latent and are blocked in the perinucleolus stage of development at this time. During sex reversal, the testicular region becomes gradually reduced as the development of the ovarian part advances until it reaches full maturity (Fig. 2). In some large males, in which sex

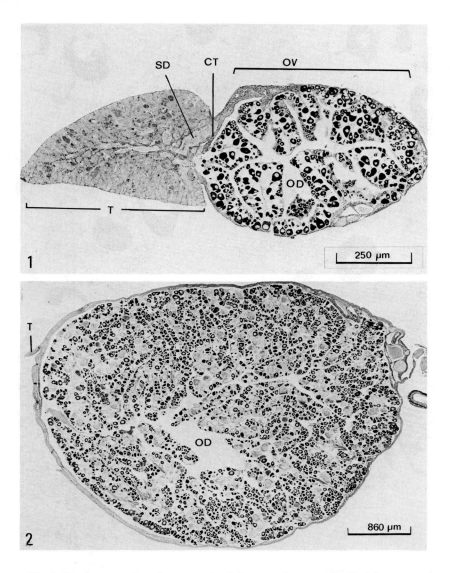

Fig. 1. Transverse section of an ovotestis of the protandrous sparid, *Rhabdosargus sarba*, showing the topographic segregation of the male and female tissue into heterosexual zones. The ovarian lobe (OV) contains ovarian lamellae projecting into the central cavity which serves as an "oviduct" (OD). The male lobe (T) is a solid structure composed of male germ cells and interstitial tissues. The lacunae next to the connective tissue (CT), which separates the heterosexual regions, form the "sperm duct" (SD).

Fig. 2. Transverse section through the gonad of *Rhabdosargus sarba* in the female phase showing the extensive ovarian development with oocytes at various stages of maturation. The remnant of the testicular lobe (T), now completely regressed, can still be discerned.

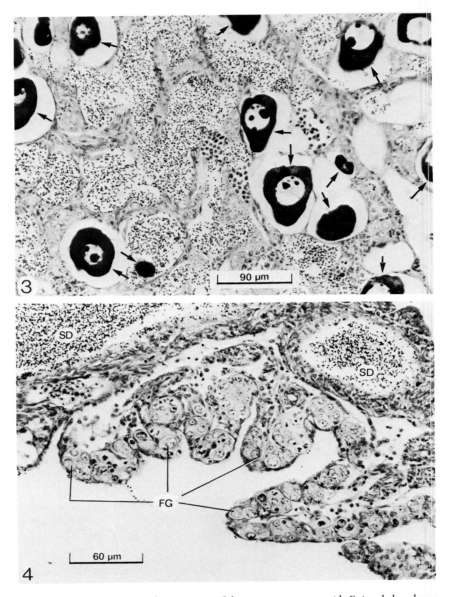

Fig. 3. Transverse section of an ovotestis of the protogynous serranid, *Epinephelus akaara*. The male and female germinal cells are intermingled in the gonad with no topographically distinct zones. The specimen is at the maturing-male phase when the male gonia undergo spermatogenesis in testicular lobules; the remnant of the female germinal cells (arrows) are scattered through the testicular tissues.

Fig. 4. Transverse section through the gonad of a large male *Rhabdosargus sarba*. The specimen is believed to have a prolonged protandrous phase and the female germ cells (FG) are found as cords of dormant cells bordering the sperm ducts (SD) of the testicular lobe. This indicates that male and female germ cells are predetermined before sex reversal.

reversal is believed to occur late or never during gonadal ontogeny, the female germ cells remain to be distinctly discernable as cords of dormant cells in the gonad (Fig. 4). In the protogynous *Monopterus*, dormant male germ cells are found in the inner region of the germinal cords, the gonadal lamellae, in the female phase (see Fig. 5); the oocytes mature along the outer edge of the same germinal cords (Chan and Phillips, 1967). A similar situation is reported in *Coris julis* (Labridae), where the origin of the testis can be traced back to particular cell groups attached to the wall of the gonad during the female phase (Reinboth, 1962a, 1970). In species where the gonad is composed of intermingled male and female germinal tissues, it is at present not possible to say whether a given type of germinal cell originates from some sexually predetermined gonia or from a stock of sexually bipotential gonocytes, because potential male and female gonocytes (or protogonia) do not have distinct differential cytological characteristic until they become differentiated into active oogonia or spermatogonia and enter active gametogenetic maturation. In some protogynous fishes such as *Thalassoma bifasciatum* (Smith, 1959, 1965; Reinboth, 1970) and *Anthias squamipinnis* (Shapiro, 1977), no trace of potential male tissue is identifiable until the onset of sex reversal when the ovary starts to degenerate and nests of male gonia within the ovarian tissue enter into the spermatogenetic cycle.

2. Somatic Elements

As in other vertebrates, somatic tissue of the gonad appears to be important in sex differentiation of fishes. Among gonochoristic species, it had been reported that somatic cells in the embryonic gonads are essential for the differentiation and morphogenesis of the testes in *Poecilia reticulata* (Takahashi, 1975). The stromal cells undergo active multiplication before the proliferation of spermatogonia and the formation of testicular lobules in the transformation of the gonad from transitory ovaries to testes during the rudimentary hermaphroditic phase in *Brachydanio rerio* and *Macropodus operculata* (Takahashi, 1977; Schwier, 1939). In the protogynous *Monopterus albus*, extensive development of the interstitial Leydig cells precedes the formation of testicular lobules during natural sex reversal (Chan and Phillips, 1967; Fig. 6). This proliferation of the Leydig cells is extensive in the interstitium of the germinal cords and highly positive in the histo-enzymological reaction for 3β-hydroxysteroid dehydrogenase; therefore, sex reversal is an event concomitant with the initiation of endocrine activities of these interstitial cells (Tang *et al.*, 1974a, 1975), whether or not this surge in steroidogenic activity forms the causative factor for the onset of sex reversal in *Monopterus* (see Section III,D).

During gonadogenesis of the synchronous hermaphrodite *Rivulus marmoratus*, the hilar stroma of the experimentally induced presumptive males

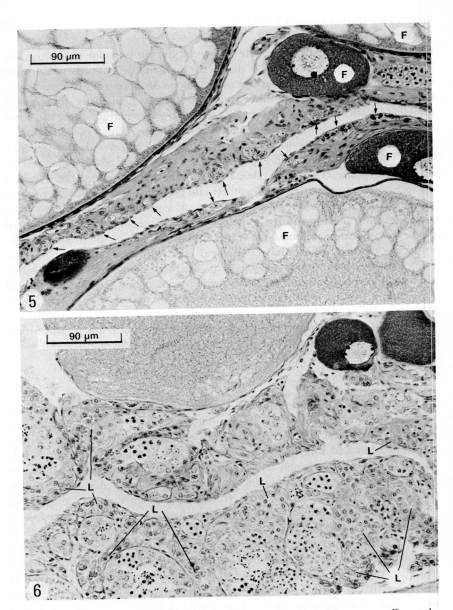

Fig. 5. Transverse section through the gonad of the protogynous *Monopterus albus* at the female phase. Prior to sex reversal, the male germ cells are found as preexisting dormant gonia (arrows) situated along the inner edges of the gonadal lamellae; however, maturing oocytes (F) are mainly confined to the outer region. Natural sex reversal in *Monopterus* is evidently a process of successive maturation of the topographically distinct female and male germ cells.

Fig. 6. Transverse section through an intersexual gonad of *Monopterus*. An extensive development of the interstitial Leydig cells (L) occurs invariably with the proliferation of the male germ cells during natural sex reversal.

increases prominently in proportion to the germ cell area; the hilar stroma is hardly recognizable in the presumptive hermaphrodites of the same genotype (Harrington, 1975). It has been suggested that the hilar stroma might retard the sexualization of the germ cells to form male cells or might decrease the germ cell mitotic rates. The exact physiological roles of the somatic tissue in the sex control of fishes remain to be substantiated by further investigations (see Section III,B).

C. Diandry and Digyny

Two types of males have been reported in a number of protogynous hermaphrodites, notably among the labrids and scarids, a phenomenon being referred to as "diandry" or "biandry" (Reinboth, 1967). The "secondary males" are believed to be derived from sex-reversed females and possess lobate testicular tissue protruding into the former ovarian cavity in a manner comparable to the ovarian lamellae. The sperm duct of a secondary male arises secondarily along the periphery of the gonad and surrounds the persisting central ovarian lumen. "Primary males" are those that are born as male without a prior existence of a female phase. They have larger testes with centrally located sperm ducts; the gonoducts in these males form simple tubes with no trace of any previous occurrence of ovarian duct or ovarian remnants. In some protogynous species, only secondary males are found in natural population, and these are referred to as "monandric" species (Reinboth, 1967). Digyny condition, the existence of both primary and secondary females in a protandrous species, has been reported in *Lates calcarifer* (Centropomidae) (Moore, 1979).

D. Dichromatism

A puzzling problem in the studies of intersexuality in the labrids and scarids is the relationship between dichromatism (i.e., the color dimorphism shown by the males) and protogynous hermaphroditism. For instance, in *Scarus sordidus*, the larger individuals are usually more brightly and distinctly colored (gaudy phase) than the small individuals (drab phase); the drab phase is exhibited by both females and males, but the gaudy fish are males only. In diandric species, both primary and secondary males contribute to the drab phase males, and studies on size-frequency distribution suggest that the primary males leave the drab phase and enter the gaudy phase via color transition at a smaller mean body size than the secondary males via sex reversal (Choat and Robertson, 1975). Changes in color patterns are also believed to be associated with sex reversal in a number of hermaphroditic wrasses. Monandric and monochromatic conditions have been reported in *Labrus turdus* and *L. merula* (Sordi, 1962), *Labroides dimidiatus* (Robertson and Choat, 1973), *Thalassoma cupido* (Reinboth, 1975a), and *Labrus bergylta* (Dipper and Pullin, 1979); diandry and dichromatism are found in *T. pavo* (Reinboth, 1962a), *Coris julis* (Reinboth, 1962a), *Halichoeres poecilopterus* (Okada, 1962; Reinboth, 1975b), *T. bifasciatum* (Reinboth, 1970, 1975b; Roede, 1972), *Labrus ossifagus* (Dipper and Pullin, 1979), and a number of species of *Scarus* (Choat and Robertson, 1975). The basis of these relationships between color phase and sex and its functional significance remain to be clarified.

III. INTRINSIC FACTORS OF SEX CONTROL AND SEX REVERSAL

A. Genetic Control

Genetic mechanisms that control sex in fishes are briefly described here; detailed reviews of this subject are found elsewhere (Dodd, 1960; Mittwoch, 1967; Yamamoto, 1969). There is little doubt that the primary factor(s) for the control of sex rests in the genetic constituents of the organisms. However, the genetic mechanism(s) for sex determination in fishes is primitive and labile (D'Ancona, 1949a, 1952; Bullough, 1947; Bertin, 1958; Dodd, 1960; Yamamoto, 1969), and many fish have no cytologically distinguishable sex chromosomes. Nevertheless, breeding experiments involving artificially sex-reversed animals indicate the presence of the heterogamete–homogamete sex determination mechanism in some species. The majority of species are gonochorists under normal, natural condition, and this may be an indication of the operation of a genetic sex control system, although such a system may be labile and subject to influences by extragenetic factors. The occurrence of "differentiated" and "undifferentiated" gonochoristic fishes offers a probable illustration of the varying degrees of genetic stability in its control of sex differentiation, although the exact genetic and biochemical mechanisms that result in such a phenomenon remain obscure.

An explanation in modern genetic terms regarding Winge's concept of multiple sex factors in sex determination in fishes (Winge, 1934) is yet to be found. However, at present this concept is probably the only logical hypothesis. In his elaboration of the concept of polygenic sex determination in fishes, Yamamoto (1969) maintained that apart from the superior sex genes in the Y and X chromosomes, multiple sex factors existed in the autosomes. In *Poecilia reticulata*, the autosomal male and female genes were believed to be more or less in balance in the majority of individuals, and sex was determined by heterosomal combinations. However, in exceptional individuals, the superior sex genes in the sex chromosomes might be overridden by many opposing autosomal genes resulting in XX males or XY females. Other undifferentiated species such as *Xiphophorus helleri*, *Poecilia vittala*, and *P. caudofasciata* also depend on this polygenic mechanism, which is presumed to be less decisive in sex determination. In these fishes, sex ratio varies widely and intersexuality is common. Sex determination in "differentiated" gonochorists, such as *Xiphophorus maculatus* and *Oryzias latipes*, is governed mainly by the superior sex genes in the sex chromosomes. Sexuality of these animals is usually stable in natural condition and intersexes are extremely rare (see Yamamoto, 1969).

With regard to the evolution of sex chromosomes in fishes, there appears to be little doubt about their autosomal origin (see Mittwoch, 1967). Apart from the fact that cross-over is possible between X and Y chromosomes in *Oryzias latipes* (Yamamoto, 1961), the unspecialized nature of sex chromosomes in teleosts is also evident by the presence of a dual system of XX:XY and WZ(Y):ZZ(YY) in the Mexican wild population and British Honduras domesticated stock of *Xiphophorus maculatus*, respectively (Gordon, 1947, 1952). Kallman (1965a,b; 1970; 1973) extended Gordon's findings and noted populations of both male and female heterogamety in a vast area of Guatamala; in some instances both types occurred in a single pool.

Little is known about the genetic mechanism that controls natural sex reversal in fishes. It is safe to assume that the established sequential sexuality in the protogynous or protandrous hermaphroditic species must be governed by some form of genetic mechanism (Chan, 1970), although it is most likely that the conventional XX:XY or ZZ:ZW mechanisms are not well established in most of the normally hermaphroditic fishes. In view of the diversity of the piscean groups and their varied genetic mechanisms of sex control, e.g., the polymorphic heterogamety systems in *X. maculatus* (Gordon, 1952) and the multiple sex chromosomes in Mexican cyprinodontid fish (Uyeno and Miller, 1971), the genetic sex determining mechanism(s) in fish may be far more complex than previously believed; further investigations will certainly be fruitful in this area.

B. Sex Inductors

In line with Witschi's inductor theory (Witschi, 1934, 1957) for amphibians, sex inductors have been proposed for the control of sex in both gonochoristic and hermaphroditic fishes. However, hypothetical inductor substances such as "androgenine and gynogenine" (D'Ancona, 1949b, 1950) and "gynotermone and androtermone" or "gynoinductor and androinductor" (Yamamoto, 1962; Yamamoto and Matsuda, 1963) provide only conceptual ideas for discussion because such sex differentiating substances have not been demonstrated or isolated in fishes. Undoubtedly, the biochemistry of sexual differentiation of the gonad most likely operates through endogenous secretions that induce the indifferent germ cells to become either male or female gonia. However, there remains an unanswered question with regard to how the teleostean gonads, which are believed to derive from a single embryonic primordium corresponding to the cortex of tetrapods (see Dodd, 1960; Atz, 1964), can elaborate two sex inductors. Indeed, many researchers believe that this apparent lack of corticomedullary organization of the teleostean gonad is related to the frequent occurrence of intersexuality and sex reversal and the instability of sex control mechanism(s) in teleosts (see Atz, 1964). It is relevant that hermaphroditism is rare among elasmobranch fishes where the embryonic gonad possesses dual components similar to that of the amphibians, and a dual inductor system is believed to operate (Chieffi, 1959).

The concept of sex inductors and antiinductors (Witschi, 1957) may offer some plausible explanation for the corticomedullary antagonism and the differentiation of the structurally bipotential embryonic gonads into either ovaries or testes in amphibians. It remains an unresolved problem with regard to how a similar sex-inductor system could operate in sex-reversing fishes to give an ovotestis with heterosexual zones or a hermaphroditic gonad with intermingled male and female sex tissues, particularly when the teleostean gonads are believed to possess a unitary stroma with no distinction between a feminizing cortex and a masculinizing medulla (D'Ancona, 1949a, 1950, 1955). Atz (1964) elaborated D'Ancona's hypothesis (1949a, 1950) that in the heterosexual gonad of sparids and serranids, gynogenine might first exceed some threshold in the protogynous species with androgenine subsequently replacing gynogenine, and vice versa in protandrous species. It is not apparent how these two antagonistic embryonic sex-inductor mechanisms could operate in the *Epinephelus* type of hermaphroditic gonads where there is no discrete segregation of the potential male and female sex tissues. Although D'Ancona (1950) speculated that a differential diffusibility of the sex differentiators might contribute to the varied gonadal patterns of hermaphroditic teleosts, it has yet to be verified if the supposedly unitary gonadal primordium could bring about two cytophysiologically distinct cell types responsible for the secretions of the male and female sex-inductor substances in the embryonic gonad of fishes. In fact, the relationship between the so-called "single embryonic origin" of the gonadal primordium and the development of hermaphroditic gonad in teleosts may require a reappraisal. As Harrington (1975) correctly noted, the hilar stroma in *Rivulus marmoratus* becomes conspicuous in the ovotestis only when it sends out the lateral ramus, at the base of which the male territory becomes manifested with male gonocytes. It may well be that in phylogenetic primitive vertebrates, because of the close spatial relationship between the nephric, interrenal, and gonadal elements in their successive differentiation from the peritoneal epithelium, the embryonic male and female primorida in teleosts and cyclostomes do not become distinct as in the amphibian medulla and cortex (see Hardisty, 1965). Harrington (1975) directed special attention to the medial and the lateral ramus in *Rivulus marmoratus, Sparus auratus, Opsanus tau,* and *Poecilia reticulatus;* the medial ramus is apparently always associated with ovarian differentiation, but the lateral ramus is related to the male territory of the teleostean gonads.

The exact nature of sex-inductor substances remains an open question. Witshci's conclusion that the inductor substances are hormonelike and that "the distribution of corticin and medullarin is by the bloodstream," together with the fact that estrogen and androgen are capable of influencing the normal course of differentiation of the gonad, led many researchers to believe that sex steroids are homologous to sex inductors (see Crew, 1952; Forbes, 1961; Yamamoto, 1962, 1969). Although androgen elaborated by the differentiating testes plays a decisive role in the differentiation of the Wolffian ducts and sex accessories in mammals, and AMH plays a role in the degeneration of the Müllerian ducts (Jost, 1965; Picard *et al.,* 1978), it appears unlikely that sex steroids are primary sex-inductor substances in the gonadal sex differentiation in mammals and fishes (Chieffi, 1959; Chan *et al.,* 1972a; see also Section III,D). Witschi (1957) proposed an antigen–antibody interrelationship between the sex inductors and antiinductors and recent studies of the H-Y antigen suggested that this H-Y antigen might play a sex-inductor role in sex control because of its apparent testis-organizing capacity in mammals (Wachtel and Koo, 1981). Although there is no conclusive evidence as to whether H-Y antigen operates in fishes, particularly in sex-reversing species (see Section III,C), it seems obvious that attention must be given to nonsteroid systems in the search for sex inductors or other biochemical factor(s) that may play a decisive role in the control of sex under natural conditions.

C. H-Y Antigen

Since the discovery of the testis-organizing role of H-Y antigen in mammals (Wachtel *et al.,* 1975b), various attempts have been made to search for the occurrence of a similar testis-inducing system in fishes. Absorption of H-Y antibodies from mouse have been reported in the male gonadal cells of *Lebistes (Poecilia) reticulatus* and *Xiphophorus helleri* (Müller and Wolf,

1979) and in the male brain cells of *Xiphophorus maculatus* (both XY and YY males), in hybrids of the genus *Tilapia*, in *Haplochromis burtoni*, and in the brain and liver cells of *Oryzias latipes* (Pechan *et al.*, 1979); female cells in the same studies gave negative results. All these fishes show an XX:XY sex determination except *H. burtoni* in which male heterogametic condition has not yet been confirmed. Evidence for the presence of structures similar to Barr bodies in female *H. burtoni* suggests the possible existence of a male heterogamety mechanism in this species (Mehl and Reinboth, 1975). However, it should be noted that Barr bodies might arise from W or Y chromosomes as in many species of snakes (Ohno, 1967) where female heterogamety was observed.

It is premature with present knowledge to draw any general conclusion regarding the role of H-Y antigen in sex control in fishes, particularly in attempts to draw any parallelism between the H-Y antigen system of XY males in mammals and that in fishes. Although H-Y antigen has been detected in XY mammals, a number of fishes (Müller and Wolf, 1979; Pechan *et al.*, 1979), and XY *Rana pipiens* (Wachtel *et al.*, 1975a), H-Y antigen is also detected in ZW female of chick (Wachtel and Koo, 1981) and ZW female of *Xenopus laevis* (Wachtel *et al.*, 1975a). Therefore, on one hand, H-Y antigen is apparently related to the heterogametic sex among various vertebrate groups, although expression of H-Y antigen is also found in XO males of the mole vole, *Ellobius lutescens* (Nagai and Ohno, 1977), and XX, Sxr/- male mice (Bennett *et al.*, 1977). On the other hand, the concept that H-Y antigen constitutes a testis-determining system in sex differentiation may only be applicable to mammals but not to all vertebrate classes, even though this antigen is capable of reorganizing dispersed newborn testicular cells *in vitro* into tubular (testicular) structure, and the H-Y antibody-treated cultures yielded follicular (ovarian) aggregates in BALB mice (Ohno *et al.*, 1978) and an inbred strain of rats (Zenzes *et al.*, 1978). It should also be noted that in the more primitive fish orders, Isopondyli (*Salvelinus alpinus, Salmo gairdneri*) and Ostariophysi (*Rutilus rutilus, Carassius auratus, Barbus tetrazona*), both male and female gonadal cells showed cross-reactivities with the H-Y antiserum and the degree of absorption was similar in both sexes (Müller and Wolf, 1979). The role of H-Y antigen as a factor in sex control of fishes remains an open question.

As yet, there have been no H-Y antigen studies on hermaphroditic teleosts, and information derived from such studies will certainly be of values. For instance, it will be interesting to know whether an expression of H-Y antigen could be found in the female or the male phase of sequential hermaphrodites, or whether cessation of production of the H-Y antigen constitutes part of the endogenous sex-control mechanism(s) in natural sex reversal in fishes. In the ovotestis of human true hermaphroditism, H-Y$^+$/H-

Y⁻ mosaicism is found; cell culture from the ovarian portion absorbed less H-Y antibody than those from the testicular portion (Winters *et al.*, 1979). Whether a similar mechanism operates in the heterosexual areas of sequential hermaphrodites in fish awaits further elucidation.

The significance of the H-Y antigen cross-reactivities in both sexes of the more primitive fish orders (Isopondyli and Ostariophysi) is not yet clear. Müller and Wolf (1979) suggested that H-Y antigen was initially expressed in both sexes in lower vertebrates and was later evolved in association with the heterogametic sex-determination mechanism(s). The questions whether H-Y antigen in these primitive fishes has any sex-inductor activity as it has in mammals, and when, during the course of vertebrate evolution, this system acquires a sex-controlling role in the differentiation of the primary sex organs have yet to be answered.

D. Gonadal Sex Steroids

1. THE OCCURRENCE AND BIOSYNTHESIS OF SEX STEROIDS

Hoar (1969) and Ozon (1972a,b) provided excellent reviews on sex steroids in fishes, particularly the gonochoristic species. However, these reports gave little consideration to the roles of gonadal steroids in sex control of hermaphroditic fishes. This aspect is considered in more detail in the following discussion.

Lupo di Prisco and Chieffi (1965) reported the occurrence of testosterone, androstenedione, adrenosterone, estradiol, estrone, estriol, and corticosteroids in freeze-dried gonadal tissues of *Serranus scriba*. Using *in vitro* incubation techniques with testosterone and progesterone as precursors, Reinboth *et al.* (1966) was unable to detect any conversion to androstenedione by the ovarian tissue of *Centropristes striatus;* pregnan-3,20-dione, 5β-androstan-3,17-dione, 5β-androstan-17β-ol-3-one, and 5β-androstan-3α,17β-diol were found among the metabolites. Chan and Phillips (1969) investigated the *in vitro* conversion of labeled pregnenolone by the gonadal tissue of *Monopterus albus* at various stages of sex reversal and found among other metabolites progesterone, 17α-hydroxyprogesterone, androstenedione, testosterone, 17β-estradiol, and estrone; a marked increase in the production of androstenedione and testosterone was found at the intersexual phase concomitant with the extensive proliferation of the interstitial Leydig cells. Similar *in vitro* studies were conducted with the protandrous *Sparus auratus* using mature testis, sex-reversing testicular tissue, sex-reversing ovarian tissue, and mature ovary (Colombo *et al.*, 1972). The precursor pregnenolone was metabolized to progesterone, 17α-hydroxyprogesterone, androstenedione, testosterone, 17β-estradiol, and estrone. Progesterone and 17α-hydroxyprogesterone were the most abundant products and were the only steroids found in the ovarian tissue during sex reversal. The yield of testosterone was only significant in mature testis, but estradiol and estrone were more abundant in mature ovarian tissue than in testicular tissue. Colombo and co-workers (1972) also reported that the gonadal tissues appeared to have a much greater metabolic potential during maturity than during sex reversal. In his study of the metabolism of testosterone in *in vitro* incubation of *Coris julis* and *Pagellus acarne*, Reinboth (1972) reported the production of 11-ketotestosterone and 11β-hydroxytestosterone; the latter was found in much larger quantity in transforming gonads than in normal ovarian tissue from the same period of the year.

Attempts to measure plasma levels of sex steroids in hermaphroditic fishes was first reported by d'Istria *et al.* (1973, quoted by Reinboth, 1979). Plasma testosterone was quantitatively assayed by radioimmunoassay (RIA) in *Serranus cabrilla*, but estrone and estradiol were undetectable in the same study. In a more extensive study of the steroid content of blood of both gonochoristic and hermaphroditic fishes, Idler *et al.* (1976) reported the presence of 11β-hydroxytestosterone in the plasma of the male phase of *Diplodus sargus, Serranus cabrilla, Pagellus acarne*, and *P. erythrinus;* 11-ketotestosterone was not detected in the same materials from these hermaphroditic fishes. However, using the same double-isotope technique, Idler and co-workers found that 11-ketotestosterone was the predominant androgen in gonochoristic species, such as *Salmo salar* and *Pseudopleuronectes americanus*, and that the plasma concentration of 11β-hydroxytestosterone was extremely low. Idler and co-workers proposed that a basic difference in the steroid enzyme complement might exist between hermaphroditic and gonochoristic teleosts.

However, the endocrine profiles related to steroid metabolism in hermaphroditic fishes are apparently more complex. In his studies of the metabolites formed by *in vitro* incubations of gonadal tissue of different sexual phases in *Serranus carbrilla, Pagellus acarne*, and *Coris julis*, Reinboth (1975b) reported that progesterone was converted to a significant amount of 5β-pregnan-3α-ol-20-one, 5β-pregnan-3β-ol-20-one, 5β-pregnan-3,20-dione, and 5β-pregnan-17α-ol-3,20-dione. 17α-Hydroxyprogesterone was absent from *S. cabrilla* and 11β-hydroxytestosterone was absent from *S. cabrilla* and *P. acarne*. Androstenedione and testosterone were absent or found only in trace amounts. In incubations where testosterone was used as precursor, 11β-hydroxytestosterone was the most abundant product in all species, and 5β-reduction appeared to be the predominant metabolic pathway, producing 5β-androstan-3β,17β-diol and 5β-androstan-3α,17β-diol; androstenedione was present in trace amount or absent. Reinboth (1975b) also reported that progesterone was metabolized more actively than testosterone and interspecific differences were important.

More recently, Reinboth (1979) provided a detailed account of the metabolites obtained by *in vitro* incubations of progesterone or testosterone with gonads of various hermaphroditic fishes. In *Pagellus acarne*, regressing testes converted progesterone to 5α- and 5β-pregnanedione in the ratio of about 1 to 2. The same tissue converted testosterone to 11-ketotestosterone and 11β-hydroxytestosterone in varying ratios, but 50% of the metabolites was androstanediols, of which 5β-androstan-3α,17β-diol formed the major constituent. Ovarian tissue of sex-reversing specimens converted testosterone to a significant amount of androstenedione, 5β-androstan-3α-ol-17-one, and 5α-androstanedione. The metabolic activities of both the regressing testis and the developing ovary were high, but the regressing testicular tissue was more active in metabolizing testosterone than was developing ovarian tissue from the same specimen.

In *Spicara maena*, testicular tissue metabolized progesterone to both 5β- and 5α-pregnanedione, but ovarian tissue yielded only the 5α isomer. With testosterone as precursor, both ovarian and testicular tissues yielded 11β-hydroxytestosterone and 11-ketotestosterone, both being present in larger quantities in male than in female, and the amount of 11β-hydroxytestosterone always exceeded that of the 11-keto derivative. Further, 5α- and 5β-androstanedione were also formed in the male, but the female yielded only the 5α isomer. Metabolic activities were higher in spring than in autumn in both sexes, and less polar steroids, such as 4-androstenedione, androstanediones, and androstanolones, were produced in greater amounts in spring than in autumn.

In *Coris julis, in vitro* incubation of labeled progesterone with male tissue yielded both 5α- and 5β-pregnan-3,20-dione, the latter being the major type in the primary and secondary males and in the sex-reversing specimens; the female tissue yielded only the 5α isomer. Conversion of testosterone to 11β-hydroxytestosterone was more active in the secondary male and the sex-reversing specimens; activity in the primary male was similar to that of the female. Conversion to 11-ketotestosterone was found in both types of males and the sex-reversing specimens, but the production was most prominent in the sex-reversing males. It is interesting to note that testes from the secondary males metabolized testosterone more actively than testes from the primary males; metabolic activity in the female appeared to be relatively low. However, it should be mentioned that Reinboth (1979) found that the patterns of metabolic activity could not be confirmed in subsequent experiments conducted under the same experimental conditions; a much lower metabolic activity was found in the sex-reversing specimens and in the secondary males.

The metabolic activity of the synchronous hermaphrodite *Serranus cabrilla* appeared to be diverse with both progesterone and testosterone as precursors. 11β-Hydroxyprogesterone, 21-deoxycortisol, 17α-hydroxyprogesterone, 5β-pregnan-3α,20α-diol, 5β- and 5α-pregnan-3,20-dione, 5β-pregnan-17α-ol-3,20-dione, and 5β-pregnan-3α,17α,20α-triol were isolated as metabolites from progesterone. 11β-Hydroxytestosterone, 11β-hydroxyandrostenedione, 11-ketotestosterone, androstenedione, 5α- and 5β-androstan-17β-ol-3-one, 5α- and 5β-androstan-3β,17β-diol, 5α- and 5β-androstan-3,17-dione, and 5β-androstan-3α,17β-diol were produced from testosterone; the 5α isomers were predominant in the male and 5β isomers were predominant in the female. Further, 11-ketotestosterone conversion outweighed 11β-hydroxytestosterone in the ovarian tissues, and 11β-hydroxytestosterone was always formed in larger amount in the testicular tissues. Conjugates of steroids, mainly sulphates, were also abundant in the metabolites accounting for approximately 30% of the radioactivity recovered, and the ovarian tissue produced more conjugates than the testicular tissues (Reinboth, 1979).

It is clear from the foregoing account that present knowledge of steroid biosynthesis and metabolism in hermaphroditic fishes remains sporadic and inconclusive. Recent studies of various sex-reversing species have yielded information that raises many new problems rather than settling old ones. Before more revealing studies of the physiological functions can be properly conducted, much more fundamental knowledge of the gonadal structure and the biology of the species is required, particularly with regard to aspects related to the species' pattern of sexuality and reproductive cycles. It remains an open question as to whether gonadal tissues taken from hermaphroditic species of different phylogenetic origin, different sexual pattern, different reproductive cycle, and different ecological background yield the same, or similar, metabolites in quality and quantity. These careful and thorough *in vitro* studies, nevertheless, are extremely essential in providing useful information regarding the probable metabolic pathway and the possible presence, or absence, of certain steroid enzyme(s) in the tissues of various hermaphroditic fishes; the results will provide meaningful clues to the key steroids to be assayed in future investigation of the *in vivo* endocrine profile of the species.

In a different approach, Eckstein *et al.* (1978) examined the effect of human chorionic gonadotropin (HCG) on the *in vitro* conversion of androstenedione by various gonadal tissues of *Sparus auratus* and found a more significant conversion to testosterone (up to 50% yield in the male) than to 11-ketotestosterone (up to only 4% yield) and 11β-hydroxytestosterone (2% yield). It is interesting to note that HCG enhanced the production of testosterone in the female, but inhibited it in the male. These findings suggest a possible role played by the pituitary gonadotropins in the trigger of natural sex reversal in fishes as previously proposed by Reinboth (1962a) and Chan *et al.* (1975; see also Section III,E).

2. THE ROLES OF SEX STEROIDS IN SEX CONTROL AND SEX REVERSAL

Sex steroids have a potent effect on the process of sex differentiation in gonochoristic fishes, such as *Oryzias latipes, Poecilia reticulata, Xiphophorus helleri, Tilapia mossambica,* and *Carassius auratus.* Yamamoto (1969) provided a thorough discussion on the subject. In general, it was found that estrogens (e.g., estriol, 17β-estradiol, estrone, and stilbestrol) could induce the production of XY females, but that androgens (e.g., dehydroepiandrosterone, testosterone propionate, androstenedione, and methyltestosterone) could result in XX male formation. In fact, sex steroids were so potent in affecting sex differentiation in fishes that Yamamoto (1962, 1969) asserted a role of natural sex inductors for sex steroids and concluded that "sex hormones act specifically as sex inducers" and that "estrogens act as female inducer and androgens as male inducer."

Information concerning the effects of sex-steroid hormones on natural sex reversal in fishes is scanty and confined to only a few species. One major problem in this type of study is the determination of the sexual status of the animals before experimentation. Biopsy is apparently the only approach by which the gonadal structure and the sexual status of a sex-reversing fish can be reliably studied both before and after hormonal treatments; however, this type of experiment is usually difficult, if not impossible, to perform. So far, only *Coris julis* (Reinboth, 1962a) and *Monopterus albus* (Chan *et al.*, 1972b; also see further discussion) have been studied with such a technique. Many investigators have conducted hormonal treatments of hermaphroditic fishes without a clear picture of the gonadal condition of the experimental animals prior to the treatments; the results derived from such experiments should be interpreted with reserve.

Reinboth (1962b) studied the effects of androgen on the control of sex in the protogynous *Coris julis.* Intramuscular injection of 2.5 mg microcrystalline testosterone-isobutyrate into females resulted in a change to the secondary male coloration in 3 weeks. After a lapse of 6–7 weeks, there was, in general, a pronounced destruction of oocytes and a development of undifferentiated gonia in the periphery of the gonads, resembling spontaneous sex change. Although complete sex reversal was observed in some experimental animals when sacrificed after a lapse of 4 months or longer, some specimen still maintained an apparently normal ovary with an increase in mitotic activities and young oocytes. Reinboth (1970) concluded that, in general, administration of androgens causes a precocious sex reversal in protogynous fishes, and estradiol injection results in considerable damage to both male and female gonadal tissues in *Diplodus aunularis, Thalassoma pavo, Coris julis,* and *Spicara maena.*

In an ill-defined study, Okada (1964a,b) found that injections of estradiol benzoate into male *Halichoeres poecilopterus* caused complete destruction of the male germ cells, but testosterone propionate induced disintegration of ovarian tissue with the onset of spermatogenesis in females. He also reported that methyltestosterone stimulated the testicular portion of the gonad in *Mylio macrocephalus* (Okada, 1965).

Studies of the actions of sex steroids on the expression of female and male sex in *Monopterus* gonad were of particular interest because natural sex reversal in this species was invariably associated with a proliferation of Leydig cells in the interstitium of the gonadal lamellae (Chan and Phillips, 1967). Marked steroidogenic ultrastructural characteristics such as smooth endoplasmic reticulum and mitochondria with tubular cristae (Fig. 7) together with an increase in 3β-hydroxysteroid dehydrogenase activities were also observed in this interstitial tissue (Tang *et al.,* 1974a, 1975). These

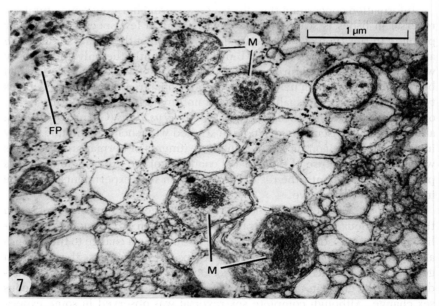

Fig. 7. Electronmicrograph of *Monopterus* Leydig cell (magnification 38,475×). The steroidogenic nature is shown by the characteristic smooth endoplasmic reticulum and mitochondria (M) with tubular cristae. Note also the close relationship of filopodia (FP) with adjacent Leydig cell.

cytological features, along with the shift to an increased androgen production during midintersexual phase (Chan and Phillips, 1969; Chan et al., 1975) indicated that a change in the steroid secretory profile was concomitant with the expression of the male phase, although it was unclear whether this endocrine alteration constituted a causative or a consequential event to the trigger of natural sex reversal. Therefore, the effects from steroids in the control of the expression of female and male phase in *Monopterus* were studied by extensive experiments involving the administration of steroid hormones, by implantation or by injection, to animals whose sexual status had been first determined by biopsy study. Various androgenic and estrogenic steroids at various dosages were used and experiments included all reproductive stages in the annual cycle. Details of the experiments, such as animals, dosages, route, and duration of treatments, and general observations were previously reported and are summarized in Tables I and II.

Therefore, in *Monopterus*, estrogens (e.g., estrone, 17β-estradiol, and estradiol benzoate) showed little effect in the female phase; they caused marked destruction of the male tissues in the intersexual and male phases, and were incapable of inducing a return to the female phase. Androgens (e.g., testosterone, methyltestosterone, and 11-ketotestosterone) at various dosages failed to enhance a precocious sex reversal of the female fish to male. The dormant male gonia was insensitive to the exogenous androgens during the female phase probably because of some inherited genetic or age-dependent factor. Even with cyanoketone to remove any possible inhibitory effect of endogenous estrogen on the male gonia, androgen treatments still failed to induce a precocious sex reversal in *Monopterus* females. However, male germ cells in the mid and late intersexual stages, i.e., after the onset of natural sex reversal, were highly sensitive to the spermatokinetic effects of the exogenous androgens. Available evidence suggests that the shift in endocrine profile that accompanies the structural sex change in the *Monopterus* gonad is a secondary event and not a causative factor for natural sex reversal.

E. Adenohypophysial Function and Sex Reversal

In higher vertebrates, early hypophysectomy of the embryo does not interfere with gonadal morphogenesis up to the stage when the gonads are fully characterized as testes or ovaries (Puckett, 1940; van Deth et al., 1956; Burns, 1961; Jost, 1970). Therefore, it appears that the pituitary is not concerned in the primary differentiation of sex. Presumably because of technical difficulties, similar research using teleosts has rarely been conducted. Hypophysectomy prevents the development of the gonad in juvenile *Gobius paganellus* (Vivien, 1941). Spermatogenesis, development of testicular en-

Table I

Summary on the Effects of Various Hormones and Drugs on the Female Gonad of *Monopterus* (with Biopsy Studies)

Hormone	Dosage per gram fish	Treatment[a]	Duration[b]	Major observation in relation to sex reversal[c]	Reference
Estrone	2 mg[e]	Implantation	40 days	No observable changes in the gonad	Chan et al. (1975)
Testosterone	5 mg[e]	Implantation	60 days	Male gonia remain dormant	Chan et al. (1972a)
Methyltestosterone	4 µg	3 Injections/wk	7 wk		
Methyltestosterone	4 µg	3 Injections/wk	7 wk		
	+8 µg[f]	3 Injections/wk	4 wk		
	8 µg	3 Injections/wk	8 wk	Male gonia remain dormant; no interstitial Leydig tissue development; some with a few "hollow lobules" but lack normal spermatogenesis	Tang et al. (1974a)
	1 µg	3 Injections/wk	7.5 wk		
	4 µg	3 Injections/wk	7.5 wk		
	4 µg	3 Injections/wk	5 wk		
Testosterone	4 µg	3 Injections/wk	8 wk		
11-Ketotestosterone	4 µg	3 Injections/wk	8 wk		
11-Ketotestosterone	4 µg	3 Injections/wk	16 wk		
Cyanoketone[d]	7 µg				
+ Cortisol	1 µg	3 Injections/wk	7.5 wk	Destruction of maturing oocytes and follicles; male gonia remain dormant	
+ Methyltestosterone	4 µg				
Cyanoketone[d]	7 µg				
+ Cortisol	1 µg	3 Injections/wk	7.5 wk	Destruction of maturing oocytes and follicles	

Progesterone	50 µg[g]	1 Injection/day	20 days	No observable changes in the gonad	
Progesterone + Estrone	50 µg[g] +2 µg[g]	1 Injection/day	20 days	No observable changes in the gonad	
NIH-LH-S$_{14}$	3 µg	3 Injections/wk	8 wk	Extensive Leydig cell development; formation of testicular lobules but with little normal spermatogenesis	Tang *et al.* (1974b)
NIH-LH-S$_{14}$	1.5 µg	3 Injections/wk	7.5 wk		
Ovine LH (Mann)	0.1 IU[h]	1 injection/12 hr	10 days	Leydig cell development, testicular lobule formation, some with spermatogenetic activities	
Ovine FSH (Sigma)	10 units[h]	1 Injection/12 hr	10 days	Leydig cells development and lobule formation; effects less marked than the LH-treated fish	

[a] Intraperitoneal.
[b] Treatments with methyltestosterone cover all seasons in the year.
[c] Based on histological comparison with the control and with the same gonad at first biopsy.
[d] 17β-ol-4,4,17-trimethyl-3-oxo-androst-5-ene-2α-carbonitrile from Sterling-Winthrop Research Institute, N.Y.
[e] One dose in pellet, following the first biopsy.
[f] Following a previous treatment and after a second biopsy on the gonad.
[g] Dosage per fish per day.
[h] Dosage per fish.

195

Table II

Summary on the Effects of Sex Steroids in Intersexual and Male Gonad of *Monopterus* (with Biopsy Studies)

Sexual status	Treatments	Dosage per gram fish[a]	Duration	Major changes in relation to sex reversal[b]	Reference
Intersex	Testosterone	5 mg[c]	60 days	Enhance spermatogenesis	Chan *et al.* (1972a)
	Estrone	2 mg[c]	40 days	Suppression of spermatogenesis and destruction of male tissue	Chan *et al.* (1975)
Early intersex	Methyltestosterone	8 µg	8 wk	No observable change	
		4 µg	7.5 wk	No observable change	
Midintersex	Methyltestosterone	4 µg	5 wk	Increase of testicular lobules and spermatogenesis	
	Methyltestosterone	4 µg	7.5 wk		
	17β-Estradiol	8 µg	2 wk		
		+4 µg	3 wk	Destruction of male tissue and suppression of spermatogenesis	Tang *et al.* (1974a)
	Estradiolbenzoate	4 µg	2 wk		
		+2 µg	5 wk		
	Estradiolbenzoate	1 µg	7.5 wk		
Male	Testosterone	10 mg[c]	60 days	Enhance spermatogenesis	Chan *et al.* (1972a)
	Estrone	5 mg[c]	60 days	Suppression of spermatogenesis and destruction of male tissue	Chan *et al.* (1975)
	17β-Estradiol	8 µg	2 wk	Suppression of spermatogenesis and destruction of male tissue	
		+4 µg	3 wk		
	Estradiolbenzoate	4 µg	2 wk	Suppression of spermatogenesis and some with no observable change	
		+2 µg	5 wk		
	Estradiolbenzoate	1 µg	4–7 wk	Suppression of spermatogenesis and destruction of male tissue	Tang *et al.* (1974a)

[a]All treatments were three intraperitoneal injections per week unless otherwise stated.

[b]Based on histological comparison with the control and with the same gonad at first biopsy.

[c]One dose as pellet implanted intraperitoneally.

docrine tissue, and differentiation of secondary sexual characters in juvenile *Poecilia recticulata* are blocked by hypophysectomy (Pandey, 1969). In the same study, methyltestosterone treatment of hypophysectomised guppies stimulated the development of secondary sexual characters. The treatment affects neither the sex ratio nor the gonadal tissues. However, higher male-to-female sex ratio and stimulation of testicular development were found in intact juvenile guppies and in embryos (carried within pregnant mothers) after methyltestosterone treatment (Eversole, 1941; Miyamori, 1961; Dzwillo, 1962; Clemens *et al.*, 1966). Pandey (1969) suggested that the exogenous steroid might act in the intact juveniles and the embryos via the pituitary, causing the early release of gonadotropins of the male type. There is also other evidence indicating that gonadotropins may exert some effect on the development of the sexual elements in teleostean gonads. Mammalian anterior pituitary powder accelerates the masculinization effect of testosterone propionate on the female gonad of *Xiphophorus helleri* (Régnier, 1937, 1938). Masculinization in the same species can be induced by chorionic gonadotropins, the effect of which is promoted by incomplete hypophysectomy (Baldwin and Li, 1942; Vivien, 1952). Atz (1964) also suggested that the effect of sex steroids on the formation of ovotestes on gonochoristic teleosts acts via the pituitary although the pituitary gland has not been directly implicated in the production of hermaphroditism.

The pituitary is essential for the proper maturation of the germ cells and the formation of the associated endocrine tissues of the gonad (see review by Pickford and Atz, 1957). In the protogynous *Monopterus albus*, natural sex reversal is accompanied by an extensive development of interstitial Leydig cells (Chan and Phillips, 1967, 1969). This offers a possible route by which the pituitary may affect sex reversal in teleosts.

Using mammalian anterior lobe extracts, Tuchmann (1936) was able to induce a change to bright body coloration from the originally dull colored individuals of *Coris julis*. In *Sparus auratus*, annual cyclical changes of oocytes in the ovotestes during the male phase has been reported (Reinboth, 1962a). This, along with the suggestion that more gonadotropin is needed for ovarian than for testicular maintenance, led to the hypothesis that in the protandrous species a gradual year-to-year increase in gonadotropin output by the pituitary finally yields a titer sufficient for the full activation of the ovarian component of the ovotestes (Reinboth, 1962a). However, the hypothesis was difficult to apply to protogynous forms. In a subsequent study, Reinboth and Simon (1963, quoted by Atz, 1964) was unable to induce sex changes in *Serranus cabrilla* with injections of fish pituitary homogenate. The only successful experiment of this sort is the one on *Monopterus albus*. In this teleost, precocious development of interstitial Leydig cells and some of the male germ cells could be induced by mammalian luteinizing hormone

(LH) and follicle stimulating hormone (FSH), although the effect of the latter was less marked than that of the former (Tang *et al.*, 1974b; Chan *et al.*, 1977). However, some of the testicular lobules formed under such experiments were partly hollow and had incomplete spermatogenic activities. This suggested that a complete sex reversal resembling the natural process was not achieved and might be attributable to either species specificity of the exogenous gonadotropins, or the absence of an additional stimulatory factor, or the lack of an intrinsic age-dependent response mechanism within the male gonia (Chan *et al.*, 1975, 1977).

There are few anatomical studies of the pituitary of hermaphroditic teleosts. Cytophysiological studies of the pituitary gland of *Monopterus* revealed seven adenohypophysial cell types (O and Chan, 1974). Two types of gonadotrops were found among the glycoprotein-secreting basophils; (O, 1973). Contrary to the reported inactivity of salmon gonadotropins on mammalian tissue (Yamazaki and Donaldson, 1968a,b; Channing *et al.*, 1974), crude pituitary extract of *Monopterus* was active in a number of bioassays including the ovarian ascorbic acid depletion test, the ovarian HCG augmentation test, ventral prostate weight test, ovarian cholesterol depletion test, ^{32}P uptake by the 1-day-old chick gonads (Ng, 1976; Chan *et al.*, 1977). The LH-like and FSH-like activity of *Monopterus* pituitary extract were assayed by the ovarian HCG augmentation and ascorbic acid depletion tests (Chan *et al.*, 1975, 1977). In later studies, the LH-like activity of the pituitary homogenate was further demonstrated by its *in vivo* stimulation of an increase in plasma testosterone level in mature male rats, increased conversion of [^3H]cholesterol to progesterone by rabbit luteal tissue *in vitro*, and elevated ovarian cyclic-AMP levels in rats (Chan and Lee, 1978). Bioassay results of gonadotropic activity at various sexual phases showed a high LH-like activity in the mature female (O, 1973; Ng, 1976; Chan *et al.*, 1975, 1977). Preliminary biochemical separation of the *Monopterus* pituitary extract yielded two fractions of different gonadotropic activities (S. T. H. Chan, T. B. Ng, and Y. H. Lee, unpublished data). This is in line with current evidence that a number of teleosts possess two different gonadotropins (Idler and Ng, 1979). These findings, together with the fact that sex reversal in *Monopterus* is a postspawning event (Chan and Phillips, 1967), suggest that a rise in the LH-like gonadotropic secretion in the mature female toward the breeding season could trigger, on the one hand, ovulation, and on the other hand, natural sex reversal by inducing the development of both the interstitial cells and the resting male gonia (Chan *et al.*, 1975; see also Fig. 8).

A hypothetical scheme for the endocrine events of natural sex reversal in *Monopterus* has been proposed (see Chan *et al.*, 1975; Chan, 1977). This scheme suggests that natural sex reversal is governed by the innate nature and the age-dependent responsiveness of the germ cells to hormonal stim-

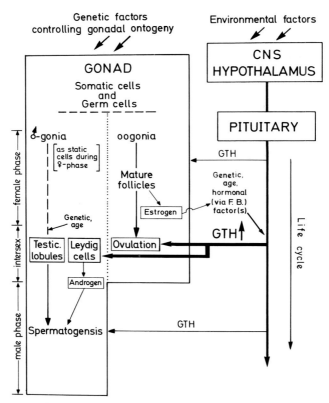

Fig. 8. A scheme illustrating the possible endocrine mechanism of natural sex reversal in *Monopterus*. The scheme involves the innate nature of the female and male germ cells, the age-dependent responsiveness of the male germ cells to extra- and intragonadal hormonal stimulations, the gonadal endocrine interactions including pathways via neuroendocrine feedback and the hypothalamic–hypophysial–gonadal axis (Modified from Chan *et al.*, 1975)

ulation(s), the gonadal endocrine interactions including feedback, and the function of the hypothalamic–hypophysial–gonadal axis (see Fig. 8). The possible involvement of the adenohypophysis in controlling the onset of natural sex reversal in *Monopterus* inevitably incorporates the hypothalamic–hypophysial–gonadal axis as part of the sex-control mechanism, and will broaden the perspective of studies of the relationship between environmental factors and natural sex reversal in fish. It has been emphasized that natural sex reversal in fishes is usually a process of successive sexual maturation of the female and male germ cells (Chan, 1970, 1977), and also that the onset time of sex change appears to be genetically governed but environmentally sensitive in *Monopterus* (Chan and Phillips, 1967; Chan, 1970; Chan and O, 1981) and in *Rivulus* (Harrington, 1971, 1975). With regard to genetic control and sexual maturation, it is worthwhile noting that, in the gonochoristic *Xiphophorus maculatus*, a sex-linked gene regulates the onset of sexual maturation of the male and a close relationship exists between the expression of this gene locus and the adenohypophysial function (Kallman and Schreibman, 1973). Kallman (1982) reported that the gene locus, *P*, determines the age of sexual maturation via its control on the hypothalamic–hypophysial axis and on the differentiation of the gonadotropic cells, and that depending on the genotype, *X. maculatus* may mature at age 8 weeks or may remain immature at 2 years. The age at which the fish spontaneously initiates sexual maturation is positively correlated with the genotype-governed responsiveness to luteinizing hormone releasing hormone (LHRH). These recent findings open new areas of research on the triggering mechanism(s) of natural sex reversal in fishes.

IV. EXTRINSIC FACTORS OF SEX CONTROL AND SEX REVERSAL

Experimental studies of extrinsic factors on sex control and sex reversal in fishes are scanty because of the presence of certain technical difficulties, e.g., the necessity of rearing fish embryos or fry under controlled laboratory conditions. Under experimental conditions, it is usually difficult to maintain fishes during early development without excessive losses; the differential mortality of one sex versus the experimental induction of the other cannot be resolved if mortality exceeds a certain limit. It is also possible that not only can one environmental factor modify the physiological effect of another, but a single factor reaching sufficient intensity may alter the effect of other physiological factors. Kinne and Kinne (1962) reported that the developmental rates of *Cyprinodon macularis* increased with oxygen content in the water and decreased with increasing salinity, the latter effect was mediated by a changing coefficient of oxygen absorption and saturation in water. Both the retardation and the acceleration were increasingly accentuated by an increase in temperature (Kinne and Kinne, 1962).

In the study of environmental influences on sex reversal, it is usually difficult to rear long-lived marine teleosts, which comprise a majority of the sex-reversing fishes; therefore, the early history of a wild-caught animal is unknown to the investigator. It has been noted that if the previous history of an animal in interaction with its total environment is imprecisely known, its subsequent responses to environmental factors can be misinterpreted (Aiken, 1969). Harrington (1971) gave a good illustration of this point. In

Rivulus marmoratus, the onset of responsiveness to short day was found to be influenced by the ontogenetic histories of the animal. The responsiveness of fish belonging to the same clone was advanced in time by early rearing at high temperature.

Despite the various difficulties encountered in the study of the problem, the fact that the environment can exert, under certain circumstances, a significant effect on the sex ratio or initiate a sex change is well established. However, almost nothing is known to date about the type of physiological events taking place in the animal under these circumstances, and even the conditions under which it occurs are not often well-defined.

A. Temperature

The study of temperature effect on sex differentiation in gonochoristic fishes involves rearing of the eggs or fry at various temperatures. Padoa (1939) reported that *Salmo irrideus* reared at 17°–20°C differentiated bisexually without the intermediate female stage that had been observed at a lower rearing temperature of 8°–13°C. It was believed that high rearing temperature was responsible for the absence of the intermediate female stage. Female *Salmo trutta* reared at 13°C showed an intersexual phase during sexual differentiation; this intersexual phase was different from that noted in trouts reared at 8°C (Ashby, 1959). However, this temperature treatment did not apparently affect the normal sex ratio after accounting for the possible occurrence of differential mortality. Lucas (1968) also reported that among various environmental factors tested on *Betta splendens*, temperature was the only one that had no effect on sex ratio; no significant variation from the 1:1 sex ratio was found in fish kept at 78° and 82°F. Mires (1974) studied the effect of cold temperature (21°C) on newly hatched tilapias in an attempt to reveal whether temperature is the cause of the production of high percentage of males in captive spawning *T. nilotica* and *T. shirana* (80–100%), and *T. aurea* and *T. valcani* (55–70%); the working procedure of captive spawning exposes the fry to a cold temperature of 20°C. No simple conclusion was reached because of the considerable discrepancy in the sex ratio of the cold-treated tilapias attributable to the high mortality rate of the experimental animals. By rearing *Anguilla anguilla* of various origins at various temperatures, D'Ancona (1959) concluded that elvers had at least "partial genotypic" determination that could undergo a sexual deviation caused by environmental factors. In particular, higher temperatures favored differentiation of males. The evidence was questioned by Sinha and Jones (1966) and regarded as inconclusive by Harrington (1967). However, temperature does have some effect on sex ratio in certain gonochoristic fishes.

Van Doorn (1962) obtained a higher percentage of male cyprinodontids, *Epilatys chaperi*, at low rearing temperature. In agreement with Van Doorn's finding, Lindsey (1962) obtained a higher proportion of female sticklebacks, *Gastersteus aculeatus*, from eggs reared at higher temperature.

The effect of environmental temperature on hermaphroditic teleost has been studied only in *Rivulus marmoratus*. The spontaneous occurrence of primary male in this normally synchronous hermaphrodite was found to be below 5% through more than 10 uniparental laboratory generations. In a study where individuals in their early stages of development were reared, each in its own jar, under eight combinations of bright or dim light, seawater or fresh water, and high or low temperature, Harrington (1967) found that over 35% primary male could be produced by low temperature treatment. In addition to the increase in males, some structural–functional abnormalities were observed to be related to different dim-light–salinity– temperature combinations. Another series of experiments were performed to determine whether the results were caused by male induction or high hermaphrodite mortality. In this second series of experiments, in which mortalities were low and structural–functional abnormalities were absent, it was concluded that primary male production in *Rivulus marmoratus* was correlated with low-temperature rearing (Harrington, 1967). A "thermolabile phenocritical" period of sex determination in *R. marmoratus* was began after stage 31b (formation of neural and hemal arches on caudal vertebrae) and ended before hatching (Harrington, 1968). The temperature threshold for the induction of primary male was found to be at or below 20°C, above which only hermaphrodites were produced. It was noted that exposure to low temperature not only controlled the direction of sex differentiation in *Rivulus* but also related to overall rate of growth and development, and might even prolong the thermolabile phenocritical period itself (Harrington, 1968). Nothing is known of the type of mechanism triggered by low temperature in the embryo of the fish that results in the delayed differentiation of the gonad.

No definitive study has yet been conducted on the effect of temperature on the timing of natural sex reversal in teleosts. Liem (1968) speculated that temperature affects sex reversal in *Monopterus*, but provided no experimental evidence. In *Rivulus marmoratus*, sex reversal from synchronous hermaphrodite to secondary male was induced by high temperatures during early rearing (Harrington, 1971). It was shown that "extraparental" and "prehatching" incubation of eggs at low temperature yielded both primary males and hermaphrodites, and continuous exposure to low temperature for 3–6 months posthatching did not favor the conversion of these hermaphrodites to secondary males. However, early rearing at high temperature was a precondition for the change from hermaphrodite to secondary male. This sex change was triggered by short day conditions and this aspect of the problem is discussed in the following section.

B. Light and Radiation

1. DAY LENGTH

Photoperiod is the most influential celestial factor regulating reproduction in vertebrates. Photoperiod is also known to affect sex reversal in fishes. In *Rivulus marmoratus*, Harrington (1971) found that synchronous hermaphrodites could change to secondary males when day length was 12 hr or less, but early rearing at high temperature was the precondition for the responsiveness to short-day treatment. Following the high-temperature treatment, the testicular zones of the ovotestes were found to increase progressively at a much faster rate than the ovarian zones. When the ratio of the testicular to ovarian tissue became great enough, the next short-day season triggered the development of a secondary testis by a rapid proliferation of the testicular tissue and an involution of the ovarian tissue. Secondary male coloration also began to develop. The number of short-day seasons required for the induction of a sex change was found to be genotype specific. Hermaphrodites exposed to moderate or low temperature at early rearing period responded to short-days only extremely late (or not at all) in their life cycle.

2. LIGHT INTENSITY

Light intensity did not apparently have any effect on sex control or sex reversal in *Rivulus marmoratus* (Harrington, 1971).

3. RADIATION

Bonham and Donaldson (1972) and Donaldson *et al.* (1972) found no alteration in sex ratios of premigratory smolt chinook salmon (*Oncorhynchus tschawytscha*) irradiated with 0.5–50 Rads/day for the first 80 days of life. Gonadal development was retarded in fry, and germ cells were lacking in fingerlings which received more than 10 Rads of radiation per day. Anders and co-workers (1969) found that pregnant female *Platypoecilus maculatus* irradiated with 1000–2500 Rads of X rays gave birth to XY females, which when mated with nonirradiated XY males produced progenies of XX females, XY males, and YY males in a 1:2:1 ratio. This indicated that the control of sex by X rays for the production of XY females was via some physiological processes rather than an induced change in the genetic sex-determining factors. Although these studies provide interesting results, it is unlikely that any fish would be exposed to such a high dose of radiation under natural conditions. Egami and Aoki (1966) showed that X rays at a dose of 2 kRads caused degeneration of large yolky oocytes in female loach, *Misgurnus guillicaudatus*. The same effect was observed even if only the anterior portion of the body was irradiated. It was concluded that this effect was attributable to a reduction of gonadotropin secretion from the pituitary gland. Therefore, it is possible that radiation may sometimes exert its influence on sex differentiation through pituitary hormones which appear to play some part in sex control of certain hermaphroditic fishes (see Section III,E).

C. Water Quality

Very little is known about the quality of water (e.g., the hardness and pH) on sex control in fish. *Betta splendens* has a highly aberrant sex ratio in nature (Lucas, 1968). Either sex may be greatly in excess in a given brood. Sex ratios deviated significantly from 1:1 among fry reared individually in both hard water and soft water; the erratic sex ratios were obtained even in replicates. It was suggested that the quality of water has little effect on sex control and that other factors are involved. Sex determination in *B. splendens* is highly labile (Lucas, 1968).

Among undifferentiated gonochorists, sex differentiation in *Anguilla* is believed to be especially susceptible to environmental influences because of the prolonged indifferent sexual condition. In his extensive embryological studies, Grassi (1919) found that salinity, and probably nutrition produced a sex change in *Anguilla anguilla*. Tesch (1928), Gandolfi-Hornyold (1931), and Bertin (1956) also believed that environmental factors influence sex in eels. In a number of cases, the vast majority of *A. anguilla* found in estuaries were males. Exceptional cases have been reported. In Italy, Grassi (1919) found that a coastal lake at Porto contained only females, but another lake at Orbetallo, similarly situated, contained only males. In Holland, a high percentage of males (70–74%) was found in two seawater lakes, the Zuider Zee and the Wadden Zee (Tesch, 1928). However, these early studies deserve careful reexaminations. Recently, Egusa and Hirose (1973) found no difference in sex ratios between *A. anguilla* in freshwater ponds and those in salt water ponds.

The effect of salinity on sex control in the European eel has long been a puzzling phenomenon. D'Ancona (1949b,1950) argued that differential migration could account for the segregation of sexes in *A. anguilla*. However, Sinha and Jones (1966, 1975) held a different view and suggested that the sex of each eel was predetermined and that the elvers were distributed at random when they reached the coast, because they found both male and female *A. anguilla* in freshwater as well as in brackish water. Egusa and Hirose

(1973) found that the female had a greater growth rate. This observation lends support to Sinha and Jones' hypothesis (1966) that the female moves out from the crowded condition in order to continue for her relatively faster growth, but the male survives in the crowded estuaries.

D. Crowding

It is interesting to note that among teleosts, most studies on the effect of crowding on sex control reported that this factor usually favored the differentiation of males. However, two reports indicated that crowded condition produce higher percentages of females (Lindsey, 1962; Kuhlmann, 1975).

Lindsey (1962) reported that crowded rearing conditions produce higher percentages of female *Gasterosteus aculeatus*. D'Ancona (1950) suggested that crowding favors the differentiation of male eels. After studying the elvers reared in ponds at varying degrees of crowding, Fidora (1951) agreed with D'Ancona that the percentage of male eels was positively correlated with increasing stocking density. Sinha and Jones (1975) believed that the duration of all these experiments on eels was too short, making the correct identification of sex difficult. Kuhlmann (1975) obtained a higher percentage of females by rearing the elvers with a stocking density 200 times higher than that in Fidora's experiments. Although the results of Fidora's and Kuhlmann's experiments were different, Kuhlmann (1975) concurred with D'Ancona (1951) that sex determination in *Anguilla anguilla* is basically genetic but can be influenced by environmental conditions.

By rearing *Betta splendens* in crowded conditions, Eberhardt (1943) obtained a statistically significant excess of males and suggested that poor space, food, and water favor male differentiation. He tried to eliminate the possibility of selective mortality of females by keeping 25 other broods on scanty food, thereby maximizing the usual high mortality in the first 2 weeks of life; the survivors were well fed after the first 2 weeks. Because experimental death was less than 1% after the first 2 weeks, it was concluded that there was no selective mortality of females in his previous experiments. Harrington (1967) commented that "under-feeding and crowding cannot *a priori* be equated with regard to selective mortality, nor can either *a priori* be assumed without influence on sex determination."

In general, males are the less viable sex, as measured by longevity and resistance to adverse environmental agents (Atz, 1964). The biological significance of developing into males under unfavorable conditions such as crowding awaits further explanation. The manner in which adverse conditions may affect sex is uncertain. It may involve the accumulation of hormones, waste products, or other biochemical substances in confined quarters. Egami (1954) demonstrated that in *Oryzias latipes*, close confinement results in the uptake of androgenic substances released by other fishes. Recently, Howell and co-workers (1980) reported that the effluent discharge of a paper mill caused phenotypical masculinization in the normally gonochoristic mosquito fish *Gambusia affinis*. The precise chemical action of these substances on sex control remains to be elucidated.

In sex-reversing *Monopterus albus*, Liem (1963) suggested that exceptionally severe ecological conditions such as periodic drought and malnutrition can cause sex reversal. The suggestion was based on no experimental evidence but rather on an assumption that malnutrition causes degeneration of ovaries in fishes (Suworow, 1959) and that, in some fishes, degeneration of the ovary is accompanied by the development of male characteristics (Bullough, 1940; D'Ancona, 1950). In a crowded condition or seasonal drought, the oxygen tension in the water is likely to be affected. Although Harrington (1967) did not intend to study the effect of oxygen on sex differentiation of *Rivulus marmoratus*, he estimated the concentration of oxygen in water from the data of Kinne and Kinne (1962), and came to the conclusion that oxygen tension has no effect on sex differentiation.

E. Social Factors

The concept of social influence on sex control and sex reversal has been suggested only recently. Fishelson (1970) postulated that social factors such as the presence or absence of a male in a social group contribute to sex reversal in coral fish. He found that when the male was removed from a group of female *Anthias squamipinnis*, one of the females would change sex, developing the typical male color and behavior. If this new male was removed, another female would develop into a male. It was concluded that sex reversal in this protogynous hermaphrodite was regulated by the presence or absence of a male fish within the social group. Fishelson's experiments have stimulated many further studies. Robertson (1972) reported that sex reversal in the protogynous *Labroides dimidiatus* is controlled by a dominant male in a group. In his observation, each group consisted of a male with a harem of females. The male in each harem suppressed the production of other males by aggressive dominance over the females. Death of the male released this suppression and the most dominant female of the harem immediately change sex. A similar phenomenon has been found in protandrous species. Fricke and Fricke (1977) presented a report on the protandrous *Amphiprion alkallopisos* and *A. bicinctus* in which females controlled the production of females by aggressive dominance over males. Other field observations on *Thalassoma bifaciatum* (Warner et al., 1975), two *Paragobiodon* species (Lassig, 1977), *Amphiprion melanopus* (Ross, 1978), six other species of *Amphiprion* (Moyer and Nakazono, 1978b), *Centropyge interruptus* (Moyer and Nakazono, 1978a), and *C. resplendens* (Bruce, 1980) also suggested that sex reversal might be initiated by a sudden alteration in the sex composition of the social groups.

Shapiro (1981) noted that most of these studies failed to provide control experiments in either field observations or in laboratory studies. Therefore, sex reversal might still be related to other factors, for instance a nonspecific disruption of the social group instigated by the predator's attack or by the experimentor's removal (spearing) of the male and a high rate of spontaneous sex reversal, which coincided with the male removal. To eliminate some of these variables, Shapiro (1981) performed a series of controlled male-removal studies on 11 single-male and 5 multiple-male social groups of *Anthias squamipinnis* under both the laboratory and the field conditions. Laboratory controls limited the likelihood that sex reversal was induced by some non-specific disruption; field observation demonstrated that sex reversal resulted from male removal and that it was not a coincidental ongoing spontaneous event. From these studies, Shapiro (1981) concluded that, in general, the removal of n males leads to the sex reversal of n females, and sex reversal that is not preceded by male disappearance occurs only infrequently.

The interesting male-removal experiments alone permit only the most general and simple conclusion, i.e., sex reversal can be initiated by certain alteration of the group composition. Shapiro (1981) argued that the experiment itself could not distinguish the effect of male removal from the effect of male absence, and that it could not distinguish a mechanism of disinhibition from a mechanism of stimulation. It did not indicate any specific male–female interaction that was the causal factor of initiating a sex change. The presence of all female groups of *A. squamipinnis* both in the field and in the established laboratory condition, and the fact that the removal of one male from a multiple-male group of *A. squamipinnis* led to the reversal of only one female despite the presence of other males in the group, suggested that at least in *A. squamipinnis*, sex reversal is not controlled simply by the absence or presence of males among females (Warner, 1975; Shapiro, 1979; 1981, 1982). The sex change phenomenon in *A. squamipinnis* was further complicated by the fact that simultaneous multiple–male removal resulted in serial sex reversals in females of the group with a mean interval of 2 days between successive onset time of sex reversal (Shapiro, 1980). Therefore, it appears that sex change of an individual can be influenced by other sex reversals within the group. These investigations involving male removal and studies of behavioral interactions among members within the social group mark only the beginning of research for a better understanding of the phenomenon of natural sex reversal in coral fishes.

Behavioral studies of sex-reversing hermaphrodites have been conducted on *Labroides dimidiatus* (Robertson and Hoffman, 1977; Robertson and Warner, 1978), *Amphiprion alkallopisos* and *A. bicinctus* (Fricke and Fricke, 1977), and in much greater detail in *Anthias squamipinnis* (Shapiro, 1979). Shapiro (1977, 1979) found that male and female *A. squamipinnis* had different behavioral profiles when six specific types of their behaviors were studied quantitatively. Male removal affected the behavioral profile of all females of the group. The magnitude of change of behavior was related directly to the dominance rank of the females. During sex reversal, the most dominant female altered her behavioral profile from that of a female to a male. At the same time, the remaining females began to treat the sex-reversing female as if she were a male. In the naturally occurring all-female groups, the most dominant female has a profile of "received behaviors" similar to that of a male, but she does not change into a male as the dominant female in male-removal experiments; this observation suggests that being suddenly treated as a male could not be in itself the cause of sex reversal (Shapiro, 1977). By comparing the behavioral profile of different sexes in *A. squamipinnis*, Shapiro (1979) proposed the "priming hypothesis" for initiating sex reversal. According to this hypothesis, a male is required to prime or to provide the behavioral preconditions for the most dominant female to undergo sex reversal after his disappearance. The most dominant female is actively stimulated to change sex by the alteration of a particular behavior or a set of behaviors at a critical magnitude after the male removal (Shapiro, 1979). This hypothesis is markedly different from the concept that the most dominant female changes sex because inhibition has been removed following the disappearance of the male (Robertson, 1972). Further studies suggested that the central nervous system of the most dominant female might play a role in the sex reversal by comparing the expected and the actual "behavioral received profile" the animal encounted; sex reversal occurred when the difference in these behavioral profiles was sufficiently great (Shapiro, 1982).

A sex ratio threshold was proposed in *A. squamipinnis* (Shapiro, 1977, 1979; Shapiro and Lubbock, 1980). According to this model, a female changed sex as soon as the ratio of females to males within a group exceeded a certain threshold value. A female "recognized" the sex ratio of its group by the amount of male–female behavioral interactions. Sex ratio might also be detected through visual cues because sex reversal in *A. squamipinnis* can be prevented for a time if a group of females sees a male behind a glass partition (Fishelson, 1970, 1975). Data from a population of *A. squamipinnis* in the Sudanese Red Sea (Shapiro and Lubbock, 1980) and *L. dimidiatus* (Shapiro, 1979) apparently fit the model well. However, the model is less satisfactory for accounting for the presence of all female groups in *A. squamipinnis*.

The effect of social condition on the sex ratio of gonochoristic teleost is uncertain. According to a report by Goedakian and Kosobutzky (1969), sex ratio in a population of *Lebistes (Poecilia) reticulatus* can be regulated by the sex ratio of the parental generation. Of four glass vessels, each containing 5 males and 5 females, two vessels were placed into a large aquarium containing 90 males, and into the other two vessels, the water from the large aquarium with 90 males was added. In the former two vessels, the sex ratios of the progeny showed significant increase toward the female sex (42.7% male), but no change in sex ratio was observed in the latter two vessels. When the apparent sex ratio of the parent was raised to 152:1, again a significant shift in the value of sex ratio in the progeny was obtained. The mechanism of this regulation of sex ratio requires further investigation.

F. Other Factors

Huxley (1923) investigated the effect of time of fertilization on the sex ratio of the brown trout, *Salmo trutta*. He found no significant effect by fertilizing eggs 7 days earlier and he regarded the slight excess of males obtained by delayed fertilization for 21 days after normal time of shedding as of doubtful significance. However, Mřsić (1923) noted that in rainbow trout a moderate delay in fertilization (4–7 days) resulted in a slight preponderance of females; with a considerable delay of 21 days, he obtained 55% males, 33% females, and 12% hermaphrodites. The sex ratio of the control was 1:1.

Harrington (1967) commented that the mortality in Mřsić's experiment was more than enough to create a dilemma of a differential mortality of one sex versus experimental induction of the other. Mřsić (1923) tried to discount the mortality as having occurred too early in ontogeny to be the deciding factor; his argument was based on the histology of differentiating gonad 121 days after hatching. This histological study was later negated by more thorough studies on the gonads of rainbow trout (Padoa, 1939) and other salmonids (Ashby, 1952; Robertson, 1953).

In a number of cases, the precise factor(s) affecting sex determination in fish was ill-defined. It might perhaps be possible that a combination of several factors form the resultant cause of the difference in the phenotypic expression. A greater proportion of males appeared in selected strains of guppies and medakas during the summer months (Winge, 1934; Aida, 1936).

The cyprinodontid fish *Rivulus marmoratus* is unique among fishes in being consistently self-fertilizing hermaphrodites (Harrington, 1961, 1963). It is the first vertebrate known to exist in nature as a homozygous clone (Harrington and Kallman, 1968). In Florida, no secondary male and only one primary male was found by Harrington (1968). However, primary males appeared to be not uncommon on the island of Curacao (Hoedman, 1958;

Kristensen, 1965). By comparing the spawning activity and embryonic development of the fish with environmental factors, such as temperature profile, salinity and photoperiod in Florida, it appeared that some, or all, of these factors might be the cause of the nonprevalence of primary male gonochorists in Florida (Harrington, 1968). The case in Curacao is not at all clear, and it would not be surprising that primary and secondary males are produced here by influence of environmental factors that are different from those found in Florida.

V. INTERACTION OF GENETIC AND ENVIRONMENTAL FACTORS IN SEX CONTROL AND SEX REVERSAL

There can be little doubt that sex phenotype is a consequence of the interaction between genetic constitution of the organism and the environment, although the extent of environmental influence varies from one species to another (see Chan and O, 1981). Because the information available at present is incomplete and fragmentary, the precise mechanism by which external factors affect sex determination and sex differentiation remains unclear.

For an extrinsic factor to exert its effect, the stimulus itself must be in some way "recognized" by the animal and translated into biological signals which in turn control the biochemistry of sex differentiation and sex maturation. It has been suggested that a neuroendocrine pathway may bridge an external stimulus and the internal events that control sex (Chan et al., 1975; Chan and O, 1981). Experimental evidence indicates that some fishes can perceive these "sex-control" stimuli through a visual channel. In the gonochoristic Lebistes (Poecilia) reticulatus, the sex ratio of a new generation is affected by the sex ratio of the population that surrounds the parental generation (Goedakian and Kosobutzky, 1969). Visual isolation, but not acoustic isolation, can block this regulatory mechanism. Among hermaphrodites, sex reversal in a group of Anthias squamipinnis after male removal can be prevented for a time if the females can view a male through a glass partition (Fishelson, 1970, 1975). The behavioral pattern perceived allows an Anthias to determine its social status. By comparing the relative proportion of different behaviors received with a template of expected proportion, the central nervous system of this species is believed to have the capability to trigger sex reversal when the difference between the actual and expected proportion is sufficiently great (Shapiro, 1982). How these signals via sensory perception can be expressed as physiological changes in sex control require further elucidation. In vertebrates, it has been established that the central nervous system and the hypothalamus intimately control the adenohypophysial–gonadal function. Therefore, it seems possible that the central nervous system may act through the pituitary in sex control and maturation of the germinal elements in the gonad. As discussed in Section III, E, gonadotropins do affect sex reversal in some teleosts.

However, it is not impossible that environmental factors may act directly on the gonad. Low temperature is known to induce the production of primary males in *Rivulus marmoratus*. Gonadal morphogenic studies show that low temperature delays sexualization and decreases the mitotic rate of the germ cells. Low temperature also induces an enlargement of the hilar stroma of the developing gonad before sexualization of the germ cells (Harrington, 1975). It has been suggested that the low-temperature induction of primary male may be induced by the lowering of germ cell mitotic rate and by the development of the hilar stroma; these effects may retard the process of male germ cells or suppress the normal differentiation of female cells.

The susceptibility of a fish to environmental influence on sex differentiation and maturation varies according to the species and its sex genotypes. Chan (1970), in reviewing the sex pattern of vertebrates, suggests that gonadal ontogeny is determined by a developmental homeostasis which is a result of the action of the male and female sex-determining genes and a developmental switch mechanism. The former controls the detailed developmental processes for the expression of different sex phenotypes. The latter, which may be of a genetic and/or environmental nature, determines the expression of alternative sex programs and the stage of sex ontogeny at which a particular sex should commence differentiation and maturation. The onset of the operation of the switch mechanism controls the duration of the critical period for sex determination and is believed to be genotype specific and age dependent. A detailed discussion on the interaction of genotype and environmental factors has been published (Chan and O, 1981).

VI. ADVANTAGES OF HERMAPHRODITISM

Many explanations have been offered for the origin of hermaphroditism. Most of them are based on the advantages obtained by an individual existing as a hermaphrodite. For example, hermaphroditism is believed to increase zygote number (Smith, 1967). In a detailed review on the evolution of hermaphroditism, Ghiselin (1969) discussed three different models: the low-density model, the size-advantage model, and the gene-disperse model. The low-density model, which offered an explanation for the occurrence of self-fertilizing hermaphroditism in an isolated habitat, suggested that hermaphroditism occurred when it was difficult to find a suitable mate in a low-density population. Consecutive hermaphroditism was explicable by the size-advantage model. According to this theory, hermaphroditism evolves when an individual reproduces most efficiently first as one sex and then as a member of the other sex when it grows older and larger in size. The gene-dispersal model suggested that hermaphroditism maintained genetic variability by reducing inbreeding and preventing genetic drift in a population. These three models assumed that an animal's reproductive organs were adapted to maximize the probability of the individual's genetic material being incorporated into the next generation.

Using population-genetic techniques and based on the same assumption, mathematical–genetic theories have been developed to explain the occurrence of sequential hermaphroditism (Warner *et al.*, 1975; Leigh *et al.*, 1976; Charnov *et al.*, 1978). A number of these models support Ghiselin's size-advantage model (Warner, 1975; Warner *et al.*, 1975; Leigh *et al.*, 1976). In general, the various models suggest that sex change is favored if one sex gains in fertility much more rapidly with age than the other. One of these models fits the reproductive biology of the protogynous *Thalassoma bifasciatum* in which the fertility of the large males may be 100 times that of the females; the same model also explain the presence and the absence of primary males in *T. bifasciatum* and *Labroides dimidiatus* respectively (Warner *et al.*, 1975). Similarly, mathematical theoretical model could also correctly predict the age of sex change in the protandrous shrimp, *Pandalus jordani*, which corresponds to the time when the reproductive value of one sex increases by a percentage exceeding the percentage loss of the other sex before sex reversal (Charnov *et al.*, 1978).

Undoubtedly, the adaptive significance of certain patterns of hermaphroditism, such as those found in *Rivulus marmoratus* and coral fish, can be reasonably understood in relation to their particular socioecological settings. However, sex control in fish under natural conditions is certainly a diverse problem and spontaneous sex reversal in fish is of wide occurrence in many unrelated groups; what is known in one species might or might not necessarily apply to the others, particularly when ecological background and selective pressure are different. Workers in this field must exercise caution and avoid the danger of oversimplification and generalization. Also, the evolutionary values of any particular character must balance the advantages against the disadvantages the character incurs. For example, the increase in fertility and reproductive success with larger size or more vivid sexual coloration may be counterbalanced by an increase in risk of predation. Most of the mathematical theories concerning the evolutionary significance of hermaphroditism do not consider the anatomical, physiological, and metabolic costs to individuals involved in sex change. In addition to these biological costs, there is also a genetic penalty for hermaphroditism if death rate and the proportionate gain of fertility with age are the same in both sexes (Warner *et al.*, 1975). It should also be noted that, as far as sexual reproduction is concerned, the selective value of hermaphroditism in some fish may not necessarily rest on any advantage it might render over gonochorism. Indeed, the sequential expression of female and male phases in separate spawning seasons among some protogynous fishes such as *Monopterus* could have provided a mechanism as advantageous as gonochorism in the prevention of self-fertilization (Chan, 1970).

The mechanisms of sex control in vertebrates vary according to the phylogenetic groups and the interactions between sex genotypes and environmental influences. Because each species under natural condition is in dynamic equilibrium with its physical and biotic environment, environmental factors would have a tremendous impact on the mode of reproduction and sex mechanism of the organism, particularly among some amphibians and fishes where the genetic sex determining mechanism is nondecisive. The adaptive significance of an environmental mechanism of sex control and a nondecisive sex-determining genotype apparently lies on its lability which allows flexibility of the animal to cope with its environmental condition, especially in certain social and ecological settings where the reproductive success of a social group, or of the species, would be strongly influenced by the dynamic environmental conditions (Chan and O, 1981). On the contrary, a decisive genetic sex-determining mechanism permits an early and definite differentiation of sex; therefore, an individual may become a better male or female. It would appear that nature favors gonochorism and decisive genetic sex determination when a highly specialized reproductive system is required for reproductive success, as in the case of mammals; hermaphroditism with its flexible environmental sex control mechanisms is advantageous to the survival of the species in the special socioecological settings of some fishes such as *Rivulus* and *Anthias* (see Chan and O, 1981).

ACKNOWLEDGMENTS

Research on *Monopterus* was supported by a research grant to S. T. H. Chan from the Nuffield Foundation, London. Research was also supported by grants from the University of Hong Kong.

REFERENCES

Aida, T. (1936). Sex reversal in *Aplocheilus latipes* and a new explanation of sex differentiation. *Genetics* **21**, 136–153.
Aiken, D. E. (1969). Photoperiod, endocrinology and the crustacean molt cycle. *Science* **164**, 149–155.

Anders, A., Anders, F., and Rase, S. (1969). XY females caused by X-irradiation. *Experientia* **25**, 871.

Ashby, K. R. (1952). Sviluppo dil sistema reproduttivo di *Salmo trutta* L., in condizioni normali e sotto l'influenza di ormoni steroidi. *Riv. Biol.* **44**, 3–18.

Ashby, K. R. (1959). L'effetto di basse concentrazioni di estradiolo e di testosterone sul differenziamento gonadico di *Salmo trutta* L. (Nota preliminare). *Riv. Biol.* **51**, 453–468.

Atz, J. W. (1964). Intersexuality in fishes. *In* "Intersexuality in Vertebrates including Man" (C. N. Armstrong and A. J. Marshall, eds.), pp. 145–232. Academic Press, New York.

Baldwin, F. M., and Li, M. H. (1942). Effects of gonadotropic hormone in the fish, *Xiphophorus helleri* Heckel. *Proc. Soc. Exp. Biol. Med.* **49**, 601–604.

Bennett, D., Boyse, E. A., Mathieson, B. J., Scheid, M., Wachtel, S. S., Yanagisawa, K., and Cattanach, B. M. (1977). Serological evidence for H-Y antigen in Sxr-XX sex-reversed phenotypic males. *Nature (London)* **265**, 255–257.

Bertin, L. (1956) . "Eels—A Biological Study." Cleaver-Hume Press, London.

Bertin, L. (1958). Sexualite et fecondation. *In* "Traité de zoologie" (P.-P. Grasse, ed.), Vol. 13, pp. 1584–1652. Masson, Paris.

Bonham, K., and Donaldson, L. R. (1972). Sex ratios and retardation of gonadal development in chronically gamma-irradiated chinook salmon smolts. *Trans. Am. Fish. Soc.* **101**, 428–434.

Bruce, R. W. (1980). Protogynous hermaphroditism in two marine angelfishes. *Copeia* pp. 353–355.

Bullough, W. S. (1940). A study of sex reversal in the minnow *Phoxinus laevis*. *J. Exp. Zool.* **85**, 475–501.

Bullough, W. S. (1947). Hermaphroditism in lower vertebrates. *Nature (London)* **160**, 9–11.

Burns, R. K. (1961). Role of hormones in the differentiation of sex. *In* "Sex and Internal Secretions" (W. C. Young, ed.), Vol. 1, pp. 76–158. Williams & Wilkins, Baltimore, Maryland.

Chan, S. T. H. (1970). Natural sex reversal in vertebrates. *Philos. Trans. R. Soc. London, Ser. B* **259**, 59–71.

Chan, S. T. H. (1977). Spontaneous sex reversal in fishes. *In* "Handbook of sexology" (J. Money and H. Musaph, eds.), pp. 91–105. Elsevier/North-Holland Biomedical Press, Amsterdam.

Chan, S. T. H., and Lee, Y. H. (1978). Some aspects of the "gonadotropic activity" in *Monopterus* pituitary homogenate on mammalian tissue. *Gen. Comp. Endocrinol.* **34**, Abstr. 52.

Chan, S. T. H., and O, W. S. (1981). Environmental and non-genetic mechanisms in sex determination. *In* "Mechanisms of Sex Differentiation in Animals and Man" (C. R. Austin and R. G. Edwards, eds.), pp. 55–113. Academic Press, New York.

Chan, S. T. H., and Phillips, J. G. (1967). The structure of the gonad during natural sex reversal in *Monopterus albus* (Pisces: Teleostei). *J. Zool.* **151**, 129–141.

Chan, S. T. H., and Phillips, J. G. (1969). The biosynthesis of steroids by the gonads of the ricefield eel *Monopterus albus* of various phases during natural sex reversal. *Gen. Comp. Endocrinol.* **12**, 619–636.

Chan, S. T. H., Tang, F., and Lofts, B. (1972a). The role of sex steroids on natural sex reversal in *Monopterus albus*. *Int. Congr. Ser.—Excerpta Med.* **256**, 348.

Chan, S. T. H., O, W. S., Tang, F., and Lofts, B. (1972b). Biopsy studies on the natural sex reversal in *Monopterus albus* (Pisces: Teleostei). *J. Zool.* **167**, 415–421.

Chan, S. T. H., O, W. S., and Hui, S. W. B. (1975). The gonadal and adenohypophysial functions of natural sex reversal. *In* "Intersexuality in the Animal Kingdom" (R. Reinboth, ed.), pp. 201–221. Springer-Verlag, Berlin and New York.

Chan, S. T. H., Ng, T. B., O, W. S., and Hui, S. W. B. (1977). Hormones and natural sex succession in *Monopterus*. *In* "Biological Research in Southeast Asia" (B. Lofts and S. T. H. Chan, eds.), pp. 101–120. University of Hong Kong, Hong Kong.

Channing, C. P., Licht, P., Papkaff, H., and Donaldson, E. M. (1974). Comparative activities of mammalian, reptilian and piscine gonadotrophins in monkey granulosa cell culture. *Gen. Comp. Endocrinol.* **22**, 137–145.

Charnov, E. L., Gotshall, D. W., and Robinson, J. G. (1978). Sex ratio: Adaptive response to population fluctuations in pandalid shrimp. *Science* **200**, 204–206.

Chieffi, G. (1959). Sex differentiation and experimental sex reversal in elasmobranch fishes. *Arch. Anat. Microsc. Morphol. Exp.* **48**, Suppl., 21–36.

Choat, J. H., and Robertson, D. R. (1975). Protogynous hermaphroditism in fishes of the family Scaridae. *In* "Intersexuality in the Animal Kingdom" (R. Reinboth, ed.), pp. 263–283. Springer-Verlag, Berlin and New York.

Clemens, H. P., McDermitt, C., and Inslee, T. (1966). The effects of feeding methyl testosterone to guppies for sixty days after birth. *Copeia* pp. 280–284.

Colombo, L., Conte, E. D., and Clemenz, P. (1972). Steroid biosynthesis *in vitro* by the gonads of *Sparus auratus* L. (Teleostei) at different stages during natural sex reversal. *Gen. Comp. Endocrinol.* **19**, 26–36.

Crew, F. A. E. (1952). The factors which determine sex. *In* "Marshall's Physiology of Reproduction" (A. S. Parkes, ed.), Vol. II, pp. 741–784. Longmans, Green, New York.

D'Ancona, U. (1949a). Ildifferenziamento della gonade e l'inversione sessuale degli sparidi. *Arch. Oceanogr. Limnol.* **6**, 97–163.

D'Ancona, U. (1949b). Condizioni ambietali e correlazioni umorali nel differenziamento sessuale e nello sviluppo dell'ánguilla. *Verh. Int.—Ver. Theor. Angew. Limnol.* **10**, 126–133.

D'Ancona, U. (1950). Détermination et différenciation du sexe chez les poissons. *Arch. Anat. Microsc. Morphol. Exp.* **39**, 274–292.

D'Ancona, U. (1951). Tentaivi di deviazione sessuale nell'anguilla. *Boll. Zool.* **18**, 97–101.

D'Ancona, U. (1952). Territorial sexualization in the gonads of teleosteans. *Anat. Rec.* **114**, 666–667.

D'Ancona, U. (1955). Osservazioni sulle gonadi giovanili di *Amia calva. Arch. Ital. Anat. Embriol.* **60**, 184–200.

D'Ancona, U. (1959). Distribution of the sexes and environmental influence in the European eel. *Arch. Anat. Microsc. Morphol. Exp.* **48**, Suppl., 61–70.

Dipper. F. A., and Pullin, R. S. V. (1979). Gonochorism and sex-inversion in British Labridae (Pisces). *J. Zool.* **187**, 97–112.

d'Istria, M., Valentino, A., and Guasco, R. (1973). Dosaggio radioimmunologica di ormoni sessauli nel plasma di *Serranus cabrilla* durante il ciclo riproduttivo. *Atti. Soc. Peloritana, Sci. Fis., Mat. Nat.* pp. 241–244.

Dodd, J. M. (1960). Genetic and environmental aspects of sex determination in cold-blooded vertebrates. *Mem. Soc. Endocrinol.* **7**, 17–44.

Donaldson, L. R., Bonham, K., and Hershberger, W. K. (1972). Low level chronic irradiation of salmon. *Annu. Prog. Rep.—NTIS Springfield Va.* RLO-2225-T-2-3, 35p.

Dzwillo, V. M. (1962). Über Künstliche Erzeugung funktioneller Männchen weiblichen Genotyps bei *Lebistes reticulatus. Biol. Zentralbl.* **81**, 575–584.

Eberhardt, K. (1943). Geschlechtbestimmung und -differenzierung bei *Betta splendens* Regan. *Z. Indukt. Abstamm- Vererbungsl.* **81**, 363–373.

Eckstein, B., Abraham, M., and Zohar, Y. (1978). Production of steroid hormones by male and female gonads of *Sparus auratus* (Teleostei, Sparidae). *Comp. Biochem. Physiol. B* **60B**, 93–97.

Egami, N. (1954). Influence of temperature on the appearance of male characters in female of the fish, *Oryzias latipes*, following treatment with methyl-dihydrotestosterone. *J. Fac. Sci., Univ. Tokyo, Sect. 4* **7**, 281–298.

Egami, N., and Aoki, K. (1966). Effects of X-irradiation of a part of the body on the ovary in the loach, *Misgurus anguillicaudatus. Annot. Zool. Jpn.* **9**, 7.

Egusa, S., and Hirose, H. (1973). Further notes on sex and growth of European eels in culture ponds. *Bull. Jpn. Soc. Sci. Fish.* **39**, 611–616.

Eversole, W. J. (1941). The effect of pregnenolone and related steroids on sexual development in the fish (*Lebistes reticulata*). *Endocrinology* **28**, 603–610.

Fidora, M. (1951). Influenza dei fattori ambietali sull'accrescemento e sul differenziamento sessuale delle anguille. *Nova Thalassia* **1**, 9.

Fishelson, L. (1970). Protogynous sex reversal in the fish *Anthias squamipinnis* (Teleostei, Anthiidae) regulated by the presence or absence of a male fish. *Nature (London)* **227**, 90.

Fishelson, L. (1975). Ecology and physiology of sex reversal in *Anthias squamipinnis* (Peters) (Teleostei: Anthiidae). *In* "Intersexuality in the Animal Kingdom" (R. Reinboth, ed.), pp. 284–294. Springer-Verlag, Berlin and New York.

Forbes, T. R. (1961). Endocrinology of reproduction in cold-blooded vertebrates. *In* "Sex and Internal Secretions" (W. C. Young, ed.), pp. 1035–1087. Baillière, London.

Fricke, H., and Fricke, S. (1977). Monogamy and sex change by aggressive dominance in coral reef fish. *Nature (London)* **266**, 830–832.

Gandolfi-Hornyold, A. (1931). Le sexe de la petite anguille de repeuplement du marais de la Grande-Brière après un séjour de trois ans dans un Aquarium du Muséum. *Bull. Mus. Natl. Hist. Nat.* [2] **3**, 423.

Geodakian, V. A., and Kosobutzky, V. I. (1969). The nature of the feedback mechanism of sex regulation. *Genetika (Moscow)* **5**, 119–126.

Ghiselin, M. T. (1969). The evolution of hermaphroditism among animals. *Q. Rev. Biol.* **44**, 189–208.

Gordon, M. (1947). Genetics of *Platypoecilus maculatus*. IV. The sex determining mechanism in two wild populations of the Mexican platyfish. *Genetics* **32**, 8–17.

Gordon, M. (1952). Sex determination in *Xiphophorus (Platypoecilus) maculatus*. III. Differentiation of gonads in platyfish from broods having a sex ratio of the three females to one male. *Zoologica (N.Y.)* **37**, 91–100.

Grassi, B. (1919). Nuvo ricerche sulla storia naturale dell'Anguilla. *Mem. R. Com. Talass. Ital.* **67**, 1–141.

Hardisty, M. W. (1965). Sex differentiation and gonadogenesis in lampreys. Parts I and II. *J. Zool.* **146**, 305–387.

Harrington, R. W., Jr. (1961). Oviparous hermaphroditic fish with internal self-fertilization. *Science* **134**, 1749–1750.

Harrington, R. W., Jr. (1963). Twenty-four-hour rhythms of internal self-fertilization and of oviposition by hermaphrodites of *Rivulus marmoratus*. *Physiol. Zool.* **36**, 325–341.

Harrington, R. W., Jr. (1967). Environmentally controlled induction of primary male gonochorists from eggs of the self-fertilizing hermaphroditic fish *Rivulus marmoratus*, Poey. *Biol. Bull. (Woods Hole, Mass.)* **132**, 174–199.

Harrington, R. W., Jr. (1968). Delimitation of the thermolabile phenocritical period of sex determination and differentiation in the ontogeny of the normally hermaphroditic fish, *Rivulus marmoratus* Poey. *Physiol. Zool.* **41**, 447–460.

Harrington, R. W., Jr. (1971). How ecological and genetic factors interact to determine when self-fertilizing hermaphrodites of *Rivulus marmoratus* change into functional males, with a reappraisal of the modes of intersexuality among fish. *Copeia* pp. 389–432.

Harrington, R. W., Jr. (1975). Sex determination and differentiation among uniparental homozygotes of the hermaphroditic fish *Rivulus marmoratus* (Cyprinodonidae: Antheriniformes). *In* "Intersexuality in the Animal Kingdom" (R. Reinboth, ed.), pp. 249–262. Springer-Verlag, Berlin and New York.

Harrington, R. W., Jr., and Kallman, K. D. (1968). The homozygosity of the self-fertilizing hermaphroditic fish, *Rivulus marmoratus* Poey (Cyprinodontidae, Antheriniformes). *Am. Nat.* **102**, 337–343.

Hoar, W. S. (1969). Reproduction. In "Fish Physiology" (W. S. Hoar and D. J. Randall, eds.), Vol. 3, pp. 1–72. Academic Press, New York.

Hoedeman, J. J. (1958). Rivulid fishes of the Antilles. Stud. Fauna Curacao Carib. Isl. 8, 112–126.

Howell, W. M., Black, D. A., and Bortone, S. A. (1980). Abnormal expression of secondary sex characters in a population of mosquitofish, Gambusia affinis holbrooki. Evidence for environmentally induced masculinization. Copeia pp. 676–681.

Huxley, J. S. (1923). Late fertilization and sex ratio in trout. Science 58, 291–292.

Idler, D. R., and Ng, T. B. (1979). Studies on two types of gonadotropins from both salmon and carp pituitaries. Gen. Comp. Endocrinol. 38, 421–440.

Idler, D. R., Reinboth, R., Walsh, J. M., and Truscott, B. (1976). A comparison of 11-hydroxytestosterone and 11-ketotestosterone in blood of ambisexual and gonochoristic teleosts. Gen. Comp. Endocrinol. 30, 517–521.

Jellyman, D. J. (1976). Hermaphroditic European perch Perca fluviatilis L. N. Z. J. Mar. Freshwater Res. 10, 721–723.

Jost, A. (1965). Gonadal hormones in the sex differentiation of the mammalian fetus. In "Organogenesis" (R. L. de Haan and H. Ursprung, eds.), pp. 611–628. Holt, New York.

Jost, A. (1970). Hormonal factors in the sex differentiation of the mammalian foetus. Philos. Trans. R. Soc. London, Ser. B 259, 119–130.

Kallman, K. D. (1965a). Sex determination in the teleost Xiphophorus milleri. Am. Zool. 5, 246–247.

Kallman, K. D. (1965b). Genetic and geography of sex determination in the poeciliid fish, Xiphophorus maculatus. Zoologica (N.Y.) 50, 151–190.

Kallman, K. D. (1970). Sex determination and the restriction of sex-linked pigment patterns to the X and Y chromosomes in populations of a poeciliid fish, Xiphophorus maculatus from the Belize and Sibun Rivers of British Honduras. Zoologica (N.Y.) 55, 1–16.

Kallman, K. D. (1973). The sex-determine mechanism of the platyfish, Xiphophorus maculatus. In "Genetics and Mutagenesis of Fish" (J. H. Schroder, ed.), p. 19. Springer-Verlag, Berlin and New York.

Kallman, K. D. (1982). Genetic control of sexual maturation in fishes of the genus Xiphophorus (Poeciliidae). Proc. Symp. Comp. Endocrinol., 9th, 1981 Abstract, p. 161.

Kallman, K. D., and Schreibman, M. P. (1973). A sex-linked gene controlling gonadotropic differentiation in determining the age of sex maturation and size of the platyfish, Xiphophorus masculatus. Gen. Comp. Endocrinol. 21, 287–304.

Kawaguchi, K., and Marumo, R. (1967). Biology of Gonostoma gracile Günther (Gonostomatidae). I. Morphology, life history, and sex reversal. Coll. Repr. Oceangr. Lab. Met. Res. Inst. Univ. Mabashi 6, 53–67.

Kinne, O., and Kinne, E. M. (1962). Rates of development of embryos of a cyrpinodont fish exposed to different temperature-salinity-oxygen combinations. Can. J. Zool. 40, 231–253.

Kristensen, I. (1965). Rivulus marmoratus (Poey) in zijin natuurlijke milieu. Het Aquarium 36, 74–76.

Kulhmann, H. (1975). Der einfluss von temperatur, futter, grasse und herkunft auf die sexuelle differenzienung von glasaalen (Anguilla anguilla). Helgol. Wiss. Meeresunters. 27, 139–155.

Lassig, B. R. (1977). Sociecological strategies adopted by obligate coral-dwelling fishes. Proc. Int. Coral Reef Symp. 3rd, 1977 pp. 565–570.

Leigh, E. G., Jr., Charnov, E. L., and Warner, R. R. (1976). Sex ratio, sex change, and natural selection. Proc. Natl. Acad. Sci. U.S.A. 73, 3656–3660.

Liem, K. F. (1963). Sex reversal as natural process in the Synbranchiform fish, Monopterus albus. Copeia pp. 303–312.

Liem, K. F. (1968). Geographical and taxonomic variation in the pattern of natural sex reversal in the teleost fish order Synbranchiformes. *J. Zool.* **156**, 225–238.

Lindsey, C. C. (1962). Experimental study of meristic variation in a population of threespine stickleback, *Gasterosteus aculeatus*. *Can. J. Zool.* **40**, 271–312.

Lucas, G. A. (1968). Factors affecting sex determination in *Betta splendens*. *Genetics* **60**, 199–200.

Lupo di Prisco, C., and Chieffi, G. (1965). Identification of steroid hormones in the gonadal extract of the synchronous hermaphrodite teleost, *Serranus scriba*. *Gen. Comp. Endocrinol.* **5**, Abstr. 71.

Mehl, J. A. P., and Reinboth, R. (1975). The possible significance of sexchromatin for the determination of genetic sex in ambisexual teleost fishes. *In* "Intersexuality in the Animal Kingdom" (R. Reinboth, ed.), pp. 243–248. Springer-Verlag, Berlin and New York.

Mires, D. (1974). On the high percent of *Tilapia* males encountered in captive spawnings and the effect of temperature on this phenomenon. *Bamidgeh* **26**, 3–11.

Mittwoch, U. (1967). "Sex chromosomes." Academic Press, New York.

Miyamori, H. (1961). Sex modification of *Lebistes reticulatus* induced by androgen administration. *Zool. Mag.* **70**, 310–315.

Moore, R. (1979). Natural sex inversion in the giant perch (*Lates calcarifer*). *Aust. J. Mar. Freshwater Res.* **30**, 803–813.

Moyer, J. T., and Nakazono, A. (1978a). Population structure, reproductive behavior and protogynous hermaphroditism in the angelfish *Centropyge interruptus* at Miyake-jima, Japan. *Jpn. J. Ichthyol.* **25**, 25–39.

Moyer, J. T., and Nakazono, A. (1978b). Protandrous hermaphroditism in six species of the anemone-fish genus *Amphiprion* in Japan. *Jpn. J. Ichthyol.* **25**, 101–106.

Mŕsić, W. (1923). Die spätbefruchtung und deren einfluss auf entwicklung und geschlechtsbildung, experimentell nachgepruft an der regenbogenforell. *Arch. Mikrosk. Anat. Entwicklungsmech.* **98**, 129–209.

Müller, U., and Wolf, U. (1979). Cross-reactivity to mammalian anti-H-Y antiserum in teleostean fish. *Differentiation* **14**, 185–187.

Nagai, Y., and Ohno, S. (1977). Testis-determining H-Y antigen in XO males of mole vole (*Ellobius lutescens*). *Cell* **10**, 729–732.

Ng, T. B. (1976). Some biochemical studies on the pituitary gonadotropin(s) of the ricefield eel, *Monopterus albus* Zuiew. M. Phil. Thesis, University of Hong Kong.

O, W. S. (1973). A study on the relationship of the pituitary gland and natural sex reversal in *Monopterus albus*. M. Phil. Thesis, University of Hong Kong.

O, W. S., and Chan, S. T. H. (1974). A cytological study on the structure of the pituitary gland of *Monopterus albus*. *Gen. Comp. Endocrinol.* **24**, 208–222.

Ohno, S. (1967). "Sex Chromosomes and Sex-linked Genes." Springer-Verlag, Berlin and New York.

Ohno, S., Ngai, Y., and Ciccarese, S. (1978). Testicular cells lysostripped of H-Y antigen organize ovarian follicle-like aggregates. *Cytogenet. Cell Genet.* **20**, 351–364.

Okada, Y. K. (1962). Sex reversal in the Japanese wrasse, *Halichoeres poecilopterus*. *Proc. Jpn. Acad.* **38**, 508–513.

Okada, Y. K. (1964a). A further note on sex reversal in the wrasse, *Hali-choeres poecilopterus*. *Proc. Jpn. Acad.* **40**, 533–535.

Okada, Y. K. (1964b). Effect of androgen and estrogen on sex reversal in the wrasse, *Halichoeres poecilopterus*. *Proc. Jpn. Acad.* **40**, 541–544.

Okada, Y. K. (1965). Bisexuality in sparid fishes. II. Sex segregation in *Mylio macrocephalus*. *Proc. Jpn. Acad.* **41**, 300–304.

Ozon, R. (1972a). Androgens in fishes, amphibians, reptiles and birds. *In* "Steroids in Nonmammalian Vertebrates" (D. R. Idler, ed.), pp. 329–389. Academic Press, New York.

Ozon, R. (1972b). Estrogens in fishes, amphibians, reptiles and birds. In "Steroids in Non-mammalian Vertebrates" (D. R. Idler, ed.), pp. 390–414. Academic Press, New York.

Padoa, E. (1939). Observations ultérieures sur la différenciation du sexe, normale et modifiée par l'administration d'hormone folliculaire, chez lat Truite iridee (Salmo irideus). Biomorphosis 1, 337–354.

Pandey, S. (1969). The role of pituitary and gonadal hormones in the differentiation of testis and secondary sex characters of the juvenile guppy, Poecilia reticulata. Biol. Reprod. 1, 272–281.

Pechan, P., Wachtel, S. S., and Reinboth, R. (1979). H-Y antigen in the teleost. Differentiation 14, 189–192.

Picard, J. Y., Tran, D., and Josso, N. (1978). Biosynthesis of labelled anti-müllerian hormone by fetal testes: Evidence for the glycoprotein nature of the hormone and for its disulfide-bonded structure. Mol. Cell. Endocrinol. 12, 17–30.

Pickford, G. E., and Atz, J. W. (1957). "The physiology of the Pituitary Gland of Fishes." N.Y. Zool. Soc., New York.

Puckett, W. O. (1940). Some effects of crystalline sex hormones on differentiation of the gonads of an undifferentiated race of Rana catesbiana tadpoles. J. Exp. Zool. 84, 39–51.

Régnier, M. T. (1937). Action des hormones sexuelles sur l'inversion du sexe chez Xiphophorus helleri Heckel. C.R. Hebd. Seances Acad. Sci. 205, 1451–1453.

Régnier, M. T. (1938). Contribution à l'étude de la sexualité des cyprinodont vivipares (Xiphophorus helleri, Lebistes reticulatus). Bull. Biol. (Woods Hole, Mass.) 72, 385–493.

Reinboth, R. (1962a). Morphologische und funktionelle zweigheschlechllichkeit bei marinen teleostiern (Serranidae, Sparidae, Centracanthidae, Labridae). Zool. Jahrb., Abt. Anat. Physiol. Morphol. Tiere 69, 405–480.

Reinboth, R. (1962b). The effects of testosterone on female Coris julis (L.), a wrasse with spontaneous sex-inversion. Gen. Comp. Endocrinol. 2, 629.

Reinboth, R. (1967). Biandric teleost species. Gen. Comp. Endocrinol. 9, Abstr. 146.

Reinboth, R. (1970). Intersexuality in fishes. Mem. Soc. Endocrinol. 18, 515–544.

Reinboth, R. (1972). Hormonal control of the teleost ovary. Am. Zool. 12, 307–324.

Reinboth, R. (1975a). Spontaneous and hormone-induced sex-inversion in wrasses (Labridae). Pubbl. Stn. Zool. Napoli. 39, Suppl., 550–570.

Reinboth, R. (1975b). In vitro studies on steroid metabolism of testicular tissue in ambisexual teleost fish. J. Steroid Biochem. 6, 341–344.

Reinboth, R. (1979). On steroidogenic pathways in ambisexual fishes. Proc. Indian Natl. Sci. Acad., Part B 45, 421–428.

Reinboth, R., Callard, I. P., and Leathem, J. H. (1966). In vitro steroid synthesis by the ovaries of the teleost fish, Centropristes striatus (L.). Gen. Comp. Endocrinol. 7, 326–328.

Robertson, D. R. (1972). Social control of sex reversal in a coral-reef fish. Science 177, 1007–1009.

Robertson, D. R., and Hoffman, S. G. (1977). The roles of female mate choice and predation in the mating system of some tropical labroid fishes. Z. Tierpsychol. 45, 298–320.

Robertson, D. R., and Warner, R. R. (1978). Sexual patterns in the labroid fishes of the Western Carribean. II. The parrotfishes (Scaridae). Smithson. Contrib. Zool. 255, 1–26.

Robertson, J. G. (1953). Sex differentiation in the Pacific salmon Oncorhynchus nerka (Walbaum). Can. J. Zool. 31, 73–79.

Robertson, D. R., and Choat, J. H. (1973). Protogynous hermaphroditism and social systems in labrid fish. Proc. 2nd Int. Symp. Coral Reefs 1, 217–225.

Roede, M. J. (1972). Color as related to size, sex and behaviour in seven Caribbean labrid fish species (genera Thalassoma, Halichoeres, Hemipteronotus). Stud. Fauna Curacao 24, 1–264.

Ross, R. M. (1978). Reproductive behaviour of the anemonefish *Amphiprion melanopus* on Guam. *Copeia* pp. 103–107.

Schwier, H. (1939). Geschlechtsbestimmung und -differenzierung bei *Macropodus opercularis, concolor, chinensis* und deren Arbastarden. *Z. Indukt. Abstamm-Vererbungsl.* **77**, 291–335.

Shapiro, D. Y. (1977). Social organization and sex reversal of the coral reef fish, *Anthias squamipinnis* (Peters). Ph.D. Thesis, University of Cambridge.

Shapiro, D. Y. (1979). Social behavior group structure, and the control of sex reversal in hermaphroditic fish. *Adv. Stud. Behav.* **10**, 43–102.

Shapiro, D. Y. (1980). Serial female sex changes after simultaneous removal of males from social groups of a coral reef fish. *Science* **209**, 1136–1137.

Shapiro, D. Y. (1981). Size, maturation and the social control of sex reversal in the coral reef fish, *Anthias squamipinnis* (Peters). *J. Zool.* **195**, 105–128.

Shapiro, D. Y. (1982). Behavioral influences on the initiation and timing of adult female-to-male sex change in a coral reef fish. *Proc. Int. Symp. Comp. Endocrinol., 9th, 1981* Abstract, p. 85.

Shapiro, D. Y., and Lubbock, R. (1980). Group sex ratio and sex reversal *J. Theor. Biol.* **82**, 411–426.

Shen, S. C., and Liu, C. H. (1976). Ecological and morphological study of the fish-fauna from the waters around Taiwan and its adjacent islands. 17. A study of sex reversal in a poma-canthid fish, *Genicanthus semifasciatus* (Kamohara). *Acta Oceanogr. Taiwan.* **6**, 140–150.

Sinha, V. R. P., and Jones, J. W. (1966). On the sex and distribution of the freshwater eel (*Anguilla anguilla*). *J. Zool.* **150**, 371–385.

Sinha, V. R. P., and Jones, J. W. (1975). "The European Freshwater Eel." Liverpool Univ. Press, Liverpool.

Smith, C. L. (1959). Hermaphroditism in some serranid fishes from Bermuda. *Pap. Mich. Acad. Sci., Arts Lett.* **44**, 111–139.

Smith, C. L. (1965). The patterns of sexuality and the classification of serranid fishes. *Am. Mus. Novit.* **2207**, 20.

Smith, C. L. (1967). Contribution to a theory of hermaphroditism. *J. Theor. Biol.* **17**, 76–90.

Smith, C. L.(1975). The evolution of hermaphroditism in fishes. *In* "Intersexuality in the Animal Kingdom" (R. Reinboth, ed.), pp. 293–310. Springer-Verlag, Berlin and New York.

Sordi, M. (1962). Ermafraditismo proteroginico in *Labrus turdus* L. e in *L. merula*. *Monit. Zool. Ital.* **69**, 69–89.

Spurway, R. (1957). Hermaphroditism with self-fertilization, and the monthly extrusion of unfertilized eggs, in the viviparous fish *Lebistes reticulatus*. *Nature (London)* **180**, 1248–1251.

Suworow, J. K. (1959). Allgemenine fischkunde. *Dtsch. Verlagder Wiss. Berlin* p. 581.

Takahashi, H. (1975). Process of function sex reversal of the gonad in the female guppy, *Poecilia reticulata*, treated with androgen before birth. *Dev. Growth Differ.* **17**, 167–175.

Takahashi, H. (1977). Juvenile hermaphroditism in the zebrafish, *Brachydanio rerio*. *Bull. Fac. Fish., Hokkaido Univ.* **28**, 57–65.

Tang, F., Chan, S. T. H., and Lofts, B. (1974a). Effect of steroid hormones on the process of natural sex reversal in the ricefield eel, *Monopterus albus* (Zuiew). *Gen. Comp. Endocrinol.* **24**, 227–241.

Tang, F., Chan, S. T. H., and Lofts, B. (1974b). Effect of mammalian luteinizing hormone in the natural sex reversal of the ricefield eel, *Monopterus albus* (Zuiew). *Gen. Comp. Endocrinol.* **24**, 242–248.

Tang, F., Chan, S. T. H., and Lofts, B. (1975). A study on the 3β- and 17β-hydroxysteroid dehydrogenase activities in the gonads of *Monopterus albus* (Pisces: Teleostei) at various sexual phases during natural sex reversal. *J. Zool.* **175**, 571–580.

Tesch, J. J. (1928). On sex and growth investigation of the freshwater eel in Dutch waters. *J. Cons., Cons. Perm. Int. Explor. Mer* **3**, 52–69.

Tuchmann, H. (1936). L'influence de l'extrait de lobe antérieur de l'hypohyse sur le tractus génital d'un Labride. *C.R. Seances Soc. Biol. Ses. Fil.* **123**, 972–975.

Uyeno, T., and Miller, R. R. (1971). Multiple sex chromosomes in a mexican cyprinodontid fish. *Nature (London)* **231**, 452–453.

van Deth, J. H. M. G., van Limborgh, J., and van Faassen, F. (1956). Le role de l'hypophyse dans la détermination du sexe de l'oiseau *Acta Morphol. Neerl. Scand.* **1**, 70–80.

van Doorn, W. A. (1962). Geschlachtsbeinvloeding door temperatur. *Het Aquarium* **32**, 208–209.

Vivien, J. H. (1941). Contribution à l'étude de la physiologie hypophysaire dans ses relations avec l'appareil génital la thyroide et les crops supraenaux chez les poissons Séleciens et Téléostéens *Scyliorhinus canicula* et *Gobius pagnellus*. *Bull. Biol. Fr. Belg.* **75**, 257–309.

Vivien, J. H. (1952). Rôle de l'hypophyse dans le déterminisme de l'involution ovarienne et de l'inversion sexuelle chez les xiphophores. *J. Physiol. (Paris)* **44**, 349–351.

Wachtel, S. S., and Koo, G. C. (1981). H-Y antigen. *In* "Mechanisms of Sex Differentiation in Animals and Man" (C. R. Austin and R. G. Edwards, eds.), pp. 255–300. Academic Press, New York.

Wachtel, S. S., Koo, G. C., and Boyse, E. A. (1975a). Evolutionary conservation of H-Y ('male') antigen. *Nature (London)* **254**, 270–272.

Wachtel, S. S., Ohno, S., Koo, G. C., and Boyse, E. A. (1975b). Possible role for H-Y antigen in the primary determination of sex. *Nature (London)* **257**, 235–236.

Warner, R. R. (1975). The adaptive significance of sequential hermaphroditism in animals. *Am. Nat.* **109**, 61–82.

Warner, R. R., Robertson, D. R., and Leigh, E. G. (1975). Sex change and sexual selection. *Science* **190**, 633–638.

Winge, O. (1934). The experimental alternation of sex chromosome into autosome and *vice versa* as illustrated by *Lebistes*. *Trav. Lab. Carlsberg, Ser. Physiol.* **21**, 1–49.

Winters, S. J., Wachtel, S. S., White, B. J., Koo, G. C., Javadpour, N., Loriaux, L., and Sherins, R. J. (1979). H-Y antigen mosaicism in the gonad of a 46,XX true hermaphrodite. *N. Engl. J. Med.* **300**, 745–749.

Witschi, E. (1934). Genes and inductors of sex differentiation in amphibians. *Biol. Rev. Biol. Proc. Cambridge Philos. Soc.* **9**, 460–488.

Witschi, E. (1957). The inductor theory of sex differentiation. *J. Fac. Sci., Hokkaido Univ.* **13**, 428–439.

Witschi, E., and Optiz, J. M. (1963). Fundamental aspects of intersexuality. *In* "Intersexuality" (C. Overzier, ed.), pp. 16–34. Academic Press, New York.

Yamamoto, T. (1961). Progenies of sex reversal females mated with sex reversal males in the medaka, *Oryzias latipes*. *J. Exp. Zool.* **146**, 163–180.

Yamamoto, T. (1962). Hormonic factors affecting gonadal sex differentiation in fish. *Gen. Comp. Endocrinol. Suppl.* **1**, 341–345.

Yamamoto, T. (1969). Sex differentiation. *In* "Fish Physiology" (W. S. Hoar and D. J. Randall, eds.), Vol. 3, pp. 117–175. Academic Press, New York.

Yamamoto, T., and Matsuda, N. (1963). Effects of estradiol, stibestrol and slkyl-carbonyl androstanes upon sex differentiation in the medaka, *Oryzias latipes*. *Gen. Comp. Endocrinol.* **3**, 101–110.

Yamazaki, F., and Donaldson, E. M. (1968a). The spermiation of goldfish (*Carassius auratus*) as a bioassay for salmon (*Oncorhynchus tschawytscha*) gonadotropin. *Gen. Comp. Endocrinol.* **10**, 383–391.

Yamazaki, F., and Donaldson, E. M. (1968b). The effect of partially purified salmon pituitary gonadotropins on spermatogenesis, vitellogenesis, and ovulation in hypophysectomized goldfish (*Carassius auratus*). *Gen. Comp. Endocrinol.* **11**, 292–299.

Zenzes, M. T., Wolf, U., Günther, E., and Engel, W. (1978). Studies on the function of H-Y antigen: Dissociation and reorganization experiments on rat gonadal tissue. *Cytogenet. Cell Genet.* **20**, 365–372.

Zohar, Y., and Abraham, M. (1978). The gonadal cycle of the captive-reared hermaphroditic teleost *Sparus auratus* (L.) during the first two years of life. *Ann. Biol. Anim., Biochim., Biophys.* **18**, 877–882.

5

HORMONAL SEX CONTROL AND ITS APPLICATION TO FISH CULTURE

GEORGE A. HUNTER AND EDWARD M. DONALDSON

West Vancouver Laboratory, Fisheries Research Branch
Dept. of Fisheries and Oceans
West Vancouver, British Columbia, Canada

I. INTRODUCTION

Studies involving experimental sex manipulation by hormones in fish have become increasingly popular. This is attributable largely to the tremendous economic potential that sex-control techniques have for the culture of economically important species of fish. For the purposes of this discussion, hormonal sex control refers only to the general control of sexual processes which can be achieved by manipulating gonadal sex. The hormones primarily involved in experimental studies of the manipulation of gonadal sex have been the sex steroids.

The first major proliferation of studies directed toward the influence of hormones on the gonads of fish occurred in the late 1930s and early 1940s.

223

FISH PHYSIOLOGY, VOL. IXB

Although in the majority of species heterosomes are not distinguishable by cytological techniques, the presence of heterosomal systems have been demonstrated by genetic techniques. Aida (1921, 1936) first demonstrated sex linked color genes and the male heterogametic–female homogametic sex chromosomal system in the medaka. Vanyakina (1969) and Yamamoto (1969) have reviewed the use of similar techniques to describe the heterosomal systems in several tropical fish, notably in the family Poeciliidae. Yamamoto (1969) also provides a brief, but interesting, review of the early work involving inter- and intraspecific matings which has provided much of our current knowledge of these systems. Of particular interest is the demonstration of both XX:XY and ZZ:WZ (YY:WY) systems in the Mexican and British Honduran races of platyfish *Xiphophorus* (*Platypoecilus*) *maculatus* (Gordon, 1947; Kallman, 1965). Similar systems have been reported for *Oreochromis mossambicus* (*Tilapia mossambica*) (Hickling, 1960). Bull and Charnov (1977) have explored the possible transition from male to female heterogamety or vice versa through a intermediate polygenic system of sex determination. More recently, analysis of the sex of progeny following gynogenesis has proven useful for the determination of female homogamety in several species including grass carp, *Ctenopharyngodon idella* (Stanley, 1976), common carp, *Cyprinus carpio* (Nagy *et al.*, 1978, 1981), and coho salmon, *Oncorhynchus kisutch* (Refstie *et al.*, 1982). A detailed examination of this technique is presented in Chapter 8, this volume. Further, the analysis of the progeny produced by the matings of hormonally sex-inverted and untreated individuals, first described in the medaka by Yamamoto (1953), has been used effectively for the identification of heterosomal systems (Table I).

The control over sex determination exerted by mechanisms associated with the sex chromosomes is relatively strict within the higher vertebrates. This also appears to be the case in a number of fish species examined. However, in several species that have an apparent heterosomal system, sex determination does not appear to be strictly bound to the sex chromosomes.

B. Models of Sex Determination

Several models of sex determination have been proposed since the discovery early in the century of the sex chromosomes. The models proposed for the vertebrates, developed primarily from studies on the amniotes have been based either on chromosomal or genic inheritance.

1. CHROMOSOMAL INHERITANCE

Sex determination based on chromosomal inheritance was proposed by Mittwoch (1971) for the higher vertebrates. The model stated that the pres-

Table I

Demonstration of Heterosomal Systems by the Mating of Untreated and Sex-Inverted Individuals

Species	Chromosomal mechanism	Reference
Oryzias latipes	XX:XY	Yamamoto (1953)
Oreochromis mossambicus	XX:XY	Clemens and Inslee (1968)
Carassius auratus	XX:XY	Yamamoto and Kajishima (1969)
Oreochromis niloticus	XX:XY	Jalabert *et al.* (1974)
Hemihaplochromis multicolor	XX:XY	Hackmann and Reinboth (1974)
Oreochromis aureus	WZ:ZZ	Guerrero (1975), Liu (1977)
Poecilia reticulata	XX:XY	Takahashi (1975a)
Salmo gairdneri	XX:XY	Okada *et al.* (1979), Johnstone *et al.* (1979a)
Oncorhynchus kisutch	XX:XY	Hunter *et al.* (1982a)
Oncorhynchus tsawytscha	XX:XY	Hunter *et al.* (1983)

ence or absence of the whole Y or W chromosome determines the respective dominant male or female sex. The resulting differential chromosome volumes along with Y- or W-linked RNA synthesis were presumed to induce gonadogenesis mediated by a higher mitotic rate in the heterogametic sex. In mammals and birds heterogamety had been previously correlated with early sex differentiation (Hamilton, 1965).

This model probably does not apply to the teleosts. First, as previously mentioned, the majority of fish species do not have heteromorphic chromosomes. Second, early differentiation of germ cells apparently does not correlate with the heterogametic sex. Eckstein and Spira (1965) reported early differentiation in the female cichlid, *Oreochromis aureus* (*Tilapia aurea*), which has been demonstrated to be heterogametic (Guerrero, 1975). However, differentiation occurs first in the homogametic female *Oryzias latipes* (Satoh and Egami, 1972; Onitake, 1972; Quirk and Hamilton, 1973), the cichlid, *Oreochromis mossambicus* (Nakamura and Takahashi, 1973), and the goldfish, *Carassius auratus* (Nakamura, 1978). Early differentiation of the female has also been reported in the cichlid, *Hemihaplochromis multicolor* (Müller, 1969), *Tilapia zillii* (Yoshikawa and Oguri, 1978), the trout, *Salmo trutta* (Ashby, 1957), *Ctenopharyngodon idella*, (Shelton and Jensen, 1979), and the salmonids, *Oncorhynchus masou, Oncorhynchus keta*, and *Salvelinus leucomaenis* (Nakamura, 1978), and the threespined stickleback, *Gasterosteus aculeatus* (Shimizu and Takahashi, 1980). Early differentiation of the male in a differentiated gonochorist teleost has not been reported. Although a correlation between hetero- or homogamety is not evident, developing germ cells of many species examined do demonstrate a differential mitotic rate. Onitake (1972) observed that following the oral administration of estrone to genetically male medaka the germ cells began an atypical rapid proliferation which preceded sex differentiation. Onitake suggested that the rapid mitotic increase was necessary to the process of differentiation. In an intensive examination of factors affecting sex inversion in the cyprinodont, *Rivulus marmoratus*, Harrington (1975) reported that rearing temperatures of 19°C or 26°C resulted in higher proportions of primary males and hermaphrodites, respectively, and these results were correlated with mitotic activity in the developing gonad. The role of differential mitotic growth in the process of sex differentiation remains to be determined.

2. GENIC INHERITANCE AND POLYGENIC SEX DETERMINATION

The concept of genic balance established by Bridges (1925, 1936) was initially modified by Winge (1934) to apply to sex determination in *Poecilia* (*Lebistes*). Winge proposed that the X and Y chromosomes contained superior male and female sex-determining genes. Minor male and female sex-determining factors were held in the autosomes. Normally the autosomal genes are maintained in balance allowing sex to be determined by the heterosomal mechanism. However, in exceptional individuals, autosomal combinations or recombinations may occur that result in an excess of autosomal factors of one sex capable of overriding the heterosomal mechanism. The outcome is an individual with a phenotypic sex differing from its heterosomal sex. Similar polygenic sex-determining systems have been hypothesized for several xiphophorin fishes (Kosswig and Oktay, 1955; Kosswig, 1964; Anders and Anders, 1963; Dzwillo and Zander, 1967), *Oryzias latipes* (Yamamoto, 1963, 1969), and *Betta splendens* (Lowe and Larkin, 1975). Exhaustive reviews of the early research on the development of the polygenic concept have been provided by Kosswig (1964) and Yamamoto (1969). Yamamoto (1969) summarized the results obtained by asserting that "sex determination in gonochorists is polyfactorial with or without epistatic sex genes in sex chromosomes." When the total strength of the male factors exceeds that of the female factors, the zygote will be male and vice versa. Therefore, depending on the control exerted by epistatic genes on the sex chromosomes the stability of sexual determination will vary from species to species.

Within the tilapia, interspecific crosses of several types result in non-Mendelian sex ratios (Pruginin *et al.*, 1975). Avtalion and Hammerman (1978) examined an extensive series of hybrid crosses between homozygous *Oreochromis* (*Tilapia*) *hornorum* males and *Oreochromis mossambicus* females and between homozygous *Oreochromis macrochir* males and *Oreochromis niloticus* (*Tilapia niliotica*) females conducted by Chen (1969) and Jalabert *et al.* (1971), respectively. Assuming that the Y and Z chromosomes were identical and that an autosomal influence exists, Avtalion and Hammerman proposed that the simplest sex-determining mechanism in tilapia would consist of three gonosomes (X, W, Y) in pairs of two (XX, XY, XW, WY, WW, and YY) similar to the system proposed for *Xiphophorus maculatus* by Gordon (1947) but with the addition of a pair of autosomes. (AA, Aa, and aa). The resulting 18 possible combinations of autosomes and gonosomes could be used to predict the results obtained by Chen, but not of Jalabert *et al.* (1971). Hammerman and Avtalion (1979) suggested that the results of Jalabert *et al.* (1971) could be explained by assigning different strengths to each of the chromosomes involved. Analysis of the sex-determining mechanism has been hampered because of the fact that in tilapia, sex chromosomes cannot be identified by karyotypic analysis and no sex-linked color markers are present. Avtalion *et al.* (1975) and Hardin (1976) have identified a male-specific electrophoretic marker in adult male *Oreochromis* (*Sarotherodon*) *aureus*. However, it has only been found in sexually mature individuals and may, therefore, be hormonally induced. The usefulness of these markers is dependent on whether they are sex linked as opposed to sex limited. For further review, the reader is referred to Wolhfarth and Hulata (1981) and Avtalion (1982).

The concept of a polygenic mechanism in which autosomal genes may play a decisive role in the process of sex determination has dominated the research on fish. However, no single polygenic system has been capable of being reconciled with all empirical data (Harrington, 1974). Despite exceptions, the concept of a polygenic system remains suited to the majority of data from studies of fish. An interesting example is the recent report by Streisinger *et al.* (1981). In this study, clones of homozygous diploid female zebrafish, *Brachydanio rerio*, were produced by hydrostatic pressure or temperature shocks administered to ova, activated by ultraviolet (UV)-treated sperm. Some of the clones were predominately male and produced high proportions of males in subsequent generations. Such results are difficult to reconcile with any of the current models of sex determination other than a polygenic system. From the analysis of single gene mutations it is clear that even in the mammals, genetic sex cannot be explained by the constitution of the sex chromosomes alone. Autosomal genes which may play an important role in gonadal differentiation have been reported in pigs (Johnston *et al.*, 1958), goats (Hamerton *et al.*, 1969), and mice (Cattanach *et al.*, 1971).

C. Models of Sex Differentiation

Several models of sex differentiation have been proposed which are compatible with single or multiple gene action.

1. H-Y Antigen

Sex differentiation based on the action of an individual gene or genes has recently been given considerable support in the aves and mammalia with the discovery of the male specific histocompatibility-Y chromosome (H-Y) antigen. The antigen, first discovered in mice by Eichwald and Silmser (1955) is ubiquitously associated with the heterogametic sex in mammalian and some nonmammalian vertebrates. These observations have led to the hypothesis that the antigen plays a major role in gonadal sex differentiation. The current hypothesis is that a gene or genes on the Y or W chromosome code for the H-Y antigen and the presence of the antigen on the surface of somatic cells of the indifferent gonad results in the development of the heterogametic gonad (Wachtel *et al.*, 1975; Zenzes *et al.*, 1978). However, the mechanics of the action or regulation of H-Y antigen with respect to gonadal differentiation have yet to be discovered. Ohno *et al.* (1978) have demonstrated that the membrane H-Y antigen receptor is present only on gonadal cells in both sexes. Ohno and co-workers used H-Y antibody to strip H-Y antigen from mice gonadal cells of known genetic constitution. Culture of the cells revealed that removal of the H-Y antigen resulted in the development of spherical aggregates that resembled ovarian follicles. Unstripped cells formed cylindrical tubular structures morphologically similar to seminiferous tubules. Ohno and co-workers concluded that there was a causal relationship between the presence of the H-Y antigen on the surface of gonadal cells and the development of the testes. They further suggested that the H-Y antigen could act as a short-range hormone-inducing specific gene expression.

Müller *et al.* (1979) inverted the sex of chicken embryo testes (ZZ) with estrogen. He found that the normally H-Y (W) antigen-negative testes were positive after sex inversion, which indicated that the gene for the antigen was expressed in the absence of the W chromosome and, therefore, must be present in both sexes, supporting the report by Ohno *et al.* (1978). Assuming 'the the presence of the H-Y (W) antigen was responsible for the formation of the ovary, Müller *et al.* (1979) suggested that the hormone-induced sex inversion was an indirect effect mediated by H-Y (W) antigen. They also indicated that there was a correlation between morphogenetic changes and H-Y (W) titer in the gonad. Therefore, Müller and co-workers suggested that the induction of the H-Y (W) antigen by estrogen did not operate as a strict on–off switch mechanism, again similar to the conclusion arrived at by Ohno *et al.* (1978). Similarily, induced H-Y antigen-positive ovarian tissue in the ovaries of sex-reversed (ZZ) individuals has been demonstrated in *Zenopus laevis* (Wachtel *et al.*, 1980) and *Pleurodeles waltlii* (Zaborski and Andrieux, 1980). In a series of experiments reviewed by Zaborski (1982) involving amphibians, H-Y expression was repressed in *Pleurodeles waltlii* ovaries (ZW) or *Rana ridibunda* testes (XY) by the administration of dihydrotestosterone or estradiol, respectively. Therefore, although in the lower vertebrates the sex hormones appear to be inducers of H-Y antigen, an inhibitory control of its expression is also suggested. Because both androgens and estrogens are produced in each sex, the nature of hormonal control of the antigen system is presumed to be quantitative. However, recent research with rainbow trout gonadal homogenates suggests that during the period of sex differentiation the gonads may not be capable of estrogen production (van den Hurk *et al.*, 1982).

Sex-specific antigens have been detected in nonmammalian vertebrates including birds (Bacon, 1970; Wachtel *et al.*, 1975; Müller *et al.*, 1980), amphibians such as *Rana pipiens*, *Xenopus laevis* (Wachtel *et al.*, 1975), and reptiles such as *Emys orbicularis* (Zabroski *et al.*, 1979). Recently, Shalev and Huebner (1980), citing unpublished work, have reported the presence of the antigen in invertebrates.

The presence of the antigen system has also been demonstrated in several species of fish. Müller and Wolf (1979) tested absorption of mammalian anti-H-Y antiserum in the teleosts *Salvelinus alpinus*, *Salmo gairdneri*, *Rutilus rutilus*, *Carassius auratus*, *Barbus tetrazona*, *Poecilia reticulata* (*Lebistes reticulata*), and *Xiphophorus helleri*. The gonads of the more primitive orders *Ostariphysi*, represented by *Rutilus*, *Carassius*, and *Barbus*, and *Isospondyli*, represented by *Salvelinus* and *Salmo*, absorbed anti-H-Y antiserum; however, a clear sex difference was not observed. In the more advanced poeciliids *Poecilia* and *Xiphophorus*, the anti-H-Y antiserum was absorbed almost exclusively by the gonadal tissues of the male but not the female. As previously mentioned, *Poecilia*, *Salmo*, and *Carassius* all have male heterogametic–female homogametic systems. The sex-determining mechanism in *Xiphophorus* does not appear to be strictly heterosomal (Kosswig, 1964). The presence of the antigen system in *Poecilia reticulata* has been recently confirmed by Shalev and Heubner (1980). Further support for a sex-specific antigen expression in advanced species comes from research by Pechan *et al.* (1979). In this study, anti-H-Y antiserum absorption was found exclusively in male cells of *Xiphophorus maculatus*, *Haplochromis burtoni*, *Oryzias latipes*, and several tilapia hybrids. The H-Y antigen was detected most readily in *Xiphophorus maculatus* males (YY). Male heterogamety has not yet been determined for the cichlid *Haplochromis burtoni*.

The presence of the sex-specific antigen system in fish is certainly interesting from an evolutionary perspective. Further examination of the antigen system in the lower vertebrates may provide an answer to the basic question of when in the course of evolution the antigen system assumed a primary role in the process of sex differentiation. The demonstration H-Y antigen expression in the ovaries of homozygous chicks and amphibians sex inverted with estradiol suggests a major role of the antigen in the process of hormonal sex inversion in these groups. Examination of the role of the antigen system in the natural or hormonally induced sex inversion of both hermaphroditic and gonochorist fish is a promising area of study.

2. THE CORTICOMEDULLARY INDUCTOR MODELS

The corticomedullary inductor model was proposed by Witschi (1929) to explain sex differentiation in amphibians. Witschi observed that the primordium of the amphibian gonad, like those of most vertebrates, is comprised of both an outer cortex and inner medulla ultimately derived from the germinal epithelium. During differentiation, either the cortex or the medulla develops at the expense of the other resulting in the development of an ovary or testis, respectively. Witschi theorized that the genetic male and female factors embodied by the balance theory of sex determination were phenotypically manifested in the dualistic character of the primordial gonad. The action of these genetic factors results in the production of an embryonic cortexin or medullarin, which in turn initiated ovarian and testicular differentiation, respectively. The existence of these hypothetical inductors remains to be demonstrated. As a result, consideration of the suitability of Wistchi's model has remained a conceptual debate.

In later publications, Witschi (1965, 1967) modified his theory to include an antagonist action of the inductors. The theory of antagonism has been illustrated in the amphibians by ablation of the dominant component of the gonad which results in differentiation of the remaining component (Haffen, 1977). Witschi (1967) has suggested that the interaction of the inductors is similar to an immune reaction. Therefore, each has the capability of inhibiting or destroying the other. Reinboth (1982) has recognized the similarity between Witschi's (1967) model and the model of mammalian sex differentiation involving the H-Y antigen system and a presumptive ovarian factor. The existence of the ovarian factor and the nature of its interaction with the H-Y antigen system remain a subject of study and debate (Wachtel and Koo, 1981).

With regard to fish, the debate surrounding the dual-inductor concept has centered on reconciliation of the model with the proposed unitary origin of the teleost gonad and the common occurrence within the teleosts of various forms of hermaphroditism.

The discrete topographical division of the primoridal amphibian gonad provided Witschi with strong support for his dual-inducer concept. This dual embryonic nature appeared ideally suited to providing the separate chemical environments necessary for the divergent development of either male or female gonia. Within the fish, a dualistic structure of the primordial gonad has been reported in the elasmobranchs (Chieffi, 1959). However, unlike this group and the rest of the vertebrates the gonads of cyclostomes and teleosts have been reported to develop from a unitary primordium homologous to the cortex of other vertebrates (D'Ancona, 1941, 1949). D'Ancona's observations have received general support (Hoar, 1969). However, the reconciliation of a dual-inductor system with a single primordial source has been difficult. First, neither D'Ancona nor later investigators have been able to explain how the development of a single primordium could provide a basis for two distinct cell lines producing two antagonistic inductors. Second, discrete male and female territories are found as early as the first indication of sex differentiation in many hermaphroditic species. Further, although the classification of many hermaphroditic species remains in doubt, it is clear that they are not restricted to a few isolated advanced families as suggested by D'Ancona (1949). Smith (1975) has provided an excellent review of their distribution. Because of these difficulties and a recognition of the problems associated with the examination of the peritoneal embryonic differentiation of interrenal, nephric, and gonadal elements as described by Hardisty (1965), Harrington (1974, 1975) has suggested that a reevaluation of the single primordium hypothesis may be in order.

An additional problem presented by the hermaphrodites is the method by which the proposed antagonistic inductors may be compartmentalized in a gonad containing both testicular and ovarian areas. The gonadal organization of some species offers a certain degree of spatial isolation in the case of protogynous or protandrous serranid or sparid species (Atz, 1964) and the synchronous hermaphrodite *Rivulus marmoratus* (Harrington, 1971). However, in the grouper *Epinephelus*, there appears to be no discrete segregation of male and female territories.

3. SEX STEROIDS

Although the two presumptive inducers have never been identified, Yamamoto (1969) concluded that based on his work with *Oryzias latipes* the two inductors, which he termed the gynotermone and androtermone, were in fact estrogens and androgens, respectively. Yamamoto's synthesis of the hormone and inductor models has dominated the work on sex control in fish. However, the origins of the hormonal model are much older.

In the six decades since Lillie (1917) postulated sex steroid involvement in his explanation of the free-martin effect in cattle, exhaustive studies have been conducted in an attempt to determine the specific role of hormones in the process of sex differentiation. Numerous experiments involving classic castration and replacement have confirmed that the sex hormones mediate the development of secondary sexual characteristics in mammals and birds (Jost, 1965; Goldstein and Wilson, 1975). In both mammals (Price, 1970) and birds (Haffen, 1975), biological, biochemical, and histochemical tests indicate hormonal activity in the indifferent gonad of the dominant sex. However, evidence for a role of the sex steroids in sex differentiation resulting from steroid administration has been inconclusive. In the marsupials, Burns (1950) working with opposum, *Didelphys virginiana*, was able to demonstrate the formation of ovotestes under the influence of estradiol administration. However, in eutherian mammals, numerous experiments involving the *in vivo* or *in vitro* administration of exogenous sex steroids have been ineffective (Burns, 1961; McCarrey and Abbott, 1979).

Feminization as a result of estrogen administration to the male embryo has been achieved in the chick (Narbaitz and De Robertis, 1970) and quail (Haffen, 1969); however, the inversions are often transitory. The administration of a variety of androgens have produced either simple masculinizing effects or masculinizing and feminizing effects. Testosterone and its esters act similarily to the natural hormones produced by the embryonic male gonad masculinizing the genital ducts but not influencing the female gonads. Some androgens including androstanedione, androstenedione, androstenediol, and *trans*-hydroandrosterone masculinize female genital ducts but feminize male genital ducts and gonads (Haffen and Wolff, 1977). Numerous studies have demonstrated that (1) feminized male gonads can secrete a hormone similar to the sex reversing hormone, (2) embryonic gonadal secretions from the medulla which have the same effect as steroid hormones, and (3) indifferent avian gonads synthesize and secrete steroids, all of which Haffen and Wolff claimed support the steroid-inductor model. Somatic sex inversion has been achieved in cultured embryonic left testes administered exogenous androgens or estrogens (Carlon and Erickson, 1978). Carlon and Erickson suggested that it is the absence of steroidogenesis during the indifferent period which is necessary for testicular development.

Similarily, androgen administration to several species of reptiles has resulted in variable results (Haffen and Wolff, 1977). Estrogen administration to the green lizard, *Lacerta viridis* resulted in partial or complete inhibition of testicular development in some individuals to produce an ovotestis and complete inhibition in others to produce an ovary (Raynaud, 1967).

Within amphibians a relatively large number of studies involving hormonally induced sex inversion have demonstrated that the administration of exogenous estrogens and androgens results in functional feminization in urodeles and masculinization of ranid anurans, respectively. However, paradoxical actions of steroid treatment have been reported in several species (Burns, 1961; Haffen and Wolff, 1977).

In teleost fish as in the amphibia, sex steroids are capable of influencing the course of sex differentiation. Yamazaki (1983) reported at least 15 species in which functional sex inversion has been achieved. However, these species are primarily gonochoristic teleosts within a small number of families. Chieffi (1967) reported the effects of androgens and estrogens on the elasmobranchs, primarily *Scyliorhinus canicula*. Both estrogens and androgens influence the differentiation of the genital ducts; however, only estrogen influences the gonad in this species producing ovotestes.

The small number of hermaphroditic species which have been treated with steroids to manipulate natural sex inversion have responded inconsistently. Reinboth (1962, 1975) used a single 2-mg injection of testosterone to mimick sex inversion in several species of protogynous hermaphrodite wrasses (Labridae). He concluded that, in general, androgens induced precocious sex inversion in protogynous species, but this evidence alone did not support the steroid-inductor model (Reinboth, 1970). Chen *et al.* (1977) reported that two of three groupers, *Epinephelus tauvina*, also a protogynous hermaphrodite, fed 80 mg methyltestosterone over 30 days initiated sex inversion. Subsequently, all 25 fish given a dosage of 1 mg methyltestosterone/kg diet 3 times per week over a 2–month period underwent sex inversion.

The rice field eel, *Monopterus albus*, has perhaps been the most intensively studied protogynous hemaphrodite (Chan, 1977; Chan *et al.*, 1977). Biochemical and histochemical techniques have demonstrated 3β-hydroxysteroid dehydrogenase and 17β-hydroxysteroid dehydrogenase activity and a large increase in the production of androgens during natural sex inversion (Chan *et al.*, 1975; Tang *et al.*, 1975). However, attempts to manipulate the course of natural sex inversion by the administration of steroids have been ineffective (Tang *et al.*, 1974; Chan *et al.*, 1977). These studies are examined in detail in Chapter 4, this volume.

The achievement of functional sex inversion in several gonochorist species remains the most compelling evidence for steroid involvement in the sex differentiation of fish. However, as Reinboth (1970) has noted, it is not possible to rule out a pharmacological rather than physiological action in this process. No studies have examined the mode of action of the steroids in the induced sex inversion of teleosts. In this regard, Vannini *et al.* (1975) examined the influence of actinomycin D and puromycin, inhibitors of DNA-dependent RNA transcription and RNA-dependent protein synthesis respectively, on testosterone-induced sex inversion in *Rana dalmatina* tadpoles. Both antibiotics suppressed testosterone action. Vannini and co-workers also reported that testosterone-induced sex inversion is linked with RNA synthesis. They have proposed that the mechanisms of testosterone action involve the derepression of latent male genes in the nuclei of somatic tissue which in turn causes them to proliferate as gonadal medullary tissue. Evidence for the direct action of steroids and their specific receptor molecules on DNA and, thereby, transcriptional processes has been established (O'Malley and Schrader, 1976). Further, no studies have examined the potential sites of genetic control of hormonal action, specifically the partially common biosynthetic pathway of the sex steroids or sex-specific receptor sites on undifferentiated germ cells.

Therefore, additional supportive evidence for Yamamoto's conclusions was confined to the specificity of sex steroids as exogenous sex inductors, the very low effective dosage of sex steroids, and the selective incorporation of sex steroids into the differentiating gonad. However, based on similar evidence, later studies have added uncertainty rather than support to Yamamoto's (1969) hypothesis of a sex-steroid-inductor model.

Yamamoto based the specificity of sex-steroid action on the absence of paradoxical effects and inactivity of corticoids as sex inducers in *Oryzias latipes*. To date, no studies have reported effective sex inversion using corticoids, although other chemicals such as N,N-dimethylformamide can modify gonadal differentiation (van den Hurk and Slof, 1981).

More recent studies have reported paradoxical effects of sex steroid administration. Gresik and Hamilton (1977) citing unpublished work by J. B. Hamilton and D. D. Kantor report that the injection of eggs containing XX or XY *Oryzias latipes* embryos with either methyltestosterone, estrone acetate, or the synthetic progestin ethynodiol diacetate at low concentrations favors testicular differentiation. High concentrations favor ovarian differentiation. They also report that paradoxical sex inversion is more difficult to achieve in YY individuals than in XY individuals. Paradoxical feminization in the cichlids was first observed in *Hemihaplochromis multicolor* fry reared in water containing testosterone propionate or methyltestosterone (Müller, 1969). Subsequently, the addition of these steroids at 500 μg/l was also found to have a paradoxical effect in other cichlids including *Chichlasoma biocellatium*, *Tilapia heudeloti*, and *Oreochromis mossambicus* (Hackmann, 1974). The effect was again demonstrated in *Oreochromis mossambicus* fed methyltestosterone at 1000 mg/kg diet for a period of 50 days after hatching (Nakamura, 1975). In genetic females, oogenesis progressed, although at a slower rate than control ovaries. The ovarian cavity was formed paradoxically in genetic males. Following the end of treatment, these fish developed intersexual gonads. Advanced spermatogenesis was observed in the inner periphery of the efferent ducts. Maturing oocytes and a definite ovarian cavity were observed on the outer side of the ducts. In the same study, administration of methyltestosterone at 50 mg/kg diet resulted in gonadal masculinization. Nakamura (1975) noted that, in contrast to his observations, Reinboth (1969) had reported that in *Hemihaplochromis multi-color* a long-term treatment at extremely high androgen dosages (30–50 g/kg diet) was required to cause gonadal masculinization, but a short-term treatment at similar dosages resulted in gonadal feminization. He suggested interspecies differences as a possible explanation but acknowledged that the major reason for these variations was unknown. More recently, paradoxical actions of testosterone and ethynyltestosterone have been reported in rainbow trout (V. J. Bye, 1980, personal communication) and channel catfish, *Ictalurus punctatus* (Goudie *et al.*, 1983).

Recent work with the amphibians has demonstrated that care must be taken in the interpretation of paradoxical steroid action. Chieffi (1965) noted the paradoxical effects found in urodele gonads following high dosages of androgens. For example, in *Pleurodeles waltlii*, testosterone propionate induced complete feminization at all dosages administered (Gallien, 1950). Chieffi concluded that the nonspecific action of the steroids argued against their putative roles as sex inducers. The administration of estrogens to several ranid species results in feminization, intersexuality, or masculinization with increasing dosages (Padoa, 1942; Gallien, 1941). Hsü *et al.* (1978a,b) have also demonstrated that high dosages of estradiol result in paradoxical masculinization of *Rana catesbeiana* tadpoles. However, this treatment also inhibits Δ^5-3β-hydroxysteroid dehydrogenase, a key enzyme in ovarian steroidogenesis. Hsü and co-workers suggested that if the paradoxical action of this steroid arose through an inhibition of ovarian steroidogenesis, the paradoxical effects would lend support to the steroid-inductor theory.

The data from the limited number of studies employing steroidogenic inhibitors or antiandrogens have not supported specific steroid action. Cyproterone acetate is the most commonly used antiandrogen in studies with fish. In mammals, it has been shown to be a potent inhibitor of endogenous or exogenous testosterone (Hamada *et al.*, 1963). The antiandrogen has been partially effective in blocking the development of secondary sexual characteristics in the three-spined stickleback (Rouse *et al.*, 1976), the Indian catfish, *Heteropneustes fossilis* (Sundararaj and Nayyar, 1969), and the guppy, *Poecilia reticulata* (Smith, 1976). Schreck (1974), citing the unpublished work of Irons and Schreck, reported a possible sex inversion in male *Oryzias latipes* fed the antiandrogen, cyproterone acetate, at 50–500 mg/kg diet for 12 weeks. Hopkins *et al.* (1979) fed cyproterone acetate in combination with various natural and synthetic estrogens to *Oreochromis aureus* (*Tilapia aurea*) fry. The results indicated that rather than potentiating the effect of the estrogens it actually lessened their effectiveness. Rastogi and Chieffi (1975) demonstrated that the same antiandrogen did not inhibit the masculinizing effects of testosterone propionate or 11-ketotestosterone in the swordtail *Xiphophorus helleri* even though it inhibited the binding of the steroid to testosterone-sensitive target tissues.

The mechanism of action of cyproterone acetate in the lower vertebrates is not clear. Chieffi *et al.* (1974) administered cyproterone acetate to *Rana esculenta* tadpoles. They hypothesized that if the androgens were indeed the sex inductors, the administration of cyproterone acetate would result in competition for androgen-receptor sites and subsequent abnormal testicular development. Instead the ovaries were masculinized. Chieffi and co-workers suggested that these results indicated that the sex inductors were not structurally similar to sex steroids. However, Hsü *et al.* (1979) have also demonstrated that cyproterone acetate administered in the rearing water at 1500 μg/l for a 7-month period can transform ovaries of *Rana catesbeiana* into testes. They suggested that the masculinizing effect of the antiandrogen was attributable to inhibition of ovarian steroidogenesis, specifically Δ^5-3β- and possibly 17β-hydroxysteroid dehydrogenase. They also hypothesized that this inhibition of young ovaries would result initially in ovarian degeneration and subsequent masculinization of the gonads with prolonged treatment. Further, Hsü and co-workers maintain that these results support the steroid-inductor theory.

The observation of parodoxical steroid action also suggests the possibility of a dominant–neutral sex mechanism of dimorphism. Studies involving the culture of isolated mammalian and avian cortex and medulla suggest that sexual dimorphism occurs as a result of the presence or absence of a single dominant sex. In the absence of the dominant sex the neutral sex develops (McCarrey and Abbott, 1979). As a general rule, the dominant sex is correlated with heterogamety (Jost, 1965). Hackmann and Reinboth (1974) noted that paradoxical sex inversion in the lower vertebrates occurs only in one direction. Specifically in the urodeles and cichlids, high doses of androgen result in feminization, but in the families *Ranidae* and *Hylidae* estrogen treatment results in masculinization. Hackmann and Reinboth proposed that the evidence from the cichlids could support a hypothesis in which the female hormone induces the female sex type while the absence of the hormone promotes the development of male sex type. The evidence from the breeding experiments conducted by Hackmann and Reinboth (1974) with *Hemihaplochromis multicolor* indicated male heterogamety. Therefore, the association between heterogamety and the dominant sex observed in birds and mammals could not be supported.

Similarly, the results presented by Gresik and Hamilton (1977) and van den Hurk and Slof (1981) suggested a dominant–neutral sex hypothesis in which interrupted testicular development results in the formation of an ovary. Van den Hurk and Slof (1981) obtained feminization in rainbow trout by administering progesterone in the rearing water at 300 μg/l over the period of sexual differentiation. Van den Hurk and Slof suggested that in addition to a possible direct feminizing effect of progesterone, there also exists the possibility that progesterone may act as a steroid-blocking agent of the androgenic pathway, resulting in ovarian differentiation. An indirect action of the progestins via a blockage of androgen biosynthesis has been previously described in mammals by Gower (1975). However, Van den Hurk and Slof also noted that progesterone has not been demonstrated to affect sex differentiation in *Oryzias latipes* (Yamamoto and Matsuda, 1963). Recently, van den Hurk and Lambert (1982) have also reported that progesterone does not influence sex differentiation in rainbow trout when administered in the diet. The evidence produced by van den Hurk *et al.* (1982), demonstrating the steroidogenic capabilities of rainbow trout testes but not ovaries at the time of sex differentiation, also supports a dominant–neutral sex hypothesis.

The limited evidence from these studies is clearly not sufficient to determine whether a dominant–neutral system of sexual dimorphism is present in fish. Presumably the unitary origin of the teleost gonad has prevented dissociation–recombination experiments which would provide more substantial evidence for a dominant–neutral sex hypothesis. The discrete male and female territories in hermaphroditic species may provide an ideal vehicle for such an examination.

Hishida (1962, 1965) by demonstrating the selective incorporation of sex steroids into the developing gonads, provided Yamamoto with considerable support for his hypothesis. Administration of 4-[^{14}C]testosterone propionate to larval *Oryzias latipes* resulted in the accumulation of the labeled steroid only by the actively differentiating gonads (Hishida, 1962). In a later study, Hishida (1965) fed larval *Oryzias latipes* 16-[^{14}C]estrone and diethylstilbestrol 1-[^{14}C]monoethyl). A 4- to 10-fold concentration of the steroids was found in developing gonads, again indicating active accumulation of the steroids. Conversion of estrone to estradiol was also demonstrated. Further, based on recovered counts, the effective oral dosages for 100% sex inversion were calculated to be 1.8×10^{-2}μg estrone and 1.1×10^{-2} μg diethylstilbestrol. Based on this evidence, Yamamoto (1969) asserted that the levels of steroids required for manipulation of sex differentiation were within the physiological capabilities of the species. Further evidence that physiological levels of steroid can influence sexual differentiation was provided by Satoh (1973). In this study, trunk regions of newly hatched *Oryzias latipes* fry were transplanted into the anterior chamber of the eyes of adult fish. The gonads of genetic females transplanted into male hosts did not differentiate into an ovary, but did form spermatogenic cells. Satoh suggested that the results supported the actions of androgens as andro-inducers. Although the results provide strong supportive evidence, they do not conclusively demonstrate a role for the steroids in primary gonadal differentiation.

Studies correlating the development of steroidogenic and differentiating tissue or identifying steroid synthesis in sexually differentiating gonads are rare. Dildine (1936) first noted the correlation between the development of a stromal region and the initiation of male influences on the sex differentiation in *Poecilia reticulata*. More recently, Takahashi (1975a) observed that, in embryonic *Poecilia reticulata*, the appearance of aggregations of stromal cells in the gonadal hilus 18 days after the last parturition are indicative of testicular differentiation. Oral administration of methyltestosterone to gravid females results in somatic aggregation in the hilar region of embryonic ovaries. Females affected in this manner contain developing oocytes and malelike stromal aggregations in the hilar region. Within 20 days of birth the ovaries completely degenerate and are replaced by testes which display precociously differentiating sperm ducts, testicular interstitium, and a concomitant initiation of spermatogenesis. The period that is most sensitive to change in the developing ovaries is at 18 days after the last parturition, and it is synchronous with the period of stromal aggregation in the hilar region of developing testes. Takahashi concluded that masculinization of the somatic element is essential for functional sex inversion of females, and that the somatic differentiation may occur prior to germinal differentiation. Nakamura (1978) has reported similar stromal aggregations in the hilar region prior to male differentiation in *Oryzias latipes* and the mosquito fish *Gambusia affinis*. However, Satoh and Egami (1972) did not observe sex differences in the histological structure of the gonadal somatic tissue prior to gametogenesis in *Oryzias latipes*.

Yoshikawa and Oguri (1979) reported that in *Oryzias latipes* the differentiation of somatic cells into interstitial cells always precedes the transformation from spermatogonia to spermatocytes. Further, interstitial cells are always found in the vicinity of germ cells in meiosis, suggesting that interstitial cells are responsible for the differentiation of germ cells. However, in the same species, Satoh (1974) reported the appearance of steroidogenic cells after the onset of gonadal sex differentiation. Similarily, van den Hurk et al. (1982) found no steroid-synthesizing structures in indifferent gonads of rainbow trout 50 days postfertilization. In this study, steroidogenic (Leydig) cells were detected in differentiated testes at 100 days postfertilization.

Takahashi and Iwasaki (1973) provided the first examination of the onset of steroidogenesis in the differentiating teleost gonad. These researchers detected Δ^5-3β-hydroxysteroid dehydrogenase activity in gonadal interstitial cells in *Poecilia reticulata* 7 days postpartum. The enzyme activity was detected concurrent with the multiplication of spermatogonia and differentiation of cells in the testicular hilus into the sperm duct analogues and testicular stroma. The enzyme was detected in the cells of the stroma but not in the sperm duct analogues. Further, enzyme activity could not be detected in fish sampled at 3 or 5 days postpartum. Takahashi and Iwasaki concluded that the enzyme activity appears and increases simultaneously with the differentiation of the testicular duct system rather than the germ cells. Recently, van den Hurk *et al.* (1982) have conducted *in vitro* assays of *Salmo gairdneri* gonadal homogenates at 50, 100, and 200 days postfertilization. The histologically indifferent gonads at 50 days postfertilization contained 3β-hydroxysteroid dehydrogenase, 5,4-isomerase, and 17α-hydroxylase, indicating the capacity to synthesize progestins. Developing testes at 100 days postfertilization also contained 17α, 20-desmolase and 11β-hydroxylase, indicating the capacity to synthesize androgens. However, sex differentiation had commenced between 50 and 100 days postfertilization. Therefore, it was not possible to determine whether these androgen-synthesizing capabilities initiated or occurred as a result of testicular development. Both testes and ovaries possessed 17β-hydroxysteroid dehydrogenase at 200 days postfertilization. Further, the ovaries possessed aromatase, indicating the capability to synthesize estrogen. Therefore, although androgen- or estrogen-synthesizing capabilities could not be demonstrated in indifferent gonads, a capability for progestin synthesis was noted. Van den Hurk and Slof (1981) had previously demonstrated that progesterone, administered at the time of sex differentiation, results in gonadal feminization. As they note, the absence of estrogen-synthesizing capabilities in differentiated gonads 100 days postfertilization is substantial evidence against a steroidal ovarian inductor. In a later study, van den Hurk and Lambert (1982) demonstrated the conversion of 11β-hydroxyandrostenedione to 11-ketoandrostenedione using testicular homogenates from 100-day-old rainbow trout. Therefore, the presence of 11β-dehydroxysteroid dehydrogenase was demonstrated. Further administration of 11β-hydroxyandrostenedione at 60 mg/kg for 8 weeks from first feeding resulted in an all male population. Van den Hurk and Lambert concluded that this steroid is endogenous to rainbow trout and responsible for the initiation of testicular development.

D. Summary

No single model of sex determination and differentiation may be reconciled with all the present evidence from the fish. The sex-determining mechanism is assumed to have a genetic basis. Although the majority of fish do not have cytologically distinguishable sex chromosomes, both male and female digametic systems have been demonstrated. Nonheterosomal chromosomal sex-determining systems have also been reported for some species. Even within species with well developed heterosomal systems, sex determination has been found to be labile to extrinsic factors. Of the currently available models for sex determination, the polygenic system described by Kallman (1965) and Yamamoto (1969) appears to fit most closely the available data.

Similarily, the mechanism of sex differentiation in fish has not been determined. The sex-specific H-Y antigen system found in higher vertebrates has been reported in several fish species. Whether it plays a significant role in the process of sex differentiation in fish remains to be determined. The dual-sex-inductor model of sex differentiation continues to dominate the conceptual approach to research in fish. However, debate continues over the ability of a unitary primordium, reported for the teleosts, to elaborate separate inducers. Evidence supporting Yamamoto's (1969) hypothesis that the sex steroids are in fact the presumptive inducers remains inconclusive. The steroids are capable of influencing sex determination in the majority of gonochoristic teleost species, but have been applied with variable success to hermaphroditic species. The evidence for steroid action based on the identification of steroid activity, the juxtaposition of steroidogenic tissues to differentiating germ cells, and the effects of antisteroidal compounds remains equivocal. However, regardless of whether the action of steroids is pharmacological or physiological, steroids can play a decidedly influential role in the process of sex differentiation. Therefore, they are a valuable tool for both further exploration of the process of sex differentiation and the manipulation of gonadal sex for purposes of fish culture.

III. HORMONAL SEX CONTROL

Although there are numerous studies concerned with the application of hormones to control gonadal sex, most of them have remained within the framework for effective treatment described by Yamamoto (1969). Based on his work with *Oryzias latipes*, Yamamoto suggested that, for effective treatment, the species-specific optimal dosage of a particular steroid should be administered from the stage of the undifferentiated gonad through the time of morphological differentiation. Although the majority of studies have confirmed these criteria as excellent general rules, exceptions have been reported. Further, within these criteria considerable scope for variation exists. Therefore, to facilitate consideration of the various components of these studies a generalized model is presented (Fig. 1).

Hormonal sex-control studies consist of three chronologically ordered phases: management, treatment, and evaluation. Within the management phase, both the species and the gonadal sex appropriate to attaining manage-

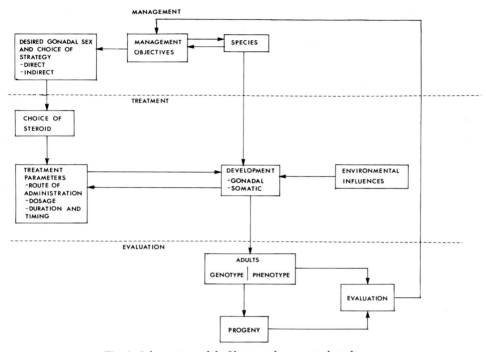

Fig. 1. Schematic model of hormonal sex-control studies.

ment objectives are chosen. The choice of gonadal sex also includes selection of the optimal strategy for achieving it. The treatment phase involves the choice of the appropriate hormone and method of application within the constraints of the gonadal and somatic development of the species. These latter constraints are in turn influenced by environmental conditions. The final phase includes the evaluation of results from either treated fish or their progeny. Where necessary, these results are used to reformulate objectives and techniques in the management and treatment phases.

A. Management

1. SPECIES CHOICE

The establishment of management objectives is clearly interactive with species choice. The initial purpose of hormonal sex-control studies was to elucidate the specific role of hormones in the process of sex differentiation. Further, the inter- and intraspecific matings of sex-inverted and untreated fish provided a means for examining basic sex-determining mechanisms. The species chosen for this research were primarily from the *Cyprinodontidae* and *Poeciliidae* for reasons that included ease of handling and breeding, the presence of a marked sexual dimorphism, and the presence of sex-linked genetic markers (Yamamoto, 1953).

The initiation of applied studies has naturally involved a greater emphasis on species which are already of economic importance and accessible for treatment. However, much of the relevant research on various treatment parameters has been conducted using the same species as in fundamental investigations. The primary concern of these studies has been the optimization of treatment procedures and the development of alternate strategies for the production of fish with a particular gonadal sex. For the purpose of this discussion the absence of phenotypic sex is included, i.e., sterility as a gonadal sex type.

2. GONADAL SEX

In general, a particular gonadal sex is chosen to enhance the value of individual fish because of sex-related morphological, physiological, or ethological characteristics. Further, the reproductive potential of the population as a whole may be modified. Reproductive potential may be enhanced by increasing the proportion of females. Reduction or elimination of this potential may be achieved by the production of sterile or monosex populations.

The ability to control gonadal sex also allows for genetic advancement through the creation of inbred lines. This may be achieved by either the production of synchronously maturing hermaphroditic fish (Jalabert *et al.*, 1975) or the production of monosex genetically homozygous individuals by gynogenetic techniques. A portion of these individuals may be sex inverted by hormonal treatment and mated with their untreated gynogenetic siblings at maturity. Repetition of this cycle over several generations results in a rapid inbreeding effect (Nagy *et al.*, 1981; Streisinger *et al.*, 1981; Donaldson and Hunter, 1982a). The various applications of genetic sex-control techniques and their usefulness to breeding programs are discussed in Chapter 8, this volume.

a. Male and Female Production. Two strategies may be employed to produce monosex male and female populations. The first involves the direct application of steroids to juvenile fish which results in the redirection of sex differentiation to the desired gonadal sex. A second, indirect method involves the sex inversion of homogametic individuals which are reared to maturity. The gametes from these sex-inverted homogametic individuals are joined with the gametes of untreated homogametic fish. If the process of sex determination in the target species is bound to a heterosomal system, the progeny should all be homogametic and, therefore, of the same sex. The monosex group produced may then be used as a reservoir of known homogametic individuals to be sex reversed and used to perpetuate the monosex stock. The use of these known homogametic individuals eliminates the necessity for further progeny testing.

A special case of the latter approach would allow for the production of a monosex heterogametic population. In this case, heterogametic individuals (XY or ZW) are sex inverted and mated with untreated heterogametic individuals. Approximately 25% of the progeny should be either YY or WW, depending on the viability of these individuals. If viable, these individuals could be raised to maturity and mated with untreated homozygous XX or ZZ individuals resulting to the production of monosex heterozygous progeny.

Yamamoto (1964a,b, 1975) has discussed the viability of YY individuals in the medaka and the goldfish. In the latter, the mating of untreated and sex-inverted XY individuals results in a male–female ratio of 3:1 indicating complete YY viability. However, in *Oryzias latipes* the $Y^R Y^R$ zygote, where R represents full xanthic coloration, is rare. However, $Y^R Y^r$ and $Y^r Y^r$ individuals, where r represents scanty coloration are common. The rarity of the $Y^R Y^R$ zygote has been attributed to the presence of an inert section on the Y^R chromosome which is usually lethal in the duplex condition. Fineman *et al.* (1974) have examined the viability of the $Y^r Y^r$ individual in the white strain of the medaka. The life expectancy of these individuals was found to be intermediate between that of the normal male XY and female XX. The viability of YY individuals appears to be high in coho salmon, *Oncorhynchus kisutch* (Hunter *et al.*, 1982a), and somewhat lower in *Hemihaplochromis multicolor* (Hackmann and Reinboth, 1974) and in *Salmo gairdneri* (Johnstone *et al.*, 1979a).

The direct strategy has the advantage that the desired phenotypic adults are produced in the same generation as the treatment. In addition, any gonadal sex may be produced. The major disadvantage of this approach has been the variability of treatment success. A further disadvantage occurs when the objective is to increase the proportions of a particular sex type which must be part of a breeding population. One-half of a population which has been sex inverted will have the genome of the opposite sex. These fish, mated with untreated fish will produce offspring in altered sex ratios. For example, male coho salmon have been demonstrated to be heterogametic. Direct treatment with estradiol produces all-female groups. However, the progeny from the XY females when mated with normal males (XY) produce offspring in a 3:1 male–female ratio (Hunter *et al.*, 1982a). Because one of the objectives of treating Pacific salmon is to increase the number of females suitable for broodstock, such a treatment may not be used. Also, this approach would not be appropriate where a desired production trait was genetically sex linked. Anderson and Smitherman (1978) sex inverted *Oreochromis aureus* and *Oreochromis niloticus* fry with ethynyltestosterone. The growth rates between the two sex-inverted groups were not significantly different. However, both sex-inverted groups grew at a significantly slower rate than a normal male control group produced by manual selection. The suboptimal growth of the sex-inverted groups was attributed to the presence of the female genotype.

The last disadvantage concerns the marketing of fish which have been treated with steroids. Steroid treatments are typically administered to juvenile fish years prior to consumption. Additionally, the amounts of steroid used are small and their half-lives are short. Johnstone *et al.* (1978) have reported that the half-life of estradiol in 1-year-old rainbow trout is less than 12 hr. Similar results have been obtained with the same steroid in coho salmon alevins (G. A. Hunter, E. M. Donaldson, G. van der Kraak, and I. Baker, unpublished). Fagerlund and McBride (1978) and Fagerlund and Dye (1979) have examined the respective depletion of [^{3}H]testosterone and 17α-1,2-[^{3}H]methyltestosterone from yearling coho salmon. These studies indicated that the steroids were rapidly taken into the blood and concentrated in the gonads. Methyltestosterone was eliminated more slowly than testosterone. The concentration of methyltestosterone 10 days from the last administration was found to be 0.01% of the dietary concentration. The authors concluded that long before the time of harvest the levels of exogenous steroid level would have decreased to nondetectable levels. It is evident that the biological basis for concern is small enough to be disregarded especially when the high concentration of endogenous sex steroids present in maturing salmon at the time of harvest are considered (Schmidt and Idler, 1962). However, marketing concerns or legislative restrictions may require the use of indirect strategies where possible.

The indirect strategy should theoretically result in the reliable production of monosex groups. However, this approach relies on the inheritance of sex as a Mendelian trait through a heterosomal system. As previously described heterosomal systems are incompletely developed in some species of fish. Therefore, this approach may not be appropriate for use in these species especially where high levels of treatment effectiveness are required. There are several other disadvantages to this approach. First, the production of fish with the desired gonadal sex requires more than one generation. Second, a means of identifying the homogametic sex-inverted adults or their gametes must be available. Johnstone *et al.* (1979b) and V. J. Bye (personal communication, 1980) noted that genetically female *Salmo gairdneri* when masculinized with androgens do not develop a functional sperm duct although they yield functional sperm when dissected. Such a characteristic has provided a valuable guide for the selection of sex-inverted female adults. However, in species that do not have sex-linked genetic markers or morphological anomalies, progeny testing is required. Donaldson and Hunter (1982a) outlined several effective strategies for the production and identification of gametes that only contain one gonosome type (X) in species, such as the Pacific salmon which exhibit a XX–XY heterosomal system (Fig. 2). In this situation, effective use may be made of cryopreservation techniques to store gametes while progeny tests are conducted (see Chapter 6, this volume). Similar screening procedures have been outlined for the tilapias (Shelton et al., 1978; Jensen and Shelton, 1979) and Salmo species (Johnstone et al., 1979b).

The identification of gametes from sex-inverted homogametic adults is not necessary when the primary objective is merely to increase the proportion of the homogametic sex type. For example, in the Pacific salmon an increase in the proportion of homogametic females is desirable for increasing the rate of stock enhancement. In such situations the use of gametes from both masculinized females and normal males resulting in the production of 75% female offspring is suitable. The need for progeny testing may also be eliminated by the use of gynogenetic techniques in species which are female homogametic (Stanley et al., 1975; Donaldson and Hunter, 1982a). The high mortality usually associated with gynogenetic techniques impedes their direct use for the production of monosex groups. However, gynogenetic individuals may be sex inverted and their gametes used for the breeding of monosex groups.

b. *Sterilization.* Numerous studies have reported the inhibitory effects of steroid treatment at high dosages or long durations on both gonadogenesis and gametogenesis (Schreck 1974). However, relatively few have attempted deliberate hormonal sterilization. The majority of these deliberate attempts have been confined to the economically important species. Further, with the notable exception of Eckstein and Spira (1965), who sterilized the gonads of *Oreochromis aureus* with stilbestrol diphosphate, all deliberate attempts at sterilization have employed androgens.

Yamamoto (1958), working with the medaka, first demonstrated that when androgens are used at dosages beyond those required to achieve masculinization, a proportional increase in percentage sterilized fish results. Several studies have demonstrated that the duration of treatment is also of importance. Oral treatment of coho salmon at three dosages and three durations of treatment demonstrated that longer durations of treatment as well as higher dosages result in a higher production of sterile fish (G. A. Hunter and E. M. Donaldson, unpublished). Van den Hurk and Slof (1981) similarly observed that the administration of methyltestosterone to juvenile *Salmo*

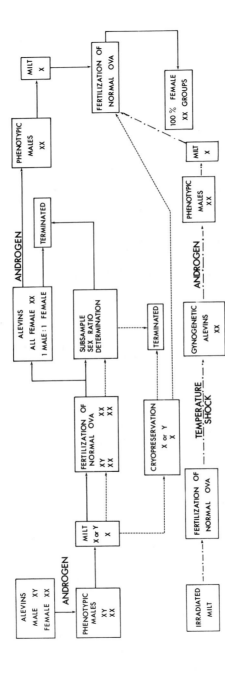

Fig. 2. Alternative strategies for the production of 100% × milt from Pacific salmon by two-stage androgen treatment with test cross (—), by single androgen-treatment cryopreservation and test cross (- - -), by radiation gynogenesis combined with androgen treatment (–·–·) (from Donaldson and Hunter, 1982a).

gairdneri at 3 μg/l in the aquarium water for 8 weeks posthatching was as effective in producing sterile fish as treatment at 300 μg/l for a 4-week period. In the same study, treatment for a 4-week period at dosages from 3 to 300 μg/l indicated a dose-dependent production of sterile fish. Van den Hurk and Slof noted that sterility was not induced when treatment was begun at 43 days postfertilization coinciding with gonadal sex differentiation. However, this time period also coincides with yolk-sac absorption and initiation of feeding. Therefore, the method of administration of this steroid may have influenced its effectiveness. The influences of route of steroid administration on effective dosage are discussed further.

Application of androgens to juveniles resulting in the sterilization of gonads and the maintenance of this gonadal state to adulthood have been demonstrated in *Oryzias latipes* (Yamamoto, 1975), *Salmo gairdneri* (Jalabert *et al.*, 1975; Yamazaki, 1976), *Oncorhynchus kisutch* (Hunter *et al.*, 1982a), and *Oncorhynchus tshawytscha* (Hunter *et al.*, 1983).

The precise mechanism of hormonally induced sterilization remains to be determined. McBride and Fagerlund (1973) reported that treatment of 3–4-month-old coho salmon fry with a diet containing 10 mg methyltestosterone/kg diet for a period of 20–32 weeks resulted in sterilization of testes, but had little or no effect on ovaries. Hirose and Hibiya (1968) administered 4-chlorotestosterone to yearling rainbow trout and found testicular degradation, restraint of male germ-cell differentiation at the spermatogonium stage, and prevention of yolk deposition in the oocytes. Ashby (1965) also reported that oocytes, once differentiated, were highly resistant to steroid influence. Hirose and Hibiya and Ashby attributed the results to a negative-feedback inhibition of gonadotropin release. In this regard, Yamamoto (1975) has reported that hypophysectomy of juvenile medaka inhibits the transformation of spermatogonia to spermatocytes. On the one hand, testes implanted into fish which had previously been sterilized by hormonal treatment underwent spermatogenesis, indicating that treatment did not adversely affect the hypothalamus or hypophysis. On the other hand, Billard *et al.* (1981) demonstrated total inhibition of testicular development in adult 2-year-old rainbow trout fed 0.5 mg/kg methyltestosterone or estradiol and partial inhibition with 0.5 mg/kg testosterone fed during the period of spermatogenesis (June–November). During this period, there was no noticeable change in plasma gonadotropin relative to controls other than the prevention of the gonadotropin rise in September. These results suggest that steroid inhibition occurs at the level of the gonad. Whether sterilization is attributable to feedback inhibition or direct action on the gonads, especially at very early stages of development, remains to be determined.

Katz *et al.* (1976) have hypothesized that the sterilization of *Oreochromis niloticus* gonads by adrenosterone added to the rearing water at 5000 μg/l for a 3-month-period involves inhibition of gonocyte migration through the splanchopleura which they suggest may be hormonally mediated. They observed that fry treated in the aforementioned manner were sterile at the end of treatment. However, at 8 months, 92% of the expected number of male fish contained testes, but no fish had ovarian tissues. Katz and co-workers suggest that the steroid action initially blocked the migration of the gonocytes, thereby preventing germinal tissue development. Ovarian tissue was completely destroyed, therefore, repopulation of gonocytes in part by mitotic division of resting cells occurred only in the testes.

Largely because of the variable effects of androgen administration on differentiated oocytes, the most effective strategy for hormonal sterilization requires the administration of steroid treatments during the period of sex differentiation at higher dosages and longer durations than those required for masculinization.

Several alternate methods for producing sterile fish have been proposed including induced polyploidy, a surgery-induced autoimmunity, and chemical or radiation treatments. These techniques have been reviewed for teleosts with special reference to the grass carp (Stanley, 1979, 1981) and the salmonids (Donaldson and Hunter, 1982a). The most promising of these techniques, induced polyploidy, is examined in detail in Chapter 8, this volume. The induction of triploidy appears to inhibit ovarian development to a much greater extent than testicular development. As a result, the use of hormonally produced all-female *Salmo gairdneri* has been suggested as a first step to producing sterile groups of this species (Lincoln and Hardiman, 1982).

The major advantage of hormonal sterilization is the ability to achieve effective treatment with genetic males and females. Therefore, the establishment of a monosex-treatment population is not required. The disadvantages of the technique include the application of steroids at an early stage of development to fish destined for human consumption and the length of treatment required.

B. Treatment

Within this phase the appropriate hormone and method of application are chosen. The objective of these choices is to ensure that the undifferentiated gonad is exposed to the hormone at a sufficient dosage and duration of treatment to redirect the course of sex differentiation. The choice of the desired gonadal sex, made in the management phase, and the strategy to obtain it determine whether androgens or estrogens are used. However, the choice of a specific steroid must involve considerations of route of administration, steroid dosage, treatment timing, and duration. These factors are in turn interactive with the gonadal and somatic development of the target species, which are influenced by environmental parameters.

1. ROUTE OF ADMINISTRATION

The choice of a particular route of administration is clearly influenced by the development of the target species. The two most popular routes have been the addition of the steroid either to the rearing water or to the diet; the choice usually depends on whether the species is capable of feeding (Table II). Using an innovative adaptation of these approaches, Dzwillo (1962) and Takahashi (1975a,b) achieved functional masculinization of embryonic guppies by androgen treatment of gravid females either in the rearing water or in the diet, respectively. Other routes of administration have included subcutaneous implantation (Egami, 1955; Okada, 1962), intraperitoneal injections (Castelnuovo, 1937; Eversole, 1939, 1941; Berkowitz, 1941; Vallowe, 1957; Okada, 1964), injection into eggs (Hishida, 1962), and implantation of steroid-containing silastic capsules (Jensen et al., 1978). The use of these alternate routes have been limited and, with the exception of the latter, used only for experimental purposes. Therefore, further discussion in this section is be confined to the more common routes of administration.

2. DOSAGE

Early research on the medaka indicated that, over the effective range of dosages, the level of controlled sex differentiation achieved was dose dependent (Yamamoto, 1969, 1975). Using genetically all-male (XY) *Oryzias latipes*, a proportional increase in the percentage of females occurred between oral estrone dosages of 10–25 mg/kg diet. At 50 mg/kg diet, 100% female fish were produced (Yamamoto, 1959b). The response of genetically all female *Oryzias latipes* to methyltestosterone was slightly different (Yamamoto, 1958). At oral dosages between 5 and 25–50 mg/kg diet a proportional increase in the percentage of males occurred, reaching a maximum level somewhat less than 100%. The 100% level was not achieved because of the production of a few sterile fish. At higher dosages of 50–300 mg/kg, the percentage of steriles increased to 100%. Yamamoto concluded that methyltestosterone both suppresses gonadogenesis in both sexes and also induces sex inversion of genetic females. At higher dosages the former effect is intensified. For both estrogens and androgens subthreshold dosage levels result in the production of intersex gonads.

The result of several studies have indicated that this dosage–effect relationship may not be as straightforward as proposed by Yamamoto. The paradoxical effects reported in various cichlid species have already been men-

Table II

Hormonal Sex-Control Studies—Economically Important Species

Species	Steroid	Dosage	Duration and timing	Treatment effect	Reference
Cichlids: Use of Androgens					
O. aureus	Testosterone Methyltestosterone	50–1000 µg/l	5–6 week starting, 4–5 week posthatching	Variable effects involuted and unaffected gonads	Eckstein and Spira (1965)
	Ethynyltestosterone Methyltestosterone	15–60 mg/kg diet	18 days, 6 days/week for 3 weeks	30–60 mg/kg produced 98–100% Males, female heterogamety demonstrated	Guerrero (1975)
O. niloticus	Ethynyltestosterone	30 mg/kg diet	32 days	Monosex males	Sanico (1975)
	Ethynyltestosterone	60 mg/kg diet	21–28 days	98.9–100% Males produced	Shelton et al. (1981)
	Methyltestosterone	40 mg/kg diet	60 days	Increased males; female homogamety demonstrated	Jalabert et al. (1974)
	Methyltestosterone	15–50 mg/kg diet	42 days	Increased males	Guerrero and Abella (1976)
	Andrenosterone	5000 µg/l	3 months postfertilization	No germ cells apparent at 3 months, 40% had testes at 8 months	Katz et al. (1976)
	Ethynyltestosterone Methyltestosterone	30–60 mg/kg diet	25–59 days	100% Males produced	Tayamen and Shelton (1978)
	Methyltestosterone	5 mg/kg diet	28 or 42 days	100% Males produced	Owusu-Frimpong and Nijhar (1981)
	Methyltestosterone	50–100 mg/kg diet	30 days from capture	100% Males produced	Nakamura and Iwahashi (1982)
T. zillii	Ethynyltestosterone	50 mg/kg diet	40 days	No sex inversion	Guerrero (1976b)
	Methyltestosterone	50–100 mg/kg diet	20 days starting 10 days post-hatching	No sex inversion; oogenesis inhibited	Yoshikawa and Oguri (1978)
T. macrochir	Methyltestosterone	50 mg/kg diet	45 days	100% Males produced	Wodiwode (1977)
	Methyltestosterone	40 mg/kg diet	2 months	100% Sterilization	Jalabert et al. (1974)
O. mossambicus	Methyltestosterone	10–50 mg/kg diet	69 days	100% Males produced at 10–40 mg/kg; female homogametic	Clemens and Inslee (1968)
	Methyltestosterone	50 mg/kg diet	19 days starting 7 days post-hatching	100% Males produced	Nakamura (1975)
	Methyltestosterone	1000 mg/kg diet	19 days starting 7 days post-hatching	Ineffective	
	Methyltestosterone	30 mg/kg diet	14–28 days	69–98% Males produced	Guerrero (1976a)
	Methyltestosterone	50 mg/kg diet	30 days	100% Males produced	Guerrero (1976b)

Species	Compound	Dose	Duration	Result	Reference
	Methyltestosterone	60 mg/kg diet	Fry (9–10 mm) for 42 days	100% Males produced	Anonymous (1979)
		30 mg/kg diet	42 days	81–85% Males produced	Anonymous (1979)
	11-Ketotestosterone	200 mg/kg diet	19 days starting 7 days posthatching	100% males produced	Nakamura (1981a)
	Methyltestosterone	50 mg/kg diet	19 days starting 7 days posthatching	100% Males produced	Nakamura (1981a)
		10 μg/l		100% Males produced	Nakamura (1981a)

Cichlids: Use of Estrogens

O. aureus

Species	Compound	Dose	Duration	Result	Reference
	Stilbestrol	50–1000 μg/l	5–6 weeks starting, 4–5 weeks posthatching	At 50 and 100 μg/l sterility induced	Eckstein and Spira (1965)
	Stilbestrol-diphosphate-Diaethyldioxystilben-diphosphate				
	Estriol	30–120 mg/kg diet	21 or 35 days	No sex inversion	Jensen (1976)
	Estrone	30–120 mg/kg diet	21 or 35 days	No sex inversion	
	Estradiol	30–120 mg/kg diet	21 or 35 days	No sex inversion	
	Ethynylestradiol } With, or without, cryproterone acetate	25–200 mg/kg diet	35 or 56 days	60–64% Females produced	Hopkins (1977)
	Diethylstilbestrol	Cryproterone acetate at 100 mg/kg diet			
	Estradiol				
	Ethynylestradiol with methallibure and cryproterone acetate	100 mg/kg diet 100 mg/kg diet 100 mg/kg diet	42 days	90% Females produced	Hopkins *et al.* (1979)

O. niloticus

Species	Compound	Dose	Duration	Result	Reference
	Diethylstilbestrol	25 and 100 mg/kg diet	25–59 days	62–90% Females produced	Tayaman and Shelton (1978)
	Estrone	100 and 100 mg/kg diet			
	Ethynylestradiol	20–60 mg/kg diet	20 days starting 10 days posthatching	No sex inversion; Some involuted gonads at 40 and 60 mg/kg	Yoshikama and Oguri (1978)

O. mossambicus

Species	Compound	Dose	Duration	Result	Reference
	Ethynylestradiol	50 mg/kg diet	19 days from 6 days posthatching	100% Females produced	Nakamura and Takahashi (1973)

Salmon: Use of Androgens

Salmo sp.

S. gairdneri

Species	Compound	Dose	Duration	Result	Reference
	4-Chlorotesterone	1.0–12.5 mg injection 6 times	30 days starting at age 1 year	Suppression of testicular differentiation and yolk deposition	Hirose and Hibiya (1968)
	Methyltestosterone	15–60 mg/kg diet	5 months starting 1 month posthatching	58% Males and 12% steriles produced	Jalabert *et al.* (1975)
	Methyltestosterone	50 mg/kg diet	5 months starting 1 month posthatching	Testes devoid of germ cells	Yamazaki (1976)
	Methyltestosterone	1 mg/kg diet	7 months starting 1 month posthatching	87.1%	Yamazaki (1976)

(continued)

253

Table II—Continued

Species	Steroid	Dosage	Duration and timing	Treatment effect	Reference
S. gairdneri (cont.)	Methyltestosterone	250 µg/l plus 3 mg/kg diet	Two 2-hr immersion of eyed and alevins, 90 days from first feeding	100% Sex inversion	Simpson (1976)
		3 mg/kg diet	90 days from first feeding	100% Sex inversion	
	Methyltestosterone	250 µg/l plus 3 mg/kg diet	Two 2-hr immersion of eyed eggs and alevins 90 days from first feeding	83% Males and 17% sterile produced	Johnstone *et al.*, (1978, 1979a)
				Demonstration of female homogamety	
	Methyltestosterone	1–10 mg/kg diet	58 days	60–68% Males produced, demonstration of female homogamety	Okada *et al.* (1979)
	Methyltestosterone	3 mg/kg diet	90 days	94% Male	V. J. Bye (personal communication, 1980)
	Methyltestosterone	25 mg/kg diet	1000°C days (9–12°C) 110 days	60% Sterility noted	V. J. Bye (personal communication, 1980)
	Methyltestosterone	30 mg/kg diet		2.9% Mature as 2 year olds versus 40% control	Harbin *et al.* (1980)
	Methyltestosterone	3–300 µg/l	28 or 56 days immersion	Sterility, high mortality	van den Hurk and Slof (1981)
	Methyltestosterone	0.5 mg/kg diet	6 months (June–November)	Gonadal inhibition	Billard *et al.* (1981)
	11β-Hydroxyandrostenedione	60 mg/kg	56 days	100% Males produced	van den Hurk and Lambert (1982)
		6.0 mg/kg	56 days	94–99% Males produced	
		0.6–6.0 mg/kg			
S. salar	Methyltestosterone	250 µg/l plus 30 mg/kg diet	Two 2-hr immersions of eyed eggs and alevins for 120 days	No female elements	Simpson (1975–1976)
	Methyltestosterone	30 mg/kg diet	120 days	Sterile	Johnstone *et al.* (1978)
	Methyltestosterone	3 mg/kg diet	90 days	100% Male or sterile	Johnstone *et al.* (1978)
S. trutta	Testosterone	50–60 µg/l	111 days starting 170 days postfertilization 6 days/week for 3 months starting 7 months postfertilization	41% Male, 35% Female	Ashby (1957)
	Testosterone	67 µg/l		24% Indeterminant testicular inhibition	Ashby (1965)

Species	Steroid	Dosage	Treatment duration	Result	Reference
Salvelinus sp.					
S. namaycush	Testosterone propionate	700 mg/kg diet	245–290°C days starting at 1250°C days	7 Males, 3 females produced	Wenström (1975)
Oncorhynchus sp.					
O. kisutch	Methyltestosterone	1–50 mg/kg diet	42 days starting 8–9 months posthatching	Severe reduction in spermatogonia	McBride and Fagerlund (1973)
	Methyltestosterone	0.2–10 mg/kg diet	504 days starting 3–4 months posthatching	Severe gonadal inhibition in group fed 10 mg/kg	Fagerlund and McBride (1975)
	Methyltestosterone	25–400 µg/l plus 20 mg/kg diet	Two 2-hr immersions of eyed eggs and alevins 70 days	94–100% Sterile	Goetz et al. (1979)
	Methyltestosterone	400 µg/l	Two 2-hr immersion of alevins	67% Males produced	G. A. Hunter and E. M. Donaldson (unpublished data)
	Methyltestosterone	400 µg/l plus	Two 2-hr immersion of either eyed eggs or alevins	83–93% Sterile	G. A. Hunter and E. M. Donaldson (unpublished data)
	Methyltestosterone	10 mg/kg diet 400 µg/l plus	3 months Two 2-hr immersion of alevins	82–92% Males produced	G. A. Hunter and E. M. Donaldson (unpublished data)
	Methyltestosterone	1 mg/kg diet, or 3–9 mg/g diet 100 or 400 µg/l plus 10 mg/kg diet	21–63 days Two 2-hr immersion of eyed eggs and alevins for 90 days	22–68% Males produced 22–78% Sterile 70–94% Sterile at age 3 years	Hunter et al. (1982a)
	Methyltestosterone	200 µg/l plus	Two 2-hr immersion of eyed eggs and alevins	97.7% Sterility at age 3 years	Donaldson and Hunter (1982b)
O. tshawytscha	Methyltestosterone	10 mg/kg diet	84 days	No reduction of spermatogonia	McBride and Fagerlund (1973)
	Methyltestosterone	0.2 or 1.0 mg/kg diet 400 µg/liter plus 3–9 mg/kg diet	28–84 days Two 2-hr immersions of alevins 21–63 days	83–93% Males; demonstration of female homogamety	Hunter et al. (1983)
O. keta and *O. gorbusha*	Methyltestosterone	50–100 mg/kg diet	14 days at age 1 year	Degeneration of spermatogonia	Yamazaki (1972)

(continued)

Table II—Continued

Species	Steroid	Dosage	Duration and timing	Treatment effect	Reference
Salmon: Use of Estrogens					
Salmo sp.					
S. gairdneri	Estrone	10–100 mg/kg diet	38–124 days posthatching	79–94% Females produced; high mortality, growth depression	Okada (1973)
	Estrone	30–120 mg/kg diet	5 months beginning 1 month posthatching	54% Females, 30% intersex produced	Jalabert *et al.* (1975)
	Estradiol	250 µg/l plus	Two 2-hr immersions of eyed eggs and alevins 30 or 56 days from feeding	100% Females produced	Simpson (1976)
		20 mg/kg diet	15 days	69% Females produced	Johnstone *et al.* (1978)
	Estradiol	20 mg/kg diet	30 days from first feeding	100% Female, growth depression; demonstration of male heterogamety	Johnstone *et al.* (1979b)
S. salar	Estradiol	250 µg/l plus	Two 2-hr immersion of eyed eggs and alevins	100% Females produced	Simpson (1976)
		20 mg/kg diet	21 days from first feeding	100% Females produced	Johnstone *et al.* (1978)
S. trutta	Estradiol	50 or 300 µg/l	70 or 111 days (560°C–888°C days) starting 170 days post-fertilization (850°C days)	High dosage produced 7 females and 3 males	Ashby (1957)
Salvelinus sp.					
S. namaycush	Estradiol	12 mg/kg	245–290°C days	8 Females and 2 males produced	Wenström (1975)
S. fontinalis	Estradiol	20 mg/kg	60 days from first feeding	99% Females produced	Johnstone *et al.* (1979a)
Oncorhynchus sp.					
O. kisutch	Estradiol	25–400 µg/l plus	Two or six 2-hr immersion of eyed eggs, two or seven immersion of alevins	Nine of 10 groups >95% female; low dosage group 60% female, increased mortality and growth depression	Goetz *et al.* (1979)
		10 mg/kg diet	70 days		
	Estradiol	100 or 400 µg/l plus	Two 2-hr immersions of eyed eggs and alevins	86–100% Females at maturity, demonstration of male heterogamety	Hunter *et al.* (1982a)
		5 mg/kg diet	90 days		

	Hormone	Dose	Treatment period	Effect	Reference
	Estradiol	400 µg/l	Two 2-hr immersions of alevins	100% Females produced	G. A. Hunter, E. M. Donaldson, and I. Baker (unpublished)
	Estradiol	200 µg/l plus	Two 2-hr immersions of eyed eggs and alevins	>99% Female at maturity	Donaldson and Hunter (1982b)
O. tshawytscha		5 mg/kg diet 400 µg/l plus	84 days Two 2-hr immersions of	96–100% Females produced	G. A. Hunter and E. M. Donaldson (unpublished)
		2 or 5 mg/kg diet 5 mg/kg diet	21–63 days 24–84 days		
O. masou	Estradiol	0.25–200 µg/l	18 days starting 5 days posthatching	All female at 0.5–5 µg/l; very high mortality at 10–200 µg/l	Nakamura (1981a)
Carp: Use of Androgens *Ctenopharyngodon idella*	Methyltestosterone	30 or 60 mg/kg diet	14–63 days from 7 days posthatching or 28 days starting 28–112 days posthatching	No effect	Stanley and Thomas (1978)
	Methyltestosterone	Silastic capsules 20.6–31.0 µg/day	303 days from 195 days posthatching; 192 days from 309 days posthatching	Growth depression: 5 fish sterile; ovarian development suppressed	Jensen et al. (1978)
	Methyltestosterone	60–120 mg/kg	255 or 410 days from 110 days posthatching	Growth and gonadal inhibition	Shelton and Jensen (1979)
	Methyltestosterone	Silastic capsules 18.7 µg/day	303 days from 195 days posthatching	Reduced germ cells	Shelton and Jensen (1979)
	Methyltestosterone	Silastic capsules 14.9–18.4 µg/day	189–461 days from 309 days posthatching	0–31.7% Females produced	Shelton and Jensen (1979)
	Methyltestosterone	Silastic capsules 16.6 µg/day	500 days from 319 days posthatching, fish stunted 123 mm	No sex inversion; 17.4% intersex	Shelton and Jensen (1979)
	Methyltestosterone	Silastic capsules 12.5 µg/day	460 days from 55 days, 128 mm length	18.5 Male, 33.3 intersex, 29.6 sterile, and 18.5% reduced ovaries produced	Shelton and Jensen (1979)
Cyprinus carpio	Methyltestosterone	100 mg/kg diet	4- to 36-day periods between 8 and 98 days posthatching	71.4–88.9% Males produced; 13.3–28.6% undifferentiated	Nagy et al. (1981)

tioned. Further, methyltestosterone treatment at very high dosages has led to decreases in masculinizing potency without sterilization. Okada *et al.* (1981) orally administered methyltestosterone for 8 weeks to genetically all-female *Salmo gairdneri* at dosages ranging from 0.01 to 500 mg/kg diet. No inversions were observed in groups fed 0.01 mg–0.1 mg/kg diet. At 0.5, 1.0, 5.0, and 10.0 mg/kg diet 84.4, 69.2, 42.0, and 24.0% males were recorded. At dosages higher than 50.0 mg/kg diet, less than 7.0% males were recorded. Intersex gonads were observed at dosages from 0.5 to 10.0 mg/kg diet.

Also related to steroid treatment are various deleterious effects usually associated with high steroid dosages. These effects include gonadal suppression and paradoxical steroid actions, which were previously described, high mortality (Ashby, 1957; Yamamoto, 1958, 1961; Yamamoto and Matsuda, 1963; Clemens and Inslee, 1968; Goetz *et al.*, 1979; Okada *et al.*, 1979; van den Hurk and Slof, 1981), and growth depression especially with estrogen administration (Ashby, 1957; Okada, 1973; Goetz *et al.*, 1979; Johnstone *et al.*, 1979b; Shelton and Jensen, 1979). Therefore, it is evident that the choice of dosage is critical to treatment efficiency. Several interrelated factors influence the optimal dosage level of a particular steroid including its biological activity, which may be related to its origin, the route of administration, the target species, and the duration of treatment. This latter factor is of considerable importance and is considered separately.

a. Biological Activity. Although a wide range of natural and synthetic steroids have been used successfully to induce sex inversion, their individual biological activities vary. Studies on the medaka reviewed by Yamamoto (1969) remain the best demonstration of relative steroid potencies. Taking advantage of the sex-linked color genes of the d-r^R strain of *Oryzias latipes* in which females $x^r x^r$ are white and males are orange-red $x^r y^R$, researchers were able to determine the 50% sex-inversion level following oral administration of various androgens and estrogens. In general, the synthetic androgens were more potent than those of natural origin. The most potent synthetic androgen was 19-*nor*-ethynyltestosterone, which produced 50% sex inversion at a dosage of 1.0 mg/kg diet. Less potent were fluoxymesterone 1.2 mg/kg, ethynyltestosterone 3.4 mg/kg, methylandrostenediol 7.8 mg/kg, methyltestosterone 15 mg/kg, and testosterone propionate 560 mg/kg. The natural androgens included androstenedione 500 mg/kg and androsterone 580 mg/kg. Using similar methods, Hishida and Kawamoto (1970) demonstrated that 11-ketotestosterone is the most potent of the natural occurring androgens at 110 mg/kg. Similarly, the synthetic estrogens hexesterol, euvastin, and ethylestradiol achieved 50% sex inversion at 0.4, 0.8, and 1.7 mg/kg, respectively (Yamamoto, 1969). To achieve 50% sex inversion, the naturally occurring estradiol, estrone, and estriol required dosages of 5.8, 20, and 130 mg/kg diet. White *et al.* (1973), observing a similar hierarchy of estrogen potency in the rat, suggested that the results were to be expected because both estrone and estriol are the metabolic products of estradiol. Similarly, 11-ketotestosterone has been demonstrated to be much more potent than testosterone in inducing sex inversion in *Poecilia reticulata* (Takahashi, 1975c).

Yamamoto (1969) and Hishida and Kawamoto (1970) have suggested that the higher potency of orally administered synthetic steroids is in part attributable to their resistance to degradation in the digestive tract. In considering the potencies of various steroids, the extremely low effective dosages of 16-[^{14}C]estrone determined by Hishida (1965) should be noted. For this steroid administered at a level of 44.6 mg/kg diet and resulting in 84% sex inversion, the effective dosage at the gonadal level was calculated to be 1.5 $\times$ 10^{-2} µg. Therefore, it is evident that the effective quantities of steroid at the site of action are low relative to those which must be administered.

Several studies have demonstrated the influence of the route of administration on steroid potency. Hishida (1965) reported a 10-fold increase in the potency of estrone when administered intraperitoneally versus orally to *Oryzias latipes* fry. Takahashi (1974) observed that when administered orally to juvenile *Poecilia reticulata*, the androgenic potency of the naturally occurring 11-ketotestosterone was low relative to synthetic steroid methyltestosterone. The equivalent dosages for the two steroids were 200 mg/kg and 20 mg/kg, respectively. However, when the steroids were added to the rearing water of the same species, 11-ketotestosterone treatment resulted in 100% functional masculinization at 10–25 µg/l, but methyltestosterone was ineffective at this dosage (Takahashi, 1975c).

In general, optimal steroid dosage appears to be species specific. However, comparisons of relative effective dosages using available data are difficult because of variability of treatment protocol. Further, observations for a particular species are often based on a limited number of dosage levels. Where comparisons may be made, the variation in effective dosage may be large. For example, similar levels of sex inversion have been achieved in genetically all-female populations of *Oryzias latipes* (Yamamoto, 1958) and *Salmo gairdneri* (Okada *et al.*, 1981) that were administered methyltestosterone at 25 mg/kg diet for 49–56 days or 0.5 mg/kg diet for 60 days, respectively. Similarly, Yamamoto and Kajishima (1969) obtained 100% sex inversion in goldfish fed estrone at 100 mg/kg diet for a period of 60 days from first feeding. A similar level of effective treatment (94%) was obtained by Okada (1973) with *Salmo gairdneri* fed the same steroid at 10 mg/kg for a period of 58 days. In general, the optimal treatment dosages for the salmonids are somewhat lower than those for other groups (Table II).

3. TIMING AND DURATION

The criteria established by Yamamoto (1969) for effective steroid treatment requires that treatment be initiated prior to the onset of normal sex differentiation and continued until the time of morphological differentiation. Failure to comply with this criterion results in ineffectual treatment. Indeed the majority of studies have indicated a specific period over which treatment must be continued. However, several studies have demonstrated treatments of shorter periods of time which are fully effective. Therefore, the relationship between the initiation and completion of sex differentiation and the timing and duration of effective treatment proposed by Yamamoto may be a conservative estimate for many species.

Treatment is ultimately dependent on an interval of time in the course of gonadal development when the gonads are labile to hormonal influences. The presumption has been that the steroids, acting as initiators of differentiation, should be administered at a time synchronous with natural differentiation. Therefore, histological techniques have been employed to determine the initiation and completion of sex differentiation. The identification of the labile period, by definition, requires the application of steroids at various times and durations within and outside of the period of histologically observable differentiation. Indicators such as age or size of fish may then be correlated with the boundaries of the labile period.

a. Determination of the Period of Sexual Differentiation. Gonadal sex differentiation has two distinct but highly interactive and interdependent processes: gonadogenesis, which is the formation of the structural and supportive elements of the gonads, and the gametogenesis, which is the formation of the gametes. As previously discussed, the physiological mechanisms governing these processes are as yet undetermined. The underlying premise of hormonal treatment is that the steroids are or have the ability to mimic the natural sex inductors. Logically, treatment should be started coincident with the release of the natural inductors and, therefore, sometime prior to the first histologically observable signs of sex differentiation. In fact, many studies have concerned themselves with gonadal differentiation in fish. However, few have examined the initial period of differentiation (Harrington, 1974). The paucity of studies and a general lack of understanding of the mechanisms involved has resulted in considerable difficulty in establishing criteria suitable for the identification of the initiation of differentiation. Gametogenesis, either oogenesis or spermatogenesis may be regarded as an indisputable sign of sex differentiation. However, the subtlety of the earliest changes in the formation of the gametes has made their individual identification difficult. The dubious identification of oogonia and spermatogonia without the benefit of somatic indicators has been stressed repeatedly (Reinboth, 1972, 1982; Harrington, 1974).

As a result, investigators have relied heavily on a variety of somatic and germinal features to demark early differentiation. Stromsten (1931) has listed a variety of such features used to identify sex differentiation in *Carassius auratus*. These features include the form and general appearance of the gonad, the means of attachment of the gonads to the wall of the coelomic cavity, the nature and disposition of the cells of the stroma and germ cells, the structure of the nuclei, the presence or absence of a nucleolus, and the vascularization of the gonad. Persov (1975) reports that toward the end of the indifferent period, the direction of development of the gonads of the *Salmo* species may be determined by the position of the germ cells in the gonad. When they are concentrated on the lateral side and large capillaries are on the medial side, ovarian development is taking place. If the germ cells are ventral to the large capillaries, testicular development is taking place. The latter assertion agrees with the data presented by Ashby (1957). Shelton and Jensen (1979) reported that anatomical changes in the ovary involving changes in its shape and development of a second point of attachment to the peritoneum are indicators of ovarian development in the grass carp, silver carp (*Hypothalmichthys molitrix*), and the goldfish.

The gonads have traditionally been considered undifferentiated during the migration of the primordial germ cells and their subsequent mitotic divisions. However, several authors have reported the possibility of using differential germ cell numbers as indicators of early sexual differentiation. The potential of this approach for the vertebrates, in general, was reviewed by Hardisty (1967). The available literature did not permit him to clarify whether differences in germ cell number were attributable to differences in the number of primoridal germ cells or an earlier or more rapid proliferation in the germ cells of one sex. However, more recent studies have indicated that a proliferation of oogonia shortly before the initiation of meiotic activity appears generalized for most teleosts (Nakamura, 1978).

In the medaka, this rapid mitotic activity occurs either shortly before (Onitake, 1972) or after hatching (Quirk and Hamilton, 1973). In several cichlids, mitotic activity of the gonia occurs simultaneously with the initiation of somatic growth leading to the formation of the ovarian cavity. These events occur at 10–16 days posthatching at 20°C in *Oreochromis mossambicus* (Nakamura, 1978; Nakamura and Takahashi, 1973), 10–15 days posthatching at 30°C in *Tilapia zillii* (Yoshikawa and Oguri, 1977), and 15–16 days posthatching at 31°C in *Oreochromis aureus* (Dutta, 1979). In the latter species, formation of the ovarian cavity was previously reported at 30 days posthatching (Eckstein and Spira, 1965). However, these fish were 10–12 mm at 30 days compared with 16–18 mm at 15–16 days reported by Dutta (1979). Growth related factors may account for the variance in observations.

In the salmonids, based on increases in the number of cysts of premeiotic gonia, female differentiation was first observed 28–25 days posthatching at

8°C in *Oncorhynchus masou* (Nakamura, 1978), 17–35 days posthatching at the same temperature in *Oncorhynchus keta* (Nakamura, 1978), 103–131 days posthatching at 1°–6°C in *Salvelinus leucomaenis* (Nakamura, 1982), and 35 days (378°C days) posthatching (Lebrun *et al.*, 1982) in *Salmo gairdneri*. In the latter species, Takashima *et al.* (1980) reported that based on the chromosomal arrangement and single nucleolus observed in some germ cells primary oocytes could be tentatively identified at 67 days from fertilization at 11.2°C (swim-up fry). At slightly higher water temperatures of 11°–13°C, van den Hurk and Slof (1981) reported cysts of oogonia in meiotic prophase at the swim-up fry stage but only 45–55 days postfertilization, 16–19 days posthatching. Ashby (1957) reported an increased number of oogonia 169 days postfertilization in *Salmo trutta* reared at 5°–8°C.

The more easily detectable ovarian characteristics have been the preferred indicators of the initiation of gonadal differentiation. However, effective indicators of testicular differentiation have been reported. Nakamura (1978) suggested that because spermatogonia remain quiescent for a variable period of time following the initiation of oogenesis, somatic differentiation is the preferred indicator of testicular differentiation. He cited as indicators of testicular differentiation formation of the efferent duct in *Oncorhynchus masou*, *Oncorhynchus keta*, and *Salvelinus leucomaenis*, differentiation of blood capillaries near the efferent duct in these species and in *Carassius auratus*, and aggregations of stromal cells in the hilar region of *Gambusia affinis* and *Poecilia reticulata*. In *Oreochromis mossambicus*, formation of the ovarian cavity and the efferent duct analoges in the gonadal stroma, indicating female and male differentiation, respectively, occur at age 20 days concurrent with the first meiotic activity of the female germ cells. These results indicate that testicular somatic differentiation can be demonstrated at a relatively early stage.

Various studies employing histological examination have provided extensive evidence for the species-specific nature of sex differentiation. In most gonochorist species examined, differentiation is initiated shortly after hatching. However, extreme variations may occur. In the cyprinodont *Poecilia reticulata*, ovarian and testicular differentiation, based on premeiotic activity of oogonia and stromal aggregations in the hilus of presumptive testes, occurs 12 and 8 days prior to parturition, respectively. In the grass carp, ovarian differentiation based on anatomical changes was initiated between age 50 and 75 days. Oogonial nests were not evident until 94–125 days and perinuclear oocytes did not appear until 240–405 days. Shelton and Jensen (1979), citing Emelyanova, report that in the silver carp, *Hypothalmichthys molitrix*, cytological differentiation of the gonads was not completed until 150–180 days posthatching.

Of additional interest is the suggestion first presented by Yamamoto

(1959a, 1962) and Yamamoto and Matsuda (1963) that the process of sex differentiation proceeds in an anterior to posterior gradient in the medaka. Yamamoto (1962) suggested that this gradient of differentiation could explain the occurrence of intersex gonads in fish administered estrogen treatments of insufficient dosage or duration. Presumably, the anterior portion of the gonad undergoes testicular differentiation influenced by the natural sex inducer, but the posterior portion of the gonad undergoes ovarian differentiation influenced by the exogenous estrogens. In the amphibians, Witschi (1967) described a similar anterior to posterior gradient of differentiation in *Xenopus laevis*. Witschi and Dale (1962) reported that a 2-day estrogen treatment of male frog larvae of various ages resulted in a gonad consisting of testicular and ovarian tissue. The ovarian mode shifted caudally with increasing age of larvae. Later, numerical examination of testicular and ovarian differentiation in *Oryzias latipes* by Yoshikawa and Oguri (1979, 1981), respectively, did not support Yamamoto's hypothesis. Yoshikawa and Oguri reported a random distribution of developing interstitial and germinal elements. In the later study, they reported that the right ovary differentiated before the left. Although numerical analysis does not support a gradient of differentiation in *Oryzias latipes*, the presence of a cephalocaudal gradient of spermatogenesis, oogenesis, and the formation of the ovarian cavity has been reported in the cichlid *Tilapia zillii* (Yoshikawa and Oguri, 1977). A similar gradient in the development of the ovarian cavity was reported for *Oreochromis mossambicus* (Nakamura and Takahashi, 1973). However, Johnstone *et al.* (1978) observed no distinct pattern in hermaphroditic rainbow trout gonads following administration of either estradiol or methyltestosterone. The question of whether a gradient of differentiation is common within the teleosts remains. However, these studies do demonstrate a variable timing of differentiation at the level of individual germinal and somatic cells, the gonad as a whole and the individual within a population. Therefore, the determination of treatment timing, treatment duration, and the assessment of treatment effectiveness based on histological analysis must account for this variability. This is particularly important where highly effective treatments are required or the treatment objective is to create simultaneously maturing intersex gonads.

   b. Determination of the Labile Period. Although histological examination may provide visual guides to the initiation and completion of sex differentiation, exact delineation of the hormone-labile period requires the administration of hormones. Only by this means can the timing and duration of an effective treatment regime be established.
   Studies involving steroid administration have demonstrated that similar to the histologically observable period of sex differentiation, the effective

treatment period varies between species. In several species, effective estrogen treatment must be started with the initiation of mitotic activity of the
primordial germ cells before sex differences are observable. In *Oreochromis
mossambicus*, the intention of mitotic activity in the germ cells is reported to
occur at 8–10 days after hatching (Nakamura and Takahashi, 1973). Meiotic
activity did not occur until age 20 days. In the same study, an effective
period of ethynylestradiol treatment starting 6 days posthatching and lasting
for 19 days was demonstrated. However, treatment exclusively within the
indifferent period 6–15 days was without effect. Similarly, Nakamura (1978)
reported the first mitotic activity of germ cells in *Oncorhynchus masou* at
14–28 days posthatching. He subsequently demonstrated an effective estradiol treatment period between 5 and 23 days posthatching.

The association between successful estrogen treatment and the earliest
mitotic activity does not appear to be universal within the teleosts. In the
guppy, ovarian differentiation, based on developing oocytes, has been reported 14 days following the preceding parturition, 12 days before birth
(Takahashi, 1975a,b). However, Takahashi (1975d) was able to obtain complete feminization of genetic males by administering ethynylestradiol at 125
mg/kg diet for 30 days after birth.

Effective androgen treatment has been demonstrated to be synchronous
with somatic sex differentiation in the guppy. Testicular differentiation in
Poecilia reticulata, based on aggregations of stromal cells in the gonadal
hilus, has been reported 18 days following the last parturition, 8 days prior to
birth (Takahashi, 1975a,b). Takahashi (1975a) was able to demonstrate that
the oral administration of methyltestosterone at 400 mg/kg diet to gravid
guppies 13–15 days after the preceding parturition, i.e., from 13–11 days
before the birth until the time of birth, resulted in the development of
stromal aggregations in the gonadal hilus of treated females. Within 20 days
of birth, these fish developed testes. Previously, Dzwillo (1962, 1966) had
demonstrated that immersion of gravid females in water containing methyltestosterone resulted in masculinization of the offspring. In the later study,
the steroid was administered at 3–4 mg/l for only 24 hr between 8–12 days
prior to parturition. Takahashi (1975a) suggested that these results indicated
an androgen-sensitive period of embryonic ovaries at about 8 days prior to
birth, corresponding to the development of the stromal aggregations in the
hilus of the normal male testis. Further, he attributed the success of the
preparturition methyltestosterone treatment to its action in initiating somatic differentiation in the hilar region.

Similar to the oral administration of estrogen, Takahashi (1975c) was able
to obtain complete masculinization of juvenile guppies postpartum by
adding 11-ketotestosterone to the rearing water at 25–50 μg/l for 35 days
following birth. Inhibition of spermatogoenesis and sperm duct formation

was observed in some individuals. Therefore, it is apparent that treatment at the initiation of differentiation is important in some species. However, in other species the developing gametes maintain their sexually bipotent nature for a considerable period and may be influenced by hormonal treatment over several discrete intervals. An excellent example of this latter case is the study by Nagy *et al.* (1981) on *Cyprinus carpio.* Using gynogenetic all-female groups of carp, these researchers were able to demonstrate that at 25°C, sex inversion could be obtained by oral administration of methyltestosterone in any of several overlaping 36-day periods initiated between 8 and 80 days posthatching.

Effective treatments of very short duration have been obtained in several species. The 24-hr treatment used by Dzwillo (1966) to masculinize embryonic guppies has been described previously. Hackmann and Reinboth (1974) masculinized *Hemihaplochromis mutlicolor* by immersing 14- to 16-day-old juveniles for 42–44 hr in water containing methyltestosterone propionate at 500 μg/l. Feminization was also achieved with a similar duration and route of administration using estradiol butyrl acetate between 11.5 and 16 days postspawning. Coho salmon have been feminized by immersion for two 2-hr periods, 4 and 11 days posthatching in water containing estradiol at 400 μg/l (G. A. Hunter, E. M. Donaldson, and I. Baker, unpublished). Typically the treatment duration for other cichlids, primarily the genus *Oreochromis* and the salmonids, genus *Salmo* have been for much longer durations (Table II).

One possible explanation for the success of unusually short treatment periods is that the gonads are exposed to the steroid for periods of time much longer than the treatment itself. Several factors including steroid dosage, rate of uptake and retention could influence the actual period of gonadal exposure relative to treatment duration. Unfortunately, few studies have examined these factors. In a recent investigation, G. A. Hunter and co-workers (unpublished) immersed newly hatched coho salmon alevins for 2.4 hr in water containing labeled estradiol at concentrations of either 0.2, 2.0, 20, or 200 μg/l. Preliminary results indicate that the uptake of the steroid is directly related to concentration. Further, at the maximum sample period of 16 days posttreatment, similar percentages of the maximum uptake of labeled estradiol were present at all dosage levels.

These results suggest that an inverse relationship may exist between dosage and duration of an effective immersion treatment. This could explain the requirement for very high immersion dosages in treatments of very short duration. This could also explain the similar effectiveness of estradiol treatment applied at 0.5–5 μg/l for 18 days starting 5 days posthatching in *Oncorhynchus masou* (Nakamura, 1981a) and a 400 μg/l treatment applied in 2-hr immersions at 4 and 11 days after hatching (G. A. Hunter, E. M. Donaldson, and I. Baker, unpublished). It is also interesting that Nakamura (1982) observed that the 18-day treatment at 10–200 μg estradiol/l resulted in a very high mortality. Goetz *et al.* (1979) observed no mortalities at 200 and 400 μg estradiol/l when confined to two 2-hr treatments in the eyed egg and alevin stages. However, high mortalities were observed when the total number of 2-hr immersions was increased to 13, seven of which were in the alevin stage.

The question remains whether a similar relationship may be found with treatments involving the oral administration of steroids. Hishida (1965) examined the retention of 18-[^{14}C]estrone fed to juvenile medaka from the 6-mm to 11-mm stages (1 month). He found that a high proportion of free estrogen remained in fish as late as the 16-mm stage, and he concluded that the metabolic clearance mechanism for free estrogen was not established at this stage. Therefore, it appears that steroids may be retained following oral treatment. However, no information on the relationship between steroid dosage and rate of uptake is available for the oral treatment of juvenile fish.

The duration of steroid treatment is of comparable importance to and may be interactive with steroid dosage for the effective sterilization of gonads. Takahashi (1977) administered dosages of 1.0–2.0 g methyltestosterone/kg diet to juvenile *Poecilia reticulata* for 35 days following birth. Treatment resulted in nearly complete sterilization of both ovaries and testes. Following hormone withdrawal, only the testes, although severely depressed, were capable of full recovery. This capability was lost when the treatment was extended to cover 70 days following birth. G. A. Hunter and E. M. Donaldson (unpublished) treated coho salmon, *Oncorhynchus kisutch,* fry, which had been previously immersed as alevins in water containing methyltestosterone, with a diet containing the steroid at 1, 3, or 9 mg/kg for either 3, 6, or 9 weeks. Increases in duration of treatment at either 3 or 9 mg/kg diet resulted in increased percentages of sterile fish.

c. Environmental Influences. The regulatory mechanism which determines the timing and duration of hormone sensitivity in gonochorist species remains to be determined. However, the labile period is assumed to be associated with specific developmental events and processes. Therefore, several researchers have suggested that factors affecting metabolic rate and growth (e.g., temperature, culture density, and feeding regimes) may influence the timing and duration of this period.

The few studies which have examined this area indicate that growth as indicated by size is a reliable criterion for the initiation of sex differentiation. This appears to be especially true for species such as the grass carp (Shelton and Jensen, 1979) and the European eel, *Anguilla anguilla* (Bieniarz *et al.,* 1981), in which differentiation occurs a relatively long time after hatching. In the latter study, differentiation commenced when eels ranging from 1.5 to 6 years and reached 14–35 cm in length. Further, the results indicated that gonadal sex differentiation was not age dependent, but was partially correlated with body length.

Size was also a more reliable indicator of the onset of sex differentiation in *Oreochromis aureus*, a species which displays a relatively early differentiation (Dutta, 1979). In this study, juveniles were reared at either 31°C or 21°C. A faster growth rate was reported at the higher temperature. Ovarian differentiation, characterized by meiotic and mitotic activity and development of the ovocoel, occurred at 15–18 mm length and 14–15 days posthatching at the high temperature, and 14–18 mm length and 24–27 days of age at the low temperature. Testicular development, characterized by mitotic activity of spermatogonia and development of the stromal lumina, occurred at 17–18 mm length and 19–20 days of age at 31°C, and 16–20 mm length and 29–30 days of age at 21°C. Irrespective of rearing conditions, when fish attained a particular length range (14–18 mm), gonadal differentiation was evident. Fish that were small for their age retained undifferentiated gonads. However, a minimum age related to growth was reported because few fish reached the critical size of 14–20 mm length before age 14–15 days. Dutta (1979) concluded that body length was better than age as an indicator of gonadal sex differentiation.

Shelton *et al.* (1981) examined the effects that factors influencing growth rate could have on the ethynyltestosterone-induced sex inversion of *Oreochromis aureus*. In this study, a decreased growth rate was observed in fry reared at 21°C compared with 30°C. However, treatment durations of 16 and 19 days and 21–28 days were, respectively, ineffective and effective in producing monosex male populations at both temperatures. Shelton and coworkers concluded that the duration of treatment within a particular age span was more critical to treatment effectiveness than factors affecting growth. This does not specifically exclude the relevance of size to treatment. In the 19-day treatment 33 and 26% of the fry were shorter than 18 mm at 21° and 30°C, respectively. Only 3.3% of the fish treated for 21 days were shorter than 18 mm, with the shortest being 14 mm long. Similar results were obtained in groups reared at high density (2600 fry /m²) to reduce growth rate. Therefore, it appears that, within the range of conditions applied in this study, duration of treatment can be used reliably as criteria for successful treatment.

To gynogenetic all-female common carp, Nagy *et al.* (1981) administered four overlapping 36-day methyltestosterone treatments initiated between 8–80 days posthatching resulting in 71.4–88.9% sex inversion. The remainder of the fish were female or contained undifferentiated gonads. However, using a similar dosage on fish of 7–21 mm, 13–39 mm, and 19–57 mm length, Nagy and co-workers were able to obtain 90, 100, and 90% males, respectively.

Therefore, it is evident that both size and age may be used effectively as indicators of developmental stage when tuned to individual species characteristics and rearing conditions. In genera such as *Oncorhynchus*, which differentiate prior to feeding, age either relative to developmental markers such as hatching or with respect to the temperature–development relationship (degree days) may be used effectively. In species which differentiate at a much later date, size criteria may be of more value.

C. Evaluation

The evaluative phase involves a quantification of treatment effectiveness relative to management objectives. In most studies, hormones are administered to a group of genetic males and females. Following treatment, histological examination is used to assess the proportions of gonadal sex types present. The occurrence of a high proportion of one gonadal sex type is usually indicative of successful sex inversion. Results that do not demonstrate a clear preponderance of one gonadal sex following treatment may be statistically compared with a control population. However, statistical nonsignificance is not conclusive evidence that sex inversion has not occurred. Sanico (1975) treated *Oreochromis aureus* with estrone obtaining an observed sex ratio not significantly different from unity. Later, Liu (1977) reported that one of these fish produced 100% male offspring indicating that sex inversion had occurred. Further, progeny testing may be required to exclude the possibility of sex-related mortality within treated groups (Yamamoto, 1953).

An alternate approach, which simplifies the analysis of treatment, involves the use of genetically monosex fish produced either by the mating of sex-inverted homogametic fish or by gynogenesis. Any deviation from the monosex gonadal type may be assumed to be a result of treatment. This approach eliminates both concerns about possible differential mortality and the need for progeny testing to confirm sex inversion.

IV. ECONOMICALLY IMPORTANT SPECIES

With the growing importance of hormonal sex-control techniques to fish culture practices, an examination of the studies involving the economically important species is in order. This discussion is primarily concerned with species from two major groups which are important to culturists. These are the cichlids, genus *Oreochromis*, known colloquially as tilapias and the salmonids, notably the genus *Oncorhynchus* and the genus *Salmo*. Mention is also made of the recent research conducted in two cyprinids *Ctenopharyngodon idella* and *Cyprinus carpio*. These groups do not represent the only economically important species to which sex-control techniques have been applied. Many of the *Poecilliidae* and *Cyprinodontidae* previously discussed are valued by pet culturists for particular sex-related characteristics such as body color and morphology. Further, these techniques have recently been applied to the turbot, *Scophthalmus maximus*. In this species, the administration of estradiol or methyltestosterone added to the weaning diet at 5 or 3 mg/kg diet, respectively, for 700°C days (58 days) induces 100% sex inversion. Methyltestosterone administered at 25 mg/kg weaning diet for 1000°C days (12°C) induces 100% sterility (Bye, 1982). As previously described, Chen *et al.* (1977) have achieved early sex inversion in the only commercially important protogynous hermaphrodite *Epinephelus tauvina*. Goudie *et al.* (1983) have achieved 100% feminization of channel catfish with estradiol or ethynyltestosterone at dosages of 6–600 mg/kg administered for 21 days following yolk sac absorption.

A. Cichlids

Central to the application of hormonal sex-control techniques to the tilapias is the desire to control breeding. In the late 1950s and early 1960s, numerous studies indicated that various cichlid species would be ideal for warm-water culture. Favorable culture attributes include rapid growth using plankton as a food source, adaptability to brackish or salt water, and high reproductive potential (Hickling, 1963). However, when held in ponds, the early maturation and high fecundity of these species rapidly results in overcrowding of the ponds with fish too small for consumption. Alternative solutions to this problem, including high-density culture, polyculture with a piscivore, cage rearing, repetitive harvesting, and monosex culture, have been reviewed by Jensen (1976).

The treatment used to achieve the monosex condition must be very effective, because the presence of the opposite sex at even very low levels will result in overbreeding. Anderson and Smitherman (1978) reported that excessive breeding occurs when females are present at levels as low as 5%. Approaches to the production of monosex populations have included hand sorting, which is time consuming and expensive, the interspecific mating of homogametic individuals, which has produced inconsistent results, and the use of hormonal sex control.

1. Androgen Treatment

Within monosex culture the production of males has been preferred. The faster growth rate of male tilapia has been reported by several researchers (Jensen, 1976). Further, the growth rate of female tilapia is reduced at maturity (Hickling, 1960).

a. Oreochromis aureus. Eckstein and Spira (1965) conducted the first experimental application of androgens to *Oreochromis aureus.* In this study, testosterone and methyltestosterone were administered to the rearing water at concentrations of 50–1000 μg/l for a period of 5–6 weeks starting at 4–5 weeks posthatching. These researchers had previously recorded the initial appearance of the genital ridge at 10–11 days posthatching with the first distinctive characteristic of the ovaries, a longitudinal groove occurring at 30 days. Sex differentiation based on germinal and somatic characteristics occurred at 7–8 weeks. The effect of the androgens was variable with some individuals having nearly involuted gonads and others being largely unaffected. Unfortunately, Eckstein and Spira did not specify the exact proportions of each gonadal sex type.

Guerrero (1975) was the first to achieve sex inversion in this species using the synthetic androgens ethynyltestosterone and methyltestosterone at 15, 30, and 60 mg/kg diet for 18 days (6 days/week for 3 weeks). Ethynyltestosterone at 60 and 30 mg/kg produced 100 and 98% males, respectively. Methyltestosterone at 30 mg/kg also produced 98% males. However, this steroid was found to be less effective at 60 mg/kg and 15 mg/kg producing 85 and 84% males, respectively. Treatments did not affect growth or survival. However, degeneration of the germinal epithelium and proliferation of the connective tissue was observed in the 60 mg/kg methyltestosterone group. Males from treated groups were mated with untreated females and the progeny analyzed. Nine families from these matings had male-to-female ratios of 1:2–3.1 indicating female heterogamety. Therefore, Guerrero was able to suggest the possibility of using the indirect method to produce monosex male fish in this species. The attempts to produce all-male stocks by the indirect method are discussed later in this section.

Recently, the pendulum has swung back in favor of the production of monosex males by the direct method. This return to the direct method is primarily attributable to the potential inconsistencies in breeding sex-inverted individuals because of the previously described complexity of the sex-determining mechanism evident in some species of tilapia. In this regard, Shelton and Jensen (1979) reported that, of 100 *Oreochromis aureus* estrogen-induced females mated with untreated males, 41 spawns were obtained. Of these spawns, only five were 100% male, one was 95% male, one was 100% female, and the rest had a 1:1 male–female ratio.

Attempting to standardize treatment for production of monosex males by the direct method, Shelton *et al.* (1981) examined the effects of temperature, stocking density, and feeding regime on treatment success. Based on Guerrero's (1975) successful treatment a dosage level of 60 mg ethynyltestosterone/kg diet was used for all tests. The dietary treatment was administered either at a fixed intake of 12% body weight per day or to satiation. Stocking densities of 160 and 2600 fry/m^2 were examined at each feeding regime. Treatment durations were between 16 and 28 days. Temperatures of either 21° or 30°C, stocking density, and feeding regime did not appear to affect treatment because 100%-male groups were produced in all but the low density, 12% feed ration, low temperature group which attained 98.9% males. Based on these results, an optimal treatment for this species was proposed consisting of a minimum treatment of 21 days, stocking densities up to 2600 fry/m^2, within a temperature range of 21–30°C and a feeding ration of 12–15% body weight.

b. Oreochromis mossambicus. Clemens and Inslee (1968) demonstrated the first successful masculinization within the tilapias, specifically *Oreochromis mossambicus (Tilapia mossambica)*. For 69 days, fry were fed a diet containing methyltestosterone at 10–50 mg/kg diet. At dosages of 0 and 50 mg/kg, the groups matured as males and females with sex ratios of 1:3.6 and 1:4.4, respectively. Treatment at 10, 30, and 40 mg/kg resulted in 100% males but one fish in the group treated at 20 mg/kg was female. However, it is notable that the sex ratios were based on 8–22 fish from an original 100 treated fish. Groups receiving similar treatments, but placed in larger tanks, which provided sample sizes of 79–128, had male-to-female ratios of 1:1.6 to 1:5.1. The discrepancy between results was attributed to an increased natural food supply available to the fish in the larger tanks. The matings of 7 of 24 males from experimental groups resulted in all-female offspring indicating that sex inversion had occurred and that the female is homogametic. Clemens and Inslee suggested that, although the results did not indicate a clearly superior dosage level, the 30 mg/kg level was promising for further study.

Subsequently, successful masculinization was obtained by treatments using 50 mg/kg diet of methyltestosterone for 19 days starting 7 days posthatching (Nakamura, 1975), thereby demonstrating a much reduced effective duration of treatment. Nakamura and Takahashi (1973) had previously demonstrated that the labile period is from 6 to 25 days of age. In this study, a dosage of 1000 mg/kg for the same period of time was ineffective. Guerrero (1976a) obtained 69, 93, and 98% male groups treated with 30 mg/kg methyltestosterone for 14, 21, or 28 days posthatching, respectively. Based on the absence of juveniles in ponds previously stocked with fry treated with 50 mg ethynyltestosterone/kg diet for 40 days, Guerrero

(1976b) concluded that 100% sex inversion had been achieved. The growth of treated fish was also superior to controls. Recently, dihydrotestosterone, testosterone propionate, and methyltestosterone have been tested on *Oreochromis mossambicus* fry. Treatment with 60 mg/kg methyltestosterone for 6 weeks achieved 100% sex inversion, and the 30 mg/kg diet treatment only produced 81–85% males (Anonymous, 1979).

Nakamura (1981) has reported the first successful sex inversion in this species using the naturally occurring androgen, 11-ketotestosterone. The effective treatment required a dose of 200 mg/kg diet for 19 days starting 7 days posthatching.

c. *Oreochromis niloticus*. Jalabert *et al.* (1974) first applied methyltestosterone treatment to juvenile *Oreochromis niloticus*. The treatment consisted of the oral administration of the steroid at 40 mg/kg diet for 2 months from first feeding. Comparison by Chi-square indicated a higher proportion of males in several groups. Sex inversion was confirmed by the mating of several treated males with untreated females which resulted in 100% female progeny, and the demonstration of female homogamety. Guerrero and Abella (1976) were able to demonstrate higher percentages of males with dosages of methyltestosterone at 15, 30, or 50 mg/kg diet for 6 weeks.

Katz *et al.* (1976) have provided the only deliberate attempts to sterilize *O. niloticus* based primarily on the early work with *Oreochromis aureus* (Eckstein and Spira, 1965) and *Oreochromis mossambicus* (Clemens and Inslee, 1968). Katz *et al.* (1976) incubated fertilized *Oreochromis niloticus* eggs in water containing 500–5000 μg/l progesterone, cyproterone acetate, androstenedione, testosterone, adrenosterone, methyltestosterone, 11 hydroxy-17α-methyltestosterone, and fluoxymetesterone, 500–1000 μg/l 11-ketotestosterone, 125–5000 μg/l estrone and estradiol, and 200–5000 μg/l diethylstilbestrol diphosphate. All treatments resulted in very high mortality with the exception of those involving adrenosterone. Fish treated at 5000 μg/l with this steroid were found to be sterile at the end of the 3-month treatment period. The gonads of these fish were described as either epithelial flaps hanging from the dorsal peritoneal lining (5.4%) or hollow ducts composed of connective tissue and blood vessels (involuted gonads). However, when sampled at 8 months, 40% of the population had developed testes, 4.4% had involuted gonads, and no gonadal development was noted in the remainder.

Tayamen and Shelton (1978) were the first to produce all-male populations of *Oreochromis niloticus*. Ethynyltestosterone and methyltestosterone were orally administered at either 30 or 60 mg/kg diet for either 25, 35, or 59 days. The high dosage of each steroid produced all-male populations during all treatment periods. The 30 mg/kg dosage produced all-male populations at the 35- and 59-day treatment periods. Recently, Owusu-Frimpong and Nij-jhar (1981) applied a 50 mg methyltestosteone/kg diet treatment to newly hatched fry for 28 or 42 days resulting in the production of all-male populations.

Recently, Nakamura and Iwahashi (1982) achieved complete masculinization by the administration of 50 or 100 mg/kg dosages of methyltestosterone for a period of 30 days, beginning the day of capture from the adult holding pond. Similar treatments begun 5 or 10 days after capture resulted in the production of intersex gonads. In most cases these intersex gonads were ovarian in the anterior region and testicular in the posterior region.

   d. Tilapia zillii. The first attempt to masculinize populations of *Tilapia zillii* was conducted by Guerrero (1976b) concurrent with his treatment of *Oreochromis mossambicus.* The treatment involved administration of 50 mg ethynyltestosterone/kg diet for 40 days, and was effective in *Oreochromis mossambicus* but not in *Tilapia zillii.* Guerrero (1976b) attributed this discrepancy to incorrect timing or dosage of treatment, which he suggested may be different for the bottom-spawning *Tilapia zillii* compared with the mouth-brooding *Oreochromis mossambicus.* However, data was not available to confirm this hypothesis. Yoshikawa and Oguri (1977) examined the period of sex differentiation in *Tilapia zillii* via histological techniques. On the basis of germ cell numbers at 15 days posthatching, they were able to distinguish between future ovaries that contained many germ cells (some of which were in meiotic prophase) and testes that contained fewer germ cells (none of which were in meiotic prophase). Following this investigation, Yoshikawa and Oguri administered an oral treatment of methyltestosterone at 50, 100, and 200 mg/kg diet for 20 days starting 10 days posthatching. The treatment resulted in inhibition of oogenesis and a proliferation of somatic elements, but did not result in sex inversion (Yoshikawa and Oguri, 1978). The researchers attributed treatment failure to the high dosage levels used. However, Woiwode (1977) achieved 100% sex inversion following oral administration of 50 mg/kg of methyltestosterone for 45 days posthatching, suggesting that treatment timing and duration may not have been optimal in the previous study. However, these results are difficult to reconcile with those of Guerrero (1976b). Given similar treatment duration and method of administration, the greater effectiveness of the methyltestosterone treatment is unexpected. Ethynyltestosterone has been demonstrated to be of greater potency than methyltestosterone (Yamamoto, 1969).

   e. Other Cichlids. Several other species have been examined, although less intensively. Hackmann (1974) reported the paradoxical feminizing action of testosterone propionate and methyltestosterone added at 500 μg/l to the rearing water of *Oreochromis mossambicus, Tilapia heudeloti,* and

Cichlasoma biocellatum. Jalabert *et al.* (1974) examined the effects of 40 mg/kg diet of methyltestosterone for 2 months on *Tilapia macrochir* concurrent with his examination of *Oreochromis niloticus.* In the latter species sex inversion was achieved. In *Tilapia macrochir,* although external male secondary characteristics were induced, the gonads were completely sterilized. Jalabert *et al.* (1971) had previously indicated male homogamety in *Tilapia macrochir.* Hackmann and Reinboth (1974) demonstrated the masculinization of *Hemihaplochromis multicolor* by immersion of the fry for 42–45 hr in water containing 50 mg testosterone propionate/l.

2. ESTROGEN TREATMENT

The production of monosex female tilapias was advocated by Bardach *et al.* (1972) as a means of alleviating the destructive nest-building activities of males.

a. Oreochromis aureus. As with the androgens, the estrogens, specifically synthetic stilbestrol and stilbestrol-diphosphate–diaethyldioxystilben-diphosphate, were first applied to *Oreochromis aureus* by Eckstein and Spira (1965). The purpose of this study was sterilization for culture purposes. The estrogens were applied to the rearing water at 50–1000 μg/l for 5–6 weeks starting at 4–5 weeks of age. At estrogen concentrations higher than 200 μg/l, mortalities were very high. At the dosages of 50 and 100 μg/l a powerful inhibition of gonadogenesis occurred, resulting, in most cases, in complete sterility.

Based on the work of Guerrero (1975), Jensen (1976) tested the effectiveness of estriol, estrone, and estradiol in inducing sex inversion in genetically male *Oreochromis aureus,* which could then be used to produce all-male progeny. In this study, these naturally occurring estrogens were administered to fry at 30, 60, or 120 mg/kg diet for either 21 or 35 days posthatching. All treatments proved ineffective. This lack of success was attributed to either insufficient dosage or duration of treatment. Following the work of Jensen, Hopkins (1977) orally administered the two synthetic estrogens, ethynylestradiol and diethylstilbestrol, and the naturally occurring estradiol alone or in combination with the antiandrogen cyproterone acetate. Steroid concentrations were 25–200 mg/kg diet, and the antiandrogen concentration was held at 100 mg/kg diet. Treatment durations were either 35 or 56 days posthatching. Ethynylestradiol at 25 or 100 mg/kg with the antiandrogen produced 60 and 63% females, respectively, but diethylstilbestrol alone resulted in 64% females. In a later experiment, Hopkins *et al.* (1979) was able to achieve 90% female groups by a 42-day oral administration of ethynylestradiol and methallibure each at 100 mg/kg diet with or without cyproterone acetate at 100 mg/kg. Hopkins and co-workers suggested that the antiandrogen may have lessened the effectiveness of treatment because of its additional antiestrogenic, progestational, and androgenic effects.

b. Oreochromis mossambicus. Nakamura and Takahashi (1973) used the oral administration of ethynylestradiol at 50 mg/kg diet to determine the optimal period of steroid treatment in *Oreochromis mossambicus.* Histological examination indicated that the primordial gonads formed 8–10 days posthatching with the histologically discernable initiation of sex differentiation occurring at 20 days of age. They were thereby able to achieve complete feminization with a treatment which lasted for 19 days from 6 days of age. The discrepancy between their determination of the time of sex differentiation and that of Clemens and Inslee (1968) (35–48 days posthatching) was attributed to the morphological criteria used.

c. Oreochromis niloticus. Tayamen and Shelton (1978) orally administered diethylstilbestrol at 25 and 100 mg/kg diet and estrone at 100 and 200 mg/kg diet for 25–35 and 59 days to newly hatched fry. The estrogen treatments produced between 62 and 90% female groups, indicating that, compared with Hopkin's (1977) treatment of *Oreochromis aureus,* estrogen-induced sex inversion was much more easily attained in *Oreochromis niloticus.*

d. Tilapia zillii. Yoshikawa and Oguri (1978) administered oral treatments of ethynylestradiol at 20, 40, and 60 mg/kg diet for 20 days starting 10 days posthatching to *Tilapia zillii* fry. Treatment did not result in sex inversion. The estrogen inhibited spermatogenesis at 20 mg/kg and at 40 and 60 mg/kg resulted in the production of few fish containing involuted gonads.

e. Other Cichlids. Monosex female groups of *Hemihaplochromis multicolor* were obtained by Hackmann and Reinboth (1974) by addition of 250 μg/l estradiol butyryl acetate to the rearing water for 42–45 hr between 11.5 and 16 days postspawning. The estrogen was effective in inducing sex inversion over a wider period of time than the androgens, testosterone propionate and methyltestosterone.

A summary of these studies is presented in Table II. The direct administration of androgens is clearly the favored strategy for producing monosex male populations. However, the value of the indirect strategy for producing monosex male populations in species such as *Oreochromis aureus,* which have male homogamety, remains to be determined. As yet, the isolation of YY males from the matings of sex-inverted XY males and untreated males, similar to that conducted by Yamamoto (1965) on *Oryzias latipes,* has not been attempted. In species such as *Oreochromis mossambicus* and *Oreochromis niloticus,* the sperm from these YY males could theoretically be used to sire monosex male populations.

B. Salmonids

Diverse objectives are associated with the hormonal sex control of salmon and trout (Goetz *et al.*, 1979; Johnstone *et al.*, 1978, 1979a; Bye and Lincoln, 1981; Hunter *et al.*, 1982a; Donaldson and Hunter, 1982a). This diversity is evident within the two major strategies employed in salmonid culture. The first, pen rearing, involves the culture of the fish in enclosures throughout the lifetime of the fish. The second, ocean ranching, involves the rearing of juvenile salmon for release into the ocean environment and subsequent harvest of these fish as adults on their anadromous migration. Hunter *et al.* (1982a) have summarized the potential applications of sex-control techniques to both strategies. Often, with pen rearing, the lack of a reliable source of gametes requires the maintenance of large numbers of nonmarketable female broodstock. The ability to produce monosex female groups could dramatically reduce the number of fish required to produce the necessary egg take, thereby reducing the costs of broodstock maintenance. Further, all-female groups are preferred in the culture of rainbow trout, because males have the propensity to mature precociously and adult males have poor growth rates, poor food conversion efficiencies, and poor survival when grown in seawater (Bye and Lincoln, 1981).

Sexual maturation restricts the period over which the fish may be marketed because of the deterioration of flesh quality and mortalities attributable to bacterial and fungal infections associated with the maturational processes. The production of sterile fish would allow for the year-round marketing of adult-sized fish of high flesh quality. The production of either female or sterile fish would alleviate the problem of sexual maturation before attainment of full adult size.

The ocean-ranching culture strategy has problems associated with sexual maturation and reproduction analogous to the pen-rearing approach. Although broodstock are not held for the full life cycle, costs are associated with the number of fish held for broodstock that must be allowed to mature sexually and are subsequently of a lower market value. Further, the number of ova which can be obtained determines the hatchery smolt output. Therefore, the production of a high proportion of females would both reduce the escapement required to meet hatchery objectives allowing a greater number of fish to be caught and provide a means of rapidly increasing the production of smolts. An increase in the proportion of females is desirable in species in which the roe is harvested and marketed separately.

The harvest of ocean-ranched stocks is also restricted in time by normal sexual maturation. In species such as the chum salmon, *Oncorhynchus keta*, the present harvest strategy, involving terminal fisheries, results in the harvest of primarily sexually mature fish which are of relatively low market value. The production of sterile fish of high flesh quality would dramatically increase the value of the catch. The production of sterile fish, which do not undergo the normal anadromous migration, would extend the active fishing season and allow alternative harvest strategies.

The prevention of the anadromous migration would eliminate the necessity for a "one-shot" harvest and would allow greater flexibility in the establishment of exploitation rates, thereby assisting the management of mixed-stock fisheries. Furthermore, by extending the effective life span of the fish the possibility of producing larger fish exists. An increase in the size of fish for harvest would provide benefits to the recreational fishery in the form of trophy-sized fish, and to the commercial fishery through an increase in total weight harvested (A higher value per unit weight is placed on large fish). This approach requires the use of stocks which remain in or return to waters accessible to the fishery. Loss of full growth potential as a result of precocious male development is also a problem for the ocean-ranching strategy. Again, the production of female or sterile fish would alleviate this problem.

1. ANDROGEN TREATMENT

a. Salmo gairdneri. The first study involving the use of androgens in rainbow trout was designed to determine the growth-promoting and go-nadal-suppressing activity of 4-chlorotestosterone acetate. The steroid was injected into 1-year-old fish at dosages of 1.0–12.5 mg/fish in each of 6 injections over 30 days (Hirose and Hibiya, 1968). Treatment suppressed both yolk deposition and testicular differentiation. Hirose and Hibiya suggested this suppression was a result of a blocking of gonadotropin release. Recently, Billard *et al.* (1981) produced gonadal inhibition in adult trout by feeding a diet containing methyltestosterone or estradiol at 0.5 mg/kg during the period of spermatogenesis (June–November). Treatment during the period of spermiation (November–February) was ineffective.

The first application of steroids during sexual differentiation was conducted by Jalabert *et al.* (1975). In this study, young fry were administered methyltestosterone orally at dosages of 15, 30, and 60 mg/kg diet for 5 months starting 1 month posthatch. Examination at 2 years indicated a combined total of 58% male, 12% sterile, 18% female, and 12% intersex fish. Working with *Horai masu*, a variant of rainbow trout, Yamazaki (1976) reported that the administration of methyltestosterone at 50 mg/kg diet for a period identical to that used by Jalabert *et al.* (1975) resulted in the production of gonads containing richly vascularized connective tissue devoid of germinal material at age 2 years. However, of three fish which survived to 3 years of age, two had testes with spermatogonial cysts and one had a threadlike ovary with a small number of yolkless ooytes. In the same study,

Yamazaki (1976) reported that treatment with methyltestosterone at 1 mg/kg diet starting 1 month posthatch and lasting for 7 months resulted in the production of 87.1% males. These results agree with Yamamoto's (1969) assertation regarding the masculinizing action of androgens at low dosages and sterilizing action at high dosages.

The first production of androgen-induced monosex male *Salmo gairdneri* was reported by Simpson (1976). Treatments involved the oral administration of methyltestosterone at 3 mg/kg diet for 90 days from first feeding with, or without, previous immersions for 2-hours of the eyed-egg and alevin stages in water containing the steroid at 250 μg/l. Immersion treatments alone were ineffective. Reduction of the duration of oral treatment to 30 days resulted in the production of 85% males at 16 months. However, in most of the male fish reported by Simpson, large areas of the gonads were devoid of germ cells and contained hypertrophied connective tissue. Johnstone *et al.* (1978) reported that following identical treatments, all animals examined at up to 12 months contained the thin filiform gonad typical of males. However, in 17% of these fish the complete gonad was composed of hypertrophied and sterile connective tissue areas. In another experiment, in which the oral treatment of 3 mg/kg was reduced to 30 days, 79% males were produced. Subsequently, milt from these mature males was used to fertilize normal ova (Johnstone *et al.*, 1979a). Of the males having mature testicular tissue, the milt from six could be expressed normally, but eight others had absent or occluded system ducts. Sperm from a total of 10 males was used to fertilize normal ova. One of these pairings resulted in the production of 100% females, indicating female homogamety in this species.

In the same year, Okada *et al.* (1979) also demonstrated female homogamety in *Salmo gairdneri*. In this study, fry were fed a diet containing 1, 5, or 10 mg methyltestosterone/kg diet for 58 days resulting in the production of 60, 80, and 88% males, respectively. Mortality ranged from 55.3 to 68.7% in these groups. Two of 11 males containing morphologically abnormal testes sired nearly 100% female offspring when mated with untreated female rainbow or steelhead trout. The few males found in these groups were attributed to contamination from other groups. The subsequent production of an F_2 generation with a normal male–female ratio indicated that the F_1 generation was genetically normal.

The observations of abnormal testicular morphology associated with sex-inverted female rainbow trout has been confirmed by V.J. Bye (personal communication, 1980; Bye and Lincoln, 1981). Bye applied an oral methyltestosterone treatment at 3 mg/kg diet for 1000°C days at 11.2°C (90 days) resulting in the production of 94% males. At age 2 years, 55% of these fish could not be stripped. On examination they were found to have abnormal testes with absent or occluded sperm ducts. At age 3 years, 70% of the fish that could not be stripped had hermaphroditic gonads consisting of ductless testes with an anterior cap of ovarian tissue. The progeny of 14 of 16 such males sired all-female progeny. All fish that could be stripped had normal tests which later sired progeny in a normal 1:1 male-to-female ratio. The duration of methyltestosterone administration necessary to produce mono-sex males without steriles was 700°C days at 9°–12°C with feeding for 10 hr/day (V. J. Bye, personal communication, 1980). Bye also reported that attempts to induce sterility with oral methyltestosterone treatment have been largely unsuccessful. Methyltestosterone when given at 25 mg/kg diet for 1000°C days inconsistently produces 60% sterility. Harbin *et al.* (1980) also attempted to induce sterilization in rainbow trout by administering methyltestosterone orally at 30 mg/kg diet for 110 days from first feeding. At age 2 years, 2.5% of the fish were found to be mature. The remainder had filiform gonads with increased vascularization and connective tissue. Serum testosterone levels of treated fish were significantly lower (0.34 ± 0.08 ng/ml) compared with controls (14.6 ± 3.63 ng/ml). Treated fish displayed none of the secondary sexual characteristics associated with maturation, presumably because of the low levels of androgen present.

In a subsequent study, van den Hurk and Slof (1981) achieved sterilization in rainbow trout by adding methyltestosterone to the rearing water 3 times per week at 3–300 μg/l for 28–56 days posthatching. They determined that sex differentiation takes place 45–55 days postfertilization. Hatching occurred around 26–29 days postfertilization. The highest percentages of steriles were obtained with treatments having the longest duration or the highest dosage. Sterility was not induced when the treatment was begun 43 or 57 days postfertilization, although the percentage of males increased. Treatments were accompanied by growth depression and increased mortality. Recently, van den Hurk and Lambert (1982) reported the production of 100 male groups of rainbow trout following treatment with 11β-hydroxyandrostenedione at 60 mg/kg diet for 56 days from first feeding. Treatment with 11β-hydroxyandrostenedione at 6 mg/kg or with methyltestosterone at 0.6 or 6.0 mg/kg produced 94–99% males.

b. Salmo Salar. Simpson (1976) first applied androgens to *Salmo salar* for the purpose of producing sex-inverted females, which could then be used to sire all-female salmon for pen culture. In this study, eyed eggs and alevins were immersed in water containing methyltestosterone at 250 μg/l for two 2-hr periods. The fry were subsequently fed a diet containing methyltestosterone at 30 mg/kg diet for 120 days. No female elements were found in the gonads up to 9 months posttreatment. Johnstone *et al.* (1978) analyzed a group of similarly treated *Salmo salar*. These fish contained sterile filiform gonads at 6 months. In another group, methyltestosterone was administered orally at 3 mg/kg diet for 90 days after first feeding producing 100% males or sterile fish.

c. Salmo trutta. Ashby (1957) reared *Salmo trutta* in water containing 50–60 μg/l testosterone for 111 days starting 170 days postfertilization (850°–1738°C days). Of 29 fish examined, 41% were females, 35% were males based on the presence of vasa deferentia, and 24% containing neither oocytes or vasa deferentia. The gonads of these latter fish were composed primarily of connective tissue with very few germ cells. Ashby (1965) again administered testosterone in the rearing water at 67 μg/l, 8 hr/day, 6 days per week for 3 months starting 7 months postfertilization. Although some ovarian inhibition was reported, testicular inhibition was far more pronounced including the complete elimination of somatic and germinal tissue.

d. Salvelinus species. Wenström (1975) fed testosterone propionate at 700 mg/kg diet to lake trout, *Salvelinus namaycush*, for either 245° or 290°C days starting at 1250°C day. Of the 10 fry examined, 7 were male and 3 were female.

e. Oncorhynchus kisutch. Coho salmon, *Oncorhynchus kisutch*, has been by far the most intensively studied of the *Oncorhynchus* species. Early experiments involved the administration of androgens after completion of sex differentiation. McBride and Fagerlund (1973) assessed the anabolic effects of orally administered methyltestosterone at 1, 10, or 50 mg/kg diet on coho salmon with an average weight of 3.79 g for up to 42 days. By the end of 28 days, a marked reduction in spermatogonia had occurred in the groups fed 10 or 50 mg/kg. At the end of 42 days the testes were nearly devoid of germ cells. The effect was more pronounced in fish receiving the higher dosage. No distinct alterations were noted in the ovary. No effect was observed on the gonads of fish fed the 1 mg/kg dosage. In a later study, Fagerlund and McBride (1975) again tested the anabolic activity of methyltestosterone on younger fish having an average weight of 0.78 gm. These fish were administered 0.2, 1.0, or 10 mg/g diet for up to 504 days. After 56 days, clear degenerative changes were observed in the gonads of those fish administered the 10 mg/kg diet. By the end of 140 days no spermatogonia could be observed in the fish in this group. By the end of 224 days the gonads consisted of thickened connective tissue. At 140 days a small number of fish in the group receiving the 10 mg/kg dosage were transferred to a control diet for 364 days. Of the 24 males examined at the end of this period, 14 were sterile and 10 had reduced spermatogonia. Of the fish fed the 1 mg/kg for 504 days, 2 of 13 appeared to be sterile, 3 had reduced spermatogonia, and 8 appeared unaffected. These results clearly suggested the possibility of producing male or sterile coho salmon by the administration of methyltestosterone. With this objective and using a modification of the procedures described by Simpson (1975–1976) for *Salmo gairdneri*, Goetz *et al.* (1979) conducted a study to determine the effects of methyltestosterone treatment at the time of sex differentiation. Sex differentiation of coho salmon and other *Oncorhynchus* species has been reported to occur earlier than that of the *Salmo* species (Persov, 1975). An initial examination of sex differentiation indicated that, based on the appearance of perinuclear oocytes, differentiation was first discernable at 880°C days, 49 days posthatching (Goetz *et al.*, 1979). Therefore, treatments were initiated prior to and including first feeding. In this study, eyed eggs were each administered 2 or 6 immersions of 2 hr duration in water containing the steroid at 25–400 μg/l. Alevins were administered either 2 or 7 immersions of 2 hr each. Fry were orally administered the steroid at 20 mg/kg diet for 70 days. All groups treated in this manner were found to be 100% sterile, with the exception of the group administered the 25 mg/kg oral treatment which was 94% sterile. A group receiving the oral treatment alone was found to be 52% sterile, 24.5% intersex, and 23.5% female.

In subsequent experiments (G. A. Hunter and E. M. Donaldson, unpublished data), the oral dosage was reduced to 10 mg/kg administered with, or without, prior immersions at 400 μg/l in the eyed-egg and alevin stages. Immersion in the alevin stage alone resulted in 67% males and no sterile fish. In groups administered the dietary treatment, 83–97% steriles were produced. In order to ascertain the effect of dosage and duration of dietary treatment, groups of coho were administered an initial immersion treatment in the alevin stage followed by dietary treatments at 1, 3, or 9 mg/kg for a period of 21, 42, or 63 days. The highest production (82–92%) of males occurred in groups receiving the 1 mg/kg dosage. The percentage of steriles increased with higher dosages or longer treatment durations. Of the 13 males that were reared to maturity in this study and mated with untreated females, one sired all-female progeny, indicating female homogamety in the species.

Hunter *et al.* (1982a) examined the development of two groups of coho salmon treated by immersion at 100 or 400 μg/l methyltestosterone in the eyed-egg and alevin stage followed by dietary treatment for 90 days. At maturity the high- and low-dosage groups contained 94 and 70% sterile fish. These sterile fish continued to live and grow for 2.5 years after the maturation and death of the control group. The fish remaining at the end of this period contained gonads composed entirely of connective tissue. During this period, these fish maintained the silvery appearance characteristic of immature fish (Fig. 3). This study first demonstrated the potential advantages of hormonal sterilization to the culture of Pacific salmon. In a recent field trial of the sterilization technique, 40,000 coho salmon were administered two 2-

Fig. 3. Comparison of three sterile (top) and two female 3-year-old coho salmon during the period of normal maturation (December).

hr immersions in water containing methyltestosterone at 200 μg/l in both the eyed-egg and alevin stages followed by an 84-day dietary treatment at 10 mg/kg (Donaldson and Hunter, 1982b). A subsample was held in a seawater net pen to determine survival, growth, and treatment success. The remaining fish, which had been nose tagged and fin clipped, were released into the ocean in May, 1980. Analysis of the fish held in the net pens indicated that 97.7% of the fish were sterile. The sterile fish survived at a similar rate, but grew at a slower rate compared to a monosex female group treated at the same time. However, the sterile group continued to grow through the period in which the females matured and ceased to grow. The steriles attained a carcass weight similar to the females by the end of the normal spawning season. Furthermore, the sterile fish continued to grow the following year, and the all-female group reached the end of their life cycle after the spawning season. No precocious males returned to the hatchery in the fall of 1980. Untreated fish released at a similar time would have been expected to return in the fall of 1981; however, no sterile fish returned to the hatchery at this time. Approximately 18 fish from the methyltestosterone-treated group did return. All contained morphologically abnormal but maturing testes. To date sterile fish have contributed to the commercial and recreational fisheries in both 1981 and 1982 and are expected to continue this contribution until normal ageing and death occur in the absence of sexual maturation.

f. *Oncorhynchus tshawytscha.* The earliest report of the actions of androgens on the gonads of chinook salmon, *Oncorhynchus tshawytscha*, is that of McBride and Fagerlund (1973). They administered 0, 0.2, or 1.0 mg methyltestosterone/kg diet to chinook fry having an average weight of 0.78 gm for either 28, 56, or 84 days. The gonads of fish treated for 84 days were visibly swollen, but no reduction in spermatogonia could be detected. Ovaries were not affected.

More recently, the effect of dietary administration of methyltestosterone at various dosages and durations of treatment have been examined (G. A. Hunter and E. M. Donaldson, unpublished data). In the study by Hunter and Donaldson, chinook alevins were administered two immersions of 2 hr each in water containing methyltestosterone at 400 μg/l. The fish were subsequently given oral treatments with methyltestosterone at 3 or 9 mg/kg diet for periods of 3, 6, or 9 weeks. All but one treatment resulted in a high (83–98%) percentage of males. In the 9 mg/kg–9 week treatment group, 43% were males and 57% appeared sterile at 8 months posthatching. At maturity, 59 males from these groups were mated with untreated female chinook or coho salmon. Greater than 92% female progeny were observed in 21 families, and 8 families had 100% females. These results indicate female homogamety in this species (Hunter *et al.*, 1983).

g. *Oncorhynchus keta and Oncorhynchus gorbuscha.* Robertson (1953) identified the period of germ cell maturation in chum salmon (*Oncorhynchus keta*) as occurring from 14 days prehatching to 42 days posthatching at 5°–6°C with the first appearance of primary oocytes at 55 days posthatching. Nakamura (1978) reported sex differentiation in chum species at 35 days posthatching at 8°C. The only report which has examined the effect of androgens on the gonads of these species is that of Yamazaki (1972). In this study 1-year-old chum and pink salmon (*Oncorhynchus gorbuscha*) were orally administered methyltestosterone at 50–100 mg/kg diet for 2 weeks. Treatment resulted in the degeneration of spermatogonia and atresia of ova.

2. Estrogen Treatment

a. Salmo gairdneri. The earliest recorded studies employing estrogens for the control of sex differentiation in rainbow trout were those of Padoa (1937, 1939). These studies involved the immersion of juvenile *Salmo gairdneri* (*irideus*) in water containing estrone. These treatments were largely without effect. Padoa (1939) determined that sex differentiation of rainbow trout takes place shortly after yolk-sac absorption is complete, during the period of first feeding. The occurrence of sex differentiation in *Salmo gairdneri* shortly after first feeding has been confirmed by Okada (1973), Takashima *et al.* (1980), and van den Hurk and Slof (1981). Okada (1973) orally administered estrone at 10, 50, 100 mg/kg diet for 58–124 days posthatching. The percentage of females obtained ranged from 79 to 94%. Treatment was accompanied by high mortalities and a dose-dependent inhibition of growth.

Jalabert *et al.* (1975) also orally administered estrone at dosages of 30, 60, and 120 mg/kg diet for 5 months beginning 1 month posthatching. When examined at age 2 years, the combined groups contained 54% females, 30% intersex, 10% males, and 6% sterile fish. Jalabert *et al.* noted the possibility that self-fertilizing hermaphroditic trout could provide a tool for producing highly homozygous strains of fish for breeding purposes.

Simpson (1976) first reported the production of 100% female trout by the administration of estradiol to the rearing water at 250 µg/l for 2-hr periods twice in both the eyed-egg and alevin stages, followed by oral administration at 20 mg/kg diet for 30 or 56 days from first feeding. Treatment for 15 days resulted in the production of 69% females. The treatment was repeated by Johnstone *et al.* (1978) with similar results. In this study, one field experiment, involving the oral administration of 20 mg/kg for 30 days without previous immersion, resulted in a 100% female group indicating that the treatment of eyed eggs and alevins was not essential. Oral treatment for shorter periods of time resulted in the production of small numbers of intersex and male fish. Some variability occurred between laboratory and field trials. Treatment for 30 days in the laboratory resulted in the production of 89% females compared with 100% female groups achieved in a similar field trial. Because the temperature regimes were similar, the discrepancy between results was attributed to photoperiod differences. During treatment, growth was depressed and fish were more susceptible to adverse conditions. Johnstone *et al.* (1979a) examined the progeny of five females from treated groups which were mated with untreated males. Of the five females examined two had progeny in a 2:1 and 2.4:1 male-to-female ratio indicating male heterogamety. The results also suggest that the YY genotype is not wholly viable.

V. J. Bye (personal communication, 1980) reported that, although repeated five times, oral treatment with 20 mg estradiol or estradiol benzoate/kg diet from the start of feeding 10 hr/day for 406°C days at a mean temperature of 9.9°C resulted in no significant alteration in the male-to-female ratio. The discrepency between these results and those already reported was attributed to the daily feeding regime. Bye suggested that, for effective treatment, administration of the treated diet be continued for 18 hr/day.

Recently, van den Hurk and Slof (1981) treated rainbow trout with progesterone at 300 μg/l for 28 days starting 27 or 43 days postfertilization. Treatment resulted in the production of 74 and 65% females. In the same study, treatment with *N*, *N*-dimethylformamide at 1 μg/l for 28 days starting 29 days postfertilization also resulted in 80 and 91% females. Significant mortalities occurred in these groups.

b. Salmo salar. Simpson (1976) reported the first production of a monosex female group of Atlantic salmon, *Salmo salar*. Treatment consisted of two immersions of the eyed eggs in water containing 250 μg estradiol/l. Alevins were treated similarly. The fry were fed a diet containing estradiol at 20 mg/kg diet for 120 days from first feeding. Johnstone *et al.* (1978) demonstrated that all-female populations of *Salmo salar* could be produced by oral administration of estradiol alone and for a shorter period of time (0–21 days). All-female groups were achieved in a group administered the immersion treatment and fed estradiol for 21 days from first feeding, and also in a group fed for 30 days starting 15 days from first feeding. Donaldson and Hunter (1982a) have noted the apparent discrepancy in treatment period overlap. Unfortunately, clearance rates of estradiol during early development are not known for this species.

c. Salmo trutta. Ashby (1957) reared juvenile brown trout, *Salmo trutta*, in water containing estradiol at 50 or 300 μg/l for 111 or 70 days (888°–568°C days), respectively, starting 170 days postfertilization (850°C days). At the end of the treatment, the group administered the lower dosage contained six males, four females, and three of indeterminant sex, but seven females and three males were found in the high dosage group. Mortalities were high in both groups.

d. Salvelinus namaycush. Wenström (1975) reported that sex differentiation in lake trout, *Salvelinus namaycush*, occurs from 750°C days to 1250°C days. Of the 10 fish examined following oral administration of estradiol at 12 mg/kg for 245–290°C days, eight were female and two were male.

e. Salvelinus fontinalis. Johnstone *et al.* (1979b) orally administered estradiol to brook trout, *Salvelinus fontinalis*, at 20 mg/kg diet for 60 days following first feeding, achieving 99% females and 1% intersex fish. Administration of estradiol for 40 days following first feeding was much less effective, achieving 67% females, 21% males, and 12% intersex fish. Significant mortalities occurred in the treated groups. Similarly to the report by Jalabert *et al.* (1975), several of the intersex fish contained testes and ovaries which matured simultaneously allowing the possibility of rapidly generating inbred stock lines. Because of the treatment period required for brook trout, Simpson *et al.* (1979) suggested that an effective estradiol treatment at 20 mg/kg diet for rainbow trout, Atlantic salmon, and brook trout include the first 60 days following first feeding.

f. Oncorhynchus kisutch. Goetz *et al.* (1979) were the first to attempt to feminize coho salmon, *Oncorhynchus kisutch*, by estrogen treatment. In this study, eyed eggs were administered two or six immersions in water containing estradiol at 25–400 μg/l. Alevins were administered two or seven immersions at identical dosages. Fry were fed a diet containing estradiol at 10 mg/kg diet for 70 days. Of the 10 experimental groups, seven contained 100% females, two were > 95% females. The lowest dosage and lowest treatment frequency group had 60% female. The group administered the treated diet alone contained 54.2% female, 18.1% male, and 27.7% intersex fish. The evidence from Goetz *et al.* (1979) suggested that treatment prior to first feeding was required for successful feminization. A dosage-dependent growth depression was also recorded. High mortalities were recorded in the two groups receiving a total of 13 immersions at 200 and 400 μg/l. Results from later tests indicate that essentially all-female groups of coho salmon may be obtained by treatment in the alevin stage alone (G. A. Hunter and E. M. Donaldson, unpublished data).

Using similar protocol, Hunter *et al.* (1982a) treated coho eyed eggs and alevins with either 100 or 400 μg/l immersions followed by oral administration of estradiol at 5 mg/kg diet. At maturity the group administered the higher dosage contained 94% mature females and 6% immature females. The group administered the lower dosage contained 92% mature females, 4% immature females, and 4% sterile fish. The ova from a total of 46 treated females and 22 control females were fertilized with sperm from untreated males. The mean male-to-female ratio (± s.d.) of the 22 control families was 1.13 ± 0.25, and the values for the treated fish fell into two categories 0.95 ± 0.17 (22 families) and 2.93 ± 0.88 (23 families). These results indicated male heterogamety in this species. Further, although variable, the mean male-to-female ratio of 2.93 indicates that the YY genotype is largely viable. Because of this altered sex ratio in the progeny, directly feminized coho salmon are not suitable for broodstock.

In the spring of 1979, approximately 40,000 coho eggs at the Capilano Salmon Hatchery were administered two 2-hr immersions in water containing estradiol at 200 µg/l both as eyed eggs and as alevins. Fry were subsequently fed estradiol at 5 mg/kg diet for 84 days. After nose tagging and fin clipping, a sample of 1000 fish were moved to a seawater net pen and the remainder released. Analysis of the fish held in the net pen indicated that over 99% of the fish were feminized. Mortalities were low and growth appeared to be normal. Tag recoveries from the ocean release indicated that this group contributed to both recreational and commercial fisheries at rates comparable to normal production fish. As expected, in the fall of 1981, approximately 500 of these fish returned to the hatchery as mature females. These fish were identical to normal production females in size and gonadosomatic index (Donaldson and Hunter, 1982b).

g. *Oncorhynchus tshawytscha.* Chinook salmon alevins were administered two 2-hr immersions in water containing estradiol at 400 µg/l. Further, the fry were orally administered estradiol at concentrations of 2 or 5 mg/kg diet for 3, 6, or 9 weeks. All treatments resulted in the production of 100% females with the exception of the 2-mg/kg-for-9-week treatment which contained 96% female and 4% sterile fish (G. A. Hunter and E. M. Donaldson, unpublished data).

h. *Oncorhynchus masou.* Nakamura (1981b) reared masu salmon, *Oncorhynchus masou*, in water containing estradiol at 0.25–200 µg/l for 18 days starting 5 days posthatching. Treatments between 0.5 and 5.0 µg/l produced essentially all-female groups when sampled at 50 or 90 days. The majority of fish treated with the 1–200 µg estradiol/l died shortly after treatment. Nakamura had previously determined the occurrence of sex differentiation at 13 days posthatching at 10°C.

A summary of the various studies on salmonids species is presented in Table II. Within the salmonids, both the production of all-female groups by the direct or indirect methods and the production of sterile groups appear to be feasible and desirable culture strategies. However, further studies are required to determine the net economic benefits of the applications of hormonal sex control to the salmonids.

C. Cyprinids

Although the various carp species are perhaps worldwide the most intensively cultured fish, remarkably little research has been directed to developing sex-control techniques in this group. However, the grass carp (*Ctenopharyngodon idella*) has been proposed as a biological control agent of aquatic weeds in the United States and in Europe. The further introduction of this species, a native of China, is dependent on the development of an effective means of preventing reproduction after its release into waterways. The production of sterile or monosex grass carp would achieve this objective. In the common carp *Cyprinus carpio* the development of sex-control techniques has been advocated by Nagy *et al.* (1981) to permit the crossing of highly inbred lines produced by gynogenesis.

1. ANDROGEN TREATMENT

a. Ctenopharyndogon idella. Stanley and Thomas (1978) attempted the indirect production of monosex stocks by the androgen-induced sex inversion of gynogenetically produced all–female groups. Methyltestosterone was administered at either 30 or 60 mg/kg. Durations for the lower dosage were 14, 28, 42, or 63 days from 7 days posthatching or for a 28-day period starting from 28, 42, 77, or 112 days posthatching. All treatments were unsuccessful. Subsequently Jensen *et al.* (1978) intraperitoneally implanted silastic capsules containing methyltestosterone into 195- or 309-day-old fish. Treatment duration was 303 and 192 days, respectively. The mean steroid diffusion rates for these two groups were 20.6 and 31.0 μg/day, respectively. The longer duration of treatment of younger fish was apparently more effective. In this group, most gonads were abnormal and smaller than controls. Further, five fish appeared to have been sterilized. Oogenesis and ovarian development were also suppressed. Survival of implanted fish was low (20%) and growth appeared depressed in the long duration treatment.

Shelton and Jensen (1979) suggested that the lack of knowledge of the critical period of sex differentiation related to the age of the grass carp was responsible for previous failures. They reported that the indifferent stage prior to gonadal sex differentiation lasted until 45–60 days (average length 43 mm) with anatomical differentiation from 50 to 75 days. Histological differentiation, marked by the appearance of the first nests of oogonia, did not occur until 94–125 days (average length 130 mm). Perinuclear oocytes appeared between 240 and 405 days (average length 150 mm). In the same study, Shelton and Jensen (1979) reported the effect of methyltestosterone treatment administered both orally and by silastic implants. Methyltestosterone treatments of 60 and 120 mg/kg diet administered to 110-day-old fingerlings lasting for either 255 or 410 days did not induce sex inversion, but both growth and gonadal development was inhibited.

Methyltestosterone administered by silastic implants to 195-day-old fingerlings (average length 107 mm) for 303 days were ineffective in altering sex ratios. Average diffusion rate was 18.7 μg/day. All treated fish had altered gonads containing extensive areas of connective tissue with reduced germ cells. In several fish, no germ cells could be identified. In an attempt to prevent the deleterious effects of steroid administration, older and larger fingerlings (309 days old, average length 158 mm) were administered methyltestosterone-filled silastic capsules for periods of 189, 202, and 461 days. The group treated for 189 days contained no females. Most gonads contained male germ cells in various stages of spermotogenesis, although the gonads anatomically resembled ovaries. Similar results were obtained in the group treated for 461 days. However, 31.7% females and 15% intersex fish were recorded in the group treated for 202 days. These results indicate that even 10-month-old grass carp remain responsive to sex manipulation with methyltestosterone. In each treatment group, gonads containing a few primary oocytes surrounded by testicular tissue undergoing spermatogenesis were observed. The occurrence of ovotests only in treated fish was attributed to the fact that anatomical differentiation had occurred prior to treatments and variable rates of development had allowed a small number of ooctyes to proceed to a point at which they were not influenced by methyltestosterone treatment.

In an attempt to determine whether the time of effective treatment could be altered by regulating the size of fish, Shelton and Jensen implanted silastic capsules containing methyltestosterone into 319-day-old fish which had been stunted (123 mm) by high-density culture. Average steroid release was calculated at 16.6 μg/day for the 500-day treatment period. Although alteration of the male-to-female sex ratio was not obtained, 17.4% of the treated fish contained intersex gonads. These gonads were dominated by perinuclear oocytes. In 40% of the fish germ cell could not be detected. Both testes and ovaries were smaller than controls. In a final experiment, 55-day-old all-female gynogenetic fish averaging 128 mm length were administered implants for 460 days. Average daily diffusion rate was 12.5 μg methyltestosterone/day. Of 27 treated fish 18.5% were male, 33.3% were intersex, 29.6% had no germinal tissue, and 18.5% had reduced ovaries. The testes of sex-inverted males were morphologically similar to ovaries. Gonadal evaluation of untreated individuals indicated that the process of oogenesis proceeded at different rates in full-sib offspring grown in similar environments. No correlation between size and rate of maturation could be determined. Shelton and Jensen suggested that the individual rates of gonadal morphogenesis observed could be the principal factor in variability of treatment outcome. This would be even more likely if normal gonadal differentiation had been activated in a few fish prior to treatment.

b. Cyprinus carpio. Similar to the grass carp, sex differentiation in the common carp, *Cyprinus carpio*, takes place a relatively long time after hatching and over an extended period of time. Davies and Takashima (1980) reported that, at 21.7°–23.5°C and based on the appearance of the ovarian cavity and "oocytelike" cells, differentiation occurred as early as 2 months posthatching in some individuals with the majority undergoing differentiation 4 months posthatching. Nagy *et al.* (1981) orally administered methyltestosterone at 100 mg/kg for 36 days starting at 8, 26, 44, 62, or 80 days posthatching to gynogenetic all-female groups of carp.

Treatments were conducted at 20° or 25°C. The most successful treatments were those conducted at 25°C. At this temperature, treatments starting from 8 to 62 days posthatching produced 71.4–88.9% males. Treatments were also conducted on fish of various length ranges. Groups composed of 90% males were produced when treatment was conducted between length 7–21 mm or 19–57 mm. A 100% male group was produced when treatment was conducted on fish of 13–39 mm length.

The use of steroids for the control of gonadal sex in the common carp has been shown to be highly effective. This technique should be of great assistance in the establishment of a selective breeding program for this species. The initial results with the grass carp are promising. However, further studies will be required to establish a treatment protocol with a level of effectiveness suitable to the management objectives associated with this species.

V. CONCLUSIONS

The use of hormonal sex-control techniques has great potential for fish culture on a world-wide basis. Despite the absence of a clear understanding of the mechanism underlying the influence of sex steroids on the process of sex differentiation, highly effective treatments have been established for the majority of species examined. This is especially true of the gonochorists and, therefore, almost all of the economically important species.

Hormonal sex-control techniques for several species of salmonids and cichlids are presently at a state of development which allows their routine use within the culture system. Further work with these species will involve an integration of the technique into existing management strategies and culture practices. Further, intraspecific differences may require some modification of existing treatment procedures. The accumulated experience from the application of hormonal sex-control techniques will facilitate the examination of species for which procedures have not been developed.

Although the biological basis for concern is negligible, legislation may constrain the use of steroids in fish destined for human consumption. This constraint may be avoided by the use of indirect methods for the production of monosex stocks. However, where highly effective treatments are required, this latter approach is constrained by the presence of autosomal influences. These constraints notwithstanding, the progress to date suggests that man will ultimately be able to control the sex ratio of all cultured fish species.

ACKNOWLEDGMENTS

Thanks is extended to P. R. Edgell, I. Baker, H. M. Dye, and A. Solmie for invaluable technical assistance in the research reported from this laboratory, J. Stoss for collaborative use of gamete cryopreservation techniques, E. T. Stone, F. K. Sandercock, and E. A. Perry for their cooperative efforts with regard to the pilot releases from the Capilano Salmon Hatchery. Thanks is also extended to Drs. W. S. Hoar, R. Reinboth, J. G. Stanley, and C. B. Schreck, for their constructive criticism of the manuscript, and M. Young for typing the manuscript.

REFERENCES

Aida, T. (1921). On the inheritance of color in a fresh-water fish *Aplocheilus latipes* Temminck and Schlegel, with special reference to the sex-linked inheritance. *Genetics* **6**, 554–573.

Aida, T. (1936). Sex reversal in *Aplocheilus latipes* and a new explanation of sex differentiation. *Genetics* **21**, 136–153.

Anders, A., and Anders, F. (1963). Genetisch bedingte XX- und XY- und XY- und YY- beim wilden *Platypoecilus maculatus* aus Mexico. *Z. Vererbungsl.* **94**, 1–18.

Anderson, C. E., and Smitherman, R. O. (1978). Production of normal male and androgen sex reversed *Tilapia aurea* and *T. nilotica* fed a commercial catfish diet in ponds. *In* "Culture of Exotic Fishes" (W. L. Shelton and J. H. Grover, eds.), pp. 34–42. Fish Cul. Sect., Am. Fish. Soc., Auburn, Alabama.

Anonymous (1979). Preliminary study of the utilization of androgen to induce the all-male tilapia. *Chin. J. Zool.* **1**, 1–3.

Ashby, K. R. (1957). The effect of steroid hormones on the brown trout (*Salmo trutta* L.) during the period of gonadal differentiation. *J. Embryol. Exp. Morphol.* **5**, 225–249.

Ashby, K. R. (1965). The effect of steroid hormones on the development of the reproductive system of *Salmo trutta* L. when administered at the commencement of spermatogenetic activity in the testis. *Riv. Biol.* **80**, No. 18, 139–169.

Atz, J. W. (1964). Intersexuality in fish. *In* "Intersexuality in Vertebrates including Man" (C. N. Armstrong and A. J. Marshall, eds.), pp. 145–232. Academic Press, New York.

Avtalion, R. R. (1982). Genetic markers in Sarotherodon and their use for sex and species identification. *In* "The Biology and Culture of Tilapias." ICLARM Conference Proceedings 7. (R. S. V. Pullin and R. H. Lowe-McConnell eds.), pp. 269–277. International Center for Living Aquatic Resources Management, Manila, Philippines.

Avtalion, R. R., and Hammerman, I. S. (1978). Sex determination in Sarotherodon (Tilapia). I. Introduction to a theory of autosomal influence. *Bamidgeh* **30**, 110–115.

Avtalion, R. R., Pruginin, Y., and Rothbard, S. (1975). Determination of allogenic and xenogenic markers in the genus Tilapia. I. Identification of sex and hybrids in Tilapia by electrophoretic analysis of serum proteins. *Bamidgeh* **27**, 8–13.

Bacon, L. D. (1970). Immunological tolerance to female wattle isographs in male chickens. *Transplantation* **10**, 124–126.

Bardach, J. E., Ryther, J. H., and McLarney, W. O. (1972). "Aquaculture." Wiley-Interscience, New York.

Berkowitz, P. (1941). The effect of oestrogenic substances in the fish (*Lebistes reticulatus*). *J. Exp. Zool.* **87**, 233–243.

Bieniarz, K., Epler, P., Malczewski, B., and Passakas, T. (1981). Development of European eel (*Anguilla anguilla* L.) gonads in artificial conditions. *Aquaculture* **22**, 53–66.

Billard, R., Breton, B., and Richard, M. (1981). On the inhibitory effect of some steroids on spermatogenesis in adult rainbow trout (*Salmo gairdneri*). *Can. J. Zool.* **59**, 1479–1504.

Bridges, C. B. (1925). Sex in relation to chromosomes and genes. *Am. Nat.* **59**, 127–137.

Bridges, C. B. (1936). Cytological and genetic basis of sex. *In* "Sex and Internal Secretions" (E. Allen, ed.), 2nd ed., pp. 15–63. Baillière, London.

Bull, J. J., and Charnov, E. L. (1977). Changes in the heterogametic mechanism of sex determination. *Heredity* **39**, 1–14.

Burns, R. K. (1950). Sex transformation in the opposum: Some new results and a retrospect. *Arch. Anat. Microsc. Morphol. Exp.* **39**, 467–483.

Burns, R. K. (1961). Role of hormones in the differentiation of sex. *In* "Sex and Internal Secretions" (W. C. Young, ed.), 3rd ed., Vol. 1, pp. 76–185. Williams & Wilkins, Baltimore, Maryland.

Bye, V. J. (1982). Steroid-induced sex reversal and sterility of the turbot *Scophthalmus maximus* L. *Gen. Comp. Endocrinol.* **46**, 392 (abstr.).

Bye, V. J., and Lincoln, R. (1981). Get rid of the males and let the females prosper. *Fish Farmer* **4**(8), 1–3.

Carlon, N., and Erickson, G. F. (1978). Fine structure of prefollicular and developing germ cells in the male and female left embryonic chick gonads *in vitro* with and without androgenic steroids. *Ann. Biol. Anim., Biochim., Biophys.* **18**, 335–349.

Castelnuovo, G. (1937). Effetti di alcuni ormoni sulla maturazione della carp. *Riv. Biol.* **23**, 365–372.

Cattanach, B. M., Pllard, C. E., and Hawkes, S. G. (1971). Sex reversed mice: XX and XO males. *Cytogenetics* **10**, 318–337.

Chan, S. T. H. (1970). Natural sex reversal in vertebrates. *Phil. Trans. Rog. Soc. Lond. B* **259**, 59–71.

Chan, S. T. H. (1977). Spontaneous sex reversal in fishes. *In* "Handbook of Sexology" (J. Money and H. Musaph, eds.), pp. 91–105. Elsevier/North-Holland Biomedical Press, Amsterdam.

Chan, S. T. H., O, W. S., and Hui, S. W. B. (1975). The gonadal and adenohypophysial functions of natural sex reversal. *In* "Intersexuality in the Animal Kingdom" (R. Reinboth, ed.), pp. 201–221. Springer-Verlag, Berlin and New York.

Chan, S. T. H., Ng, T. B., O, W. S., and Hui, W. B. (1977). Hormones and natural sex succession in Monopterus. *In* "Biological Research in Southeast Asia" (B. Lofts and S. T. H. Chan, eds.), pp. 101–120. University of Hong Kong, Hong Kong.

Chen, F. Y. (1969). Preliminary studies on the sex determining mechanism of *Tilapia mossambica* Peters and *T. hornorum* Trewavas. *Verh.—Int. Ver. Theor. Angew. Limnol.* **17**, 719–724.

Chen, F. Y., Chow, M., Chao, T. M., and Lim, R. (1977). Artificial spawning and larval rearing of the grouper, *Epinephelus tauvina* (Forskal) in Singapore. *Singapore J. Primary Ind.* **5**(1), 1–21.

Chieffi, G. (1959). Sex differentiation and experimental sex reversal in elasmobranch fishes. *Arch. Anat. Microsc. Morphol. Exp.* **48**, 21–36.

Chieffi, G. (1965). Onset of steroidogenesis in the vertebrate embryonic gonads. In "Organogenesis" (R. L. deHaan and H. Urspring, eds.), pp. 653–671. Holt, New York.

Chieffi, G. (1967). The reproductive system of Elasmobranchs: Developmental and endocrinological aspects. *Pharmacol. Endocrinol. Immunol.* **37**, 553–580.

Chieffi, G., Iela, L., and Rastogi, R. K. (1974). Effect of cyproterone, cyproterone acetate and ICI 46474 on gonadal sex differentiation in *Rana esculenta*. *Gen. Comp. Endocrinol.* **22**, 532–534.

Clemens, H. P., and Inslee, T. (1968). The production of unisexual broods of *Tilapia mossambica* sex-reversed with methyltestosterone. *Trans. Am. Fish. Soc.* **97**, 18–21.

D'Ancona, U. (1941). Ulteriori osservazione e considerazioni sull'ermafroditismo dell'orato (*Sparus auratus* L.). *Pubbl. Stn. Zool. Napoli* **18**, 314–336.

D'Ancona, U. (1949). Ermafroditismo ed intersessualitz nei Teleostei. *Experientia* **5**, 381–389.

Davies, P. R., and Takashima, F. (1980). Sex differentiation in common carp, *Cyprinus carpio*. *J. Tokyo Univ. Fish.* **66**(2), 191–199.

Dildine, G. C. (1936). Studies on telostean reproduction. I. Embryonic hermaphroditism in *Lebistes reticulatus*. *J. Morphol.* **60**, 261–277.

Donaldson, E. M., and Hunter, G. A. (1982a). Sex control in fish with particular reference to salmonids. *Can. J. Fish. Aquat. Sci.* **39**, 99–110.

Donaldson, E. M., and Hunter, G. A. (1982b). The ocean release and contribution to the fishery of all-female and sterile groups of coho salmon (*Oncorhynchus kisutch*). *Proc. Int. Symp. Reprod. Physiol. Fish, 1982* p. 78.

Dutta, O. K. (1979). Factors influencing gonadal sex differentiation in *Tilapia aurea* (Steindachner). Ph. D. dissertation, Auburn University, Auburn, Alabama (mimeo.).

Dzwillo, M. (1962). Über künstliche Erzeugung funktioneller Männchen weiblichen Genotypus bei *Lebistes reticulatus*. *Biol. Zentralbl.* **81**, 575–584.

Dzwillo, M. (1966). Über den Einfluss von Methyltestosteron auf primäre und sekundäre Geschlechtsmerkmale während verschiedener Phasen der Embryonalentwicklung von *Lebistes reticulatus*. *Zool. Anz., Suppl.* **29**, 471–476.

Dzwillo, M., and Zander, C. D. (1967). Geschlechtsbestimmung und Geschlechtsum stimmung bei Zahnkarpfen (Pices). *Mitt. Hamb. Zool. Mus. Inst.* **64**, 147–162.

Eckstein, B., and Spira, M. (1965). Effect of sex hormones on the gonadal differentiation in a cichlid, *Tilapia aurea*. *Biol. Bull.* (*Woods Hole, Mass.*) **129**, 482–489.

Egami, N. (1955). Production of testis-ova in adult males of *Oryzias latipes*. I. Testis-ova in the fish receiving estrogens. *Jpn. J. Zool.* **11**, 21–34.

Eichwald, E. J., and Silmser, C. R. (1955). Communication. *Transplant. Bull.* **2**, 148–149.

Eversole, W. J. (1939). The effects of androgens upon the fish (*Lebistes reticulatus*). *Endocrinology* **25**, 328–330.

Eversole, W. J. (1941). Effect of pregneninolone and related steroids on sexual development in fish *Lebistes reticulatus*. *Endocrinology* **28**, 603–610.

Fagerlund, U. H. M., and Dye, H. M. (1979). Depletion of radioactivity from yearling coho salmon (*Oncorhynchus kisutch*) after extended ingestion of anabolically effective doses of 17α-methyltestosterone-1,2-^{3}H. *Aquaculture* **18**, 303–315.

Fagerlund, U. H. M., and McBride, J. R. (1975). Growth increments and some flesh and gonad characteristics of juvenile coho salmon receiving diets supplemented with 17α-methyltestosterone. *J. Fish Biol.* **7**, 305–314.

294   GEORGE A. HUNTER AND EDWARD M. DONALDSON

Fagerlund, U. H. M., and McBride, J. R. (1978). Distribution and disappearance of radioactivity in blood and tissues of coho salmon (*Oncorhynchus kisutch*) after oral administration of ³H-testosterone. *J. Fish. Res. Board. Can.* **35,** 893–900.

Fineman, R., Hamilton, J., and Siler, W. (1974). Duration of life and mortality rates in male and female phenotypes in three sex chromosomal genotypes (XX, XY, YY) in the killifish, *Oryzias latipes. J. Exp. Zool.* **18,** 35–39.

Gallien, L. (1941). Recherches expérimentales sur l'action amphisexuelle de l'hormone femelle (oestradiol) dans le différenciation du sexe chez *Rana temporaria*, L. *Bull. Biol. Fr. Belg.* **75,** 369–397.

Gallien, L. (1950). Inversion du sexe et effet paradoxal (féminisation) chez l'urodele *Pleurodeles waltlii*, M., traité par le propionate de testostérone. *C. R. Hebd. Seances Acad. Sci.* **231,** 1092–1094.

Goetz, F. W., Donaldson, E. M., Hunter, G. A., and Dye, H. M. (1979). Effects of estradiol-17β and 17α-methyltestosterone on gonadal differentiation in the coho salmon, (*Oncorhynchus kisutch*). *Aquaculture* **17,** 267–178.

Goldstein, J. L., and Wilson, J. D. (1975). Genetic and hormonal control of male sexual differentiation. *J. Cell. Physiol.* **85,** 365–378.

Gordon, M. (1947). Genetics of *Platypoecilus maculatus*. IV. The sex determining mechanism in two wild populations of Mexican platyfish. *Genetics* **32,** 8–17.

Goudie, C. A., Redner, B. D., Simco, B. A., and Davis, K. B. (1983). Feminization in channel catfish by oral administration of sex steroid hormones. *Trans. Am. Fish. Soc.* **112,** 670–672.

Gower, P. B. (1975). Regulation of steroidogenesis. *In* "Biochemistry of Steroid Hormones" (H. L. J. Mankin, ed.), pp. 127–147. Blackwell, Oxford.

Gresik, E. W., and Hamilton, J. B. (1977). Experimental sex reversal in the teleost fish, *Oryzias latipes. In* "Handbook of Sexology" (J. Money and H. Musaph, eds.), pp. 107–126. Elsevier/North-Holland Biomedical Press, Amsterdam.

Guerrero, R. D. (1975). Use of androgens for the production of all-male *Tilapia aurea* (Steindachner). *Trans. Am. Fish. Soc.* **2,** 342–348.

Guerrero, R. D. (1976a). Culture of male *Tilapia mossambica* produced through artificial sex reversal. *FAO Tech. Conf. Aquacult.,* 1976, E.15, pp. 1–3.

Guerrero, R. D. (1976b). *Tilapia mossambica* and *T. zillii* treated with ethynyltestosterone for sex reversal. *Kalikasan* **5,** 187–192.

Guerrero, R. D. (1979). Use of hormonal steroids for artificial sex reversal of Tilapia. *Proc. Indian Natl. Sci. Acad., Part B* **45**(5), 512–514.

Guerrero, R. D., and Abella, T. A. (1976). Induced sex reversal of *Tilapia nilotica* with methyltestosterone. *Fish. Res. J. Philipp.* **1,** 46–49.

Guerrero, R. D. (1974). The use of synthetic androgens for production of monosex male *Tilapia aurea* (Steindachner). Ph.D. Dissertation, Auburn University, Auburn, Alabama.

Hackmann, E. (1974). Einfluss von Androgenen auf die Geschlechtsdifferenzierung verschiedener Cichliden (Teleostei). *Gen. Comp. Endocrinol.* **24,** 44–52.

Hackmann, E., and Reinboth, R. (1974). Delimination of the critical stage of hormone-influenced sex differentiation in *Hemiphaplochromis multicolor* (Hilgendorf) (Cichlidae). *Gen. Comp. Endocrinol.* **22,** 42–53.

Haffen, K. (1969). Quelques de l'intersexualite expérimentale chez le poulet (*Gallus gallus*) et la caille (*Caturnix caturnix japonica*). *Biol. Bull. (Woods Hole, Mass.)* **103,** 401–417.

Haffen, K. (1975). Sex differentiation of avian gonads *in vitro. Am. Zool.* **15,** 157–272.

Haffen, K. (1977). Sexual differentiation of the ovary. *In* "The Ovary" (S. Zuckerman and B. J. Weir, eds.), 2nd ed., pp. 69–112. Academic Press, New York.

Haffen, K., and Wolff, E. (1977). Modification of the ovarian development. *In* "The Ovary" (S. Zuckerman and B. J. Weir, eds.), Vol. 1, pp. 393–446. Academic Press, New York.

Hamada, H., Neumann, F., and Junkmann, K. (1963). Intrauterine antimaskuline Beeinflussung von Pattenfeten durch ein stark Gestagen wirksames steroid. *Acta Endocrinol. (Copenhagen)* **44**, 380–388.

Hamerton, J. R., Dickson, J. M., Pollard, C. E., Grieves, S. A., and Short, R. V. (1969). Genetic intersexuality in goats. *J. Reprod. Fertil.* **7**, Suppl. 25–51.

Hamilton, J. L. (1965). "Lillie's Development of the Chick," 3rd ed. Holt, New York.

Hammerman, I. S., and Avtalion, R. R. (1979). Sex determination in Sartherodon (Tilapia). Part 2. The sex ratio as a tool for the determination of genotype—A model of autosomal and gonosomal influence. *Theor. Appl. Genet.* **55**, 177–187.

Harbin, R., Whitehead, C., Bromage, N. R., Smart, G., Johnstone, R., and Simpson, T. (1980). Sterilization and other effects of methyltestosterone in the rainbow trout. *J. Endocrinol.* **87**, 66–67. Abstr.

Hardin, S. (1976). Electrophoresis of serum proteins from three species of Tilapia and one of their F_1 hybrids. M.Sc. Thesis, Auburn University, Alabama.

Hardisty, M. W. (1965). Sex differentiation and gonadogenesis in lampreys (Parts I and II). *J. Zool.* **146**, 305–387.

Hardisty, M. W. (1967). The numbers of vertebrate primordial germ cells. *Biol. Rev. Cambridge Philos. Soc.* **42**, 265–287.

Harrington, R. W. (1971). How ecological and genetic factors interact to determine when self-fertilizing hermaphrodites of *Rivalus mormoratus* change into functional males, with a reappraisal of the modes of intersexuality among fish. *Copeia* pp. 389–432.

Harrington, R. W., Jr. (1974). Sex determination and differentiation in fishes. *In* "Control of Sex in Fishes" (C. B. Schreck, ed.), VPI-SG-74-01, pp. 4–12. Extension Division, Virginia Polytechnic Institute and State University, Blacksburg.

Harrington, R. W., Jr. (1975). Sex determination and differentiation among uniparental homozygotes of the hermaphroditic fish, *Rivulus marmoratus* (Cyprinodontidae, Atheriniformes). *In* "Intersexuality in the Animal Kingdom" (R. Reinboth, ed.), pp. 249–262. Springer-Verlag, Berlin and New York.

Hickling, C. F. (1960). The malacca Tilapia hybrids. *J. Genet.* **57**, 1–10.

Hickling, C. F. (1963). The cultivation of tilapia. *Sci. Am.* **208**(5), 143–152.

Hirose, K., and Hibiya, T. (1968). Physiological studies on growth-promoting effect of protein-anabolic steroids on fish. II. Effects of 4-chlorotestosterone acetate on rainbow trout. *Bull. Jpn. Soc. Sci. Fish.* **34**, 473–481.

Hishida, T. (1962). Accumulation of testosterone-4-C^{14} propionate in larval gonad of the medaka, *Oryzias latipes*. *Embryologia* **7**, 56–67.

Hishida, T. (1965). Accumulation of estrone-16-C^{14} and diethylstilbestrol-(Monoethyl-1-C^{14}) in larval gonads of the medaka, *Oryzias latipes*, and determination of the minimum dosage of estrogen for sex reversal. *Gen. Comp. Endocrinol.* **5**, 137–144.

Hishida, T., and Kawamoto, N. (1970). Androgenic and male inducing effects of 11-ketotestosterone on a teleost, the medaka (*Oryzias latipes*). *J. Exp. Zool.* **173**, 279–283.

Hoar, W. S. (1969). Reproduction. *In* "Fish Physiology" (W. S. Hoar and D. J. Randall, eds.), Vol. 3, pp. 1–72. Academic Press, New York.

Hopkins, K. D. (1977). Sex reversal of genotypic male *Sarotherodon aureus* (Cichlidae). M.S. Thesis, Auburn University, Auburn, Alabama.

Hopkins, K. D., Shelton, W. L., and Engle, C. R. (1979). Estrogen sex-reversal of *Tilapia aurea*. *Aquaculture* **18**, 263–268.

Hsü, C. Y., Liang, H. M., Yu, N. W., and Chiang, C. H. (1978a). A histochemical study of the adrenogenital syndrome in frog tadpoles. *Gen. Comp. Endocrinol.* **35**, 347–354.

Hsü, C. Y., Liang, H. M., and Hsü, L. H. (1978b). Further evidence for decreased Δ⁵3β-hydroxysteroid dehydrogenase activity in female tadpoles after estradiol treatment. Gen. Comp. Endocrinol. **35**, 451–545.

Hsü, C. Y., Hsü, L. H., and Liang, H. M. (1979). The effect of cyproterone acetate on the activity of Δ⁵-3β-hydroxysteroid-dehydrogenase in tadpole sex transformation. Gen. Comp. Endocrinol. **39**, 404–410.

Hunter, G. A., Donaldson, E. M., Goetz, F. W., and Edgell, P. R. (1982a). Production of all female and sterile groups of coho salmon (Oncorhynchus kisutch) and experimental evidence for male heterogamety. Trans. Am. Fish. Soc. **111**, 367–372.

Hunter, G. A., Donaldson, E. M., Stoss, J., and Baker, I. (1983). Production of monosex female groups of chinook salmon (Oncorhynchus tshawytscha) by the fertilization of normal ova with sperm from sex reversed females. Aquaculture **33**, 335–364.

Jalabert, B., Kammacher, P., and Lessent, P. (1971). Déterminisme du sexe chez les hybrids entre Tilapia macrochir et Tilapia nilotica. Etude de la sex-ratio dans les recroisements des hybrids de première génération par les epèces parentes. Ann. Biol. Anim., Biochim., Biophys. **11**, 155–165.

Jalabert, B., Moreau, J., Planquette, P., and Billard, R. (1974). Determinisme du sexe chez Tilapia macrochir et Tilapia nilotica: Action de la methyltestostérone dans l'alimentation des alévins sur la différenciation sexuelle; proportion des sexes dans la descendance des males "inverses." Ann. Biol. Anim., Biochim., Biophys. **14**(4-B), 729–739.

Jalabert, B., Billard, R., and Chevassus, B. (1975). Preliminary experiments on sex control in trout: Production of sterile fishes an simultaneous self-fertilization hermaphrodites. Ann. Biol. Anim., Biochim., Biophys. **15**, 19–28.

Jensen, G. L. (1976). The effects of several naturally occurring estrogens on Sarotherodon aureus (Steindachner) and their potential application to yield monosex genetic males. M.S. Thesis, Auburn University, Auburn, Alabama.

Jensen, G. L., and Shelton, W. L. (1979). Effects of estrogens on Tilapia aurea: Implications for production of monosex genetic male tilapia, Aquaculture **16**, 233–242.

Jensen, G. L., Shelton, W. L., and Wilken, L. O. (1978). Use of methyltestosterone silastic implants to control sex in grass carp. In "Culture of Exotic Fishes" (R. O. Smitherman, W. L. Shelton, and J. H. Grover, eds.), pp. 200–219. Fish Cult. Sect., Am. Fish. Soc., Auburn, Alabama.

Johnston, E. R. Zeller, J. H., and Cantwell, G. (1958). Sex anomalies in swine. J. Hered. **49**, 255–261.

Johnstone, R., Simpson, T. H., and Youngson, A. F. (1978). Sex reversal in salmonid culture. Aquaculture **13**, 115–134.

Johnstone, R., Simpson, T. H., Youngson, A. F., and Whitehead, C. (1979a). Sex reversal in salmonid culture. Part II. The progeny of sex-reversed rainbow trout. Aquaculture **18**, 13–19.

Johnstone, R., Simpson, T. H., and Walker, A. F. (1979b). Sex reversal in salmonid culture. Part III. The production and performance of all-female populations of brook trout. Aquaculture **18**, 241–252.

Jost, A. (1965). Gonadal hormones in the sex differentiation of mammalian foetus. In "Organogenesis" (R. L. deHaan and H. Ursprung, eds.), pp. 611–628. Holt, New York.

Kallman, K. D. (1965). Genetics and geography of sex determination in the poeciliid fish, Xiphophorus maculatus. Zoologica (N.Y.) **50**, 151–190.

Katz, Y., Abraham, M., and Eckstein, B. (1976). Effects of adrenosterone on gonadal and body growth in Tilapia nilotica (Teleostei, Cichlidae). Gen. Comp. Endocrinol. **29**, 414–418.

Kosswig, C. (1964). Polygenic sex determination. Experientia **20**, 1–10.

Kosswig, C., and Oktay, M. (1955). Kie Geschlechtsbestimmung bei den Xiphophorini (Neue tatesachen und neue bedentung). *Istanbul Univ. Fen. Fak. Hidrobiol.* **2**, 133–156.

Lebrun, C., Billard, R., and Jalabert, B. (1982). Changes in the number of germ cells in the rainbow trout (*Salmo gairnderi*) during the first 10-post-hatching weeks. *Reprod. Nutr. Dev.* **22**(2), 405–412.

Lillie, R. F. (1917). The freemartin; a study of the action of the hormones in the foetal life of cattle. *J. Exp. Zool.* **23**, 371–452.

Lincoln, R. F., and Hardiman, P. A. (1982). The production of female diploid and triploid rainbow trout. *Int. Symp. Genet. Aquacult., 1982* Abstract, p. 36.

Liu, C.-Y. (1977). Aspects of reproduction and progeny testing in *Sarotherodon aureus* (Steindachner). M.S. Thesis, Auburn University, Auburn, Alabama.

Lowe, T. P., and Larkin, J. R. (1975). Sex reversal in *Betta splendens* Regan with emphasis on the problem of sex determination. *J. Exp. Zool.* **191**, 25–32.

McBride, J. R., and Fagerlund, U. H. M. (1973). The use of 17-α-methyltestosterone for promoting weight increases in juvenile Pacific salmon. *J. Fish. Res. Board Can.* **30**, 1099–1104.

McCarrey, J. R., and Abbott, U. K. (1979). Mechanisms of genetic sex determination, gonadal sex differentiation and germ-cell development in animals. *Adv. Genet.* **20**, 217–290.

Mittwoch, U. (1971). Sex determination in birds and mammals. *Nature (London)* **231**, 432–434.

Müller, R. (1969). Die Einwirking von Sexualhormonen auf die Geschlechtsdifferenzierung von *Hemihaplochromis multicolor* (Hilgendorf-Cichlidae). *Zool. Jahrb., Abt. Allg. Zool. Physiol. Tiere* **74**, 519–562.

Müller, U., and Wolf, U. (1979). Cross reactivity to mammalian anti-H-Y antiserum in teleostean fish. *Differentiation* **14**, 185–187.

Müller, U., Zenzes, M. T., and Wolf, U. (1979). Appearance of H-W (H-Y) antigen in the gonads of oestradiol sex-reversed male chicken embryos. *Nature (London)* **280**, 142–144.

Müller, U., Guichard, A., Reyss-Brian, M., and Scheib, D. (1980). Induction of H-Y antigen in the gonads of male quail embryos by diethylstilbestrol. *Differentiation* **16**, 129–133.

Nagy, A., Rajki, K., Horvath, L., and Csanyi, V. (1978). Investigation on carp *Cyprinus carpio* L. gynogenesis. *J. Fish Biol.* **13**, 215–224.

Nagy, A., Beresenyi, M., and Csany, V. (1981). Sex reversal in carp (*Cyprinus carpio*) by oral administration of methyltestosterone. *Can. J. Fish. Aquat. Sci.* **38**, 725–728.

Nakamura, M. (1975). Dosage-dependent changes in the effect of oral administration of methyltestosterone on gonadal sex differentiation in *Tilapia mossambica*. *Bull. Fac. Fish., Hokkaido Univ.* **26**(2), 99–108.

Nakamura, M. (1978). Morphological and experimental studies on sex differentiation of the gonad of several teleost fishes. Ph.D. Thesis, Faculty of Fish, Hokkaido University, Hokkaido, Japan.

Nakamura, M. (1981a). Effects of 11-ketotestosterone on gonadal sex differentiation in *Tilapia mossambica*. *Bull. Fac. Fish., Hokkaido Univ.* **47**, 1323–1327.

Nakamura, M. (1981b). Feminization of masu salmon, *Oncorhynchus masou*, by administration of estradiol-17-β. *Bull. Japan. Soc. Sci. Fisheries* **47**(11), 1529.

Nakamura, M. (1982). Gonadal sex differentiation in whitespotted char, *Salvelinus leucomaenis*. *Jpn. J. Ichthyol.* **28**, 431–436.

Nakamura, M., and Iwahashi, M. (1982). Studies on the practical masculinization in *Tilapia nilotica* by the oral administration of androgen. *Bull. Jpn. Soc. Sci. Fish.* **486**, 763–769.

Nakamura, M., and Takahashi, H. (1973). Gonadal sex differentiation in *Tilapia mossambica* with special regard to the time of estrogen treatment effective in inducing feminization of genetic fishes. *Bull. Fac. Fish., Hokkaido Univ.* **24**, 1–13.

Narbaitz, R., and De Robertis, E. M. (1970). Steroid producing cells in chick intersexual gonads. *Gen. Comp. Endocrinol.* **14**, 164–169.

Ohno, S. (1967). "Sex Chromosome and Sex Linked Genes." Springer-Verlag, Berlin and New York.

Ohno, S., Nagai, Y., and Ciccarese, S. (1978). Testicular cells lysostripped of H-Y antigen organize ovarian follicle-like aggregates. *Cytogenet. Cell Genet.* **20**, 351–364.

Okada, H. (1973). Studies on sex differentiation of salmonidae. I. Effects of estrone on sex differentiation of the rainbow trout (*Salmo gairdnerii irideus* Gibbons). *Sci Rep. Hokkaido Fish Hatchery* **28**, 11–21.

Okada, H., Matsumoto, H., and Yamazaki, F. (1979). Functional masculinization of genetic females in rainbow trout. *Bull. Jpn. Soc. Sci. Fish.* **45**, 413–419.

Okada, H., Matsumoto, H., and Murakami, Y. (1981). Ratio of induced males from genetic females at various dietary concentrations of methyltestosterone. *Annu. Meet. Jpn. Soc. Sci. Fish.* Abstract, p. 33.

Okada, Y. K. (1962). Sex reversal in the Japanese wrasse, *Halichoeres poecilopterus*. *Proc. Jpn. Acad.* **38**(8), 508–513.

Okada, Y. K. (1964). Effects of androgen and estrogen on sex reversal in the wrasse, *Halichoeres poecilopterus*. *Proc. Jpn. Acad.* **40**, 541–544.

O'Malley, B. W., and Schrader, W. T. (1976). The receptors of steroid hormones. *Sci. Am.* **234**, 32–43.

Onitake, K. (1972). Morphological studies of normal sex differentiation and induced sex reversal process of gonads in the medaka *Oryzias latipes*. *Annot. Zool. Jpn.* **45**, 159–169.

Owusu-Frimpong, M., and Nijjhar, B. (1981). Induced sex reversal in *Tilapia nilotica* (Cichlidae) with methyltestosterone. *Hydrobiologia* **78**, 157–160.

Padoa, E. (1937). Differenziazone e inversione sessuale (feminizzazione) di avanotti di Trota (*Salmo irideus*) trattati con ormone follicolare. *Monit. Zool. Ital.* **48**, 195–203.

Padoa, E. (1939). Observations ultérieures dur la différenciation du sexe, normale et modifiée par l'administration d'hormone folliculaire, chez la truite iridée (*Salmo irideus*). *Biomorphosis* **1**, 337–354.

Padoa, E. (1942). Femminizzazione e mascolinizzazione di girmi de *Rana esculenta* in firnzione della dose di diidrofollicolina loro somministrata. *Monit. Zool. Ital.* **53**, 210–213.

Pechan, P., Wachtel, S. S., and Reinboth, R. (1979). H-Y antigen in the teleosts. *Differentiation* **14**, 189–192.

Persov, G. M. (1975). The period of gonadal differentiation in fishes. *In* "Sex Differentiation in Fishes," p. 118. University of Leningrad Publ. House, Leningrad (Fish. Mar. Serv. Can., Transl. Ser. No. 4069).

Price, D. (1970). *In vitro* studies on differentiation of the reproductive tract. *Philos. Trans. R. Soc. London, Ser. B* **259**, 133–139.

Pruginin, Y., Rothbard, S., Wohlfarth, G., Halevy, R., Moav, R., and Hulata, G. (1975). All-male broods of *Tilapia nilotica* × *T. aurea* hybrids. *Aquaculture* **6**, 11–21.

Quirk, J. G., and Hamilton, J. B. (1973). Number of germ cells in known male and known female genotypes of vertebrate embryos (*Oryzias latipes*). *Science* **180**, 963–964.

Rastogi, R. K., and Chieffi, G. (1975). The effects of antiandrogens and antiestrogens on non-mammalian vertebrates. *Gen. Comp. Endocrinol.* **26**, 79–91.

Raynaud, A. (1967). Effets d'une hormone oestrogène sur le développement de l'appareil génital de l'embryon de lezard vert. *Arch. Anat. Microsc. Morphol. Exp.* **56**, 63–122.

Refstie, T., Stoss, J., and Donaldson, E. M. (1982). Production of all-female coho salmon (*Oncorhynchus kisutch*) by diploid gynogenesis using irradiated sperm and cold shock. *Aquaculture* **29**, 67–82.

Reinboth, R. (1962). Morphologische und funktionelle Zweigeschlechtlichkeit bei marinen Telostiern (Serranidae, Sparidae, Centracanthidae, Labridae). *Zool. Jahrb., Abt. Allg. Zool. Physiol. Tiere* **69**, 405–580.

Reinboth, R. (1969). Varying effects with different ways of hormone administration of gonad differentiation of a teleost fish. *Gen. Comp. Endocrinol.* **13**, 527–528.

Reinboth, R. (1970). Intersexuality in fishes. *Mem. Soc. Endocrinol.* **18**, 515–544.

Reinboth, R. (1972). Hormonal control of the teleost ovary. *Am. Zool.* **12**, 307–324.

Reinboth, R. (1975). Spontaneous and hormone-induced sex-inversion in wrasses (Labridae). *Pubbl. Stn. Zool. Napoli* **39**, Suppl., 550–573.

Reinboth, R. (1982). The problem of sexual bipotentiality as exemplified by teleosts. *Reprod. Nutr. Dev.* **22**(2), 397–403.

Robertson, J. G. (1953). Sex differentiation in the Pacific salmon *Oncorhynchus keta* (Walbaum). *Can. J. Zool.* **31**, 73–79.

Rouse, E. F., Coppenger, C. J., and Barnes, P. R. (1976). The effect of an androgen inhibitor on behaviour and testicular morphology in the stickleback, *Gasterosteus aculeatus*. *Reg. Conf. Comp. Endocrinol.*, *Oreg. State Univ.* Abstract, p. 42.

Sanico, A. F. (1975). Effects of 17α-ethynyltestosterone and estrone on sex ratio and growth of *Tilapia aurea* (Steindachner). M.S. Thesis, Auburn University, Auburn, Alabama.

Satoh, N. (1973). Sex differentiation of the gonad of fry transplanted into the anterior chamber of the adult eye in the teleost *Oryzias latipes*. *J. Embryol. Exp. Morphol.* **30**, 345–340.

Satoh, N. (1974). An ultrastructural study of sex differentiation in the teleost *Oryzias latipes*. *J. Embryol. Exp. Morphol.* **32**, 192–215.

Satoh, N., and Egami, N. (1972). Sex differentiation of germ cells in the teleost, *Oryzias latipes*, during normal embryonic development. *J. Embryol. Exp. Morphol.* **28**, 385–395.

Schmit, P. J., and Idler, D. R. (1962). Steroid hormones in the plasma of salmon at various states of maturation. *Gen. Comp. Endocrinol.* **2**, 204–214.

Schreck, C. B. (1974). Hormonal treatment and sex manipulation in fishes. *In* "Control of Sex in Fishes" (C. B. Schreck, ed.), VPI-SG-74-01, pp. 84–106. Extension Division, Virginia Polytechnical Institute and State University, Blacksburg.

Shalev, A., and Huebner, F. (1980). Expression of H-Y antigen in the guppy (*Lebistes reticulatus*). *Differentiation* **16**, 81–83.

Shelton, W. L., and Jensen, G. L. (1979). Production of reproductively limited grass carp for biological control of aquatic weeds. *WRRI Bull.* (*Auburn Univ.*)–**39**, 1–174.

Shelton, W. L., Hopkins, K. D., and Jensen, G. L. (1978). Use of hormones to produce monosex tilapia for Aquaculture. *In* "Culture of Exotic Fishes" (R. D. Smitherman, W. L. Shelton, and J. H. Glover, eds.), pp. 10–33. Fish Cult. Sect., Am. Fish. Soc., Auburn, Alabama.

Shelton, W. L., Rodriguez-Guerrero, D., and Lopez-Macias, J. (1981). Factors affecting androgen sex reversal of *Tilapia aurea*. *Aquaculture* **25**, 59–65.

Shimizu, M., and Takahashi, H. (1980). Process of sex differentiation of the gonad and gonaduct of the three spined stickleback, *Gasterosteus aculeatus* L. *Bull. Fac. Fish.*, *Hokkaido Univ.* **31**(2), 137–148.

Simpson, T. H. (1975–1976). Endocrine aspects of salmonid culture. *Proc. R. Soc. Edinburgh*, *Ser. B* **75**, 241–252.

Simpson, T. H., Johnstone, R., and Youngson, A. F. (1979). Female stocks less vunerable. *Fish Farmer* **2**(3), 20–21.

Smith, C. L. (1975). The evolution of hermaphroditism in fishes. *In* "Intersexuality in the Animal Kingdom" (R. Reinboth, ed.), pp. 293–310. Springer-Verlag, Berlin and New York.

Smith, H. T. (1976). Various aspects of reproductive control in fresh water fishes. M. S. Thesis, Virginia Polytechnic Inst. and State University, Blacksburg.

Stanley, J. G. (1976). Female homogamety in grass carp (*Ctenopharyngodon idella*) determined by gynogenesis. *J. Fish. Res. Board Can.* **33**, 1372–1374.

Stanley, J. G. (1979). Control of sex in fishes with special reference to the grass carp. *In* "Proc. of the grass carp conference" (J. V. Shireman, ed.) pp. 201–242. Univ. Florida. Gainesville, FL.

Stanley, J. G. (1981). Manipulation of developmental events to produce monosex and sterile fish. *In* "Early life history of fish. II. Second International Symposium held at Woods Hole, 2–5 April, 1979" (R. Lasher and K. Sherman, ed.) *Rapp, P.-V. Reun Cons. Int. Explor. Mer 178.*

Stanley, J. G., and Thomas, A. E. (1978). Absence of sex reversal in unisex grass carp fed methyltestosterone. *In* "Culture of Exotic Fishes" (R. O. Smitherman, W. L. Shelton, and J. H. Glover, eds.), pp. 194–199. Fish Cult. Sect., Am. Fish. Soc., Auburn, Alabama.

Stanley, J. G., Martin, J. M., and Jones, J. B. (1975). Gynogenesis as a possible method for producing monosex grass carp (*Ctenopharyngodon idella*). *Prog. Fish-Cult.* **37**, 25–26.

Streisinger, G., Walker, C., Dower, N., Knauber, D., and Singer, F. (1981). Production of clones of homozygous diploid zebra fish (*Brachydanio rerio*). *Nature (London)* **291**, 293–296.

Stromsten, F. A. (1931). The development of the gonads in the goldfish *Carassius auratus* (L.). *Univ. Iowa Stud. Nat. Hist.* **13**(7), 3–45.

Sundararaj, B. I., and Nayyar, S. K. (1969). Effects of estrogen, SU-9055, and cyproterone acetate on the hypersecretory activity in the seminal vesicles of the castrate catfish, *Heteropneustes fossilis* (Bloch). *J. Exp. Zool.* **172**, 399–408.

Takahashi, H. (1974). Modification of the development of female reproductive organs in the guppy, *Poecilia reticulata*, following an androgen treatment in their juvenile period. *Bull. Fac. Fish., Hokkaido Univ.* **25**(3), 174–199.

Takahashi, H. (1975a). Functional masculinization of female guppies, *Poecilia reticulata*, influenced by methyltestosterone before birth. *Bull. Jpn. Soc. Sci. Fish.* **41**, 499–506.

Takahashi, H. (1975b). Process of functional sex reversal of the gonad in the female guppy, *Poecilia reticulata*, treated with androgen before birth. *Dev. Growth Differ.* **17**, 167–175.

Takahashi, H. (1975c). Masculinization of the gonad of juvenile guppy, *Poecilia reticulata*, induced by 11-ketotestosterone. *Bull. Fac. Fish., Hokkaido Univ.* **26**(1), 11–22.

Takahashi, H. (1975d). Functional feminization of genetic males of the guppy, *Poecilia reticulata*, treated with estrogen after birth. *Bull. Fac. Fish., Hokkaido Univ.* **26**(3), 223–234.

Takahashi, H. (1977). Effects of large doses of methyltestosterone on the development of reproductive organs of juvenile guppy, *Poecilia reticulata. Bull. Fac. Fish., Hokkaido Univ.* **28**, 6–19.

Takahashi, H., and Iwasaki, Y. (1973). The occurrence of histochemical activity of 3β-hydroxysteroid dehydrogenase in the developing testes of *Poecilia reticulata. Dev. Growth Differ.* **15**, 241–253.

Takashima, F., Patino, R., and Nomura, M. (1980). Histological studies on the sex differentiation in rainbow trout. *Bull. Jpn. Soc. Sci. Fish.* **46**, 1317–1322.

Tang, F., Chan, S. T. H., and Lofts, B. (1975). A study on the 3β and 17β-hydroxysteroid dehydrogenase activities in the gonad of *Monopterus albus* (Pisces: teleostei) at various sexual phases during sex reversal. *J. Zool.* **175**, 571–580.

Tang, F., Chan, S. T. N., and Lofts, B. (1974). Effect of steroid hormones on the process of natural sex reversal in the ricefield eel, *Monopterus albus* (Zuiew). *Gen. Comp. Endocrinol.* **24**, 227–241.

Tayamen, M. M., and Shelton, W. L. (1978). Inducement of sex reversal in *Sartherodon niloticus* L. *Aquaculture* **14**, 349–354.

Thorgaard, G. H. (1977). Heteromorphic sex chromosomes in male rainbow trout. *Science* **196**, 900–902.

Vallowe, H. H. (1957). Sexual differentiation in the teleost fish, *Xiphophorus helleri* as modified by experiental treatment. *Biol. Bull. (Woods Hole, Mass.)* **112**, 422–429.

van den Hurk, R., and Lambert, J. G. D. (1982). Temperature and steroid effects on gonadal sex differentiation in rainbow trout. *Proc. Int. Symp. Reprod. Physiol. Fish, 1982* pp. 69–72.

van den Hurk, R., and Slof, G. H. (1981). A morphological and experimental study of gonadal sex differentiation in the rainbow trout, *Salmo gairdneri*. *Cell Tissue Res.* **218**, 487–497.

van den Hurk, R., Lambert, J. G. D., and Peute, J. (1982). Steroidogenesis in the gonads of rainbow trout fry (*Salmo gairnderi*) before and after the onset of gonadal sex differentiation. *Reprod. Nutr. Dev.* **22**(2), 413–425.

Vannini, E., Stagni, A., and Zaccanti, F. (1975). Autoradiographic study on the mechanism of testosterone-induced sex-reversal in Rana tadpoles. *In* "Intersexuality in the Animal Kingdom" (R. Reinboth, ed.), pp. 318–331. Springer-Verlag, Berlin and New York.

Vanyakina, E. D. (1969). Genetics of sex determination and some problems of hormonal regulation of sex in teleosts. *In* "Genetics, Selection and Hybridization of Fish" (I. B. Cherfas, ed.), pp. 25–41. Akad. Nauk SSSR (Transl. by Isr. Prog. Sci. Transl., 1972).

Vorontsov, N. N. (1973). The evolution of sex chromosomes. *In* "Cytotaxonomy and Vertebrate Evolution" (A. B. Chiarelli and E. Capanna, eds.), pp. 619–657. Academic Press, New York.

Wachtel, S. S., and Koo, G. C. (1981). H-Y antigen in gonadal differentiation. *In* "Mechanisms of Sex Differentiation in Animals and Man" (C. R. Austin and R. G. Edwards, eds.), pp. 255–299. Academic Press, New York.

Wachtel, S. S., Koo, G. C., and Boyse, E. A. (1975). Evolutionary conservation of H-Y (male) antigen. *Nature (London)* **254**, 270–272.

Wachtel, S. S., Bresler, P., and Koide, S. S. (1980). Does H-Y antigen induce the heterogametic ovary? *Cell* **20**, 859–864.

Wenström, J. C. (1975). Sex differentiation and hormone directed sex determination in the lake trout (*Salvelinus namaycush*). M. A. Thesis, Dept. Biol., Northern Michigan University, Marquette.

White, A., Handler, P., and Smith, E. L., eds. (1973). "Principles of Biochemistry," 5th ed. McGraw-Hill, New York.

Wilson, J. D. (1978). Sexual differentiation. *Annu. Rev. Physiol.* **40**, 279–306.

Winge, O. (1934). The experimental alteration of sex chromosomes into autosomes and vice versa, as illustrated by Lebistes. *C. R. Trav. Lab. Carlsberg, Ser. Physiol.* **21**, 1–49.

Witschi, E. (1929). Bestimmung und Vererbung des Geschlechts bei Tieren. *Handb. Vererbungswiss.* **2**, 1–115.

Witschi, E. (1965). Hormones and embryonic induction. *Arch. Anat. Microsc. Morphol. Exp.* **54**, 601–611.

Witschi, E. (1967). Biochemistry of sex differentiation in vertebrate embryos. *In* "The Biochemistry of Animal Development" (R. Weber, ed.), Vol. 2, pp. 193–223. Academic Press, New York.

Witschi, E., and Dale, E. (1962). Steroid hormones at early developmental stages of vertebrates. *Gen. Comp. Edocrinol. Suppl.* **1**, 356–361.

Wohlfarth, G. W. and Hulata, G. I. (1981). "Applied genetics of tilapias." ICLARM Studies and Reviews 6, International Center for Living Aquatic Resources Management, Manila.

Woiwode, J. G. (1977). Sex reversal of *Tilapia zillii* by injestion of methyltestosterone. *Tech. Pap. Ser. Bur. Fish. Aquat. Resour. (Phillips.)* **1**(3), 1–5.

Yamamoto, N. K. (1975). Effects of hypophysectomy on implanted testes of the neuter medaka *Oryzias latipes*. *Dev. Growth Differ.* **17**, 253–263.

Yamamoto, T. (1953). Artificially induced sex reversal in genotypic males of the medaka (*Oryzias latipes*). *J. Exp. Zool.* **123**, 571–194.

Yamamoto, T. (1955). Progeny of artificially induced sex reversals of male genotype (XY) in the medaka (*Oryzias latipes*) with special reference to YY-male. *Genetics* **40**, 406–419.

Yamamoto, T. (1958). Artificial induction of functional sex-reversal in genotypic females of the medaka (*Oryzias latipes*). *J. Exp. Zool.* **137**, 227–262.

Yamamoto, T. (1959a). A further study of induction of functional sex reversal in genotypic males of the medaka (*Oryzias latipes*) and progenies of sex reversals. *Genetics* **44**, 739–757.

Yamamoto, T. (1959b). The effects of estrone dosage level upon the percentage of sex-reversals in genetic male (XY) of the medaka (*Oryzias latipes*). *J. Exp. Zool.* **141**, 133–154.

Yamamoto, T. (1961). Progenies of induced sex-reversal females mated with induced sex-reversal males in the medaka, *Oryzias latipes*. *J. Exp. Zool.* **146**, 163–179.

Yamamoto, T. (1962). Hormonic factors affecting gonadal sex differentiation in fish. *Gen. Comp. Endocrinol., Suppl.* **1**, 341–345.

Yamamoto, T. (1963). Induction of reversal in sex differentiation of YY zygotes in the medaka, *Oryzias latipes*. *Genetics* **48**, 293–306.

Yamamoto, T. (1964a). The problem of viability of YY zygotes in the medaka, *Oryzias latipes*. *Genetics* **50**, 45–58.

Yamamoto, T. (1964b). Linkage map of sex chromosomes in the medaka, *Oryzias latipes*. *Genetics* **50**, 59–64.

Yamamoto, T. (1965). Estriol-induced XY females of the medaka (*Oryzias latipes*) and their progenies. *Gen. Comp. Endocrinol.* **5**, 527–533.

Yamamoto, T. (1967). Estrone-induced white YY females and mass production of white YY males in the medaka, *Oryzias latipes*. *Genetics* **55**, 329–336.

Yamamoto, T. (1968). Effects of 17β-hydroxyprogesterone and androstenedione upon sex differentiation in the medaka, *Oryzias latipes*. *Gen. Comp. Endocrinol.* **10**, 8–13.

Yamamoto, T. (1969). Sex differentiation. *In* "Fish Physiology" (W. S. Hoar and D. J. Randall, eds.), Vol. 3, pp. 117–175. Academic Press, New York.

Yamamoto, T. (1975). "Medaka (killifish) Biology and Strains." Keigaku Publ. Co., Tokyo, Japan.

Yamamoto, T., and Kajishima, T. (1969). Sex-hormonic induction of reversal of sex differentiation in the goldfish and evidence for its male heterogamety. *J. Exp. Zool.* **168**, 215–222.

Yamamoto, T., and Matsuda, N. (1963). Effects of estradiol, stilbestrol and some alkyl-carbonyl androstanes upon sex differentiation in the medaka, *Oryzias latipes*. *Gen. Comp. Endocrinol.* **3**, 101–110.

Yamamoto, T., and Susuki, H. (1955). The manifestation of the urinogenital papillae of the medaka (*Oryzias latipes*) by sex hormones. *Embryologia* **2**, 133–144.

Yamamoto, T., Takeuchi, K., and Takai, M. (1968). Male-inducing action of androsterone and testosterone propionate upon XX zygotes in the medaka, *Oryzias latipes*. *Embryologia* **10**, 142–151.

Yamazaki, F. (1972). Effects of methyltestosterone on the skin and the gonads of salmonids. *Gen. Comp. Endocrinol., Suppl.* **3**, 741–750.

Yamazaki, F. (1976). Application of hormones in fish culture. *J. Fish. Res. Board Can.* **33**, 948–958.

Yamazaki, F. (1983). Sex control and manipulation in fish. *Aquaculture* **33**, 329–354.

Yoshikawa, H., and Oguri, M. (1977). Sex differentiation in a cichlid, *Tilapia zillii*. *Bull. Jpn. Soc. Sci. Fish.* **44**, 313–318.

Yoshikawa, H., and Oguri, M. (1978). Effects of steroid hormones on the sex differentiation in a cichlid fish, *Tilapia zillii*. *Bull. Jpn. Soc. Sci. Fish.* **44**, 1093–1097.

Yoshikawa, H., and Oguri, M. (1979). Gonadal sex differentiation in the medaka, *Oryzias latipes*, with special regard to the gradient of the differentiation of the testis. *Bull. Jpn. Soc. Sci. Fish.* **45**, 1115–1121.

Yoshikawa H, and Oguri, M. (1981). Ovarian differentiation in the medaka, *Oryzias latipes*, with special reference to the gradient of the differentiation. *Bull. Jpn. Soc. Sci. Fish.* **47**, 43–50.

Zaborski, P. (1982). Expression of H-Y antigen in non-mammalian vertebrates and its relation to sex differentiation. *Proc. Int. Symp. Reprod. Physiol. Fish, 1982* pp. 64–68.

Zaborski, P., and Andrieux, B. (1980). H-Y antigen in sexual organogenesis of amphibians recent studies on its expression in some experimental conditions. *Proc. Int. Embryol. Conf., 14th, 1979* p. 114.

Zaborski, P., Dorizzi, M., and Pieau, C. (1979). Sur l'utilisation de serum anti-H-Y de Souris pour la détermination du sexe génétique chez *Emys orbiculais*, L. (Testudines, Emydiadae). *C. R. Hebd. Seances Acad. Sci., Ser. D* **288**, 351–354.

Zenzes, M. T., Wolf, U., Gunther, E., and Engel, W. (1978). Studies on the function of H-Y antigen: Dissociation and reorganization experiments on rate gonadal tissue. *Cytogenet. Cell Genet.* **20**, 365–372.

6

FISH GAMETE PRESERVATION AND SPERMATOZOAN PHYSIOLOGY

JOACHIM STOSS

Department of Fisheries and Oceans
Fisheries Research Branch
West Vancouver Laboratory
West Vancouver, British Columbia, Canada

I. INTRODUCTION

As fish farming expands and harvesting of wild stocks becomes more intense, there is a growing need for techniques for storage of gametes to facilitate artificial reproduction procedures and to preserve desirable gene

FISH PHYSIOLOGY, VOL. IXB

pools. In this discussion, short-term storage of unfrozen gametes is differentiated from long-term storage which usually infers cryopreservation.

Short-term storage, ranging from hours to weeks, is applied mostly in hatcheries to overcome problems such as asynchrony in maturation, transportation of gametes, or selective use of spawners. Cryopreservation makes possible the almost indefinite storage of desirable genes. This can substantially increase the efficiency of selective breeding of cultured species as is done with domesticated animals. For wild fish, gamete cryopreservation may provide one of the last possibilities to save unique characteristics among stocks. Reports of loss of genetic variability within populations or disappearance of entire stocks (Ryman, 1981; Hynes *et al.*, 1981) argue the need for the creation of gene banks for fish similar to those that presently exist for cultivated plants (Shirkie, 1982) and are under development for endangered species of birds (G. F. Gee, personal communication, 1982).

Cryogenic gene banking of fish gametes is now possible for sperm cells from several species, but not for ova and embryos. This poses some problem because both male and female gametes are required to preserve a population. However, the reestablishment of a lost population based on cryopreserved sperm cells only will after several generations of back-crossing, lead to a new stock with a high similarity to the original one. An alternative method of preserving endangered stocks is to rear a limited number of individuals in captivity. To cope with genetic problems associated with small populations, cryopreserved sperm cells collected from many males in previous generations, can be used to increase the effective size of the breeding population (Gjedrem, 1981).

Aspects of fish gamete preservation have been extensively reviewed (Blaxter, 1969; Shehadeh, 1975; Horton and Ott, 1976; Pullin and Kuo, 1981; Billard, 1980c; Sundararaj, 1981; Stoss and Donaldson, 1982). Tropical species have received special consideration from Harvey and Hoar (1979) and Withler (1981) and salmonids from Scott and Baynes (1980). In this chapter, all available information is reviewed. Teleosts of importance for aquaculture, particularly salmonids, dominate the literature. For this reason, a separate treatment of salmonids is presented in some sections.

Compared to egg cells, sperm cells are preferred for preservation. The reasons are the large number of available spermatozoa, the ease and repeatability of collection, and the suitability for cryopreservation. Because information about fish spermatozoa morphology, metabolism, and motility is rather scarce and scattered in the literature, these subjects are given particular attention herein. Further, general aspects of cryobiology are discussed. The inclusion of these topics may facilitate critical review of previous work and indicate new preservation procedures for species where adequate information is now lacking.

II. MORPHOLOGY OF SPERMATOZOA

Teleostean and, possibly, holostean spermatozoa lack, in contrast to other groups of fishes, a head cap, the acrosome (Afzelius, 1978). Because the absence of an acrosome coincides with the presence of a micropyle in the egg, a head-cap structure may not be necessary to enter the egg during the process of fertilization (Ginsburg, 1972).

The hypothesis that sperm cells of species employing external fertilization have a simple structure, in contrast to more developed structures associated with internal fertilization, holds true for teleostean spermatozoa (see Franzen, 1970). The sperm cell of the common carp, *Cyprinus carpio*, as described by Billard (1970), serves as an example of the primitive type, which is released during spawning into an aqueous environment. The head is ovoid with a diameter of 2.5 μm, and attached to its proximal side is a low, collarlike midpiece formed by an extrusion of the plasmalemma and an indentation of the nucleus. The midpiece contains a few mitochondria and the centrioles. The typical 9 + 2 arrangement of nine pairs of peripheral microtubules and one pair of central tubules is found in the flagellum. An undulated membrane envelops the entire cell. In other species, the shape of the head may differ and fusion of the mitochondria is common (Billard, 1970; Mattei and Mattei, 1975; Jaspers *et al.*, 1976). The plasma membrane often forms one or two finlike ridges along the tail, which are on a horizontal axis with the central microtubules (Nicander, 1970; Billard, 1970; Stein, 1981). This modification of the flagellum is believed to improve the efficiency of flagellar propulsion, although this suggestion has been questioned recently (Afzelius, 1978). In contrast to the primitive cell type, spermatozoa from fish employing internal fertilization have both an elongated head and a midpiece. Further, the mid-piece contains extensive mitochondrial structures and intercentriolar material. This general form has been demonstrated in members of the families *Poeciliidae, Jenysiidae, Pantodontidae,* and *Embiotocidae* (Billard, 1970; Dadone and Narbaitz, 1967; Stanley, 1969; Van Deurs, 1975; Gardiner, 1978a).

Peculiarities in the tail have been reported for some species. Remarkable is the absence of flagella in spermatozoa from two families belonging to the *Mormyriformes* (Mattei *et al.*, 1972). Biflagellated cells were found in *Porichtys notatus* (Stanley, 1965), and a low frequency of biflagellated spermatozoa was reported for the channel catfish, *Ictalurus punctatus* (Jaspers *et al.*, 1976), and in the guppy, *Poecilia reticulata* (Billard and Féchon, 1969). In cross sections of sperm flagella from *Anguilliformes* and *Elopiformes*, no central microtubules were found presenting a 9 + 0 pattern (Billard and Ginsburg, 1973; Mattei and Mattei, 1975). Structures, which connect the peripheral and central tubules in the 9 + 2 system, were also missing, and only one instead of two dynein arms (see also Section IV) could be found at each peripheral microtubule doublet. Nevertheless, these cells are motile (Baccetti *et al.*, 1979).

III. METABOLISM BY SPERMATOZOA

The energy for motility and basic cell metabolism is derived from the breakdown of exogenous or endogenous nutrients in the presence or absence of oxygen. In the case of external fertilization, fish spermatozoa are shed into an aqueous medium that lacks metabolic substrates. These cells depend entirely on cellular reserves, such as phospholipids, as determined in the alewife *Alosa pseudoharengus*, and the rainbow trout, *Salmo gairdneri* (Minnassian and Terner, 1966). Glycolipids were identified in several salmonids (Levine *et al.*, 1976) and glycogen in brown trout, *Salmo trutta* (Baccetti *et al.*, 1975). Significant amounts of glycogen are present in the midpiece of spermatozoa of fish from the genera *Poecilia* and *Opsanus* (Anderson and Personne, 1970; Billard and Jalabert, 1973; Baccetti *et al.*, 1975). With internal fertilization, endogenous nutrients are probably supplemented by those present in fluids of the female reproductive tract.

Areobic breakdown of substrate has been demonstrated to be the dominant metabolic pathway for spermatozoa of rainbow trout, sucker, *Catostomus commersonnii*, sunfish, *Lepomis* sp., Atlantic salmon, *Salmo salar*, and Atlantic cod, *Gadus morhua* (Terner and Korsh, 1963a; Mounib, 1967). Extracellular acetate and glyoxylate were oxidized by Atlantic salmon and Atlantic cod spermatozoa, but did not, in contrast to pyruvate, affect the rate of O_2 uptake (Mounib, 1967). Oxidation of added glucose by rainbow trout and sunfish spermatozoa was low but measurable. Further, in the presence of added substrate, oxidation of intracellular components prevailed (Terner and Korsh, 1963a; Mounib, 1967).

Resynthesis of endogenous lipids, organic acids, proteins, and nucleic acids from added substrate has been demonstrated in sperm cells from rainbow trout, alewife, and Atlantic salmon (Terner and Korsh, 1963b; Minassian and Terner, 1966; Mounib and Eisan, 1968a). The ability to fix carbon dioxide and to incorporate it into organic acids was noted in sperm cells from Atlantic cod (Mounib and Eisan, 1968b).

Under anaerobic conditions, the ability to metabolize extracellular components was reduced in Atlantic cod and Atlantic salmon spermatozoa. Simultaneously, no increase of lactate, resulting from glycolysis of carbohydrates, was observed as was the case under aerobic conditions (Mounib, 1967). The detection of any lactate by Mounib was of significance, because Terner and Korsh (1963a) could not measure lactate production by rainbow trout spermatozoa under anaerobic conditions. Later identification of lactate dehydrogenase (LDH) in spermatozoa from Atlantic salmon (Baccetti *et al.*, 1975) and rainbow trout (Billard and Breton, 1976) further indicated that glycolysis takes place. Its importance appears to be limited; unlike mammalian spermatozoa, Atlantic salmon sperm cells do not use energy provided by extracellular glucose to restore electrolyte balances in *in vitro* stored sperm (Hwang and Idler, 1969). However, this finding was obtained by adding glucose in an aqueous solution, thereby reducing the osmotic pressure of the semen. In two species with internal fertilization (*Cymatogaster aggregata* and *Poecilia reticulata*), glycolysis reached a level comparable to that observed in mammalian spermatozoa, and the sperm cells utilized extracellular glucose extensively (Gardiner, 1978b).

Data on O_2 uptake by fish spermatozoa is limited. *Lepomis* and *Catostomus commersonnii* use $110-140$ μl O_2 per 10^{10} cells per hr. Corresponding values for rainbow trout, Atlantic salmon, and Atlantic cod spermatozoa were $20-40$ μl O_2 per 10^{10} cells per hr (Terner, 1962; Terner and Korsh, 1963a; Mounib, 1967). Comparison of the results is inappropriate because of possible differences in cell size among species, relatively high incubation temperatures for salmonid spermatozoa (25°C) very different motility characteristics, and the possibility that oxygen uptake was measured in motile sunfish, sucker, or cod spermatozoa but measured in immotile salmonid cells (see also Section IV).

In summary, the well-developed mitochondrial sheath in the midpiece of spermatozoa from fish employing internal fertilization indicates the need for extensive metabolic activity, including glycolysis. Some of these spermatozoa are capable of living up to several months in the female reproductive tract (Jalabert and Billard, 1969). In contrast, sperm cells from many fish that reproduce by external fertilization are more primitive in structure and have a comparatively short life span after being released into an aqueous environment. Provided there is sufficient dissolved oxygen in the water, aerobic metabolism of intracellular substrates may dominate.

IV. MOTILITY OF SPERMATOZOA

In preservation of spermatozoa, motility is a useful parameter for assessing the viability of sperm cells. Observation is easy and can be done under the microscope using a hemocytometer (Holtz *et al.*, 1977). In general, motility and fertility are well correlated. However, because different parts of the cell are responsible for motility and fertility, there are examples where motile, γ-irradiated (Lasher and Rugh, 1962; J. Stoss, unpublished), or cryopreserved (Kossmann, 1973; Mounib *et al.*, 1968; Stein and Bayrle, 1978) cells were not fertile.

Before discussing details of fish spermatozoan motility, the mechanism of flagellar movement is briefly described. The system of microtubules in the flagellum represents the motile apparatus of a sperm cell. One of each of the nine peripheral double tubules (the A tubule) carries two arms which consist of an ATPase called dynein. Under hydrolysis of ATP, these dynein arms interact with the protein tubulin of the B tubule from the adjacent double set, causing a slide between them. Because of the presence of interconnecting elements between all tubules, continuous sliding between some of them creates tension and results in oscillation of the flexible flagellum (see Satir, 1974; Stebbings and Hyams, 1979). Recent studies demonstrated that the axoneme of demembranated rainbow trout or chum salmon flagella had to be exposed to both cyclic AMP (cAMP) and ATP-Mg^{2+} in order to become functionally motile (Morisawa and Okuno, 1982; Morisawa et al., 1982). The environmental factors which activate this cAMP–ATP-Mg^{2+} system and initiate motility vary among fishes and are discussed in more detail in later sections.

A. Induction of Motility

Fish spermatozoa are immotile in the testis, and, in many species, in the seminal plasma. During natural reproduction, motility is induced after the discharge of sperm into the aqueous environment or the female genital tract. The particular factors that suppress motility are, as a general rule, neutralized by the environmental conditions during spawning. Therefore, environmental factors, such as ions, pH, or osmolality, may depolarize the cell membrane, stimulating motility.

1. CATIONS

Potassium in the seminal plasma prevents motility in salmonid sperm. This was suggested by Scheuring (1925) who was unable to induce motility in *Salmo gairdneri* spermatozoa after dilution with potassium solutions. Schlenk and Kahmann (1938) directly related potassium in the seminal fluid to immotility of spermatozoa. Dilution of potassium or its removal by dialysis from the seminal plasma (Benau and Terner, 1980) induces motility. As shown by Kusa (1950) in chum salmon, *Oncorhynchus keta*, sperm, potassium inhibits both motility and fertility. The action of potassium is still apparent in 1 mM solutions, which fail to initiate motility in *Salmo gairdneri* and *Salmo trutta* sperm (Scheuring, 1925; Schlenk and Kahmann, 1938; Stoss et al., 1977; Baynes et al., 1981). Because motility is not inhibited in physiological solutions, such as ovarian fluid, which contain more than 1 mmole of K^+ (Hwang and Idler, 1969; Stoss et al., 1977), interactions with other components must occur.

Scheuring (1925) first reported that Na^+, Ca^{2+}, and Mg^{2+} reduced the inhibitory action of K^+, the bivalent cations being more effective than Na^+. Seminal plasma to which NaCl is added also induces motility. Schlenk and Kahmann (1938) observed motility with combined Na^+-K^+ solutions as long as the Na^+-to-K^+ ratio was 16:1 or greater. This observation was basically confirmed by other researchers (Scheuring, 1925; Stoss *et al.*, 1977; Baynes *et al.*, 1981) who estimated somewhat lower relationships. Baynes *et al.* (1981) demonstrated that solutions of K^+ or K^+ and Na^+ which did not induce motility, completely activate sperm cells when small amounts of Ca^{2+} are added. A similar, but slightly reduced, effect has been noted with Mg^{2+}.

Occasionally, substances in the seminal plasma, termed androgamones, are believed to control spermatozoan motility in salmonids (Runnström *et al.*, 1944; Hartmann *et al.*, 1947). However, these androgamones are nothing else than K^+ (Baynes *et al.*, 1981). The effect of K^+ on the motility of spermatozoa in other teleostei is less clear, but it is reported not to inhibit motility in some species (Terner and Korsh, 1963a; Morisawa and Suzuki, 1980; Chao, 1982).

2. pH

The pH of an activating solution also affects motility. Buffered solutions (including ovarian fluid) did not induce motility in rainbow trout spermatozoa when the pH was adjusted to values below 7.8 (Baynes *et al.*, 1981; Schlenk, 1933). Alkaline conditions similar to or greater than those of seminal plasma or ovarian fluid (see Scott and Baynes, 1980) apparently enhance the motility and fertility of salmonid spermatozoa (Petit *et al.*, 1973; Billard *et al.*, 1974; Billard, 1981). Although Gaschott (1928), Inaba *et al.* (1958), and Pautard (1962) reported the effectiveness of an alkaline pH, they did not find complete inhibition of motility following dilution of rainbow trout sperm with buffered electrolyte solutions in a range of acid pH values. Unbuffered solutions or water of acidic pH do not inhibit motility in salmonid spermatozoa (J. Stoss, unpublished observations). Apparently the effect of a low pH on salmonid spermatozoa requires further clarification.

A few observations are reported from other fishes. Spermatozoa from the mullet, *Mugil capito*, were active in buffered seawater between pH 5.5 and 10.0, with a distinct optimum at pH 7 (Hines and Yashouv, 1971). Long-jawed goby (*Gillichthys mirabilis*) and sea bass (*Dicentrarchus labrax*) spermatozoa were motile in diluted and buffered seawater between pH 5 and 10, with the optimum for sea bass being around pH 9 (Weisel, 1948; Billard, 1980c).

3. OSMOLALITY

The induction of motility by a change in the osmotic pressure of the suspending medium has been reported repeatedly. According to Morisawa and Suzuki (1980), hypotonic suspension media initiates motility in spermatozoa from freshwater fishes such as *Cyprinus carpio*, *Carassius auratus*, and *Tribolodon hakonensis*. They used either NaCl, KCl, or mannitol solutions. Earlier observations by Suzuki (1959) or those summarized by Ginsburg (1972) confirm these observations, but there are also examples where isotonic media effectively activate motility in freshwater spawners. Pike, *Esox lucius*, spermatozoa are motile in an isotonic NaCl solution (250 mOsm, pH 9.0; Billard and Breton, 1976), and channel catfish, *Ictalurus punctatus*, spermatozoa are motile in a 0.65% saline (Guest *et al.*, 1976). Consequently, hypotonicity is not the only factor explaining motility induction in freshwater spawners.

Hypertonicity induces the motility of spermatozoa in marine teleosts, as demonstrated in cod, *Gadus morhua macrocephalus*, flounders, *Limanda yokohamae* and *Kareius bicoloratus* (Morisawa and Suzuki, 1980), sea bass, *Dicentrarchus labrax*, sea bream, *Sparus auratus* (Billard, 1978a), grey mullet, *Mugil cephalus* (Chao *et al.*, 1975), and the goby, *Gillichthys mirabilis* (Weisel, 1948). In the latter species, hypertonic sugar solutions also were effective.

In salmonid fishes, motility can be initiated by hypotonic, isotonic, and, to a certain degree, hypertonic media (Gaschott, 1928; Ellis and Jones, 1939; Pautard, 1962; Stoss *et al.*, 1977; Morisawa and Suzuki, 1980). Solutions of NaCl or mannitol below approximately 400 mOsm initiate motility of sperm cells in *Salmo gairdneri* and *Oncorhynchus keta* (Stoss *et al.*, 1977; Morisawa and Suzuki, 1980). According to Billard (1980a), sucrose solutions between 200–300 mOsm (pH 9) do not activate rainbow trout spermatozoa; however, Van der Horst *et al.* (1980) observed activation at 300 mOsm when the sperm dilution rate was high.

Mechanisms for the induction of motility in the viviparous fishes are unclear. Mosquito fish, *Gambiusa affinis* and *Poecilia reticulata*, spermatozoa, which are released in aggregates (spermatozeugma), become motile after exposure to NaCl and KCl solutions between 50 and 300 mOsm (Morisawa and Suzuki, 1980). Breakdown of the spermatozeugma follows, resulting in release of the sperm cells. Because mannitol was ineffective, Morisawa and Suzuki concluded that rather than osmolality, a change of ionic concentrations in the environment, particularly in K^+, assists in the initiation of the motility of spermatozoa. A rapid breakdown of spermatozeugma occurs also in Ringer's solution, but no information has been provided on motility (Ginsburg, 1972). However, Billard (1978a) reported that spermatozoa of *Poecilia reticulata* achieve motility spontaneously after dissociation of the spermatozeugma, without any dilution. Subsequently, the cells stay motile for an extended period (see Table I). Also in killifish *Fundulus heteroclitus* and the cichlid *Oreochromis mossambicus* sperm dilution is not required. Spermatozoa become motile during stripping or exposure to air. Very careful collection and preventing agitation of samples maintains immotility in *O. mossambicus* sperm cells (Kuchnow and Foster, 1976; B. Harvery, 1983, personal communication).

It is obvious that environmental conditions during spawning activate the motility of spermatozoa. However, not all the factors involved are well understood.

B. Duration and Prolongation of Motility

The duration of motility in the natural environment varies greatly between species of fish and coincides in general with the fertile period of spermatozoa. A detailed listing of a number of salt, brackish, and freshwater spawners has been given by Ginsburg (1972), and some selected data are included in Table I. Scott and Baynes (1980) summarized data for salmonid fishes. The chemical characteristics of the motility inducing medium essentially determine the duration of spermatozoan motility. Temperature alters the motile period, i.e., low temperatures result in prolonged duration of locomotion at a reduced speed of cells (Schlenk and Kahmann, 1937; Lindroth, 1947; Turdakov, 1971; Hines and Yashouv, 1971).

In fresh water, spermatozoa show an immediate outburst of motility upon dilution that ceases after 15 sec in *Salmo gairdneri* and may last for 2–3 min in *Esox lucius* (Lindroth, 1947; Billard and Breton, 1976). Swelling and lysis of the cells in the hypotonic water limits the duration of motility (Huxley, 1930; Schlenk and Kahmann, 1938; Billard, 1978a). Motility in salt water may last from 2 min as in *Sparus auratus* (Billard, 1978a) to 2 days as in the herring *Clupea harengus pallasi* (Yanagimachi, 1957). In the viviparous fish *Poecilia reticulata*, motility in the seminal fluid was observed for 48 hr (Billard, 1978a).

Neither fresh water nor full strength sea water have ever been shown to be particularly suited for maintaining spermatozoan motility. Only diluted sea water has been used with some success (Ellis and Jones, 1939; Hines and Yashouv, 1971; Billard, 1978a). Artificial media, which induce good activation of spermatozoa without exposing them to extreme osmotic conditions, prolong spermatozoan motility and the period of fertility. These media are of interest for the development of incubation or dilution media for short-term and cryopreservation storage of spermatozoa. Data on the duration of motility and fertility of fish spermatozoa in such media are summarized in Table I. Trout spermatozoa (*S. trutta, S. gairdneri*) are motile between 1 to 5 min in various isotonic media (Ginsburg, 1963; Scheuring, 1925; Baynes *et al.*, 1981) and highly fertile for 8 min when suspended in a buffered NaCl

Table I

Duration of Motility and Fertility of Fish Spermatozoa in Motility-Inducing Media

Species	Medium	Temperature (°C)	Dilution ratio[a] (semen:medium)	Duration of motility[b]	Fertility[c] (%)	Insemination at minutes after dilution with medium	Reference
Salmo trutta	Ovarian fluid	3.0–9.2	1/30	60 sec	78.8	5	Ginsburg (1963)
	Fish Ringer's: 112 mM NaCl 3.4 mM KCl, 2.4 mM NaHCO$_3$, 2.7 mM CaCl$_2$			62 sec	2.0	5	
S. gairdneri	0.1 M NaHCO$_3$	6–11	1‰	115–165	—	—	Scheuring (1925)
S. trutta	0.1 M NaCl, 0.1 M CaCl$_2$	6–11	1‰	>300 sec	—	—	
S. gairdneri	0.14 M NaCl, 0.02 M KCl, 2.5 mM Ca^{2+} (pH 9.0)	15	1‰	>200 sec	—	—	Baynes et al. (1981)
S. gairdneri	NaCl (250 mOsm) 0.05 M glycine, 0.02 M Tris (pH 9.0)	10	1‰	—	100	8	Billard (1977)
S. gairdneri	0.12 M NaCl (1 mM phosphate buffer, pH 7.4), 5 mM 3-isobutyl-1 methylxanthine	11	1/2	90 sec	—	—	Benau and Terner (1980)
Oncorhynchus gorbuscha	0.12 M NaHCO$_3$, 1-5 mM 3-isobutyl-1-methylxan-thine	15	1‰	>10 min	—	—	Stoss et al. (1983)

Species	Medium	Temperature (°C)	Dilution	Duration of motility	Survival	Fertility[b,c]	Reference
Esox lucius	NaCl (255 mOsm, pH 9.4)	10	—	—	106	3	Billard and Breton (1976)
Esox lucius	NaCl (250 mOsm, pH 9.0)	10	1‰[a]	3–5 min	29	8	Billard (1978a)
Catostomus commersonnii, Lepomis (sp.)	0.12 *M* KCl	4	—	1440 min	—	—	Terner and Korsh (1963a)
Stizostedion vitreum	Ovarian fluid	—	1‰[a]	45 min	—	—	
Fundulus heteroclitus	Seminal plasma	2	—	240 min	—	—	Kuchnow and Foster (1976)
Dicentrarchus labrax	334 mM NaCl, 83 mM glycine, 1.05 mM MgSO$_4$, 1.7 mM CaCl$_2$, 20 mM Tris, HCl to pH 8.5	20	1/100	6 min	—	—	Billard (1978a)
Sparus auratus	Ringer's	20	1/100	6 min	—	—	
Poecilia reticulata	Seminal plasma	20	1/100	60 min	—	—	
Poecilia reticulata		4	—	2880 min	—	—	
Poecilia reticulata	207 mM NaCl, 5.4 mM KCl, 1.3 mM CaCl$_2$, 0.49 mM MgCl$_2$, 0.41 mM MgSO$_4$, 10 mM Tris, addition of glucose (1 mg/1 ml)	18	1/50	60 min	—	—	Gardiner (1978b)
		18	1/50	1140 min	—	—	

[a] 1‰ indicates that semen was diluted in excess with medium without exact control of dilution rate.

[b] Active locomotion.

[c] Fertility is expressed in % of control values if given.

solution (Billard, 1977). The phosphodiesterase inhibitor 3-isobutyl-1-methylxanthine (IBMX) extended *Salmo gairdneri* spermatozoan motility to 90 sec as compared to 30 sec in a saline-lacking IBMX (Benau and Terner, 1980). Using the same substance (1–5 mmole) in 1% NaHCO$_3$, more than 10 min of intensive motility in pink salmon, *Oncorhynchus gorbuscha* sperm (Stoss *et al.*, in preparation) was observed. IBMX inhibits the degradation of cAMP to AMP and causes a distinct increase in cAMP levels in the sperm of *Salmo gairdneri* (Benau and Terner, 1980) as it does in mammalian spermatozoa (Hoskins *et al.*, 1975). In contrast, the addition of cAMP or ATP to sperm of rainbow trout (Billard, 1980b) or to sperm from several other oviparous teleosts (Pautard, 1962) had little or no effect on spermatozoan motility.

In the viviparous fish *Poecilia reticulata* and *Cymatogaster aggregata*, motility is enhanced by the presence of reducable exogenous sugar (Gardiner, 1978b).

Motility enhancing substances can be of natural origin and represent an integral part of the fertilization process. The eggs from the herring (*Clupea harengus pallasi*), several bitterlings (*Acheilognathus lanceolata, A. tabira, Rhodeus ocellatus*), and the fat minnow (*Sarcocheilichthys variegatus*) (Yanagimachi, 1957; Suzuki, 1958) are well-known examples. In the proximity to an egg or upon physical contact with the microplyar region, the swimming speed of the spermatozoa of these species increases, and they are attracted to the micropyle where they aggregate for a few minutes. In the herring, the necessity for such a stimulation seems to be related to the typical mass spawning where gametes from both sexes are deposited rather independently. To ensure maximum fertilization under these conditions, ova and sperm cells have to stay fertile for some time. Indeed, herring spermatozoa are fertile in seawater for several days (compare Section VI, B), probably due to their sluggish motility. This may extend the life-span by preserving energy but requires stimulation to actively fertilize.

Motility of salmonid spermatozoa is enhanced by ovarian fluid, which is released with the eggs. The duration is at least doubled compared to that in fresh water, and the period of fertility is prolonged (Ginsburg, 1963; Billard, 1977; see Table I). This effect has been attributed to motility enhancing factors in the ovarian fluid such as astaxanthine, beta carotene (Hartmann *et al.*, 1947), or an unspecified substance of low molecular weight (Yoshida and Nomura, 1972). Carotenoid pigments such as astaxanthine or a synthetic cantaxanthine had, according to a more recent study (Quantz, 1980), no effect on spermatozoan motility. To explain the good motility of salmonid spermatozoa in ovarian fluid, its isotonicity, combination of ions, and alkaline pH are sufficient reasons (Hwang and Idler, 1969; Holtz, *et al.*, 1977; Baynes *et al.*, 1981).

The secretory product of the seminal vesicles present in some teleosts (Hoar, 1969) does not affect spermatozoan motility (Weisel, 1948). Its absence, however, reduces the fertilizing ability of spermatozoa in the catfish *Heteropneustes fossilis* (Sundararaj and Nayyar, 1969).

C. Reactivation of Motility

Once activated, salmonid spermatozoa lose their fertilizing capability very quickly, owing to the brief duration of motility. If activated in a physiological solution, however, motility and fertility can be re-initiated. This was first demonstrated by Schlenk and Kahmann (1938) who immersed activated trout spermatozoa in a K^+-rich solution. After 37 min, transfering the spermatozoa into any solution with a lower K^+ concentration (also seminal plasma) induced motility. A similar finding was made by Nomura (1964) for *Salmo gairdneri* spermatozoa. Upon dilution of milt in ovarian fluid, good reactivation with water was possible approximately 1–18 hr later. Kusa (1950) reported that *Oncorhynchus keta* milt lost its fertility in Ringer's solution after 90 min, but had regained it after 24 h. Diluting *Salmo trutta* milt in Ringer's, Ginsburg (1963) noted a drop of fertility to almost zero after 20 min, but a complete restoration after 90 and 120 min. The same tendency, but at a lower level of fertility, was also found in ovarian fluid (compare also Fig. 3 of Billard and Jalabert, 1974). Dilution of milt from *Oncorhynchus keta* in Ringer's and redilution with the same medium when inseminating eggs at consecutive intervals, resulted in a gradual decrease in fertility to 10% after 120 min (Yamamoto, 1976). Without redilution, fertility was almost nil after 2 min, but showed a slight restoration between 60 and 80 min. Reactivation has also been induced in spermatozoa from *Ictalurus punctatus* (Guest *et al.*, 1976). Carp spermatozoa, having once been activated, regained motility spontaneously after a period of rest (Sneed and Clemens, 1956).

All these examples demonstrate that spermatozoa can become reactivated after a certain period of rest. Since previously activated spermatozoa remain metabolically active, some time is required to restore initial levels of available energy. This resting period is not necessary when phosphodiesterase inhibitors such as IBMX or theophylline are being used. IBMX induced immediate reactivation when added at 5 mM in a 0.12 M NaCl solution (pH 7.4) to trout spermatozoa only a few minutes after motility had ceased (Benau and Terner, 1980). Similarly reactivated brook trout, *Salvelinus fontinalis* spermatozoa, were capable of fertilization. Benau and Terner concluded from these results that once the motility-suppressing K^+ had been removed during the first activation, other factors such as ATP or cAMP were responsible for the initiation and maintenance of motility. This observation was confirmed by Morisawa and Okuno (1982). Whereas Benau and Terner (1980) used saline only for the first activation, Billard (1980b) maintained fertility most effectively when first diluting *Salmo gairdneri* spermatozoa with 0.01 *M* theophylline and saline (250 mOsm, pH 9.0) and subsequent redilution 30 or 60 min later with saline only. For redilution, theophylline was not necessary.

Some cases have been reported where salmonid spermatozoa maintained motility and fertility for extended periods, i.e., hours or days, when sperm were immersed in saline, ovarian fluid, or diluted sea water (Ellis and Jones, 1939; Nomura, 1964; Terner and Korsh, 1963a). Reactivation may have taken place in these cases, leading to the interpretation that motility had never ceased. Since the immersion medium can also induce reactivation (Yamamoto, 1976), it may have done so when aliquots of previously diluted milt were transferred onto a slide for microscopic examination. This possibility is further supported by the observation that when these samples are examined microscopically, motility ceases very quickly (Terner and Korsh, 1963a).

V. GAMETE QUALITY

Gametes from different fish of the same species may show very different suitability for preservation (Ott and Horton, 1971a; Billard *et al.*, 1977; Stoss and Holtz, 1983a). The aging of spermatozoa *in vivo* has been identified as a cause for reduced keeping quality. This observation applies to species in which spermatogenetic processes precede the spawning season and spermatozoa are being stored during the spermiation period in the testis (see Billard and Breton, 1978). Thus a continuous reduction in the duration of motility was noted in sperm cells from sea bass *Dicentrarchus labrax* when milt was collected at the beginning, middle, or end of the spawning season. When these samples were being stored at 4°C, motility could be induced after 70 hr in milt collected early in the spawning season. Cells collected 2 months later lost the ability to become activated after only 9 hr of storage (Billard *et al.*, 1977).

Similarly, Legendre and Billard (1980) reported that, after cryopreservation, rainbow trout sperm demonstrated reduced fertility with progressing spermiation stages. However, this reduction was not confirmed in a study with coho salmon (*Oncorhynchus kisutch*). Repeated sperm sampling from the same group of males or sampling from different males at the same spermiation stage from beginning to end of spawning, resulted in high post-thaw fertility throughout (Stoss *et al.*, 1984). Because the spermiation period in the coho salmon was limited to 3 weeks by the natural death of these fish, the situation in the trout, a repeat spawner, may be very different. Further, the composition of fatty acids in rainbow trout spermatozoa has been related to postthaw survival (Baynes and Scott, 1982).

For preservation purposes, only the elimination of aged spermatozoa may presently provide a useful quality criterion. Other sperm characteristics, such as spermatozoan motility, cell density, and the concentration of organic and inorganic components, have not been related to the suitability of a particular sample for storage. Interactions between sperm quality and cryopreservation techniques should also be considered. Therefore, the quality effects may be less pronounced when optimized techniques are employed.

Little is known regarding the quality of ova as related to preservation. Aging, usually referred to as overripening, is documented in various species (Nomura *et al.*, 1974; Escaffre *et al.*, 1977; De Montalembert *et al.*, 1978). The observation that eggs from certain females are readily fertilizable with fresh spermatozoa, but very poorly fertilized with short-term stored or cryopreserved cells, requires interpretation (Billard, 1981; Stoss and Holtz, 1981b; Harvey and Stoss, in preparation.).

VI. SHORT-TERM PRESERVATION OF SPERMATOZOA

Short-term storage of sperm or eggs is beneficial in situations when male and female gametes are being collected at different times or locations, or when the collection site and incubation facility are some distance apart and delayed fertilization is necessary. Preservation techniques are designed to reduce the metabolic activity of the cells in order to extend their life span. Most fish spermatozoa have the advantage of being quiescent in the seminal plasma; therefore, no energy is consumed for motility. This characteristic makes them most suitable for short-term storage, and, not surprisingly, successful storage for a few hours or even several days was reported as early as the 1800's. Reviews of the earlier literature can be found in Scheuring (1925), Barrett (1951), Blaxter (1955), and Ginsburg (1972).

A. Storage of Undiluted Sperm

1. SALMONIDS

The main factors which influence storage are (1) temperature, (2) gaseous exchange, (3) sterile conditions, and (4) prevention of desiccation.

A reduction of the storage temperature to levels just above freezing prolongs the storage capability of salmonid sperm (Withler and Morley, 1968; Hiroi, 1978; Stoss *et al.*, 1978). Although a common phenomenon in mammals, a detrimental effect of low temperatures, known as thermal shock has not been reported. However, in *Salmo gairdneri*, fertility is reduced at a fertilization temperature of 0°C when a low density of spermatozoa is used. Raising the temperature to 5°C or increasing the density of cells overcomes this effect (Billard and Gillet, 1975).

The need for gaseous exchange must be considered during storage. Unlike mammalian spermatozoa where inert gases such as CO_2 reduce the metabolic activity, storage of *Salmo gairdneri* or *Salvelinus fontinalis* milt under a CO_2 atmosphere kills the cells (Scheuring, 1925; Henderson and Dewar, 1959; Büyükhatipolgu and Holtz, 1978). Further, N_2 or a mixture of N_2, H_2, and CO_2 reduce the storage capability (Büyükhatipoglu and Holtz, 1978). This inability to survive under anaerobic conditions agrees with the finding of poor glycolytic activity in these cells.

Air and, preferably, pure oxygen are most suitable for maintaining cell viability. This fact was first demonstrated by Scheuring (1925) who stored trout milt under air and oxygen at 0°C. He was able to induce motility in sperm samples stored 4 days under an O_2 atmosphere, but this was not possible in corresponding air-stored samples after 24 hr. Later, Truscott *et al.* (1968) demonstrated the importance of sufficient gas exchange in stored *Salmo salar* milt. When kept in vials at 2–3°C under air, full fertility was maintained for at least 5 days; however, samples kept in sealed vials showed reduced fertility after 1 day. By providing air, Withler and Morley (1968) successfully kept sperm from *Oncorhynchus gorbuscha* and *Oncorhynchus nerka* for 4 days (3°C) before noticing a reduction in fertility. Because desiccation is a problem during prolonged storage, Büyükhatipoglu and Holtz (1978) kept *Salmo gairdneri* milt under moisture-saturated O_2 or air. Motility was retained on average for 12 and 8 days, respectively, and the fertilizing capacity after 15 days of storage under oxygen was 81% (control 98%). In subsequent studies using 4° and -2°C storage temperatures, full fertilizing capacity was maintained for at least 23 days when oxygen was used and 17 days when air was used (Stoss *et al.*, 1978). Full fertility was retained in further experiments at 0°C under O_2 for 34 days (Stoss and Holtz, 1983b). The superiority of oxygen over air was also confirmed by Billard (1981).

Because oxygen enters samples by diffusion, its availability within a sample decreases with increasing distance from the surface. For this reason, fertility is best maintained in samples approximately 6-mm deep. Increasing the sample depth reduces storage ability drastically (J. Stoss and W. Holtz, unpublished). This observation may partly explain the enormous variability in the reported storage capability of salmonid sperm lasting from hours to weeks (compare also Barrett, 1951; Nomura, 1964; Plosila and Keller, 1974; Carpentier and Billard, 1978).

Collection of milt under sterile conditions is difficult, and the occurrence of bacterial growth often limits storage to a few days (Withler and Morley, 1968; Hiroi et al., 1973; Hiroi, 1978). A combination of 9000 IU penicillin and 9000 μg streptomycin does not affect motility of rainbow trout spermatozoa, and a much lower concentration (125 IU/125 μg per ml of sperm) is sufficient during storage (Stoss and Refstie, 1983). Induction of motility is prevented by dissolving antibiotics in seminal plasma and subsequently adding a quantity equal to 5% of the sample volume. Instead of seminal plasma, nonactivating sperm diluents (see further discussion) may be used.

Under field conditions, milt may be kept in oxygenated plastic bags. When stored at a liquid-to-gas ratio (v/v) between 1:50 and 1:120 (0°C), antibiotic-protected Atlantic salmon and rainbow trout milt maintain their fertility for 10 and 20 days, respectively (Stoss and Refstie, 1983; Stoss and Holtz, 1983b). A similar technique also was successfully applied by Billard (1981).

In a recent study (D. F. Alderdice and J. O. T. Jensen, personal communication), the effects of temperature and storage time were examined in *Oncorhynchus keta* gametes. Sperm were kept in polyethylene bags under air at a liquid-to-gas ratio (v/v) exceeding 1:30. Initial sperm fertility was reduced by 10% and 50% after 147 hr and 192 hr, respectively, at 3°C, the lowest tested storage temperature. At 15°C, 90% survival was observed after 23 hr of storage and 50% after 41 hr. When sperm fertility was tested with stored instead of fresh eggs, the decrease in sperm fertility occurred earlier.

2. OTHER FISHES

Dry storage of sperm has been practiced with success in a variety of nonsalmonid fish. Again, no cold shock attributable to abrupt cooling or to a low preservation temperature has been reported and the lowest temperature tested has always been the most successful. This was apparent in the herring, *Clupea harengus*, where fertility was about 90% after 2 days of storage at 4°C, however, the corresponding value at 7°C was 7% (Blaxter, 1955). By keeping Pacific herring, *Clupea pallasi*, spermatozoa at 0.8°–1.0°C, high fertility was retained for 3 weeks (Dushkina, 1975). In killifish, *Fundulus heteroclitus*, storage at 2°C was superior to that at 4°C, and fertility was maintained for 4 and 2 hr, respectively. Spermatozoa of *Fundulus* are motile in the seminal plasma, which explains their rather brief period of keeping quality. Storage of the whole testis (2°C) leaves the cells immotile and fertile for 72 hr (Kuchnow and Foster, 1976).

The role of gaseous exchange is not clear. In some cases, anaerobic storage conditions appear favorable, as in the white bass, *Roccus chrysops*. Spermatozoa stored at 3°C in vials or in oxygenated plastic bags were motile following induction after 6 and 8 days, respectively. Storage in syringes that probably provide little airspace allowed motility induction after 38 days of storage (Clemens and Hill, 1969). Milkfish (*Chanos chanos*) sperm was kept in polyethylene syringes at 4°C, and motility and fertility were retained up to 14 and 10 days, respectively (Pullin and Kuo, 1981). Atlantic cod milt kept in covered test tubes at 0°C retained full motility until day 7. Storage was extended by penicillin (6000 IU/ml/week) to 10 days (Mounib *et al.*, 1968). These examples indicate that a large liquid-to-gas interface may not be necessary for these species.

There are several other reports demonstrating that fertility or motility can be retained for a period of hours or a few days. These include the mullets, *Mugil capito* and *Mugil cephalus* (Hines and Yashouv, 1971; Chao *et al.*, 1975), the catfish, *Clarias lazera* (Hogendoorn and Vismans, 1980), common carp (Hulata and Rothbard, 1979), and sea bass, *Dicentrarchus labrax* (Billard *et al.*, 1977). Because storage conditions were not always varied in these reports, improvements may be possible.

B. Storage of Diluted Sperm

The use of diluents for the storage of fish spermatozoa may provide better control of the physiochemical conditions during storage than would be possible in undiluted milt. According to Mann (1964), an ideal diluent is (1) isotonic, (2) has a good buffering capacity, (3) contains nutrients, stabilizing colloids, and antioxidants, (4) is antibacterial, and (5) has, in general, a good keeping quality. For fish sperm it is also a requirement that the diluent does not activate motility of spermatozoa. Diluents that correspond to the ionic composition of seminal plasma, have often been preferred to media, such as fish Ringer's or Cortland's medium, which are based on blood plasma (Randall and Hoar, 1971; Truscott and Idler, 1969; Billard and Jalabert, 1974; Büyükhatipoglu and Holtz, 1978). Fresh water as a sperm diluent, because of reasons mentioned previously, is unsuitable, although sometimes proposed (Poon and Johnson, 1970; Plosila *et al.*, 1972; Plosila and Keller, 1974).

1. SALMONIDS

Only a few attempts with limited success have been reported in storing diluted sperm from salmonid fishes. Henderson and Dewar (1959), who studied brook trout, obtained inferior results using frog Ringer's versus undiluted storage. Truscott and Idler (1969) retained motility in Atlantic salmon spermatozoa for 6 days at 4°C using their diluent Hfx#1 which was based on inorganic components of seminal plasma. Compared to the storage ability of undiluted sperm for several weeks, storage diluents appear to have no advantage. Because motility is not inhibited by isotonic media, but by the balance of ions (which changes during storage), any dilution may affect the capability of spermatozoa to reestablish ionic equilibrium with the extracellular fluid. By contrast, so-called insemination diluents, which are added to sperm prior to the addition to the eggs, may increase the efficiency of artificial insemination, as has been demonstrated in rainbow trout. Duration of sperm fertility in such nonactivating diluents is also limited to a few minutes (Billard, 1975, 1980a).

2. OTHER FISHES

In fishes other than salmonids, diluents have been applied more successfully, often extending storage time beyond that for undiluted milt. Diluted in seawater (7°C), *Clupea harengus* spermatozoa maintained 23% fertility after 24 hr (control = 90%). Buffering seawater to pH 8 and adding egg yolk reduced the decrease to 65% and a low level of fertility in both media was still observed after 5 days (Blaxter, 1955). Yanagimachi (1957) observed high fertility (92%) in seawater-diluted *Clupea pallasi* sperm (6–9°C) after 14 hr, and 62% fertility after 2 days. Dilution to isotonicity appeared advantageous as compared to full-strength seawater. Alderdice and Velsen (1978), working with the same species, found high fertility in diluted seawater (17% salinity, 4°C) for up to 7 hr (85%), and a decrease to 35% after 48 hr. Dushkina (1975) reported that *Clupea pallasi* spermatozoa maintain a high level of fertility in diluted seawater (17–18‰ salinity) for 2 days at 2°C, and 6 days at 0.8°C. The relatively long survival of herring spermatozoa in their natural spawning medium may be related to their sluggish motility, ensuring a slow use of cell substrate. In the goby, *Gillichthys mirabilis*, motility was prolonged in isotonic dilutions of seawater and lasted for 2 weeks at a storage temperature between 2° and 4°C (Weisel, 1948).

Using artificial diluents, Guest *et al.* (1976) stored macerated pieces of ripe channel catfish (*Ictalurus punctatus*) testis at 4°C. After 9 weeks, motility could still be induced after storage in Truscott and Idler's (1969) Hfx#1 solution, in a modified Cortland's solution (Truscott *et al.*, 1968) or in 0.5% NaCl. Motility after storage in the balanced salt solutions was slightly better than in the saline. It is not clear whether these diluents activated spermatozoa motility within the testis. Furthermore, no undiluted milt was stored and no fertility tests were reported.

Catfish, *Clarias lazera*, spermatozoa stored at diluted rates of 10^{-1}–10^{-3} in 0.8% NaCl for 24 hr (5°C) maintained a higher level of fertility than under undiluted storage (Hogendoorn and Vismans, 1980). Further, a dilution of common carp sperm with a 0.3% urea − 0.4% NaCl solution maintained full fertility for at least 45 hr (0–5°C) (Hulata and Rothbard, 1979), and motility was observed after 6 days of storage at 4°C (Kossmann, 1973). Carp sperm stored in frog Ringer's (3°–5°C) retained motility for 30 days (Sneed and Clemens, 1956). White bass, *Roccus chryptus*, spermatozoa kept the capability of becoming motile for 14 days in Ringer's at 3°C (Clemens and Hill, 1969). Milkfish (*Chanos chanos*) blood serum was a superior storage medium as opposed to cow serum, Ringer's, 400 mM glucose, or 150 mM NaCl. In this species, sperm dilution was clearly advantageous to undiluted storage (Hara *et al.*, 1982), contradicting information given by Kuo (1982). *Mugil cephalus* milt diluted (1:1) and stored at 5°C in marine teleost Ringer's showed motile spermatozoa following activation after 23 days (Chao *et al.*, 1975). Finally, Mounib *et al.* (1968) had better success storing Atlantic cod sperm in a diluent (330 mmole NaCl, 83 mmole glycine, 26 mmole NaHCO$_3$) than undiluted. With the addition of penicillin (5000 I.U./ml per week), spermatozoa were fully motile after 20 days and still showed a low level of motility after 25 days. In *Oreochromis mossambicus*, diluents were superior to undiluted storage (B. Harvey, personal communication).

In conclusion, sperm diluents have the potential to prolong storage periods. However, a number of diluents used in the aforementioned studies acted only as isotonic media. Attempts to more closely adapt diluents to the requirements of the cells will likely improve storage success and result in techniques which are reliably applicable by the aquaculturist.

C. Supercooling

An aqueous medium is supercooled when the theoretical freezing point is just passed, but the medium remains unfrozen unless ice seeding is induced. Addition of cryoprotectants reduces the nucleation temperature of both the suspension medium and the cell liquid (compare under section cryopreservation) and allows the storage of cells in a supercooled state at several degrees below 0°C. This has been attempted in fish spermatozoa. Truscott *et al.* (1968) tested the effect of temperatures between −3° and −6.5°C on the storage of Atlantic salmon sperm using either dimethylsulfoxide (DMSO) or ethylene glycol (EG) as cryoprotectants. In a Cortland's solution, which was modified by supplementing potassium and 5% DMSO, fertility was 96 and 81% after storage at −4.5°C for 11 and 28 days, respectively. With EG, 70% fertility was observed after 38 days of storage at −3°C. All samples were flushed with O$_2$. Sanchez-Rodriguez and Billard (1977) followed up these results by also including two of their own diluents which had been developed as dilution and insemination media for trout sperm. Supercooled rainbow trout milt (−4°C) suspended in modified Cortland's with either 5% DMSO or 5% EG retained a high proportion of motile spermatozoa for at least 35 days. Motility in the two other media ceased within 1–2 weeks. Fertility tested after 9 days was retained completely when using modified Cortland's and DMSO, but zero in the case of EG. Also, Zell (1978) supercooled brook trout and Atlantic salmon sperm in two diluents using DMSO or polyvinylpyrrolidone (PVP). Samples which were kept for 2–3 min at either −6° or −8°C, subsequently showed fertility. A temperature of −20°C was not tolerated, indicating that intracellular freezing may have taken place.

In general, the choice of the suspension medium, the concentration and the nature of the cryoprotectant used, and the storage temperature are the most critical variables during supercooled storage. Because Truscott *et al.* (1968) and Sanchez-Rodriguez and Billard (1977) discontinued storage before fertility or motility were extinguished, supercooling may have the potential to store semen successfully for longer periods than those reported to date.

D. Postmortem Storage

The duration of sperm viability when left in the testis after the death of the fish may have some practical importance. In *Clupea hargenus*, Blaxter (1955) found a rapid loss of fertility within 18 hr of storage, but Dushkina (1975) achieved high fertility for over 2 days when dead *Clupea pallasi* were kept at 0.8°C. In chum salmon, *Oncorhynchus keta*, Okada *et al.* (1956) observed good fertility for up to 90 min of storage at 11°–12°C, but a decrease to 10% after 3 hr. Higher temperatures reduced the storage ability drastically. Brook trout, *Salmo trutta*, spermatozoa kept in killed fish, either submerged in water or stored dry (4°C), showed a linear loss in fertility to almost zero within 18 hr (Billard *et al.*, 1981).

All reports demonstrate that the drastic physiological changes taking place postmortem affect the viability of the sperm cells within a short period. This underlines the superiority of *in vitro* storage.

In summary, sperm viability is prolonged by a low storage temperature. The availability of oxygen is essential, particularly in salmonid spermatozoa. However, anaerobic conditions may prove to be superior in cells which perform glycolysis. Storage diluents were applied successfully, but are of no advantage compared to undiluted storage in salmonids. The limited knowledge regarding metabolism in spermatozoa hinders the formulation of proper diluents. Supercooled storage may develop into an interesting alternative. Postmortem storage is most unsuitable.

VII. SHORT-TERM PRESERVATION OF OVA

Mature ovulated ova remain arrested in metaphase II of meiosis until they become activated. Activation removes the development block and is associated with a number of physiological changes (see Ginsburg, 1972; Gilkey, 1981).

Short-term preservation is only feasible with eggs from species in which activation is controllable, e.g., in salmonid eggs which become activated following insemination and exposure to fresh water. In the genera *Carassius*, *Cyprinus*, *Tribolodon*, *Hypomesus*, and *Pecoglossus*, autoactivation occurs in isotonic Ringer's solution soon after collection of ova (Yamamoto, 1961). For this reason, attempts to preserve eggs in ovarian fluid from the common carp *Cyprinus carpio*, Indian carp, *Labeo rohita*, and catfish, *Pangasius sutchi*, for only a few hours were unsuccessful because of parthenogenetic development (Withler, 1980). Parthenogenetic zygotes are usually haploid and, therefore, are not viable. Data reported by Yamamoto (1944, 1961) from medaka *Oryzias latipes* eggs indicate that autoactivation can be reduced when eggs are collected by dissection of the ovary instead of by stripping.

The duration of egg fertility in the natural spawning medium is comparable to that of sperm cells and, in general, is of more limited duration in fresh water than in salt water. In salmonid eggs, fertility decreases sharply in fresh water within 30 sec (Szöllösi and Billard, 1974); however, cod (*Gadus morhua*) eggs remain highly fertile in seawater (34‰) for 15 min (Davenport *et al.*, 1981).

Loss of fertility in fresh water coincides with the sealing of the internal orifice of the micropyle (Szöllösi and Billard, 1974), or the disconnection of the micropyle from the yolk membrane (Suzuki, 1959). Inability to maintain osmoregulation, incomplete activation, or parthenogenetic development are reasons for time-limited fertility in salt water (Davenport *et al.*, 1981; Dushkina, 1975). A tabulation of the durabilities of ova from various fish in their natural spawning environment was provided by Ginsburg (1972).

Salmonid ova can be stored successfully in ovarian fluid. Storage at temperatures between 0° and 4°C has always been superior to that at a higher temperature. Therefore, little change from initial fertility was observed in pink and sockeye salmon after 2 and 3 days of storage (3°C), respectively (Withler and Morley, 1968). Chum salmon eggs, also kept at 3°C, maintained 90% of their initial fertility for approximately 6 days (D. F. Alderdice and J. O. T. Jensen, personal communication), and Ginsburg (1972) observed 70% fertility in *Salmo trutta* eggs after 10 days of storage (0.4°–1.0°C). *Clupea harengus* eggs retained high fertility for 2 days when stored at 4°C (Blaxter, 1955).

Eggs from *Fundulus heteroclitus* showed reduced or no viability when stored at temperatures lower than 6°C. At the optimal temperature range (6°–10°C) fertility remained high for 24 hr. When exposed to low or high temperatures, the proportion of abnormally developed eggs increased with storage time (Kuchnow and Foster, 1976). Parthenogenetic development may have taken place. Sturgeon (*Acipenser güldenstadti*) eggs did not become fertilized at 0°C, indicating the unsuitability of a low storage temperature (Ginsburg, 1972).

In rainbow trout eggs, storage under gases, such as N_2, or a mixture of 95% O_2 and 5% CO_2, was inferior to air or pure oxygen. Storage of more than 4 layers of eggs above each other reduced the durability drastically. This may have been caused by reduced gaseous exchange to eggs in the lower layers, or by the weight pressure exerted by the eggs above. Rainbow trout eggs, to which antibiotics (125 IU penicillin + 125 μg streptomycin per gram eggs) were added, maintained a high level of fertility for 10 days and showed a reduction in fertility to 70% after 20 days of storage (1°C) (H. Pueschel, W. Holtz, and J. Stoss, unpublished data; see also Stoss and Donaldson, 1982). However, in repeated tests, the period of storage was reduced in some cases, indicating differences between eggs from various females. Very likely, the highest quality can be expected when eggs are collected shortly after ovulation.

Dilution media during storage have shown no advantage. Although T. Yamamoto (1939) found that *Oryzias latipes* eggs retained their fertility in an isotonic electrolyte solution (128 mM NaCl, 2.6 mM KCl, 1.8 mM CaCl$_2$, pH 7.3) for a few hours, chum salmon eggs that were kept for 24 hr (10°C) in a similar solution showed incomplete breakdown of cortical alveoli after subsequent activation with fresh water (T. S. Yamamoto, 1976). Yanagimachi (1956), using a medium of greater molarity (207 mM NaCl, 7.3 mM KCl, 2.1 mM CaCl$_2$, 3.3 mM MgCl$_2$, pH 7.6 adjusted with NaHCO$_3$) for *Clupea pallasi* eggs, observed a high level of fertility for only 5 hr (5°–12°C). Using either 222 mM NaCl or KCl, fertility decreased only slightly after 100 hr of storage. Balanced salt solutions which preserve fertility for at least a period of hours have been developed for salmonids (Billard, 1980a; Stoss and Donaldson, 1983).

Because of the role of Ca^{2+} during the process of activation, high levels of Ca^{2+} or Mg^{2+} in a suspension medium may activate eggs under isotonic conditions (Kusa, 1953; Yanagimachi, 1956; Gilkey, 1981).

Supercooling of eggs, as has been done with sperm cells, has not been attempted for prolonged storage. As discussed in Section VIII,D, eggs can withstand temperatures below 0°C in the presence of cryoprotectants, but subsequent survival may be affected. Low permeability of cryoprotectants through the egg membranes may pose problems in removing them after

storage. Unhatched embryos from the capelin, *Mallotus villosus*, are protected under natural conditions and can be supercooled to −11.9°C before
freezing intracellularly (Davenport *et al.*, 1979).

An interesting form of natural egg storage is found in some cyprinidonts.
"Annual fishes" that inhabit water bodies which dry out seasonally survive
the dry season as zygotes. Very unique features such as the formation of an
extraembryonic membrane and the formation of dispersed blastomeres,
which all retain the ability to later develop independently into an embryo,
make these eggs very resistant (Wourms, 1971).

VIII. CRYOPRESERVATION OF GAMETES

In 1949, Polge *et al.*, discovered that fowl spermatozoa retained full
motility after freezing and thawing in the presence of glycerol. This finding
initiated extensive cryobiological research, which also led to the first successful frozen storage of mammalian embryos in 1972 (Whittingham *et al.*,
1972; Wilmut, 1972). Although successful cryopreservation of spermatozoa
from herring (*Clupea harengus*) already had been achieved by Blaxter
(1953), it was not until recently that significant success in the frozen storage
of fish sperm has been reported. The most consistent data now available are
from salmonids, but studies on a variety of other fishes also demonstrate the
feasibility of sperm preservation.

A few unsuccessful attempts have been made to cryopreserve fish ova or
embryos.

A. General Aspects of Cryopreservation

The physical events during the freezing and thawing of cells have been
reviewed by, among others, Mazur (1977) and Farrant (1980). Cells suspended in a medium can be supercooled to temperatures below 0°C.
Eventually ice forms in the extracellular medium, but since the cell is protected by the cell membrane, ice crystals do not grow into the cell and the
cytoplasm remains unfrozen. Because the supercooled water within the cell
has a higher chemical potential than the frozen water outside, the cell dehy-
drates and shrinks. Because the concentration of extracellular solutes increases as the solvent progressively crystalizes, osmotic forces support dehydration. If there is enough time, the cell will lose all of its free water
before the temperature is reached at which intracellular water nucleates.
This may be somewhere between −10°C and −40°C (see Mazur, 1977).
However, if cooling proceeds too fast, remaining intracellular water
eventually freezes.

A two factor hypothesis has been postulated to explain the type of injury the cell may undergo during temperature reduction.

1. During slow freezing, increased concentrations of extracellular solutes expose the cells to osmotic stress. Depending on the concentration as well as duration and temperature of exposure the cell membrane may collapse (see Meryman, 1971b; Meryman *et al.*, 1977).

2. During fast freezing, formation of intracellular ice injures the cell; however, small amounts of ice are not necessarily detrimental (see Leibo *et al.*, 1978).

In the absence of cryprotectants, the two aforementioned effects mostly overlap, allowing no survival. Cryoprotectants open or widen this window by buffering the effect of concentrated solutes and by lowering the nucleation temperature of the intracellular water. The exact mechanisms involved remain unclear, but cryoprotectants are believed to act by altering the chemical and physical properties of the extracellular medium rather than by affecting the cells directly (Mazur, 1977). Therefore, cryoprotectants protect against injury from slow freezing (Mazur, 1977).

The main requirements for a cryoprotectant are good solubility in water and nontoxicity. There are permeating and nonpermeating cryoprotectants. Permeating ones are methanol, DMSO, EG, and glycerol, with the first showing fastest and the last showing slowest permeation. Nonpermeating agents include mono- and polysaccharides, polyvinylpyrrolidone (PVP), hydroxyethyl starch (HES), dextrans, and proteins (Meryman, 1971a). Permeating agents provide better protection at relatively slow cooling rates, and nonpermeating components are more suitable in connection with fast freezing. It has been questioned whether permeating agents provide cryoprotection only after permeation (Mazur and Miller, 1976a,b).

To find the ideal conditions for sufficient cell dehydration to ensure optimal postthaw survival, the investigator has to consider a number of interacting variables. Obviously, cell dehydration during cooling depends directly on the speed of temperature reduction. Optimal cooling rates vary considerably because of differences between various types of cells in relative amounts of intracellular water, cell size, membrane permeability for water, and temperature coefficient. Based on these variables, Mazur (1963, 1977) developed a mathematical model to estimate optimum cooling rates for a given cell, which minimizes the chance of intracellular ice formation. Therefore, small cells with a high water permeability, such as red blood cells, must be frozen very quickly, but large cells with an intermediate water permeability coefficient, such as mammalian eggs, require slow cooling. Optimal cooling rates further depend on type and concentration of the cryoprotectant used (Leibo and Mazur, 1971; Polge, 1980).

When freezing has been done quickly leading to some intracellular ice formation, thawing also must be conducted quickly to prevent recrystallization of the ice to larger crystals at warmer temperatures. The critical range would be approximately −60°C and above.

Once the intracellular nucleation temperature has been passed, the cells can be stored safely in liquid nitrogen at −196°C. The transfer should not be done before the samples have reached approximately −70°C. Storage at −79°C in dry ice may show some limitation, particularly when glycerol is used (see Polge, 1980; Pullin, 1972, 1975). In contrast, storage in liquid nitrogen is only affected by background radiation, limiting storage to somewhere between 200 and 32,000 years (see Whittingham, 1980; Ashwood-Smith, 1980).

B. Techniques

General aspects of freezing techniques were discussed by Farrant and Ashwood Smith (1980). Maurer (1978) reviewed special techniques related to mammalian ova and embryo preservation.

The freezing of spermatozoa is mostly done with either dry ice (−79°C) or liquid nitrogen (−196°C). By keeping ampules or straws with diluted sperm at a certain distance above the surface of liquid nitrogen or by immersing them into the liquid, variable freezing rates can be obtained. Touching the nitrogen surface, for example, with the warm sample holder, causes evaporation, and the samples become uniformly surrounded by vapor. Samples can also be buried in dry ice or the ice may be used to cool an alcohol bath in which the samples are immersed. More sophisticated equipment is commercially available.

The cooling rate depends also on the type of sample container and the sample volume. Small volumes ensure more uniform cooling rates within the sample. The only way to establish a precise cooling rate is by measuring the change in temperature within a reference sample by a thermocouple while a certain procedure is applied.

The pellet technique, developed by Nagase and Niwa (1964) provides a simple method with a good repeatability of freezing rates. Sperm mixed with extender are dropped into small holes which were melted or drilled in solid dry ice; the droplet freezes instantly at a rate of approximately 20°–30°C/min. The pellets obtained can easily be removed for transfer into liquid nitrogen.

Sperm samples are usually thawed in a temperature-controlled water bath or in air. Pellets are preferably thawed in a temperature-controlled thawing solution.

C. Preservation of Spermatozoa

Considerable progress has been reported in the cryopreservation of salmonid spermatozoa (Mounib, 1978; Stein and Bayrle, 1978; Büyükhatipoglu and Holtz, 1978; Legendre and Billard, 1980; Kurokura and Hirano, 1980; Erdahl and and Graham, 1980; Stoss and Holtz, 1981a,b, 1983a; Stoss and Refstie, 1983). These results indicate that species differences within salmonids are minor or nonexistent and do not warrant a separate discussion.

Studies involving tropical fresh and saltwater species with potential for aquaculture have demonstrated the feasibility of cryopreservation with some very promising results. Species included are silver carp, *Hypophthalmichthys molitrix* (Sin, 1974), grey mullet, *Mugil cephalus* (Chao *et al.*, 1975; Chao, 1982), Indian carp, *Labeo rohita*, tawes carp, *Puntius gonionotus*, bighead carp, *Aristichthys nobilis*, catfish, *Pangasius sutchi*, (Withler, 1982), grouper, *Epinephelus tauvina* (Withler and Lim, 1982), grass carp *Ctenopharyngodon idella* (Durbin *et al.*, 1982), and milkfish, *Chanos chanos* (Hara *et al.*, 1982). Variable postthaw fertility results were reported from pike, *Esox lucius*, common carp, *Cyprinus carpio* (Stein and Bayrle, 1978; De Montalembert *et al.*, 1978; Moczarski, 1977), and striped bass (Kerby, 1983); however, rather high fertility was obtained with cryopreserved spermatozoa from herring, *Clupea harengus*, cod, *Gadus morhua*, plaice, *Pleuronectes platessa*, sea bream, *Sparus auratus*, sea bass, *Dicentrarchus labrax*, whitefish, *Coregonus muksun*, and zebrafish, *Brachydanio rerio* (Blaxter, 1953; Mounib *et al.*, 1968; Pullin, 1972, 1975; Billard, 1978b; Piironen and Hyvärinen, 1983; Harvey *et al.*, 1982). In the following, various effects of the preservation procedure on fish spermatozoa are discussed.

1. PREFREEZING EFFECTS

Between sperm collection and freezing, a period of short-term preservation is inevitable. In spite of the good durability of undiluted milt, prefreezing storage at 0°C for only 60 min reduces the fertility of frozen–thawed rainbow trout sperm, as compared to 15 min of storage (Stoss and Holtz, 1983a). A similar observation was also made by H. Stein (personal communication).

Permeating cryoprotectants have been used at concentrations which increase the osmotic pressure of the basic diluent severalfold. This may cause osmotic shock after addition as observed for DMSO (6.8–12.5% final concentration) in rainbow trout semen when the exposure exceeded 1 min at 0°C (Stoss and Holtz, 1983). A similar observation was made with glycerol in carp semen (Sneed and Clemens, 1956). Gradual addition of the cryoprotectants reduced the detrimental effect (Stoss *et al.*, 1983).

Glycerol has been reported to be toxic to salmonids (Ott and Horton, 1971a; Erdahl and Graham, 1980), even when added gradually (Truscott *et al.*, 1968), and also to *Epinephelus tauvina* sperm (Withler and Lim, 1982). In contrast, *Mugil cephalus* and *Gadus morhua* spermatozoa tolerate glycerol (Chao *et al.*, 1975; Mounib *et al.*, 1968). Ethylene glycol added at 8% to the diluent affects coho sperm fertility (Ott and Horton, 1971a), but a final concentration of 12.5% is not detrimental to *Salmo salar* sperm (Truscott *et al.*, 1968). Propylene glycol at increasing concentrations reduces fertility of unfrozen *Salmo salar* sperm (Truscott and Idler, 1969).

The need for equilibration of spermatozoa in the diluent prior to freezing is sometimes postulated in order to provide good penetration of the cryoprotectant. Data presented by Bayrle (1982) showed an inconsistent response of rainbow trout spermatozoa to equilibration. However, according to other researchers, equilibration is not necessary for salmonid sperm, and it may, in fact, reduce subsequent fertility (Ott and Horton, 1971a; Stein and Bayrle, 1978; Legendre and Billard, 1980; Bayrle, 1980; Stoss and Holtz, 1983). This observation applies also to spermatozoa from sea bream (Billard, 1978b), common carp (Moczarski, 1977), and channel catfish (Guest *et al.*, 1976). In salt-water spawners, hypertonic extenders probably activate spermatozoa motility leading to a depletion of energy reserves with increasing equilibration time.

Dilution rates of sperm-to-extender ranging from 1:1 to 1:19 have had no effect on postthaw fertility of salmonid sperm (Truscott and Idler, 1969; Ott and Horton, 1971a; Büyükhatipoglu and Holtz, 1978). In contrast, Legendre and Billard (1980) tested dilution rates of 1:1, 1:3, and 1:9 in rainbow trout spermatozoa, and found the 1:3 dilution to be superior when a constant number of spermatozoa for subsequent fertilization tests were used. In pike as well as in sea bream and sea bass, any dilution exceeding 1:2 is disadvantageous (De Montalembert *et al.*, 1978; Billard, 1978b).

2. EXTENDERS

Diluents which resembled the inorganic composition of seminal plasma have been used with some success in salmonids (Truscott and Idler, 1969; Büyükhatipoglu and Holtz, 1978; Stoss and Holtz, 1981a; Erdahl and Graham, 1980). Seminal plasma from rainbow trout, obtained by centrifugation of semen has not been an ideal diluent for freezing (Table II). It has been demonstrated that extenders with few components perform as well as more complex media. Horton and Ott (1976) reported that, compared to an earlier rather complex extender (Ott and Horton, 1971a,b), one consisting of only NaCl, $NaHCO_3$, and lecithin was sufficient. Further addition of mannitol or fructose did not improve subsequent fertility when freezing Pacific salmon spermatozoa (F. C. Withler and R. B. Morley, personal communication).

By optimizing the concentrations of sucrose, glutathione, and $KHCO_3$, Mounib (1978) achieved postthaw fertility in Atlantic salmon and Atlantic cod, which was identical to fresh sperm. Wide variations of sucrose and KCl concentration in the freezing medium did, in contrast, not affect postthaw fertility of rainbow trout spermatozoa (Stoss and Holtz, in preparation).

A few examples of the performance of various diluents applying the same freezing and thawing technique are given in Table II. Obviously DMSO diluted in distilled water resulted in fair success in rainbow trout. Combining glycerol or DMSO with 0.3 M glucose gave good survival in Atlantic salmon and *Coregonus* spermatozoa, respectively. The suitability of an isotonic glucose–DMSO solution was reconfirmed, including five species of Pacific salmon (Stoss *et al.*, in preparation). It should be noted that this extender induced motility in salmonids. Later reactivation, however, is possible. Similar glucose–cryoprotectant solutions were also effective for spermatozoa from grass carp and grey mullet (Durbin *et al.*, 1982; Hara *et al.*, 1982). Van der Horst *et al.* (1980) found sucrose solutions (250 and 280 mOsm/kg) combined with various levels of DMSO to be effective in maintaining postthaw motility. Clearly, more complex extenders like the ones used by Truscott and Idler (1969), Stein and Bayrle (1978), or Stoss and Holtz (1981a) provide no advantage (Table II). The brief exposure of spermatozoa to the various extenders prior to freezing and after thawing may limit the importance of any diluent. Unsuitable media possibly interfere with fertility by affecting prefreezing or postthawing motility (Mounib, 1978; Legendre and Billard, 1980).

According to Stoss and Holtz (1981b), no buffer is required in an extender for rainbow trout spermatozoa, but if included, any resultant pH below 7.0 is detrimental. The addition of proteins or egg yolk to extenders improved postthaw survival in sperm from salmonid fishes (Büyükhatipoglu and Holtz, 1978; Legendre and Billard, 1980; Stoss and Holtz, 1983a). S. Baynes (personal communication) observed that rainbow trout spermatozoa frozen in Mounib's (1978) medium with 10% DMSO and 10% egg yolk yielded a higher postthaw survival rate than did the diluent when free of egg yolk. The cell density required to achieve satisfactory fertilization was reduced by the yolk. The usefulness of simple extender media was also found in a variety of other fish. A NaCl–NaHCO$_3$–glycine medium has been used successfully in Atlantic cod sperm (Mounib *et al.*, 1968) and in plaice (Pullin, 1972, 1975). In silver and bighead carp, a NaCl solution was superior to a more complex medium (Sin, 1974). Isotonic NaCl solutions have also been successfully used in grass carp and grey mullet sperm (Durbin *et al.*, 1982; Hara *et al.*, 1982).

Mixtures of diluted seawater and cryoprotectant have been reported to provide postthaw motility or fertility in the salt-water spawners *Clupea harengus* and *Mugil cephalus* (Blaxter, 1953; Pruginin and Cirlin, 1975); however, media without cryoprotectant resulted in expectedly poor survival of *Clupea harengus* spermatozoa (Rosenthal *et al.*, 1978). In *Cyprinus carpio*, mammalian semen extenders were unsuitable, but some fertility was observed with an extender consisting of 0.75% KCl, 10% lecithin, and 5% DMSO (Pavlovici and Vlad, 1976). Modified Cortland's medium was suitable

Table II

Fertility of Spermatozoa Frozen in Various Diluents[a]

Diluent (sperm-to-diluent ratio 1:3)	Cryoprotectant and concentration in final dilution (%)		Species	Fertility[b] (% eyed eggs)		Replicates, eggs per replicate (n/n)	References
				X	SD		
Distilled water	DMSO	7.5	Salmo gairdneri	67.4	4.0	(5/155)	Stoss (1979)
0.3 M Glucose	DMSO	7.5	S. salar	91.3	7.6	(4/125)	Stoss and Refstie (1982)
0.3 M Glucose	Glycerol	15.0	Coregonus muksun	98.6	—	(1/2156)	Piironen and Hyvärinen (1983)
Distilled water + 20% egg yolk	DMSO	7.0	S. gairdneri	82.1	7.9[c]	(6/118)	Bayrle (1980)
Seminal plasma	DMSO	7.5	S. gairdneri	6.4	2.3	(5/155)	
Truscott and Idler (1969) Hfx#10[d]	DMSO	7.5	S. gairdneri	83.3	4.8	(5/229)	
Mounib (1978)[e]	DMSO	7.5	S. gairdneri	93.9	3.7	(5/229)	Stoss (1979)
Stoss and Holtz (1981a)[f]	DMSO	7.5	S. gairdneri	87.7	4.1	(5/229)	
Stein and Bayrle (1978)[g]	DMSO	7.5	S. gairdneri	88.2	2.9	(5/229)	

[a]Pellet technique, thawing in 120 mmole NaHCO$_3$ (10°C) (Stein, 1975) resulting in a sperm dilution of approximately 1/80.

[b]Fresh sperm control is 100.

[c]Sperm dilution after thawing 1/115.

[d]Media contains 103 mM NaCl, 22 mM KCl, 1.3 mM CaCl$_2$, 0.5 mM MgSO$_4$, 3.3 mM fructose, 79.9 mM glycine.

[e]Media contains 125 mM sucrose, 6.5 mM glutathione, 100 mM KHCO$_3$.

[f]Media contains 101 mM NaCl, 23 mM KCl, 5.4 mM CaCl$_2$, 1.3 mM MgSO$_4$, 200 mM Tris-citric acid (pH 7.25), 0.4% bovine serum albumin, 0.75% Promine D.

[g]Media contains 128.3 mM NaCl, 23.8 mM NaHCO$_3$, 3.7 mM Na$_2$HPO$_8$, 0.4 mM MgSO$_4$, 5.1 mM KCl, 2.3 mM CaCl$_2$, 5.6 mM glucose, 66.6 mM glycin, 20% egg yolk.

for *Cyprius carpio* and *Ctenopharyngodon idella* spermatozoa (Moczarski, 1976, 1977).

3. CRYOPROTECTANTS

DMSO has been employed mostly in cryopreservation of salmon sperm and has provided good protection. Optimum concentrations may vary among the applied freezing techniques or species (Ott and Horton, 1971a,b; Stein, 1979). According to data obtained by Stoss and Holtz (1983), there was no difference in postthaw fertility when 6.8–12.5% (v/v) DMSO (final concentration) was used to preserve rainbow trout sperm. A concentration of 3.3% DMSO in the sperm-extender suspension also was found to be sufficient (Erdahl and Graham, 1980). Glycerol has been used with much less success in salmonids, possibly because of its deleterious effect on unfrozen spermatozoa (see Section VIII,C,1).

Ethylene glycol, tested over a wide range of concentrations, provided little protection in brown trout spermatozoa (Stein, 1979). However, Erdahl and Graham (1980) reported good postthaw fertility using either EG or DMSO for the same species. Propylene glycol provided as much protection as DMSO for Atlantic salmon sperm (Truscott and Idler, 1969). Polyvinylpyrrolidone has been employed unsuccessfully in coho salmon and brown trout (Ott and Horton, 1971a; Stein, 1979).

Dimethylsulfoxide has also been used successfully for a variety of species, e.g., *Mugil cephalus, Sparus auratus, Dicentrarchus labrax, Ictalurus punctatus, Cyprinus carpio, Ctenopharyngodon idella,* and *Morone saxatilis* (Chao *et al.*, 1975; Billard, 1978b; Guest *et al.*, 1976; Moczarski, 1976, 1977). Glycerol has been applied with very good success in *Clupea harengus, Pleuronectes platessa,* and *Gadus morhua* (Blaxter, 1953; Pullin, 1972, 1975; Mounib *et al.*, 1968; Kerby, 1983). The ability of glycerol to cryoprotect *Coregonus muksun* spermatozoa (Table II) has been related to the high glycerol concentration found in seminal plasma in the genus *Coregonus* (Piironen and Hyvärinen, 1983). Methanol was effective for *Brachydanio rerio* spermatozoa (Harvey *et al.*, 1982).

4. FREEZING AND THAWING

As already noted, freezing and thawing rates are the most critical variables, but surprisingly, little systematic research has been done on fish spermatozoa. Only Billard (1978b) has investigated the effect of various freezing rates in combination with cryoprotectant concentration and dilution rates in sea bass and sea bream spermatozoa. The optimum freezing rate for sea bass was about 10°–20°C/min. Further, DMSO at 10% was superior in each case to 5, 15, or 20%, and postthaw fertility reached levels around 90%.

Comparison of data from the literature often is complicated either by insufficient recording of freezing rates or by termination of controlled freezing at too high temperatures. Slow freezing rates in the range of 1°–5°C/min are insufficient for rainbow trout spermatozoa, but a rate of 30°C/min provided some survival (Graybill and Horton, 1969). Freezing of sperm pellets on dry ice results in cooling velocities around 20–30°C/min (depending on pellet size), and this range is suitable for spermatozoa of salmonids (Ott and Horton, 1971a; Büyükhatipoglu and Holtz, 1978; Stein and Bayrle, 1978; Legendre and Billard, 1980; Erdahl and Graham, 1980; Stoss and Holtz, 1981a,b; Stoss and Refstie, 1983), of the whitefish, *Coregonus muksun* (Piironen and Hyvärinen, 1983), and of the pike, *Esox lucius* (Stein and Bayrle, 1978; De Montalembert *et al.*, 1978).

There is indirect evidence that some intracellular ice forms while pellets are freezing on dry ice, which causes damage during slow thawing. By increasing the thawing rate from approximately 120°C/min to 1500°C/min, postthaw fertility has been improved in chum salmon spermatozoa (Stoss *et al.*, 1984). Compared to this finding, thawing in a number of instances as reported in the literature, was performed at a rather slow rate.

Fast freezing rates as achieved by immersing samples directly into liquid nitrogen were unsuccessful in spermatozoa from *Ictalurus punctatus* (Guest *et al.*, 1976), *Cyprinus carpio* (Moczarski, 1977), and *Salmo salar* (Hoyle and Idler, 1968). However, Mounib (1978) obtained excellent postthaw fertility after freezing 1-ml samples from *Salmo salar* and *Gadus morhua* in liquid nitrogen. Cell injury by water recrystallization may have been prevented by the fast thawing procedure applied. Freezing at a rate of approximately 100°C/min provided excellent postthaw survival in milkfish (Hara *et al.*, 1982).

Spermatozoa from some saltwater spawners, such as *Clupea harengus* and *Gadus morhua*, tolerated slow freezing rates in the range of 1°–5°C/min (Blaxter, 1953; Mounib, *et al.*, 1968). Because cell injury during slow freezing is often related to prolonged exposure to highly concentrated solutes, spermatozoa which are adapted to seawater may be less susceptible to increased salt concentrations during freezing.

5. POSTTHAWING EFFECTS

Frozen–thawed spermatozoa can differ from unfrozen cells in motility characteristics. Motility, although not induced by the extender after dilution, can begin spontaneously during thawing (Stein, 1975; Bayrle, 1980; Stoss and Holtz, 1981a; B. Harvey, personal communication). Attempts to suppress spontaneous motility by thawing rainbow trout sperm in nonactivating media failed to maintain subsequent fertility. Only isotonic media with good activation properties are suitable (Stoss and Holtz, 1981a).

Changes in the membrane potential, possibly by leaking electrolytes (Kurokura and Hirano, 1980) may stimulate motility induction.

The period of motility can be drastically reduced. *Brachydanio rerio* spermatozoa were motile for 10–15 sec after thawing, but fresh cells show motility in fish Ringer's for 1 hr (Harvey *et al.*, 1982). Thawing pink salmon (*Oncorhynchus gorbuscha*) spermatozoa in an IBMX-supplemented solution (compare Table I), which considerably prolonged motility in fresh cells, showed no such effect on thawed spermatozoa (Stoss *et al.*, 1984). Also, motility of frozen–thawed grouper (*Epinephelus tauvina*) sperm in seawater was reduced from 30 min in fresh cells to 1 min (Withler and Lim, 1982). Stein (1979) reported motility of 1–2 sec in thawed *Salmo trutta* spermatozoa. In spite of the brief motility, postthaw fertility in most cases was reported to be high. In contrast, the postthaw percentage of motile cells was correlated with fertility by Mounib (1978) and Harvey *et al.* (1982).

Because of the brief duration of motility, it is not surprising that a delay of only 30 sec between thawing in an activating solution and insemination resulted in reduced fertility in rainbow trout (Stoss and Holtz, 1981a). Erdahl and Graham (1980) thawed milt from brown and rainbow trout in the freezing extender devoid of cryoprotectant, and kept the thawed samples for an unspecified period prior to insemination of eggs. However, in this case, subsequent fertility seemed not affected by postthaw storage. A postthaw incubation period was suggested by Zell (1978).

As demonstrated by Billard and Legendre (1980) and Stoss and Holtz (1981a), higher densities of frozen spermatozoa than of fresh cells are required to achieve maximum fertility. Because fertility success with cryopreserved sperm cells has reached an acceptable level, particularly for several salmonid fishes, further improvements will focus on increasing the rate of cell survival, thereby allowing the use of a given number of spermatozoa more efficiently.

Inconsistent results with cryopreserved milt have been related to the particular male or female employed in the cryopreservation and fertility test (Ott and Horton, 1971b; Stoss and Holtz, 1981b; Harvey *et al.*, 1982). Pooling milt from several rainbow trout prior to cryopreservation reduced the variability of fertility results (Stoss and Holtz, 1983a). However, Legendre and Billard (1980) reported that the fertility of the pooled milt was much higher than the mean of all individual fish involved. Further, gamete quality effects are discussed in Section V.

6. Freeze Drying

Freeze drying, i.e., dehydrating cells by lyophilization, has been conducted with limited success using mammalian spermatozoa (compare Jeyendran *et al.*, 1981). In rainbow trout, Zell (1978) obtained a maximum of

64% fertility (control, 70%) with one sample of vacuum-dried spermatozoa, but was unable to repeat this result. Some degree of fertility also was maintained when samples were stored at −11°C for up to 1 year. Zell (1978) related the inconsistent results primarily to technical difficulties. Because acrosomal damage is related to infertility in freeze-dried mammalian spermatozoa (Saacke and White, 1972), teleostean sperm cells lacking an acrosome may be less susceptible to freeze-drying damage. If this technique can be improved, it may provide an alternative storage technique in regions without a continuous supply of coolants. However, currently, cryopreservation is the superior technique.

D. Cryopreservation of Ova and Embryos

Cryopreservation of fish ova and embryos appears to be more complicated than the freezing of spermatozoa. Several factors interfere with the removal of intracellular water during cooling: (1) the large egg volume, (2) the presence of two different membranes (the outer capsule and the perivitelline membrane which surrounds the yolk), and (3) the different water permeability of both membranes (Loeffler and Løvtrup, 1970). There is little information about cryopreservation attempts. Most of it refers to the very large egg of salmonids which also has a low permeability for water, especially after the formation of the pervivitelline space after water activation (Prescott, 1955; Potts and Rudy, 1969; Loeffler and Løvtrup, 1970). Penetration of cryoprotectants such as glycerol, DMSO, and methanol has been found to be extremely slow in unactivated ova, and cooling rates as low as 0.01°C/min are still too high to pervent intracellular freezing (Harvey and Ashwood-Smith, 1982; see also Harvey, 1982).

The effect of pre- and postfreezing treatments requires careful investigation because concentration of cryoprotectant, the mode of its addition, or duration of exposure may affect subsequent development, as demonstrated in herring embryos and rainbow trout zygotes (Whittingham and Rosenthal, 1978; Haga, 1982; Stoss and Donaldson, 1983). Furthermore, the choice of the developmental stage is critical because temperature tolerance may change, as reported for plaice embryos (Pullin and Bailey, 1981). Further, water permeability varies during embryogenesis as demonstrated in the zebrafish, *Brachydanio rerio* (Harvey and Chamberlain, 1982). Although a high proportion of cells may remain unfrozen in the ice-seeded suspension medium, subsequent development of herring and rainbow trout embryos or coho salmon zygotes was increasingly inhibited with reduction of the temperature (Whittingham and Rosenthal, 1978; Hara, 1982; Stoss and Donaldson, 1983). This is in agreement with findings by Harvey and Ashwood-Smith (1982) indicating that mechanical damage occurs in supercooled cells with progressive temperature reduction.

Cryoprotectants reduce the temperature for intracellular freezing, and ice formation takes place in salmonid eggs between ice seeding in the medium (−4° to −5°C) and approximately −20°C (Harvey and Ashwood-Smith, 1982; Stoss and Donaldson, 1983). Zell's (1978) finding that inseminated salmonid ova withstand −55°C could not be confirmed either by the former researchers or by Erdahl and Graham (1980).

The basic problem, sufficient dehydration during cooling, has not been solved. The salmonid egg is probably the least suited to conduct cryopreservation studies. Further reserach may be more successful if smaller cells with a higher water permeability and less yolk are chosen, applying an approach similar to that in mammalian ova (Leibo, 1980). Because the amount of experimental work has been very limited, the field is open for challenging research.

As mentioned in the introduction, preservation techniques for ova are urgently needed for stock conservation purposes. Because it is the goal to preserve the genes, but not necessarily the entire egg, possibilities for nuclei transplants into sterile donor eggs are worth exploring. One way could be by inducing androgenesis, i.e., inseminating sterile (irradiated) eggs with viable sperm cells. Subsequent destruction of the first mitotic spindle would result in diploid organisms. Techniques similar to those used to clone zebrafish may be suitable (Streisinger et al., 1981). Haploid androgenetic embryos have been obtained in chum salmon (Arai et al., 1979), but no attempts have been made so far to produce diploids. However, a high degree of homozygosity in resulting offspring may limit application of this technique.

IX. FINAL REMARKS

The mode of fertilization (external or internal) as well as the spawning environment (salt or fresh water) provide some indication about morphology, metabolism, and motility of the sperm cells from a particular species, and can be taken into consideration when applying storage procedures.

For short-term storage of unfrozen gametes, temperature and gaseous exchange are the most critical but easily controllable variables. The induction of motility and autoactivation must be avoided in sperm cells and ova, respectively.

Although fish ova or embryos have not been successfully cryopreserved, freezing and thawing of sperm cells poses no major problems. In summarizing Section VIII of this chapter, a brief guideline on freezing techniques for fish spermatozoa may be of use to those who have the need to preserve gametes from a species for which information is still inadequate.

Sperm collected during the peak of the spawning season, possibly pooled from several males, usually responds best. It is apparently advisable to chill the milt immediately and to process it as quickly as possible. Requirements for a dilution medium are isotonicity, and preservation of the cell's ability to become motile after induction. Adding organic components such as purified proteins, egg yolk, or sugars can be advantegous.

From all cryoprotectants, DMSO has been used mostly at concentrations between 1 and $2M$. Glycerol and methanol may also produce satisfactory results; glycerol is effective particularly for salt-water species. These spermatozoa are generally hardier than those from freshwater fish.

To establish a freezing rate, relatively simple techniques can be applied (see Section VIII,B). Various ranges between 1° and 5°C/min, 10° and 50°C/min, or above 100°C/min can be tested roughly to determine the optimal range. Control of freezing rate is necessary between approximately 0° and −70°C. If the rate can only be controlled to, for example, −40°C, "two-step" procedures which halt freezing somewhere between −20° and −40°C for various intervals, before transfer to the storage temperature, may be tried. Subsequent thawing of the sample is probably most successful when applying fast thawing rates.

To assess cell viability after freezing and thawing, motility tests are possible, but will often only indicate whether or not any spermatozoa survived. To fertilize eggs with frozen–thawed spermatozoa, sufficient sperm density, which may be well above the requirements for fresh cells, must be established. Further, duration of motility and fertility in cryopreserved cells can be extremely short, making an immediate insemination after thawing or after motility induction necessary.

In general, one must realize that all variables may be highly interactive, emphasizing the need for precise standardization of the freezing, thawing, and insemination procedure to achieve consistent results.

ACKNOWLEDGMENT

While writing this manuscript, J. Stoss received a postdoctoral fellowship from the Natural Sciences and Engineering Research Council of Canada, funded by the Department of Fisheries and Oceans.

REFERENCES

Afzelius, B. A. (1978). Fine structure of the garfish spermatozoan. *J. Ultrastruct. Res.* **64**, 309–314.

Alderdice, D. F., and Velsen, F. P. J. (1978). Effects of short-term storage of gametes on fertilization of Pacific herring eggs. *Helgol. Wiss. Meeresunters.* **31**, 485–498.

Anderson, W. A., and Personne, P. (1970). Recent cytochemical studies on spermatozoa of some invertebrate and vertebrate species. *In* "Comparative Spermatology" (B. Baccetti, ed.), pp. 431-449. Academic Press, New York.

Arai, K., Onozato, H., and Yamazaki, F. (1979). Artificial androgenesis induced with gamma irradiation in masu salmon, *Oncorhynchus masou*. *Bull. Fac. Fish., Hokkaido Univ.* 30, 181–186.

Ashwood-Smith, M. J. (1980). Low temperature preservation of cells, tissues and organs. In "Low Temperature Preservation in Medicine and Biology" (M. J. Ashwood-Smith J. Farrant, eds.), pp. 19–44. Pitman Medical Ltd., Tunbridge Wells, Kent.

Baccetti, B., Pallini, V., and Burrini, A. G. (1975). Localization and catalytic properties of lactate dehydrogenase in different sperm models. *Exp. Cell Res.* 90, 183–190.

Baccetti, B., Burrini, A. G., Dallai, R., and Pallini, V. (1979). The dynein electrophoretic bands in axonemes naturally lacking the inner or the outer arm. *J. Cell. Biol.* 80, 334–340.

Barrett, I. (1951). Fertility of salmonid eggs and sperm after storage. *J. Fish. Res. Board Can.* 8, 125–133.

Baynes, S. M., and Scott, A. P. (1982). Cryopreservation of rainbow trout spermatozoa: Variation in membrane composition may influence spermatozoan survival. *Proc. Int. Symp. Reprod. Physiol. Fish 1982* p. 128.

Baynes, S. M., Scott, A. P., and Dawson, A. P. (1981). Rainbow trout *Salmo gairdnerii* Richardson, spermatozoa: Effects of cations and pH on motility. *J. Fish Biol.* 19, 259–267.

Bayrle, H. (1980). Untersuchungen zur Optimierung der Befruchtungsfähigkeit von gefrierkonserviertem Fischsperma. Dissertation, Tech. Univ. München.

Bayrle, H. (1982). Cryopreservation of rainbow trout sperm: Effect of equilibration. *Proc. Int. Symp. Reprod. Physiol. Fish, 1982* pp. 129–130.

Benau, D., and Terner, C. (1980). Initiation, prolongation, and reactivation of the motility of salmonid spermatozoa. *Gamete Res.* 3, 247–257.

Billard, R. (1970). Ultrastructure comparée de spermatozoïdes de quelques poissons téléostéens. In "Comparative Spermatology" (B. Baccetti, ed.), pp. 71–79. Academic Press, New York.

Billard, R. (1975). L'insémination artificielle de la truite *Salmo gairdneri* Richardson. IV. Effets des ions K+ et Na+ sur la conservation du pouvoir fécondant des gamètes. *Bull. Fr. Piscic.* 265, 88–100.

Billard, R. (1977). Utilisation d'un systeme tris-glycocolle pour tamponner le diluent d'insémination pour truite. *Bull. Fr. Piscic.* 264, 102–112.

Billard, R. (1978a). Changes in structure and fertilizing ability of marine and freshwater fish spermatozoa diluted in media of various salinities. *Aquaculture* 14, 187–198.

Billard, R. (1978b). Some data on gametes preservation and artificial insemination in teleost fish. *Actes Colloq. Cent. Nat. Exploitation Oceans (CNEXO)* 8, 59–73.

Billard, R. (1980a). Survie des gamètes de truite arc-en-ciel après dilution dans des solutions salines ou de sucrose. *Reprod. Nutr. Dev.* 20, 1899–1905.

Billard, R. (1980b). Prolongation de la durée de motilité et du pouvoir fécondant des spermatozoïdes de truite arc-en-ciel par addition de théophylline au milieu de dilution. *C. R. Hebd. Seances Acad. Sci., Ser. D* 291, 649–652.

Billard, R. (1980c). Reproduction and artificial insemination in teleost fish. *Int. Congr. Anim. Reprod. Artif. Insemin. [Proc.], 9th, 1980* RT-H-3, pp. 327–337.

Billard, R. (1981). Short-term preservation of sperm under oxygen atmosphere in rainbow trout (*Salmo gairdneri*). *Aquaculture* 23, 287–293.

Billard, R., and Breton, B. (1976). Sur quelques problèmes de physiologie du sperme chez les poissons téléostéens. *Rev. Trav. Inst. Peches Marit.* 40, 501–503.

Billard, R., and Breton, B. (1978). Rhythms of reproduction in teleost fish. In "Rhythmic Activity of Fishes" (J. E. Thorpe, Ed.), pp. 31–53. Academic Press, New York.

Billard, R., and Fléchon, J. E. (1969). Spermatogonies et spermatocytes flagelles chez *Poecilia reticulata* (Téléostéens Cyprinodontiformes). *Ann. Biol. Anim., Biochim., Biophys.* **9,** 281–286.

Billard, R., and Gillet, C. (1975). Effets de la température sur la fécondation et la survie des gamètes chez la truite arc-en-ciel, *Salmo gairdneri. Bull. Fr. Piscic.* **259,** 53–65.

Billard, R., and Ginsburg, A. S. (1973). La spermiogenese et le spermatozoïdes d'*Anguilla anguilla.* L'études ultrastructurale. *Ann. Biol. Anim., Biochim., Biophys.* **13,** 523–534.

Billard, R., and Jalabert, B. (1973). Le glycogène au cours de la formation des spermatozoïdes et de leur transit dans les tractus genitaux male et femelle chez le guppy (poisson poecilide). *Ann. Biol. Anim., Biochim., Biophys.* **13,** 313–320.

Billard, R., and Jalabert, B. (1974). L'insémination artificielle de la truite (*Salmo gairdneri* Richardson). II. Comparaison des effets de differents dilueurs sur la conservation de la fertilité des gamètes avont et après insemination. *Ann. Biol. Anim., Biochim., Biophys.* **14,** 601–610.

Billard, R., Jalabert, B., and Breton, B. (1974). L'insémination artificielle de la truite (*Salmo gairdneri* Richardson). III. Definition de la nature et de la molarité du tampon a employer avec les dilueurs d'insémination et de conservation. *Ann. Biol. Anim., Biochim., Biophys.* **14,** 611–621.

Billard, R., Dupont, J., and Barnabé, G. (1977). Diminution de la motilité et la durée de conservation du sperme de *Dicentrarchus labrax* L. (poisson, téléostéen) pendant la periode de spermiation. *Aquaculture* **11,** 363–367.

Billard, R., Marcel, J., and Matei, D. (1981). Survie *in vitro* et *post mortem* des gamètes de truite fario (*Salmo trutta* fario). *Can. J. Zool.* **59,** 29–33.

Blaxter, J. H. S. (1953). Sperm storage and cross-fertilization of spring and autumn spawning herring. *Nature (London)* **172,** 1189–1190.

Blaxter, J. H. S. (1955). Herring rearing. I. The storage of herring gametes. *Mar. Res. Dep. Agric. Fish. Scotl.* **3,** 1–12.

Blaxter, J. H. S. (1969). Development: Eggs and larvae. In "Fish Physiology" (W. S. Hoar and D. J. Randall, eds.), Vol. 3, pp. 177–252. Academic Press, New York.

Büyükhatipoglu, S., and Holtz, W. (1978). Preservation of trout sperm in liquid or frozen state. *Aquaculture* **14,** 49–56.

Carpentier, P., and Billard, R. (1978). Conservation à court terme des gamètes de Salmonides a des températures voisines de 0°C. *Ann. Biol. Anim., Biochim., Biophys.* **18,** 1083–1088.

Chao, N. H. (1982). New approaches for cryopreservation of sperm of grey mullet, *Mugil cephalus. Proc. Int. Symp. Reprod. Physiol. Fish,* 1982 pp. 132–133.

Chao, N. H., Chen, H. P., and Liao, I. C. (1975). Study on cryogenic preservation of grey mullet sperm. *Aquaculture* **5,** 389–406.

Clemens, H. P., and Hill, L. G. (1969). The collection and short-term storage of milt of white bass. *Prog. Fish-Cult.* **31,** 26.

Dadone, L., and Narbaitz, R. (1967). Submicroscopic structure of spermatozoa of a Cyprinodontiform teleost, *Jenynsia lineata. Z. Zellforsch. Mikrosk. Anat.* **80,** 214–219.

Davenport, J., Vahl, O., and Lønning, S. (1979). Cold resistance in the eggs of the capelin *Mallotus villosus. J. Mar. Biol. Assoc. U.K.* **59,** 443–453.

Davenport, J., Lønning, S., and Kjørsvik, E. (1981). Osmotic and structural changes during early development of eggs and larvae of the cod, *Gadus morhua* L. *J. Fish Biol.* **19,** 317–331.

De Montalembert, G., Bry, C., and Billard, R. (1978). Control of reproduction in northern pike. *Am. Fish. Soc., Spec. Publ.* **11,** 217–225.

Durbin, H., Durbin, F. J., and Stott, B. (1982). A note on the cryopreservation of grass carp milt. *Fish. Manage.* **13,** 115–117.

Dushkina, L. A. (1975). Viability of herring (*Clupea*) eggs and fertilizing capacity of herring sperm stored under various conditions. *J. Ichthyol.* **15**, 423–429.

Ellis, W. G., and Jones, J. W. (1939). The activity of the spermatozoa of *Salmo salar* in relation to osmotic pressure. *J. Exp. Biol.* **16**, 530–534.

Erdahl, D. A., and Graham, E. F. (1980). Preservation of gametes of freshwater fish. *Int. Congr. Anim. Reprod. Artif. Insem. [Proc.]*, 9th, 1980 RT-H-2, pp. 317–326.

Escaffre, A. M., Petit, J., and Billard, R. (1977). Evolution de la quantité d'ovules récoltes et conservation de leur aptitude a être fécondes au cours de la periode post ovulatoire chez la truite arc-en-ciel. *Bull. Fr. Piscic.* **265**, 134–142.

Farrant, J. (1980). General observations on cell preservation. *In* "Low Temperature Preservation in Medicine and Biology" (M. J. Ashwood-Smith and J. Farrant, eds.), pp. 1–18. Pitman Medical Ltd., Tunbridge Wells, Kent.

Farrant, J., and Ashwood-Smith, M. J. (1980). Practical aspects. *In* "Low Temperature Preservation in Medicine and Biology" (M. J. Ashwood-Smith and J. Farrant, eds.), pp. 285–310. Pitman Medical Ltd., Tunbridge Wells, Kent.

Franzen, A. (1970). Phylogenetic aspects of the morphology of spermatozoa and spermiogenesis. *In* "Comparative Spermatology" (B. Baccetti, ed.), pp. 29–46. Academic Press, New York.

Gardiner, D. M. (1978a). Fine structure of the spermatozoon of the viviparous teleost *Cymatogaster aggregata. J. Fish Biol.* **13**, 435–438.

Gardiner, D. M. (1978b). Utilization of extracellular glucose by spermatozoa of two viviparous fishes. *Comp. Biochem. Physiol.* **59**, 165–168.

Gaschott, O. (1928). Beiträge zur Reizphysiologie des Forellenspermas. I. Die optimalen Konzentrationen einiger Salzlösungen. II. Beeinflussung der Spermatozoenbewegung durch wechselnde pH-Konzentrationen. *Arch. Hydrobiol., Suppl.* **4**, 441–478.

Gilkey, J. C. (1981). Mechanisms of fertilization in fishes. *Am. Zool.* **21**, 359–375.

Ginsburg, A. S. (1963). Sperm-egg association and its relationship to the activation of the egg in salmonid fishes. *J. Embryol. Exp. Morphol.* **11**, 13–33.

Ginsburg, A. S. (1972). "Fertilization in Fishes and the Problem of Polyspermy." Keter Press, Jerusalem (translated from Russian, available from U.S. Dept. of Commerce, NTIS, Springfield, Virginia).

Gjedrem, T. (1981). Conservation of fish populations in Norway. *Ecol. Bull.* **34**, 33–36.

Graybill, J. R., and Horton, H. F. (1969). Limited fertilization of steelhead trout eggs with cryopreserved sperm. *J. Fish. Res. Board Can.* **26**, 1400–1404.

Guest, W. C., Avault, J. W., and Roussel, J. D. (1976). Preservation of channel catfish sperm. *Trans. Am. Fish. Soc.* **105**, 469–474.

Haga, Y. (1982). On the subzero temperature preservation of fertilized eggs of rainbow trout. *Bull. Jpn. Soc. Sci. Fish.* **48**, 1564–1572.

Hara, S., Canto, J. T., and Almendras, J. M. E. (1982). A comparative study of various extenders for milkfish, *Chanos chanos* (Forsskål), sperm preservation. *Aquaculture* **28**, 339–346.

Hartmann, M., Medem, F. G., Kuhn, R., and Bielig, H. J. (1947). Untersuchungen über die Befruchtungsstoffe der Regenbogenforelle. *Z. Naturforsch B: Anorg. Chem., Org. Chem., Biochem., Biophys., Biol.* **2b**, 330–349.

Harvey, B. (1982). Cryobiology and the storage of teleost gametes. *Proc. Int. Symp. Reprod. Physiol. Fish, 1982* pp. 123–127.

Harvey, B., and Ashwood-Smith, M. J. (1982). Cryoprotectant penetration and supercooling in the eggs of salmonid fishes. *Cryobiology* **19**, 29–40.

Harvey, B., and Chamberlain, B. J. (1982). Water permeability in the developing embryo of the zebra fish, *Brachydanio rerio. Can. J. Zool.* **60**, 268–270.

Harvey, B., and Hoar, W. S. (1979). "The Theory and Practice of Induced Breeding in Fish, IDRC-TS21e. Int. Dev. Res. Cent. Ottawa.

Harvey, B., Kelly, R. N., and Ashwood-Smith, M. J. (1982). Cryopreservation of zebrafish spermatozoa using methanol. *Can. J. Zool.* **60**, 1867–1870.

Henderson, N. E., and Dewar, J. E. (1959). Short-term storage of brook trout milt. *Prog. Fish-Cult.* **21**, 169–171.

Hines, R., and Yashouv, A. (1971). Some environmental factors influencing the activity of spermatozoa of *Mugil capito* Cuvier, a grey mullet. *J. Fish Biol.* **3**, 123–127.

Hiroi, O. (1978). Studies on the retention of gametes of salmonid fishes. 3. Change in fry-liberation rate of stored chum salmon eggs inseminated with sperms preserved in same conditions. *Sci. Rep. Hokkaido Salmon Hatchery* **32**, 19–26.

Hiroi, O., Yasukawa, M., and Suetake, T. (1973). Studies on the retention of gametes in salmonid fishes - 2. On the storage of chum salmon (*Oncorhynchus keta*) sperm. *Sci. Rep. Hokkaido Salmon Hatchery* **27**, 39–44.

Hoar, W. S. (1969). Reproduction. *In* "Fish Physiology" (W. S. Hoar and D. J. Randall, eds.), Vol. 3, pp. 1–72. Academic Press, New York.

Hogendoorn, H., and Vismans, M. M. (1980). Controlled propagation of the African catfish, *Clarias lazera* (C. & V.). II. Artificial reproduction. *Aquaculture* **21**, 39–53.

Holtz, W., Stoss, J., and Büyükhatipoglu, S. (1977). Beobachtungen zur Aktivierbarkeit von Forellenspermatozoen mit Fruchtwasser, Bachwasser und destilliertem Wasser. *Zuchthygiene* **12**, 82–88.

Horton, H. F., and Ott, A. G. (1976). Cryopreservation of fish spermatozoa and ova. *J. Fish. Res. Board Can.* **33**, 995–1000.

Hoskins, D. D., Hall, M. L., and Munsterman, D. (1975). Induction of motility in immature bovine spermatozoa by cyclic AMP phosphodiesterase inhibitors and seminal plasma. *Biol. Reprod.* **13**, 168–176.

Hoyle, R. J., and Idler, D. R. (1968). Preliminary results in the fertilization of eggs with frozen sperm of Atlantic salmon (*Salmo salar*). *J. Fish. Res. Board Can.* **25**, 1295–1297.

Hulata, G., and Rothbard, S. (1979). Cold storage of carp semen for short periods. *Aquaculture* **16**, 267–269.

Huxley, J. S. (1930). The maladaption of trout spermatozoa in fresh water. *Nature (London)* **125**, 494.

Hwang, P. C., and Idler, D. R. (1969). A study of major cations, osmotic pressure, and pH in seminal components of Atlantic salmon. *J. Fish. Res. Board Can.* **26**, 413–419.

Hynes, J. D., Brown, E. H., Helle, J. H., Ryman, N., and Webster, D. A. (1981). Guidelines for the culture of fish stocks for resource management. *Can. J. Fish. Aquat. Sci.* **38**, 1867–1876.

Inaba, D., Nomura, M., and Suyama, M. (1958). Studies on the improvement of artificial propagation in trout culture. II. On the pH values of eggs, milt, coelomic fluid and others. *Bull. Jpn. Soc. Sci. Fish.* **23**, 762–765.

Jalabert, B., and Billard, R. (1969). Etude ultrastructurale du site de conservation des spermatozoïdes dans l'ovaire de *Poecilia reticulata* (Poisson téléostéen). *Ann. Biol. Anim., Biochim., Biophys.* **9**, 273–280.

Jaspers, E. J., Avault, J. W., and Roussel, J. D. (1976). Spermatozoal morphology and ultrastructure of channel catfish, *Ictalurus punctatus*. *Trans. Am. Fish. Soc.* **105**, 475–480.

Jeyendran, R. S., Graham, E. F., and Schmehl, M. K. L. (1981). Fertility of dehydrated bull semen. *Cryobiology* **18**, 292–300.

Kerby, J. H. (1983). Cryogenic preservation of sperm from striped bass. *Trans. Am. Fish. Soc.* **112**, 86–94.

Kossmann, H. (1973). Versuche zue Konservierung des Karpfenspermas (*Cyprinus carpio*). *Arch. Fischereiwiss.* **24**, 125–128.

Kuchnow, K. P., and Foster, R. S. (1976). Thermal tolerance of stored *Fundulus heteroclitus* gametes: Fertilizability and survival of embryos. *J. Fish. Res. Board Can.* **33**, 676–680.

Kuo, C. M. (1982). Progress in artificial propagation of milkfish. *ICLARM Newsl.* **5**, 8–10.

Kurokura, H., and Hirano, R. (1980). Cryopreservation of rainbow trout sperm. *Bull. Jpn. Soc. Sci. Fish.* **46**, 1493–1495.

Kusa, M. (1950). Physiological analysis of fertilization in the egg of the salmon, *Oncorhynchus keta*. I. Why are the eggs not fertilized in isotonic Ringer solution. *Annot. Zool. Jpn.* **24**, 22–28.

Kusa, M. (1953). Significance of the cortical change in the initiation of development of the salmon egg (physiological analysis of fertilization in the egg of the salmon, *Oncorhynchus keta*, II). *Annot. Zool. Jpn.* **26**, 73–77.

Lasher, R., and Rugh, R. (1962). The "Hertwig effect" in teleost development. *Biol. Bull. (Woods Hole, Mass.)* **123**, 582–588.

Legendre, M., and Billard, R. (1980). Cryopreservation of rainbow trout sperm by deep-freezing. *Reprod. Nutr. Dev.* **20**, 1859–1868.

Leibo, S. P. (1980). Water permeability and its activation energy of fertilized and unfertilized mouse ova. *J. Membr. Biol.* **53**, 179–188.

Leibo, S. P., and Mazur, P. (1971). The role of cooling rates in low-temperature preservation. *Cryobiology* **8**, 447–452.

Leibo, S. P., McGrath, J. J., and Cravalho, E. G. (1978). Microscopic observations of intra-cellular ice formation in unfertilized mouse ova as a function of cooling rate. *Cryobiology* **15**, 257–271.

Levine, M., Bain, J. Narashimhan, R., Palmer, B., Yates, A. J., and Murray, R. K. (1976). A comparative study of the glycolipids of human, bird and fish testes and of human sperm. *Biochim. Biophys. Acta* **441**, 134–145.

Lindroth, A. (1947). Time of activity of freshwater fish spermatozoa in relation to temperature. *Zool. Bidr. Uppsala* **25**, 164–168.

Loeffler, C. A., and Løvtrup, S. (1970). Water balance in the salmon egg. *J. Exp. Biol.* **52**, 291–298.

Mann, T. (1964). "The Biochemistry of Semen and of the Male Reproductive Tract." Methuen, London.

Mattei, C., and Mattei, X. (1975). Spermiogenesis and spermatozoa of the *Elopomorpha* (teleost fish). *In* "The Functional Anatomy of the Spermatozoon" (B. A. Afzelius, ed.), pp. 211–221. Pergamon, Oxford.

Mattei, X., Mattei, C., Reizer, C., and Chevalier, J. L. (1972). Ultrastructure des spermatozoïdes aflagelles des *Mormyres* (poissons téléostéens). *J. Microsc. (Paris)* **15**, 67–78.

Maurer, R. R. (1978). Freezing mammalian embryos: A review of the techniques. *Theriogenology* **9**, 45–68.

Mazur, P. (1963). Kinetics of water loss from cells at subzero temperatures and the likelihood of intracellular freezing. *J. Gen. Physiol.* **47**, 347–369.

Mazur, P. (1977). The role of intracellular freezing in the death of cells cooled at supraoptimal rates. *Cryobiology* **14**, 251–272.

Mazur, P. (1977). The role of intracellular freezing in the death of cells cooled at supraoptimal rates. *Cryobiology* **14**, 251–272.

Mazur, P., and Miller, R. H. (1976b). Survival of frozen-thawed human erythrocytes as a function of the permeation of glycerol and sucrose. *Cryobiology* **13**, 523–536.

Meryman, H. T. (1971a). Cryoprotective agents. *Cryobiology* **8**, 173–183.

Meryman, H. T. (1971b). Osmotic stress as a mechanism of freezing injury. *Cryobiology* **8**, 489–500.

Meryman, H. T., Williams, R. J., and Douglas, M. S. J. (1977). Freezing injury from "solution effects" and its prevention by natural or artificial cryoprotein. *Cryobiology* **14**, 287–302.

Minassian, E. S., and Turner, C. (1966). Biosynthesis of lipids by human and fish spermatozoa. *Am. J. Physiol.* **210**, 615–618.

Moczarski, M. (1976). Cryobiological factors in grass carp preservation. *Proc. Int. Congr. Anim. Reprod. Artif. Insemin. 8th, 1976, Vol. 4*, pp. 1030–1033.

Moczarski, M. (1977). Deep freezing of carp *Cyprinus carpio* L. sperm. *Bull. Acad. Pol. Sci., Ser. Sci. Biol.* **15**, 187–190.

Morisawa, M., and Okuno, M. (1982). Cyclic-AMP induces maturation of trout sperm axoneme to initiate motility. *Nature (London)* **295**, 703–704.

Morisawa, M., and Suzuki, K. (1980). Osmolality and potassium ion: Their roles in initiation of sperm motility in teleosts. *Science* **10**, 1145–1147.

Morisawa, M., Okuno, M., and Morisawa, S., (1982). Physiological study on the initiation of sperm motility in chum salmon, *Oncorhynchus keta. Proc. Int. Symp. Reprod. Physiol. Fish, 1982*, p. 131.

Mounib, M. S. (1967). Metabolism of pyruvate, acetate and glyoxylate by fish sperm. *Comp. Biochem. Physiol.* **20**, 987–992.

Mounib, M. S. (1978). Cryogenic preservation of fish and mammalian spermatozoa. *J. Reprod. Fertil.* **53**, 13–18.

Mounib, M. S., and Eisan, J. S. (1968a). Biosynthesis of lipids by salmon sperm from pyruvate, acetate and glyoxylate. *Comp. Biochem. Physiol.* **25**, 193–200.

Mounib, M. S., and Eisan, J. S. (1968b). Carbon dioxide fixation by spermatozoa of cod. *Comp. Biochem. Physiol.* **25**, 703–709.

Mounib, M. S., Hwang, P. C., and Idler, D. R. (1968). Cryogenic preservation of Atlantic cod (*Gadus morhua*) sperm. *J. Fish. Res. Board Can.* **25**, 2623–2632.

Nagase, H., and Niwa, T. (1964). Deep freezing bull semen in concentrated pellet form. I. Factors affecting survival of spermatozoa. *Proc. Int. Congr. Anim. Reprod. Artif. Insemin., 4th, 1964, Vol.* 3, pp. 410–415.

Nicander, L. (1970). Comparative studies on the fine structure of vertebrate spermatozoa. *In* "Comparative Spermatology" (B. Baccetti, ed.), pp. 47–56. Academic Press, New York.

Nomura, M. (1964). Studies on reproduction of rainbow trout, *Salmo gairdneri* with special reference to egg taking. VI. The activities of spermatozoa in different diluents, and preservation of semen. *Bull. Jpn. Soc. Sci. Fish.* **30**, 723–733.

Nomura, M., Sakai, K., and Takashima, F. (1974). The over-ripening phenomenon of rainbow trout. I. Temporal morphological changes of eggs retained in the body cavity after ovulation. *Bull. Jpn. Soc. Sci. Fish.* **40**, 977–984.

Okada, S., Ishikawa, Y., and Kimura, G. (1956). On the viability of the sperm and the egg left in the dead body of dog salmon, *Oncorhynchus keta* (Walbaum). *Sci. Rep. Hokkaido Fish Hatchery* **11**, 7–17.

Ott, A. G., and Horton, H. F. (1971a). Fertilization of chinook and coho salmon eggs with cryo-preserved sperm *J. Fish. Res. Board Can.* **28**, 745–748.

Ott, A. G., and Horton, H. F. (1971b). Fertilization of steelhead trout (*Salmo gairdneri*) eggs with cryo-preserved sperm. *J. Fish. Res. Board Can.* **28**, 1915–1918.

Pautard, F. G. E. (1962). Biomolecular aspects of spermatozoan motility. *In* "Spermatozoan Motility" (D. W. Bishop, ed.), Publ. No. 72, pp. 189–232. Am. Assoc. Adv. Sci., Washington, D.C.

Pavlovici, I., and Vlad, C. (1976). Some data on the preservation of carp (*Cyprinus carpio* L.) seminal material by freezing. *Rev. Cresterea Anim.* **4**, 45–48. (Can. Fish. Mar. Serv. Transl. Ser. 3965).

Petit, J., Jalabert, B., Chevassus, B., and Billard, R. (1973). L'insémination artificielle de la truite (*Salmo gairdneri* Richardson). I. Effets du taux de dilution, du pH et de la pression osmotique du diluteur sur la fécondation. *Ann. Hydrobiol.* **4**, 201–210.

Piironen, J., and Hyvärinen, H. (1983). Cryopreservation of spermatozoa of the whitefish (*Coregonus muksun* Pallas). *J. Fish. Biol.* **22**, 159–163.

Plosila, D. S., and Keller, W. T. (1974). Effects of quantity of stored sperm and water on fertilization of brook trout eggs. *Prog. Fish-Cult.* **36**, 42–45.

Plosila, D. S., Keller, W. T., McCartney, T. J., and Robson, D. S. (1972). Effects of sperm storage and dilution on fertilization of brook trout eggs. *Prog. Fish-Cult.* **34**, 179–181.

Polge, C. (1980). Freezing of spermatozoa. *In* "Low Temperature Preservation in Medicine and Biology" (M. J. Ashwood-Smith and J. Farrant, eds.), pp. 45–64. Pitman Medical Ltd., Tunbridge Wells, Kent.

Polge, C., Smith, A. U., and Parkes, A. S. (1949). Revival of spermatozoa after vitrification and dehydration at low temperature. *Nature (London)* **164**, 666.

Poon, D. C., and Johnson, A. K.(1970). The effect of delayed fertilization on transported salmon eggs. *Prog. Fish-Cult.* **32**, 81–84.

Potts, W. T. W., and Rudy, P. P. (1969). Water balance in the egg of the Atlantic salmon *Salmo salar. J. Exp. Biol.* **50**, 223–237.

Prescott, D. M. (1955). Effect of activation on the water permeability of salmon eggs. *J. Cell. Comp. Physiol.* **45**, 1–12.

Pruginin, Y., and Cirlin, B. (1975). Techniques used in controlled breeding and production of larvae and fry in Israel. *EIFAC Tech. Pap.* **EIFAC/T25**, 90–97.

Pullin, R. S. V. (1972). The storage of plaice (*Pleuronectes platessa*) sperm at low temperatures. *Aquaculture* **1**, 279–283.

Pullin, R. S. V. (1975). Preliminary investigations into methods for controlling the reproduction of captive marine flatfish. *Pubbl. Stn. Zool. Napoli* **39**, Suppl., 282–296.

Pullin, R. S. V., and Bailey, H. (1981). Progress in storing marine flatfish eggs at low temperatures. *Rapp. P.-V. Reun,. Cons. Int. Explor. Mer.* **178**, 514–517.

Pullin, R. S. V., and Kuo, C. M. (1981). Developments in the breeding of cultured fishes. *Adv. Food-Prod. Syst. Arid Semiarid Lands [Proc. Symp.], 1980* papers, (in press).

Quantz, G. (1980). Ueber den Einfluss von carotinoidreichem Trockenfutter auf die Eibe-fruchtung der Regenbogenforelle (*Salmo gairdneri* R.). *Arch. Fischereiwiss.* **31**, 29–40.

Randall, D. J., and Hoar, W. S. (1971). Special techniques. *In* "Fish Physiology" (W. S. Hoar, and D. J. Randall, eds.), Vol. 6, pp. 511–528. Academic Press, New York.

Rosenthal, H., Alderdice, D. F., and Velsen, F. P. J. (1978). Cross-fertilization experiments using Pacific herring eggs and cryopreserved Baltic herring sperm. *Can. Fish. Mar. Serv., Tech. Rep.* **844**.

Runnström, J., Lindvall, S., and Tiselius, A. (1944). Gamones from the sperm of sea-urchin and salmon. *Nature (London)* **153**, 285–286.

Ryman, N. (1981). Fish gene pools. *Ecol. Bull.* (Stockholm) **34**.

Saacke, R. G., and White, J. M. (1972). Semen quality test and their relationship to fertility. *Proc. Tech. Conf. Artif. Insem. Reprod. NAAB, 4th, 1972*, pp. 2–7.

Sanchez-Rodriguez, M., and Billard, R. (1977). Conservation de la motilité et du pouvoir fécondant du sperme de truite arc-en-ciel maintenu à des températures voisines de 0°C. *Bull. Fr. Piscic.* **265**, 143–152.

Satir, P. (1974). The present status of the sliding microtubule model of ciliary motion. *In* "Cilia and Flagella" (M. A. Sleigh, ed.), pp. 131–141. Academic Press, New York.

Scheuring, L. (1925). Biologische und physiologische Untersuchungen an Forellensperma. *Arch. Hydrobiol., Suppl.* **4**, 181–318.

Schlenk, W. (1933). Spermatozoenbewegung und Wasserstoffionenkonzentration. Versuche mit dem Sperma der Regenbogenforelle. *Biochem. Z.* **265**, 29–35.

Schlenk, W., and Kahmann, H. (1937). Reaktionskinetische Untersuchung der Bewegung der Forellenspermatozoen. Z. Vergl. Physiol. **24**, 518–531.

Schlenk, W., and Kahmann, H. (1938). Die chemische Zusammensetzung des Spermaliquors und ihre physiologische Bedeutung. Untersuchung am Forellensperma. Biochem. Z. **295**, 283–301.

Scott, A. P., and Baynes, S. M. (1980). A review of the biology, handling and storage of salmonid spermatozoa. J. Fish Biol. **17**, 707–739.

Shehadeh, A. H. (1975). Induced breeding techniques—a review of progress and problems. EIFAC Tech. Paper **EIFAC/T25**, 72–89.

Shirkie, R. (1982). Science in service, the consultive group on international agricultural research. IDRC Rep. **11**(3), 24–26.

Sin, A. W. (1974). Preliminary results on cryogenic preservation of sperm of silver carp (Hypophthalmichthys molitrix) and bighead (Aristichthys nobilis). Hong Kong Fish. Bull. **4**, 33–36.

Sneed, K. E., and Clemens, H. P. (1956). Survival of fish sperm after freezing and storage at low temperatures. Prog. Fish-Cult. **18**, 99–103.

Stanley, H. P. (1965). Electron microscopic observations on the biflagellate spermatids of the teleost fish Porichtys notatus. Anat. Rec. **151**, 477.

Stanely, H. P. (1969). An electron microscope study of spermiogenesis in the teleost fish Oligocottus moculosus. J. Ultrastruct. Res. **27**, 230–243.

Stebbings, H., and Hyams, J. S. (1979). "Cell Motility." Longmans, Green, New York.

Stein, H. (1975). Spezielle Untersuchungen am Fischsperma unter besonderer Berücksichtigung der Spermakonservierung. Dissertation, Tech. Universität München.

Stein, H. (1979). Zum Einfluss von Schutzsubstanz und Konfektionierung auf das Befruchtungsergebnis mit tiefgefrorenem Bachforellensperma (Salmo trutta forma fario L.). Berl. Meunch. Tieraerztl. Wochenschr. **92**, 420–421.

Stein, H. (1981). Licht- und elektronenoptische Untersuchungen an den Spermatozoen verschiedener Süsswasserknochenfische (Teleostei). Z. Angew. Zool. **68**, 183–198.

Stein, H., and Bayrle, H. (1978). Cryopreservation of the sperm of some freshwater teleosts. Ann. Biol. Anim., Biochim., Biophys. **18**, 1073–1076.

Stoss, J. (1979). Spermakonservierung bei der Regenbogenforelle (Salmo gairdneri). Dissertation, Universität of Göttingen.

Stoss, J., and Donaldson, E. M. (1982). Preservation of fish gametes. Proc. Int. Symp. Reprod. Physiol. Fish, 1982 pp. 114–122.

Stoss, J., and Donaldson, E. M. (1983). Studies on cryopreservation of eggs from rainbow trout (Salmo gairdneri) and coho salmon (Oncorhynchus kisutch). Aquaculture **31**, 51–65.

Stoss, J., and Holtz, W. (1981a). Cryopreservation of rainbow trout (Salmo gairdneri) sperm. I. Effect of thawing solution, sperm density and interval between thawing and insemination. Aquaculture **22**, 97–104.

Stoss, J., and Holtz, W. (1981b). Cryopreservation of rainbow trout (Salmo gairdneri) sperm. II. Effect of pH and presence of a buffer in the diluent. Aquaculture **25**, 217–222.

Stoss, J., and Holtz, W. (1983). In preparation.

Stoss, J., and Holtz, W. (1983a). Cryopreservation of rainbow trout (Salmo gairdneri) sperm. III. Effect of proteins in the diluent, sperm from different males and interval between collection and freezing. Aquaculture (in press).

Stoss, J., and Holtz, W. (1983b). Successful storage of chilled rainbow trout (Salmo gairdneri) spermatozoa for up to 34 days. Aquaculture **31**, 269–274.

Stoss, J., and Refstie, T. (1983). Short-term storage and cryopreservation of milt from Atlantic salmon and sea trout. Aquaculture **30**, 229–236.

Stoss, J., Büyükhatipoglu, S., and Holtz, W. (1977). Der Einfluss bestimmter Elektrolyte auf die Bewegungsauslösung bei Spermatozoen der Regenbogenforelle (*Salmo gairdneri*) *Zuchthygiene* 12, 178–184.

Stoss, J., Büyükhatipoglu, S., and Holtz, W. (1978). Short-term and cryopreservation of rainbow trout (*Salmo gairdneri* Richardson) sperm. *Ann. Biol. Anim., Biochim., Biophys.* 18, 1077–1082.

Stoss, J. *et al.* (1984). In preparation.

Streisinger, G., Walker, C., Dower, N., Knauber, D., and Singer, F. (1981). Production of clones of homozygous diploid zebra fish (*Brachydanio rerio*). *Nature (London)* 291, 293–296.

Sundararaj, B. I. (1981). Reproductive physiology of teleost fishes—a review of present knowledge and needs for future research. *Aquacult. Dev. Coord. Programme, FAO-UN, Rome* **ADCP/REP 16.**

Sundararaj, B. I., and Nayyer, S. K. (1969). Effect of extirpation of "seminal vesicles" on the reproductive performance of the male catfish, *Heteropneustes fossilis* (Bloch). *Physiol. Zool.* 42, 429–437.

Suzuki, R. (1958). Sperm activation and aggregation during fertilization in some fishes. I. Behavior of spermatozoa around the micropyle *Embryologia* 4, 93–102.

Suzuki, R. (1959). Sperm activation and aggregation during fertilization in some fishes. II. Effect of distilled water on the sperm-stimulating capacity and fertilizability of eggs. *Embryologia* 4, 359–367.

Szöllösi, D., and Billard, R. (1974). The micropyle of trout eggs and its reaction to different incubation media. *J. Microsc. (Paris)* 21, 55–62.

Terner, C. (1962). Oxydative and biosynthetic reactions in spermatozoa. *In* "Spermatozoan Motility" (D. W. Bishop, ed.), Publ. No. 72, pp. 89–98. Am. Assoc. Adv. Sci. Washington, D.C.

Terner, C., and Korsh, G. (1963a). The oxidative metabolism of pyruvate, acetate and glucose in isolated fish spermatozoa. *J. Cell. Comp. Physiol.* 62, 243–249.

Terner, C., and Korsh, G. (1963b). The biosynthesis of fatty acids of glycerides and phosphatides by isolated spermatozoa of the rainbow trout. *J. Cell. Comp. Physiol.* 62, 251–256.

Truscott, B., and Idler, D. R. (1969). An improved extender for freezing Atlantic salmon spermatozoa. *J. Fish. Res. Board Can.* 26, 3254–3258.

Truscott, B., Idler, D. R., Hoyle, R. J., and Freeman, H. C. (1968). Sub-zero preservation of Atlantic salmon sperm. *J. Fish. Res. Board Can.* 25, 363–372.

Turdakov, A. F. (1971). The effect of temperature conditions on the speed and fertilizing capacity of the spermatozoa of some Issyk-kul fishes. *J. Ichthyol.* 11, 206–215.

Van der Horst, G., Dott, H. M., and Foster, G. C. (1980). Studies on the motility and cryopreservation of rainbow trout (*Salmo gairdneri*) spermatozoa. *S. Afr. J. Zool.* 15, 275–279.

Van Deurs, B. (1975). The sperm cells of Pantodon (Teleostei) with a note on residual body formation. *In* "The Functional Anatomy of the Spermatozoon" (B. A. Afzelius, ed.), pp. 311–318. Pergamon, Oxford.

Weisel, G. F. (1948). Relation of salinity to the activity of the spermatozoa of *Gillichthys*, a marine teleost. *Physiol. Zool.* 21, 40–48.

Whittingham, D. G. (1980). Principles of embryo preservation. *In* "Low Temperature Preservation in Medicine and Biology" (M. J. Ashwood-Smith and J. Farrant, eds.), pp. 65–83. Pitman Medical Ltd., Tunbridge Wells, Kent.

Whittingham, D. G., and Rosenthal, H. (1978). Attempts to preserve herring embryos at subzero temperatures. *Arch. Fischereiwiss.* 29, 75–79.

jection of pituitary glands from *Prochilodus platensis* (Houssay, 1930, 1931). This technique was soon applied in Brazil to induce spawning of *Prochilodus argenteus* (von Ihering and Azevedo, 1934; von Ihering, 1935, 1937). These studies were followed by successes with the induced spawning of the sturgeon in the Union of Soviet Socialist Republics (USSR) (Gerbil'skii, 1938) and with induced spawning of North American species in the United States (Hasler *et al.*, 1939, 1940). These and subsequent early studies were thoroughly reviewed by Pickford and Atz (1957) in their comprehensive treatise on the fish pituitary gland. Since that time, literally thousands of publications have dealt with reproductive processes in fish in general (Atz and Pickford, 1964; Donaldson, 1977; Swann and Donaldson, 1980), and a significant proportion of these relate directly or indirectly to the control of reproductive processes. A number of reviews have been published which deal either exclusively or in large part with induced spawning. These include Pickford and Atz (1957), Atz and Pickford (1959), Clemens and Sneed (1962), Donaldson (1973, 1975), Shehadeh (1975), Fontaine (1976), Harvey and Hoar (1979), Chondar (1980), Woynarovich and Horvath (1980), Pullin and Kuo (1981), Davy and Chouinard (1981), Sundararaj (1981), Lam (1982), Billard (1983), and Fostier and Jalabert (1982).

II. INDUCED MATURATION IN FISH CULTURE

The application of induced-maturation techniques in fish falls into two basic categories. The first involves the induction of maturation and spawning in fish which would not otherwise reproduce in captivity; the second involves the manipulation of spawning time in fish which do normally reproduce in captivity.

A. Species That Do Not Reproduce Spontaneously in Captivity

A number of species that are currently of great economic significance or are of potential significance for freshwater, brackish water, or seawater aquaculture do not reproduce spontaneously in captivity. There are two reasons for this failure to reproduce.

1. The species is cultured outside of its natural range where environmental conditions are not conducive to the completion of normal sexual maturation. For example, the grass carp, *Ctenopharyngodon idellus*, which is indigenous to China, is now cultured in many different regions including the USSR, Europe, North America, Central America, and southeast Asia.

2. The environmental conditions in the culture container are dissimilar from those prevailing in the natural environment during spawning. The environmental factors which militate against normal spawning may induce specific physiological responses or fail to entrain the culmination of the normal reproductive cycle, e.g., inappropriate salinity, temperature, photoperiod, hydrostatic pressure, or lack of spawning substrate. Alternatively, the culture environment may induce a generalized stress response in the fish which inhibits or disables the reproductive process, e.g., density, water quality, flow rate, extremes of temperature, photoperiod, or salinity. The subject of stress evaluation in cultured fish has recently been reviewed (Donaldson, 1981). In some situations, it is possible to alter the environment in the broodstock enclosure to permit completion of sexual development and normal spawning. In other situations, it is the normal practice to induce final maturation, ovulation, and spermiation by hormone treatment.

B. Species That Reproduce in Captivity

Although the primary use of induced-spawning techniques is often thought to be directed toward those species that fail to reproduce in captivity, there is an increasing realization that the technique can be used to alter or synchronize the spawning time of species that do mature and produce gametes in captivity (e.g., the common carp, *Cyprinus carpio*, and the salmonids, *Salmo* sp. and *Oncorhynchus* sp.). Reproduction in these and other cultured species is usually restricted to a specific period of the year when environmental conditions are optimal. Hormone treatments can be used to (1) obtain viable gametes outside the normal spawning season and thereby permit intraspecific hybridization of genetically distinct stocks that have differing maturity dates, (2) permit the interspecific hybridization of related species that have different spawning seasons, (3) obtain gametes earlier and thereby lengthen the period available for rearing prior to stocking, transplanting, or releasing (Donaldson *et al.*, 1981a,b; Hunter *et al.*, 1981), (4) improve production efficiency by synchronization of groups of broodstock to ovulate and spermiate on predetermined dates (Donaldson *et al.*, 1981a,b; Hunter *et al.*, 1981), (5) facilitate the collection of gametes from wild broodstock in remote locations where facilities for holding adults are minimal, (6) minimize gamete losses attributable to prespawning mortality by accelerating maturation in captive broodstock which are being held in hatchery facilities under suboptimal environmental conditions (Hunter *et al.*, 1978, 1979; Sower *et al.*, 1982), (7) maximize survival of progeny by facilitating gamete production, fertilization, and incubation under relatively aseptic hatchery conditions rather than under seminatural pond conditions,

(8) stimulate spawning behavior in a small number of individuals which are then placed with a larger number of untreated fish to induce spawning, and (9) produce ova with a known ovulation time for experimental studies such as induced gynogenesis or triploidy (Donaldson and Hunter, 1982; Refstie *et al.*, 1982).

III. INDUCED FINAL MATURATION AND OVULATION

A. Levels of Intervention

Reproduction in female teleosts is controlled by the hypothalamic– pituitary–ovarian axis (Donaldson, 1973). This sequential mechanism allows for intervention at several levels to promote or interfere with the maturation process. Until the mid-1970s, procedures for induced ovulation in fish were limited to what can be referred to as the first generation technique. This consisted of hypophysation (von Ihering, 1937), i.e., the injection of homologous or heterologous pituitary extracts which contained gonadotropin. During the 1970 s, this first generation technique was refined by the production at the experimental (Donaldson *et al.*, 1972b; Donaldson, 1973), pilot, and, finally, application scale of partially purified piscine gonadotropin expressly for the purpose of induced final maturation and ovulation in teleosts. During this period, successful spawning was induced by mammalian gonadotropins especially human chorionic gonadotropin (HCG) either alone or in conjunction with fish pituitary extracts. The use of fish or mammalian pituitary preparations involves the collection and processing of materials from biological sources. The size and complexity of the typical gonadotropin molecule makes it unlikely that synthetic piscine gonadotropin will be available in the foreseeable future, unless it can be produced in quantity by genetic engineering in microorganisms as has recently been achieved for human growth hormone and insulin.

The search for synthetic alternatives to gonadotropin in the 1970s has led to the development of second generation techniques for the induction of final maturation, ovulation, and spermiation in fish. These techniques either operate at a higher level in the axis by stimulating the production and/or release of gonadotropin in the pituitary gland (e.g., antiestrogens and gonadotropin releasing hormones) or they operate at a lower level of the axis by supplying ovarian hormones (e.g., 17α,20β-dihydroprogesterone and prostaglandins) which would normally be stimulated by endogenous or exogenous gonadotropin. These second generation ovulation inducers are all

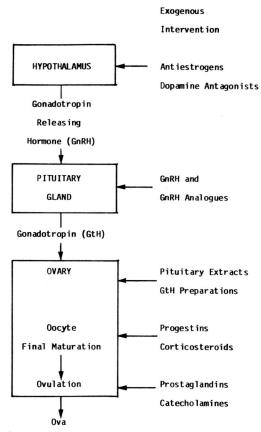

Fig. 1. Levels of external intervention in the hypothalamic–pituitary–ovarian axis which can be utilized to induce maturation and ovulation in teleosts.

small molecules relative to gonadotropin, and can be produced synthetically. They can also be expected to be relatively free of species specificity. A number of the possible levels of intervention in the hypothalamic–pituitary–ovarian axis are illustrated in Fig. 1.

B. Use of Specific Compounds

1. ANTIESTROGENS

A significant amount of evidence has been accumulated which indicates that gonadotropin secretion in teleosts, as in higher vertebrates, is regulated by negative feedback of gonadal steroids. In female catfish, *Heteropneustes fossilis,* injection of estradiol or testosterone inhibited vitellogenesis and caused atresia of mature oocytes (Sundararaj and Goswami, 1968), and estradiol inhibited compensatory ovarian growth after unilateral ovariectomy (Goswami and Sundararaj, 1968). In the sockeye salmon, *Oncorhynchus nerka,* bilateral gonadectomy caused an activation of the gonadotrops (McBride and van Overbeeke, 1969) which was reversed by treatment with estradiol or estradiol cypionate (van Overbeeke and McBride, 1971). The presence of this negative-feedback control of gonadotropin secretion by estrogen led to spawning trials with antiestrogens. Antiestrogens are synthetic compounds that are capable of competing with estrogen for binding sites on estrogen receptors. Two of these nonsteroidal compounds from the triphenylethylene series, which have been used in both higher and lower vertebrates, are clomiphene citrate [a racemic mixture of zuclomiphene citrate and euclomiphene citrate (2-[*p*-(2-chloro-1,2-diphenylvinyl)phenoxy] triethylamine dihydrogen citrate)] and tamoxifen [*trans*-1-(*p*-β-dimethyl-amineethoxyphenyl)-1,2-diphenylbut-1-ene] (Fig. 2a). The successful induction of ovulation in women with clomiphene was reported by Greenblatt *et al.* (1961) and with tamoxifen by Klopper and Hall (1971). The action of these antiestrogens in higher vertebrates has not been fully elucidated partly because of the fact that these compounds have both estrogenic and antiestrogenic effects, the proportion varies with differences in estrogen receptors between species and tissues within individual species (Patterson, 1981; Adashi *et al.*, 1981).

The first studies of induced ovulation with an antiestrogen in teleosts

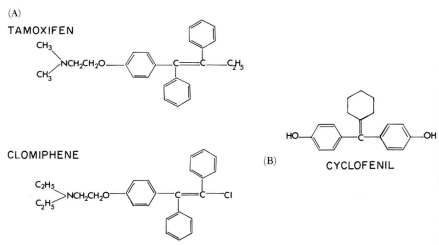

Fig. 2. A. Structure of the antiestrogens tamoxifen and clomiphene. B. Structure of cyclofenil. (free diol F6060 is shown; F6066 is diacetate).

were conducted on goldfish (*Carassius auratus*). Gravid females held at 13°C on a 16-hr photoperiod were injected daily ip with 1 mg/kg or 10 mg/kg clomiphene in 0.6% saline. In both groups, 90% of the fish ovulated after between 1 and 4 days of treatment, but no controls ovulated (Pandey and Hoar, 1972). Goldfish hypophysectomized 4–7 hr prior to injection failed to ovulate after clomiphene treatment, but sham-operated goldfish ovulated normally, indicating that the action of clomiphene in the goldfish is at the hypothalamic or pituitary level (Pandey *et al.*, 1973). Clomiphene appears to lack estrogenic as opposed to antiestrogenic activity in goldfish because it failed to induce vitellogenesis in goldfish hypophysectomized 12 weeks previously (Pandey and Stacey, 1975). Forced feeding of approximately 50 mg clomiphene per fish every 48 hr to a total of 250 mg in late November and early December failed to alter the spawning date of maturing plaice (*Pleuronectes platessa*); both treated and control fish ovulated in March and early April (Pullin, 1975). Treatment of intact *Heteropneustes fossilis* with 150 μg clomiphene per fish (approximately 3 mg/kg) per day ip induced ovulation in all fish between day 3 and day 6; similar treatment of hypophysectomized fish did not induce ovulation (Singh and Singh, 1976). Further evidence for the release of gonadotropin by clomiphene was provided by the fact that ovarian ^{32}P uptake, indicating RNA synthesis, was significantly increased in intact catfish which received a single 150-μg injection of clomiphene, but not in hypophysectomized fish. Also, in clomiphene-treated fish, plasma and pituitary gonadotropin levels, determined by ^{32}P-uptake bioassay, were significantly higher than those in control fish (Singh and Singh, 1976). Another compound cyclofenil (sexovid, F6066), *bis*-(-*p*-acetoxyphenyl)-cyclohexylidene methane (Fig. 2b), which is a nonsteroidal compound with endocrine properties and has been used in mammals as an ovulation stimulant, was also tested in the aforementioned series of experiments on catfish. At the same dose as clomiphene (150 μg/fish/day), ovulation was induced in all fish between days 2 and 4. Hypophysectomy inhibited ovulation in only 22% of sexovid-treated fish, suggesting that it has an effect at the ovarian level as well as the hypothalamic–pituitary level. Sexovid also increased ovarian ^{32}P uptake in both intact and in hypophysectomized catfish, and increased bioassayable gonadotropin levels in the pituitary and plasma. However, the increase in pituitary gonadotropin was less than in clomiphene-treated fish (Singh and Singh, 1976).

Injection of the loach, *Misgurnus anguillicaudatus*, with clomiphene, 1 or 10 mg/kg daily at 17°–22°C, resulted in ovulation in 2–4 days at the low dosage and 2–3 days at the higher dosage. Especially at the high dosage, ovulation was incomplete in the anterior portion of the ovary (Ueda and Takahashi, 1976). Injection of clomiphene, 1 mg/kg at 13°–15°C, into loach in the prespawning period also induced spawning, but after a longer period

(6–9 days at 1 injection per day and 6–15 days at 1 injection every 3 days). Gonadotrops in these clomiphene-treated fish contained fewer small granules, and the cisternae of the rough endoplasmic reticulum were dilated indicating gonadotropin release (Ueda and Takahashi, 1977a). In a further experiment, Ueda and Takahashi (1977b) examined the effect of clomiphene, 1 mg/kg every 3 days for 30 days, at 23–25°C and 18-hr photoperiod, on immature female goldfish. In this period, fish with ovaries in the early vitellogenic phase advanced to the secondary or tertiary yolk stage; fish in the primary yolk stage advanced to the tertiary yolk stage. Oocytes in the late perinucleolar stage advanced to the yolk vesicle stage. However, clomiphene had no effect on oocytes in the early perinucleolus stage. Therefore, clomiphene is promising as a stimulant of ovarian development providing that vitellogenesis has been initiated.

To partially elucidate the mechanism of action of clomiphene in female teleosts, Breton *et al.* (1975) gave a single ip injection of 0.1, 1.0, or 10.0 mg/kg clomiphene to carp (*Cyprinus carpio*) of varying maturity held at 18°C on a 12-hr photoperiod and measured plasma gonadotropin. At 0.1 mg/kg, all females responded between 16 and 28 hr, but the plasma gonadotropin maxima were less than 50 ng/ml. At 1.0 mg/kg, two females responded between 32 and 40 hr, and others responded between 60 and 72 hr. About 50% of the fish exhibited gonadotropin maxima in the 140–290 ng/ml range. At 10.0 mg/kg only two of six fish gave a significant response at 52–56 hr. It is difficult to interpret the data without additional information about the state of maturity of the fish; however, the data do clearly indicate that the antiestrogen clomiphene is capable of stimulating gonadotropin release in the female teleost, and that the gonadotropin maxima can exceed the concentration measured in ovulating goldfish (Breton *et al.* 1972).

To localize the site of estrogen negative feedback in the teleostean hypothalamic–hypophyseal axis, Billard and Peter (1977) implanted 4.2 μg clomiphene or tamoxifen in cocoa butter pellets in the hypothalamus and pituitary of sexually mature goldfish and determined serum gonadotropin concentration. Clomiphene implanted in the pituitary gland induced a significant increase in plasma gonadotropin to 80 ng/ml at 24 and 72 hr. Tamoxifen implanted in the pituitary induced a greater response, 145–200 ng/ml, and also induced a response when implanted in the nucleus lateral tuberis (NLT) (40 ng/ml at 72 hr). These data indicate that the pituitary is the primary site of action of antiestrogens in the goldfish with the NLT involved to a lesser extent. In fact, the NLT is very close to the pituitary gland, and Billard and Peter (1977) note that the antiestrogen may have diffused or been transported from this location to the pituitary. Some, but not all of the goldfish in the aforementioned study that had a serum GtH concentration above 75 ng/ml 24 hr after implantation ovulated. Specifically, five of six fish which received pituitary implants of clomiphene ovulated, but only one of five of the tamoxifen pituitary-implant group ovulated; however, three of this latter group were found to have atretic ovaries. If a simple method could be devised for implantation of antiestrogen in the pituitary gland, this might be an effective technique for inducing ovulation.

Bieniarz *et al.* (1979) attempted to induce ovulation in carp by injecting clomiphene 1 mg/kg/day for three days. This treatment increased the proportion of ova having the germinal vesicle in the peripheral position, but did not induce ovulation. Other fish in the same experiment did ovulate after treatment with 1 mg/kg/day carp hypophyseal homogenate for 3 days, suggesting that the clomiphene dosage was too low. In another study on the carp, Kapur and Toor (1979) successfully induced ovulation with clomiphene in fish which had been pretreated with indomethacin, 5 or 10 mg/kg, to block spontaneous ovulation by inhibition of prostaglandin synthesis. The effective dosage of clomiphene was 10 mg/kg given 7 hr and 31 hr after indomethacin treatment. Single doses of 5 mg/kg clomiphene at 7 hr or 31 hr did not induce ovulation. Control carp, which did not receive indomethacin, all ovulated after 23 hr in the spawning hapas. In the absence of clomiphene effects in hypophysectomized fish (Pandey *et al.*, 1973), the ability of the clomiphene to override the effect of indomethacin suggests that prostaglandin may be involved both at the hypothalamic–hypophyseal level and at the ovarian level.

Donaldson *et al.* (1978b, 1981b) have examined the feasibility of inducing final maturation and ovulation in Pacific salmon with the antiestrogen tamoxifen. In the first study, conducted in November, 1977, coho salmon (*Oncorhynchus kisutch*) received a primer injection of 0.1 mg/kg partially purified salmon gonadotropin (SG-G100) followed on day 3 by 1 or 10 mg/kg tamoxifen ip in peanut oil. Ova in both treated groups underwent 100% germinal vesicle breakdown (GVBD) and 85% ovulation compared to an eventual 84.6% GVBD and 77% ovulation in control fish (Fig. 3). Mean times to ovulation at 10°C were 11.4 days in the 0.1-mg SG-G100–10-mg tamoxifen group, 16.1 days in the 0.1-mg SG-G100–1-mg tamoxifen group, and 19.8 days in the peanut-oil control group. Ovulation in the high dosage tamoxifen group was much more closely synchronized than in the control group; all fish ovulated on days 9–13 compared to days 12–32 in the control group. Further studies were conducted during the fall of 1978 to determine whether tamoxifen could be used to induce ovulation in salmon prior to the initiation of the normal spawning season and without the SG-G100 primer (Donaldson *et al.*, 1981b). In this study, coho salmon were injected with 4 mg tamoxifen on day 0, 10 mg on day 3, and 10 mg on day 6. No treated fish ovulated before November, but control fish started to ovulate in late October. Preovulatory ova in treated fish sampled 41 days after treatment were

saline of 50, 100, or 200 μg LHRH/fish (body weight 17.5–36.4 g). After 2 days, 40%, 50%, and 80% of the fish had ovulated at the low, medium, and high dosages, respectively. Treatment with LHRH was also accompanied by an increase in the water and sodium content of the ovary. Again using synthetic LHRH, Lam *et al.* (1975) were able to demonstrate ovulation in LHRH-injected mature female goldfish at 12°C. Daily ip injection of 10 mg/kg or intracranial injection (not into the brain) at 2 mg/kg resulted in 75% and 100% ovulation, respectively, after 4–7 days of treatment. In a subsequent experiment, the effect of lower dosages of LHRH was investigated in goldfish. At 0.1–1.0 mg/kg ip daily for 5 days, 33–80% of the fish ovulated, but after intracranial injection over the same dosage range 0–100% of the goldfish ovulated (Lam *et al.*, 1976).

In 1975, the first report appeared of the use of LHRH to spawn five species of cultured fish indigenous to the People's Republic of China (Conference on Application of Hormones to Economic Fish, 1975). The study compared pure versus "crude" LHRH, intraperitoneal and intramuscular versus intracranial injection, single versus multiple injection, and LHRH alone versus LHRH supplemented with piscine pituitary extract or HCG. The variable size of the treatment groups and variety of treatment regimes utilized does not permit statistical comparison; however, some generalizations are possible. Spawning was induced in females of all five species with an overall success rate of 64.5%, and fertilization and hatching were normal. Crude LHRH was less effective than pure LHRH. *Ctenopharyngodon idellus* (grass carp), *Hypophthalmichthys molitrix* (silver carp), and *Megalobrama ambylocephala* (blunt-snout bream) spawned after receiving a minimum total of 1.6–2.5 mg/kg pure LHRH in two or three injections; however, the grass carp and *Aristichthys nobilis* (spotted silver carp) spawned after receiving a minimum total of 5 mg/kg crude LHRH in one or two injections. Intracranial injection required a lower dosage of LHRH; a minimum total of 68 μg/kg pure LHRH over three injections was effective in one silver carp, and higher dosages up to 1 mg/kg were effective in others. Most of the silver carp that received intracranial injections of LHRH did not spawn. Intracranial injection of a total of 0.3 mg/kg or more crude LHRH in two injections induced spawning in grass carp, spotted silver carp, and blunt-snout bream. Intracranial injection of LHRH, with or without pituitary extract, was slightly effective in *Mylopharyngodon piceus* (black carp). The Chinese researchers concluded that LHRH is effective over a wide dosage range, and that its use at high dosages does not cause prespawning mortality by abdominal distension similar to that caused by high dosages of HCG. In a subsequent study on the mechanism of action of LHRH in the grass carp, spawning was induced by injection of 0.5 mg/kg and 6.5 mg/kg LHRH spaced 8 hr apart (Anonymous, 1978).

In the Japanese medaka, *Oryzias latipes*, ovarian maturation and ovulation has been induced with LHRH in females held under conditions of warm temperature and short photoperiod, which normally inhibit gonadal development (Chan, 1977). Fish received 1, 10, or 100 μg/kg or 1 mg/kg LHRH ip in saline twice per week for 6 weeks. The two higher dosages caused a significant increase in gonadosomatic index and in the number of oocytes containing yolk. Ovulation occurred in 5 out of 12 fish at the highest dosage (1 mg/kg). These results indicate clearly that LHRH can be used both to stimulate ovarian development and to induce ovulation.

The Japanese plaice (*Limanda yokohamae*) has been demonstrated to respond to LHRH when injected im at a dosage of 1 mg/fish emulsified in complete adjuvant, but not when injected with a similar dosage in saline (Aida *et al.*, 1978). In this experiment, the plaice (average weight 189 g) received 5.3 mg/kg LHRH in saline on days 1 and 3. No fish ovulated, and 50% of the fish were given 5.3 mg/kg LHRH in adjuvant on days 6 and 9, as were an additional group that had received saline only on days 1 and 3. All fish ovulated between days 11 and 13, but the fish that had received LHRH in saline on days 1 and 3 generally ovulated sooner. In a subsequent examination of the dose response to a single im injection of LHRH in adjuvant, all fish ovulated within 12 days (average 4.2 days) at 5.3 mg/kg, 80% ovulated (average 6.75 days) at 1.1 mg/kg, and 40% ovulated (average 8.5 days) at 0.2 mg/kg. No controls ovulated during the 12-day period. These researchers also succeeded in inducing ovulation with LHRH in the goby, *Acanthogobius flavimanus* (average weight 46.9 g). A single im injection of 2.1 mg/kg in complete adjuvant induced ovulation in 55% of the gobies in 3–6 days, and 7.5 mg/kg induced ovulation in 100% of the gobies in 3–5 days. A dosage of 2.1 mg LHRH/kg in saline only induced ovulation in 10% of the gobies (Aida *et al.*, 1978). Aida *et al.* propose that the function of emulsion in adjuvant is to prolong the absorption of the LHRH.

In the common carp, repeated daily intrahypophyseal injection of 1 μg/kg LHRH caused germinal vesicle migration and GVBD, but not ovulation, although injection into the third brain ventricle or into muscle tissue had no effect (Sokolowska *et al.*, 1978). In the tench (*Tinca tinca*), 60% ovulation was obtained by injection of LHRH at 100 μg/kg (Kouril and Barth, 1981).

Studies on the induction of final maturation and ovulation in Pacific salmon with LHRH were initiated in November, 1977. A primer injection into mature coho salmon of 0.1 mg/kg partially purified salmon gonadotropin (SG-G100) on day 0 (Jalabert *et al.*, 1978a) was followed by ip injection of 1.6 mg/kg LHRH on days 3, 4, and 5. Germinal vesical breakdown was complete in all fish and ovulation occurred in 85.7% after a mean interval of 11.7 days from the first injection (Fig. 3) (Donaldson *et al.*, 1978b, 1979, 1981a).

In the following year, treatment in October using a similar protocol, 0.1 mg/kg SG-G100 followed by 2 mg/kg LHRH, was again successful. By day 11, 60% of the treated group had ovulated compared to 14% of the saline control group. In November, 1978, an attempt was made to accelerate spawning in coho salmon using either ip or im injection of a primer consisting of 0.1 mg/kg LHRH followed on days 3 and 5 or days 3, 4, and 5 by injection of 1.0 mg/kg LHRH. None of these groups ovulated significantly sooner than the control group. However, a group which received a 0.1 mg/kg SG-G100 ip primer followed by 1.0 mg/kg LHRH did ovulate sooner than the control group (Donaldson et al., 1979, 1981a).

   b. Luteinizing Hormone Releasing Hormone Analogues. This discussion has focussed on the use of LHRH to induce ovulation in teleosts. However, biochemical research on LHRH did not end with its isolation and characterization in 1971. Soon after the amino-acid sequence of the natural LHRH decapeptide was described the des-Gly[10]-LHRH(1–9) ethylamide nonapeptide analogue was synthesized and demonstrated to have a higher biological activity than LHRH in a mammalian assay (Fujino et al., 1972). Substitution in position six with D-alanine also provided an analogue (D-Ala[6]LHRH) with increased potency (Monahan et al., 1973). When these two substitutions were combined to produce the D-Ala[6], des-Gly[10]-LHRH(1–9) ethylamide analogue (LHRHa D-Ala[6]) (Fig. 5) it was found to have a biological activity 35–50 times that of the native LHRH in mammals (Coy et al., 1974; Fujino et al., 1974). Other highly active analogues have been synthesized by substituting the des-Gly[10]-LHRH(1–9) ethylamide in position six with D-serine (Bu[t]) (Fig. 5) (Konig et al., 1975) or with D-tryptophane.

   A number of proposals have been made to explain the increased potency of the highly active analogues in mammals. The receptor binding conformation may be improved by the alkylamine substitution in position 10 or the D amino acid in position 6 (Coy et al., 1979). Further, the analogues may be subject to a slower rate of enzymic degradation, thereby increasing their half life (Buckingham, 1978). In this regard, the [D-Ser(Bu[t])[6]]-LHRH(1–9) eth-

pyroGlu-His-Trp-Ser-Tyr-D-Ala-Leu-Arg-Pro-NHC$_2$H$_5$

   1   2   3   4   5   6   7   8   9

D-Ala[6]desGly[10]-LH-RH(1-9)ethylamide

pyroGlu-His-Trp-Ser-Tyr-D-Ser(Bu[t])-Leu-Arg-Pro-NHC$_2$H$_5$

   1   2   3   4   5   6   7   8   9

[D-Ser(Bu[t])[6]]desGly[10]-LH-RH(1-9)ethylamide

Fig. 5. Amino-acid sequences of two high potency LHRH analogues.

ylamide analogue [LHRHa D-Ser(But)^6] has been shown to be degraded 27 times more slowly than LHRH (Clayton and Shakespear, 1978).

The availability of LHRH analogues (LHRHa), which are highly active in mammalian assays, combined with the knowledge that LHRH itself can induce ovulation in teleosts, if given at a sufficient dose and in an appropriate manner, has led to spawning trials with LHRHa in teleosts.

Chinese scientists were the first to examine the feasibility of using a potent LHRHa to induce ovulation in cultured fish. The overall spawning success in Chinese carps injected with the LHRHa D-Ala^6 nonapeptide was 78.5%, and the success rate in carps treated with a combination of LHRHa D-Ala^6 and fish pituitary extract was 75% (Cooperative Team for Hormonal Application in Pisciculture, 1977). In grass carp a single injection dose of 1–100 μg/kg was used. The response interval was 12–22 hr within a temperature range of 20–27°C, and the spawning rate was 86.3%. When two injections were used in this species, the initial dose of 1–7 μg/kg was followed by a second dosage of 5–63 μg/kg after 7–12 hr. The grass carp spawned 3–14 hr after the second injection at a success rate of 82%. The recommended treatment for grass carp is a single dose of 5–10 μg/kg at a temperature not exceeding 28°C. In the black carp, LHRHa D-Ala^6 at a dosage of 9–500 μg/kg divided between two injections was only 43% effective. A priming dose of LHRHa D-Ala^6 followed by two combination treatments with fish pituitary extract was found to be 75% effective, and a recommended treatment of 10 μg/kg LHRHa D-Ala^6 combined with 1–2 mg fish pituitary extract in a single injection at 20°–28°C was proposed. In the silver carp and spotted silver (bighead) carp, LHRHa D-Ala^6 was administered at a dosage of 1.3–300 μg/kg to produce a spawning rate of 71.9% and 78.7% in the two species. The recommended treatment for these species was a priming dose of 2 μg/kg LHRHa D-Ala^6 followed after 7 or 26 hr by a second dose of 10 μg/kg at 20°–28°C. The response interval after this treatment, 8–9 hr, is similar to that obtained with fish pituitary extract or HCG. It should be noted that there were no control groups in this series of experiments.

In a recent study on induced ovulation of Chinese carps (Freshwater Commercial Fish Artificial Propagation Work Groups in Fujian, Jaingsu, Zhejiang, and Shanghai, 1977), further attempts were made to optimize LHRHa D-Ala^6 administration especially in fish which had been treated in previous years with HCG. Silver carp that had not spawned previously were very sensitive to a single injection of LHRHa D-Ala^6 at 1–100 μg/kg. The recommended dosage was 10 μg/kg. The response time of 22–24 hr was longer than that observed in the grass carp. Silver carp that had been induced with HCG in previous years were less sensitive to LHRHa D-Ala^6 and required two doses totaling 10–25 μg LHRHa D-Ala^6 24 hr apart. Mature black amur received a primer injection of 1–3 μg LHRHa D-Ala^6 followed

24–40 hr later by 6–9 µg LHRHa D-Ala[6] and 0.4–2 mg/kg fish pituitary extract. Less mature black amur received 15–23 µg/kg LHRHa D-Ala[6] and 2 mg/kg fish pituitary extract divided between three injections. In black amur, a preparatory injection of 2–5 µg LHRHa D-Ala[6] per fish in a 1:1 tea oil–aqueous emulsion was given 2–20 days before induced ovulation to accelerate the maturation process. In grass carp, a single injection of 10 µg LHRHa per kg was recommended. If the female grass carp is given two injections and the male a single injection maturation is not synchronous. In the bighead carp the recommended treatment is 1 µg/kg LHRHa D-Ala[6] followed 12 hr later in mature carp by 8–9 µg/kg LHRHa D-Ala[6]. In less mature bighead carp the interval between injections is 24 hr.

In the silver carp, a number of additional investigations were conducted on the use of LHRHa D-Ala[6] (Freshwater Commerical Fish Artificial Propagation Work Groups in Fujian, Jaingsu, Zhejiang, and Shanghai, 1977). Injection of silver carp with 2 µg LHRHa D-Ala[6] and 500 µg cAMP/kg followed by 8 µg LHRHa D-Ala[6] and 500 µg cAMP/kg was reported to be effective, and lower doses of 100–150 µg cAMP combined with LHRHa D-Ala[6] were also reported to be effective. Silver carp, which had been treated with HCG in previous years, developed antibodies to HCG, and induced ovulation was more difficult in these fish. It was also noted that HCG-treated fish required increasing dosages in successive years. The slow response time observed at low temperatures was more noticeable with LHRHa D-Ala[6] than with HCG or fish pituitary extract, but no statistical data were presented. The mortality rate in spawners was reported to be much lower when LHRHa D-Ala[6] was used compared to HCG or fish pituitary extract. The LHRHa D-Ala[6] was also successfully used in the People's Republic of China to induce spawning in the Japanese eel (*Anguilla japonica*) (Research Group of Eel Reproduction, 1978).

In the coho salmon, it has been demonstrated that the high potency analogues of LHRH are very effective for induction of final maturation and ovulation when used either alone or in conjunction with a gonadotropin primer (Donaldson *et al.*, 1979, 1981a, 1982, 1984; Van Der Kraak *et al.*, 1982, 1983b; Sower *et al.*, 1982). In the initial study in November, 1977, a primer of 0.1 mg/kg SG-G100 followed by 0.2 mg/kg LHRHa D-Ala[6] on days 3, 4, and 5 was found to be more effective than the same primer followed by an eightfold larger amount of LHRH. Moreover, all the treated salmon ovulated in the single sampling interval prior to day 10 (Fig. 3). Survival to the eyed-egg stage was 98% in this group compared to 93% in a control group (Donaldson *et al.*, 1981a). In November, 1978, a second LHRHa study was conducted using coho salmon, this time with LHRHa D-Ser(But)[6]. The 0.1 mg/kg SG-G100 primer was followed on days 3, 4, and 5 with 100, 33, or 11 µg/kg LHRHa D-Ser(But)[6]. The response appeared to be dose related and the 100-μg/kg group ovulated after a mean interval of 7.9 days, which was significantly sooner than the saline control group. Survival to hatching in the 100-μg/kg group was 97.6% compared to 93% in the control group. Although the effect of the 100 μg/kg LHRHa D-Ser(But)[6] was significant, the utility of the lower dosages was difficult to interpret because of the lack of a control group which received the SG-G100 primer alone (Donaldson et al., 1981a). Therefore, in the following year, again using coho salmon, comparison was made between the SG-G100 primer alone, and SG-G100 primer followed by 300 and 33 μg/kg LHRHa D-Ala[6] or LHRHa D-Ser(But)[6]. These two analogues were also compared at 300, 100, and 33 μg/kg preceded by an LHRHa primer at 20% of these dose levels instead of the SG-G100 primer. In this experiment, in which the salmon were relatively close to maturity, all treatment groups using either of the analogues with or without SG-G100 ovulated earlier than control fish, indicating for the first time in the salmonids that ovulation could be induced using LHRHa alone (Donaldson et al., 1982, 1984). In 1980, LHRHa D-Ala[6] from two different sources was compared using injections of 100, 33, or 11 μg/kg on days 2, 3, and 4 preceded by a primer injection on day 0 of 0.1 mg/kg SG-G100. The pooled 100-μg/kg group was 100% ovulated by day 10, but the saline control was only 14% ovulated at this time. The 11-μg/kg LHRHa D-Ser(But)[6] group did not ovulate sooner than the group that received the SG-G100 primer alone; however, both of these treatments were effective relative to the control group (Donaldson et al., 1982, 1984).

In the fall of 1980, the reason for the greater effectiveness of LHRHa compared to LHRH was determined in the Pacific salmon. When maturing female coho salmon received a single injection of 0.2 or 1.0 mg/kg LHRH, there was a dose-related increase in plasma gonadotropin, as determined by radioimmunoassay (RIA), which peaked at 1.5–3 hr and declined rapidly to close to basal levels at 24 hr (Fig. 6). However, single injections of 200 or 20 μg/kg LHRHa D-Ala[6] resulted in dose-related increases in plasma gonadotropin that were sustained for 96 hr (Van Der Kraak et al., 1983a). Ova in fish which had received LHRHa D-Ala[6] underwent GVBD in the 96-hr (Fig. 6) experimental period, but those in fish which received LHRH did not. Furthermore, in those fish that underwent GVBD, final maturation was preceded by a remarkable increase in the plasma concentration of 17α-hydroxy-20β-dihydroprogesterone (Van Der Kraak and Donaldson, 1983). These results explain the ineffectiveness of LHRH as a primer and support the finding that LHRHa D-Ala[6] is an effective inducer of ovulation in Pacific salmon.

In the fall of 1981, additional tests were conducted on coho salmon to further investigate the ability of LHRHa D-Ala[6] to induce ovulation (Van Der Kraak et al., 1982). A single injection of 0.2 mg/kg LHRHa D-Ala[6], or

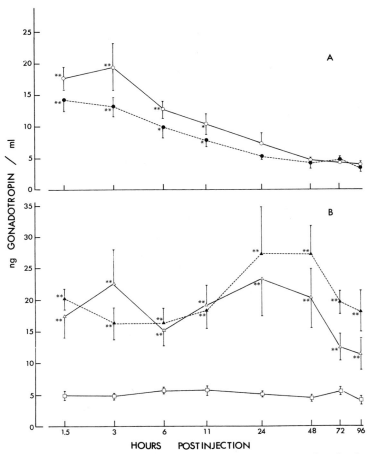

Fig. 6. Plasma gonadotropin concentrations in intact coho salmon (*Oncorhynchus kisutch*) at intervals after the injection of LHRH (A): $\bigcirc$, 1.0 mg/kg; $\bullet$, 0.2 mg/kg or (B) LHRHa D-Ala[6] ($\blacktriangle$, 0.2 mg/kg; $\triangle$, 0.02 mg/kg), or saline ($\square$). Reproduced with permission from Van Der Kraak *et al.* (1983a).

injection of 0.2 or 0.02 mg/kg LHRHa D-Ala[6] combined with 0.1 mg/kg SG-G100, was more effective than single injections of 0.02 mg/kg LHRHa D-Ala[6] or 0.1 mg/kg SG-G100. Two separate injections on days 0 and 3 of either 0.2 and 0.2 mg/kg, or 0.02 and 0.02 mg/kg LHRHa D-Ala[6] or 0.1 mg/kg SG-G100 and 0.2 mg/kg LHRHa D-Ala[6], were all somewhat more effective in inducing ovulation than the single injection treatments. These latest results indicate that LHRHa D-Ala[6] is an effective practical inducer of ovulation in Pacific salmon when one or, at most, two injections are given either alone or in combination with salmon gonadotropin.

Recently, ovulation has been induced in landlocked Atlantic salmon (*Salmo salar*) implanted in the perivisceral region with cholesterol pellets containing 125 μg (D-Trp[6]-des-Gly[10]) LHRH-ethylamide (Crim et al., 1983). Ovulation occurred between days 7 and 12 after implantation, a latent period similar to that observed in Pacific salmon (Donaldson et al., 1981a).

Not all attempts at causing ovulation in teleosts with LHRHa have been successful. In the pike (*Esox lucius*), a single ip injection of 1 μg/kg (D-Trp[6]-des-Gly[10]) LHRH-ethylamide emulsified in olive oil failed to induce ovulation (Billard and Marcel, 1980). This single low dose of the LHRHa was probably insufficient to elicit an ovulatory surge of gonadotropin.

In goldfish held at 12°C (T. J. Lam and S. Pandey, unpublished, in Lam et al., 1978; Lam, 1982) LHRHa D-Ala[6] is partially effective in inducing ovulation at a dose (10 μg/kg) at which LHRH is ineffective. Lin et al. (1984) were able to stimulate gonadal growth in regressed goldfish by injection of 10 μg or 1 mg/kg LHRHa D-Ala[6] daily for 10 days; however, ovulation was not induced (R. Peter, unpublished, in Lin et al. 1983). Recently, 50% of a group of female goldfish were induced to ovulate after two injections of 100 μg/kg LHRHa D-Ala[6] spaced 9 hr apart. These fish had been held at 12°C on a 16-hr photoperiod and the temperature was increased to 20°C 4 days before injection (R. Peter, unpublished, 1982). In cyprinids and perhaps other groups of teleosts, factors such as time of injection in relation to circadian rhythms, self-potentiation effects at certain injection intervals, gonadal stage at time of administration, and environmental effects (e.g., temperature) require further investigation.

To summarize progress made in the use of LHRHa to induce ovulation in teleosts of commercial significance, it can be said that the hope held out at the FAO World Conference on Aquaculture in Kyoto for the use of LHRHa as a practical tool (E. M. Donaldson, unpublished, 1976) is well on the way to being fulfilled. Further research is required to establish optimal dosages, modes of administration (e.g., in solution, in oil emulsion, or by pellet), and location of treatment. With regard to treatment costs, multiple injections may require less LHRHa; however, single injections or pellet implantations although requiring more hormone involve less handling of the fish. Further studies are also required to establish which are the most effective analogues.

A major step forward in the study of controlled reproduction in fish occurred recently with the isolation and sequencing of an LHRH-like decapeptide from the hypothalamus of the Pacific salmon having the sequence shown in Fig. 7 (Sherwood et al., 1982, 1983). This peptide has considerably less activity than LHRH in an *in vitro* rat bioassay. Recent tests in Pacific salmon have demonstrated that the salmon GRH can induce ovulation.

pyroGlu-His-Trp-Ser-Tyr-Gly-Trp-Leu-Pro-Gly-NH$_2$

    1   2   3   4   5   6   7   8   9   10

Fig. 7. Amino-acid sequence of GnRH isolated from the Pacific salmon (Sherwood *et al.*, 1982; 1983).

However, analogues based on the salmon GRH structure are more effective than the natural salmon GnRH (Donaldson *et al.*, 1983). The potency of salmon GnRH in other teleosts and the possibility of variations in GRH structure among various teleost genera remains to be explored. It is interesting to note that the differences between salmon and mammalian GRH can be explained on the basis of two single base pair changes in the DNA sequence of the GnRH gene. Synthetic analogues based on modifications of the teleost GnRH structure may be at least as effective and perhaps more effective in fish than those based on the mammalian LHRH structure which have been used to date.

3. DOPAMINE ANTAGONISTS

The recent accumulation of evidence from preoptic lesioning experiments in favor of the presence of a gonadotropin release-inhibitory factor (GRIF) in the goldfish (Peter *et al.*, 1978; Peter and Paulencu, 1980; Peter, 1982), suggests another possible mechanism for induced ovulation in those teleosts where GRIF is present. The catecholamine, dopamine, and its agonist apomorphine reduce the increase in plasma gonadotropin concentration associated with preoptic lesions, which suggests that GRIF in goldfish may be dopamine (Chang and Peter, 1982, 1983a; Chang *et al.*, 1983). Furthermore, injection of the dopamine antagonist pimozide into intact goldfish caused an increase in plasma gonadotropin concentration (Chang and Peter, 1982, 1983a). Injection of dopamine into LHRHa D-Ala[6]-treated goldfish blocked the normal increase in plasma gonadotropin associated with LHRHa D-Ala[6] injection; however, injection of pimozide into LHRHa D-Ala[6]-treated goldfish potentiated the effect of the LHRHa D-Ala[6] on plasma gonadotropin and induced GVBD and ovulation (Chang and Peter, 1982, 1983b). Therefore, in teleosts, where GRIF plays a significant role in the regulation of gonadotropin release from the pituitary gland, injection of the dopamine antagonist pimozide either alone or in combination with a suitable potent LHRH analogue may prove to be an effective means of ovulation induction.

4. GONADOTROPINS

Gonadotropins are hormonal glycoproteins of pituitary or placental origin which stimulate gonadal development and function. Gonadotropins used for induced ovulation in fish fall into two categories according to whether they are of piscine or mammalian origin. Mammalian gonadotropins can be further segregated depending on whether they are of pituitary origin, i.e., luteinizing hormone (LH) and follicle stimulating hormone (FSH), or of placental origin, i.e., human chorionic gonadotropin (HCG) and pregnant mare serum gonadotropin (PMSG). From a practical point of view, piscine gonadotropin preparations may be subdivided according to their degree of purification. These preparations range from the aqueous extracts of whole pituitary glands used in the traditional hypophysation procedure through acetone-dried pituitary powders, lyophilized pituitaries, lyophilized aqueous extracts of acetone-dried pituitary powders, precipitated ethanol extracts of pituitary glands, to partially purified gonadotropins prepared by gel filtration, with or without ion-exchange chromatography, or by affinity chromatography on concanavalin A–Sepharose.

Teleost gonadotropin has a molecular weight of about 30,000 (Burzawa-Gerard, 1971, 1974; Donaldson et al., 1972b; Idler et al., 1975a), and consists of two subunits (Donaldson, 1973; Burzawa-Gerard et al., 1976; Pierce et al., 1976; Lo et al., 1981), as do the gonadotropins in other vertebrates (Licht et al., 1977). Besides this normal carbohydrate-containing fish gonadotropin, there may be a second fish pituitary hormone with a lower carbohydrate content having gonadotropic functions such as the ability to accelerate yolk uptake into the oocyte during vitellogenesis (Idler et al., 1975b; Idler and Ng, 1979, also this volume; Ng and Idler, 1978b, 1979). Further studies are required to define whether this protein is in fact a gonadotropin, and whether it has a physiological role in oocyte development that could be exploited by fish culturists.

a. Fish Pituitary Extracts. As indicated in the introduction, aqueous extracts of homoplastic or heteroplastic pituitary glands freshly dissected from sexually maturing donor fish were the first materials to be used for the induced spawning of cultured fish (Houssay, 1931; von Ihering, 1937). This procedure has stood the test of time and in many regions of the world is still the practical means by which spawning is induced in many types of cultured fish (Harvey and Hoar, 1979; Davy and Chouinard, 1981; Rothbard, 1981). The advantages of the technique are that (1) it is simple, requiring little in the way of capital facilities or materials, (2) it does not require refrigerated storage, (3) dosage can be simply calculated based on the relationship between body weight of donor and recipient, and (4) other hormones present in the extract may have a synergistic effect. Disadvantages are (1) the lack of potency standardization, (2) that donor fish of equivalent or preferably lower value must be killed at the spawning location to obtain pituitary glands, (3) the fish may develop an immune reaction to the gonadotropin or other

proteins in the extract over a period of time, especially in the case of heteroplastic pituitaries, (4) nongonadotropic hormones present in the crude extract may cause deleterious effects and negate or modify the effect of the
gonadotropin, and (5) it is conceivable that injection of a fresh pituitary
extract could possibly transmit disease organisms from the donor to the
recipient. In the common carp, storage of heads for 24 hr at 4°C prior to
removal of the pituitary gland did not effect gonadotropic potency, also
pituitaries from small ripe fish had the same gonadotropin content as large
ripe fish (Yaron et al., 1982).

The hypophysation procedure has recently been capably reviewed
(Harvey and Hoar, 1979; Sundararaj, 1981; Rothbard, 1981; Lam, 1982). The
normal methodology involves two intramuscular injections of aqueous pituitary extract. Procedural variations between hatcheries include the following:
(1) choice of donor species and degree of maturity, (2) evaluation of maturity
of recipient, (3) homogenization technique and temperature, (4) separation
of extract from residue by centrifugation, filtration, or settling, (5) addition of
other compounds, e.g., glycerine to increase viscosity and thereby prevent
loss after injection or to prolong uptake, (6) location of injection, (7) injection
volume, (8) water temperature at the time of injection, (9) division of dosage
between first (priming) and second (resolving) dose, (10) time of day when
the first injection is given, (11) interval between the first and second injections, (12) determination of spawning readiness, (13) use of an anesthetic,
and (14) decision between natural spawning and stripping. The existence of
this great number of variables has meant that in practice no single study or
series of studies has investigated all the relevant variables in a single species.
Variables of major importance are (1) donor species, (2) maturity of spawners, (3) dosage, (4) division of dosage, (5) timing, and (6) temperature.

The next step beyond the basic hypophysation procedure with fresh
pituitary extract involves the use of stored pituitary glands or extracts.
Whole glands can be dehydrated and defatted by several changes of cold
acetone and then stored in whole or powdered form, or they can be
lyophilized or stored frozen at −20°C or below for a number of years without
significant loss of potency (Donaldson et al., 1978a). Alternatively an aqueous extract can be prepared from the pituitary glands or acetone powder and
then stored either in the frozen lyophilized form or in the presence of a
preservative such as glycerol. In the latter case, the material is stored in
glass ampules. An advantage of the acetone powders or preserved extracts is
that they can be made in a homogeneous form from a large number of
pituitaries, and then standardized by bioassay of a small aliquot. Other
advantages are that the pituitaries of donor fish can be collected and pro-
cessed at various times and/or locations eliminating the need for on-site
donor fish. Acetone-dried pituitaries were first used by von Ihering and

Azevedo (1936) to spawn *Astyanax taeniatus* and *A. bimaculatus,* and the procedure has been used extensively since then (see Table 46 in Pickford and Atz, 1957). Acetone-dried carp pituitaries have been successfully used after 10 years of storage (A. D. Hasler, R. K. Meyer, and W. J. Wisby, unpublished, 1950, in Pickford and Atz, 1957). Fresh pituitary glands can be placed directly into dry-ice cooled acetone (Burrows *et al.,* 1952; Palmer *et al.,* 1954); however, it may be more convenient to freeze fresh pituitary glands on a slice of dry ice in a styrofoam container and then conduct the drying or extraction procedure at the laboratory. The use of dry ice obviates the need for electrical refrigeration equipment in the field. In general, it is recommended that the acetone be changed two or three times to ensure complete dehydration and defatting.

An alternative to acetone drying is dehydration in absolute alcohol (de Azevedo and Oliveira, 1939; Fontenele, 1955; Pickford and Atz, 1957; Ibrahim and Chaudhuri, 1966; Chondar, 1980; Rothbard, 1981). The defatted dry weight of pituitary glands is 13–17% of the wet weight (Pickford and Atz, 1957), and the dry weight of Pacific salmon pituitary glands after lyophilization is 17% of the wet weight (E. M. Donaldson, unpublished).

Aqueous extracts of fresh or dried pituitaries can be preserved by addition of glycerine 2:1 (Ribeiro and Neto, 1944; Ibrahim and Chaudhuri, 1966; Ibrahim, 1969a,b) or by lyophilization (Syndel Laboratories, unpublished). In the goldfish, injection of solutions containing more than 10% glycerine cause peritoneal inflammation (Clemens and Sneed, 1962). Extraction of ethanol-preserved pituitary glands with trichloroacetic acid has not been successful (Bhowmick and Chaudhuri, 1968).

In general, aqueous extracts of fresh or preserved pituitary glands from mature fish are all that is required for induced ovulation on a practical basis. A number of hypophysation studies from a range of species have been tabulated recently by Harvey and Hoar (1979) and Lam (1982).

b. Purified Fish Gonadotropin Preparations. Since the 1970s, partially purified and purified fish gonadotropin preparations have been produced that have a higher specific activity than whole-pituitary preparations. All purification procedures developed to date involve some loss of gonadotropic activity, and as a consequence, the end product has a higher gonadotropic activity per unit weight, but contains less total activity than the starting material.

Therefore, these preparations are more suited to research and development studies rather than direct application in fish culture. The advantage from a research point of view is that purified gonadotropins (1) are more or less free of other pituitary hormones, (2) can be standardized by *in vivo* or *in vitro* bioassay, radioreceptor assay, or radioimmunoassay, (3) can be stored for many years, and (4) are readily soluble in water or saline with no residue. Partially purified or purified gonadotropins have been prepared from the pituitary glands of only a small number of teleosts using a combination of extraction and chromatographic procedures. These include the common carp, *Cyprinus carpio* (Burzawa-Gerard, 1971, 1974; Idler and Ng, 1979), chinook salmon, *Oncorhynchus tshawytscha* (Donaldson *et al.*, 1972b; Donaldson, 1973; Pierce *et al.*, 1976), chum salmon, *Oncorhynchus keta* (Idler *et al.*, 1975a,b; Yoneda and Yamazaki, 1976; Idler and Hwang, 1978; Ng and Idler 1978a; Idler and Ng, 1979), pink salmon, *Oncorhynchus gorbuscha* (Federov and Smirnova, 1976), Indian catfish, *Heteropneustes fossilis* (Sundararaj and Samy, 1974), rainbow trout, *Salmo gairdneri* (Breton *et al.*, 1976) tilapias, *Sarotherodon mossambicus* (Farmer and Papkoff, 1977) and *Sarotherodon spirulus* (Hyder *et al.*, 1979), American plaice, *Hippoglossoides platessoides*, and winter flounder, *Pseudopleuronectes americanus* (Ng and Idler, 1978b, 1979), and pike eel, *Muraenesox cinereus* (Huang *et al.*, 1981).

To date, most of the partially purified and purified teleost gonadotropins have been produced in very limited quantities, and this has restricted their availability for *in vivo* studies on induced ovulation. An exception to this generalization is the partially purified gonadotropin of Pacific salmon, *Oncorhynchus* sp., SG-G100 (Donaldson and Yamazaki, 1968; Donaldson *et al.*, 1972b; Donaldson, 1973), which was made available to investigators at an early stage, and has since been produced in relatively large quantities. Species in which ovulation has been induced with SG-G100 include the goldfish, *Carassius auratus* (Yamazaki and Donaldson, 1968b; Stacey *et al.*, 1979a), Indian catfish, *Heteropneustes fossilis* (Sundararaj *et al.*, 1972), grey mullet, *Mugil cephalus* (Shehadeh *et al.*, 1973a; Kuo *et al.*, 1974), ayu, *Plecoglossus altivelis* (Ishida *et al.*, 1972), milkfish, *Chanos chanos* (Nash and Kuo, 1976; Vanstone *et al.*, 1976), Japanese flounder, *Limanda yokohamae* (Hirose *et al.*, 1976, 1979), pacu *Colossoma mitrei* (Castagnolli and Donaldson, 1981), coho salmon, *Oncorhynchus kisutch* (Jalabert *et al.*, 1978a; Hunter *et al.*, 1981 (Fig. 3); Sower *et al.*, 1982), and the chinook salmon, *Oncorhynchus tschawytscha* (Hunter *et al.*, 1978). With a crude preparation the chum salmon, *Oncorhynchus keta*, steelhead and rainbow trout, *Salmo gairdneri*, and cutthroat trout, *Salmo clarki*, have also been ovulated (E. M. Donaldson and G. A. Hunter, unpublished). Another partially purified salmon gonadotropin preparation (PPSG) has been used to induce ovulation in the pike, *Esox lucius* (De Montalembert *et al.*, 1978a; Billard and Marcel, 1980), and a third salmon gonadotropin preparation, Salmon Con AII, MW 40,000 induces ovulation in the winter flounder, *Pseudopleuronectes americanus* (Idler and Ng, 1979).

Regarding the dosage of partially purified gonadotropin required to induce ovulation, two factors are significant. First, species specificity; salmonids (Hunter *et al.*, 1981) require a lower dose of SG-G100 than some nonsalmonids, e.g., grey mullet (Kuo *et al.*, 1974). Second, the ovarian maturity stage at the time of treatment is important. The more mature the fish is, the lower is the dosage required; in the grey mullet this has been clearly correlated with egg diameter (Fig. 8) (Kuo and Nash, 1975; reviewed by Nash and Shehadeh, 1980). In pike, salmon gonadotropin was most effective when given on the day of capture. The ovulation rate dropped from 96% to 40% over the 3 days after capture and was attributed to oocyte degeneration associated with the stress of capture and confinement (De Montalembert *et al.*, 1978b).

c. *Mammalian Gonadotropins.* Mammalian gonadotropins are of either pituitary or placental origin; LH and FSH fall into the former category and HCG and PMSG fall into the latter category. These hormones are available either individually under the generic name or a trade name or as a combination of placental and pituitary gonadotropin. The commercial preparation, Synahorin, falls into this latter category. These mammalian hormones potentially have several advantages over fish pituitaries. They (1) are readily available, (2) can be stored for a long period, (3) are uniform and standardized, (4) are competitive on a cost basis, (5) eliminate the need to kill donor fish, and (6) eliminate the collection, processing, and preservation of fish

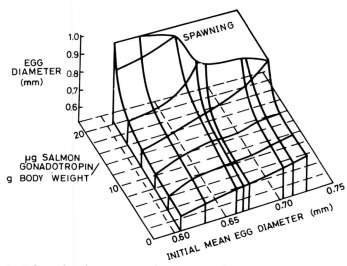

Fig. 8. Relationship between initial mean egg diameter, spawning dose of salmon gonadotropin SG-G100, and egg development in the grey mullet, *Mugil cephalus.* Reproduced with permission from Kuo and Nash (1975).

pituitary extracts (Shehadeh, 1975). However, since 1975 standardized salmon pituitary extracts which have similar advantages have become commercially available (Syndel Laboratories). Of the mammalian gonadotropins HCG has been used effectively on the widest range of species, either alone, e.g., in the freshwater catfish, *Clarias macrocephalus* (Carreon *et al.*, 1976) or *Clarias lazera* (Eding *et al.*, 1982; see Table II in Lam, 1982, for additional examples), or in combination with fish pituitary preparations, e.g., in the grass carp, *Ctenopharyngodon idella* (Schoonbee *et al.*, 1978; see Table III in Lam, 1982, for additional examples). In addition to being commercially available in clinical grade, HCG can also be prepared locally from human pregnancy urine (Katzman and Doisy, 1932; Pullin and Kuo, 1980), and crude HCG has been used to induce ovulation in the Indian carp, *Labeo rohita* (Bhowmick, 1979). A negative aspect of HCG is that repeated use can lead to fish becoming refractory as a result of an immune response (Freshwater Commerical Fish Artificial Propagation Work Groups in Fujian, Jaingsu, Zhejiang, and Shanghai, 1977). The other placental gonadotropin preparation PMSG has been used successfully alone in only a limited number of species including *Heteropneustes fossilis* (Sundararaj and Goswami, 1966) and the gulf croaker, *Bairdiella icistia* (Haydock, 1971), and was ineffective when tested alone in *Clarias* and *Heteropneustes* (Ramaswami and Lak-

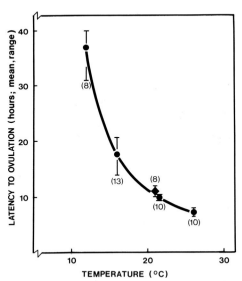

Fig. 9. Effect of temperature on the latency to ovulation in goldfish following injection of HCG (●) at 10 μg/g, or SG-G100 (◆) at 20 μg/g body weight. Data are presented as mean ± range of ovulation times, with numbers of ovulating fish in parentheses. Reproduced with permission from Stacey *et al.* (1979a).

shman, 1959), the white sucker, *Catostomus commersoni* (Tao, 1973), or the cichlid, *Tilapia nilotica* (Babiker and Ibrahim, 1979). In the latter two studies, HCG was effective alone under similar conditions. The latent period to ovulation after a single injection of HCG in the goldfish was negatively correlated with temperature over the range 12°–26°C (Fig. 9; Stacey *et al.*, 1979a). At 21°C the latency period of 9–12 hr was the same as that after SG-G100 injection, and correlates well with the time course of the surge in the plasma concentration of endogenous gonadotropin associated with natural spawning (Stacey *et al.*, 1979b). The effect of temperature on the interval between sturgeon pituitary injection and ovulation in the beluga (*Huso huso*) has been described by Igumnova (1974).

Of the two mammalian pituitary gonadotropins, only LH has been demonstrated to be an effective inducer of ovulation in fish (Sundararaj and Goswami, 1966, 1969; Anand and Sundararaj, 1974; Kapoor, 1976), and it offers no advantage over HCG on the basis of cost or effectiveness. However, preparations which contain a mixture of placental and pituitary gonadotropins, e.g., Synahorin, have been used with success either alone, as in the goldfish (Yamazaki, 1965) and in *Tilapia mossambica* (Khalil, 1974), or together with homologous fish pituitary extract, as in the grey mullet, *Mugil cephalus* (Shehadeh and Ellis, 1970).

5. STEROIDS

The actions of gonadotropin at the ovarian level are regarded as being largely mediated by the steroid hormones (Jalabert, 1976). Therefore, a number of studies have been conducted to determine the feasibility of inducing final maturation and ovulation with steroids, rather than with gonadotropin or hypothalamic factors. In general, steroids have the advantage of commerical synthesis into a pure product of indefinite storage life. The biochemistry and biology of steroids in fish reproduction are the subject of Chapter 7, Volume IXA, this series and their *in vitro* effects are described in Chapter 3, this volume. In this section they are dealt with according to their application in induced ovulation.

a. Progestins. Progesterone (Prog) derivatives 17α-hydroxyprogesterone (17α-Prog) and 17α-hydroxy-20β-dihydroprogesterone (17α,20β-Prog) (Fig. 10) were detected in significant quantities in the plasma of maturing sockeye salmon, *Oncorhynchus nerka*, by Idler *et al.* (1960) and Schmidt and Idler (1962). These progestins and in particular 17α,20β-Prog were later shown to be effective inducers of final maturation (germinal vesicle migration and breakdown) in teleost oocytes (Jalabert, 1976; Goetz, Chapter 3, this volume). *In vivo* ovulation was successfully induced in mature trout containing oocytes with the germinal vesicle at the subperipheral stage using

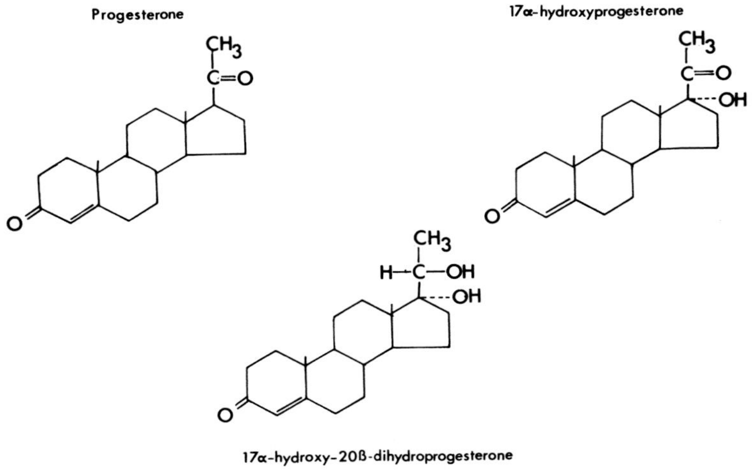

Fig. 10. Structures of the progestins progesterone, 17α-hydroxyprogesterone, and 17α-hydroxy-20β-dihydroprogesterone.

2 mg/kg 17α,20β-Prog or Prog. After 9 days, approximately 80% of the 17α-20β-Prog, 55% of the Prog group, and 30% of the control group had ovulated. Progesterone may have formed a substrate for 17α,20β-Prog biosynthesis (Jalabert et al., 1976). When rainbow trout were injected twice with 3 mg/kg 17α,20β-Prog 4–8 weeks before spawning and before germinal vesicle migration, 94% underwent final maturation, but only 25% ovulated. Priming with trout pituitary extract 2 days before a single injection of 17α,20β-Prog increased the ovulation rate to 58% (Jalabert et al., 1978b). These results (Jalabert et al., 1976, 1978b) suggest that in trout, which are close to maturity and have high endogenous levels of plasma gonadotropin, injection of 17α,20β-Prog is sufficient to induce final maturation and ovulation, but in less mature fish, characterized by lower endogenous gonadotropin levels, it is necessary to supplement 17α,20β-Prog with gonadotropin to ensure that both final maturation and ovulation occur. In subsequent studies (Jalabert et al., 1980; Bry, 1981), ovulation was again induced in trout with oocytes containing the germinal vesicle in the subperipheral position. Monitoring of plasma gonadotropin indicated a fall in gonadotropin concentration 24 hr after injection, perhaps as a result of negative feedback, followed by a gradual increase which continued after ovulation especially in fish which were not stripped.

In coho salmon, *Oncorhynchus kisutch*, containing oocytes in which the germinal vesicle had not reached the peripheral position, injection of 2 mg of 17α,20β-Prog/kg followed after 10 days by a further dosage of 4 mg/kg induced 100% final maturation after 20 days, but failed to induce ovulation and, in fact, inhibited ovulation relative to the control group. There was a significant decline in plasma gonadotropin concentration after the first injection from 4 ng/ml to less than 0.2 ng/ml, and a lesser decline after the second injection (Jalabert *et al.*, 1978a). In the same experiment, a primer of 0.25 mg/kg SG-G100 induced 100% final maturation after 8 days and 60% ovulation after 12 days. Injection of 2 mg/kg 17α,20β-Prog 2 days after the SG-G100 did not improve the response. In a later trial using salmon closer to maturity, injection of 3 mg/kg 17α,20β-Prog after a lower dosage of primer, 0.1 mg/kg SG-G100, provided a better ovulation response than primer alone (Jalabert *et al.*, 1978a). This indicates that in Pacific salmon, 17α,20β-Prog can be used to reduce the amount of fish gonadotropin required to induce ovulation in females which are close to maturity.

In the carp (*Cyprinus carpio*) held at low temperatures (13°–15°C), it was also necessary to prime, in this case with carp pituitary extract 0.6 mg/kg, prior to injection of 2 mg/kg 17α,20β-Prog to obtain ovulation on the following day in 50–60% of the fish. In the same experiment, hypophysation, which is normally conducted at 20°C or above, was only partially effective, and 17α,20β-Prog alone, 11-desoxycorticosterone (DOC) alone, or 17α,20β-Prog and DOC combined were ineffective (Jalabert *et al.*, 1977). In *Crenimugil labrosus*, ovulation was induced by injection of progesterone alone 20 mg/kg/day for 6 days (Cassifour and Chambolle, 1975).

In the northern pike, *Esox lucius*, the effectiveness of treatments in inducing ovulation fell in the following order: 0.1 mg/kg salmon GtH > 0.03 mg/kg salmon GtH + 3 mg/kg 17α,20β-Prog > 0.03 mg/kg salmon GtH + 3 mg/kg 17α-Prog > 0.02 mg/kg salmon GtH + 3 mg/kg 17α,20β-Prog > 3 mg/kg 17α,20β-Prog (De Montalembert *et al.*, 1978a). As in the rainbow trout, 17α,20β-Prog only induced final maturation and ovulation of ova that contained germinal vesicles in the peripheral or subperipheral stage.

Clearly, progesterone derivatives can be used to induce final maturation; however, ovulation will only occur when the endogenous gonadotropin concentration is sufficient, or when a minimum of exogenous gonadotropin is supplied. In other words, the progestins can partially replace gonadotropin. In the future, it may be possible to combine progestin treatment with factors directly involved in the ovulation process.

b. Corticosteroids. The corticosteroids are largely of interrenal origin, but may also be synthesized in gonadal tissue (Colombo *et al.*, 1973; Fostier *et al.*, Chapter 7, Vol. 9A, this series). Plasma corticosteroid concentrations and secretion rates have been demonstrated to increase in a number of teleosts during reproductive development and spawning (Donaldson and Fagerlund, 1970; Fagerlund and Donaldson, 1970; Katz and Ecksein, 1974;

Wingfield and Grimm, 1977; Cook *et al.*, 1980). The *in vitro* effects of corticosteroids on the ovary are discussed elsewhere (Chapter 3, this volume). Studies on the induction of ovulation in fish using corticosteroids followed those on amphibia (see Pickford and Atz, 1957). Corticosteroids that induce ovulation are shown in Fig. 11. The first successful use of a corticosteroid to spawn a teleost was the induction of ovulation in *Misgurnus fossilis* using a single dosage of 1 mg DOCA per fish (Kirshenblatt, 1952, in Pickford and Atz, 1957; Kirshenblatt, 1959). Ovulation has also been induced with DOCA in the hypophysectomized intact Indian catfish (Ramaswami, 1962; Sundararaj and Goswami, 1966, 1972). In the latter study, intact catfish held at 25°C started to ovulate 6 hr after a single im injection of 5 mg DOCA per fish compared to 8 hr after injection of LH. Cortisol acetate and cortisone acetate were also effective in the catfish at 25 mg/fish (Sundararaj and Goswami, 1966). Recently the African catfish, *Claria lazera,* has been induced to reproduce in ponds by ip injection of 50 mg/kg DOCA in oil suspension into females (Hogendorn, 1979).

In goldfish injected ip with 100 mg/kg DOC, cortisone or corticosterone, cortisone was the most effective corticosteroid inducing ovulation in 80% after 1 day (Khoo, 1974). However, injection of mature goldfish held at 12°C with the 11β-hydroxylation inhibitor metopirone also induced ovulation. Metopirone would be expected to inhibit cortisol biosynthesis leading to a build-up of 11-deoxycortisol (Pandey *et al.*, 1977). In comparison, injection of metopirone into hypophysectomized catfish inhibited the ovulatory re-

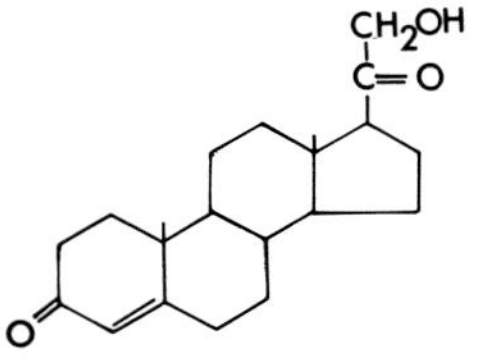

11- deoxycorticosterone

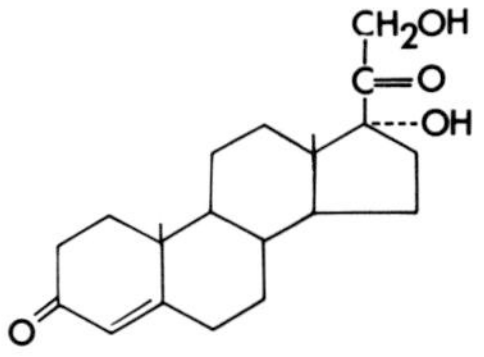

11- deoxycortisol

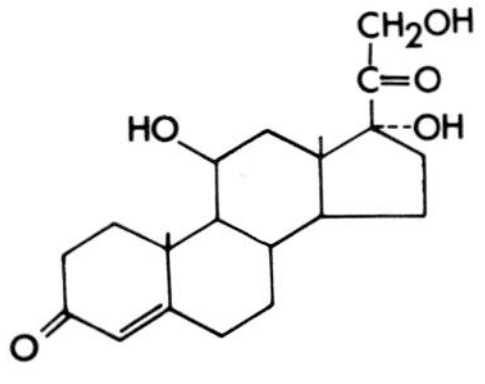

Cortisol

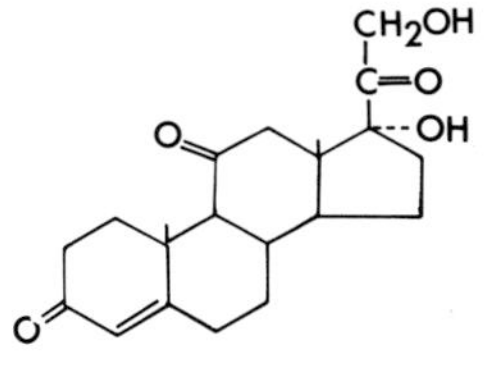

Cortisone

Fig. 11. Structures of the corticosteroids 11-deoxycorticosterone, 11-deoxycortisol, cortisol, and cortisone.

sponse to LH and DOCA (Sundararaj and Goswami, 1969). In the ayu, ip injection of 5, 10, or 20 mg cortisol per fish body weight (40–50 gm) induced ovulation. The 10-mg dosage was more effective than 1000 IU HCG. A combination treatment of 300 IU HCG followed after 1 day by 5 mg cortisol was demonstrated to be as effective as 10 mg cortisol and more effective than HCG alone. Injection of cortisol or HCG increased both the water content and the sodium concentration of the ovary (Hirose and Ishida, 1974b). Similar effects on ovarian water and sodium after injection of corticosteroids have been described in *Tilapia nilotica*. In this species, cortisol at a level of 175 mg/kg, induced ovulation in all fish, but corticosterone at a level of 195 mg/kg induced ovulation in 75% of the fish. Higher or lower doses of both steroids were less effective (Babiker and Ibrahim, 1979). Injection of carp with DOC at 1 mg/kg was ineffective (Jalabert *et al.*, 1977); however, this dosage was much lower than the effective levels used by other researchers. Recently, ovulation of grey mullet, *Mugil cephalus*, in aquaria followed by natural fertilization was achieved on a reproducible basis by priming with either 50 mg/kg carp pituitary homogenate, 5 mg/kg SG-G100, or 16,700 IU HCG, followed by a secondary injection of 120 mg/kg DOC (Pullin and Kuo, 1981).

The corticosteroids are clearly effective in inducing ovulation in most teleosts when they have been administered in high enough dosages alone or in combination with gonadotropins. Further research is required to elucidate their physiological or pharmacological role in the maturation and ovulation process. Jalabert (1976) has suggested that DOC and 11-deoxycortisol may displace 17α,20β-Prog bound to plasma protein, and cortisol and cortisone may lower oocyte sensitivity to 17α,20β-Prog.

c. Estrogens and Androgens. Plasma estrogen concentrations in female teleosts decline after the completion of vitellogenesis and prior to final maturation and ovulation (Wingfield and Grimm, 1977). This decline in estrogen concentration coincides with an increase in plasma 17α,20β-Prog concentration, and suggests that estrogens would be inappropriate for induction of final maturation and ovulation. This contention is supported by data from *in vitro* final maturation studies which indicate that 17β-estradiol slightly inhibits the effect of gonadotropin (Jalabert, 1975, 1976). However, in *Tilapia nilotica*, 17β-estradiol at 190 mg/kg did increase ovarian sodium concentration and induced ovulation in 66% of the females. Administration of estrone or estriol over the dosage range 65–245 mg/kg failed to induce ovulation (Babiker and Ibrahim, 1979). In the grass and silver carps, an attempt to induce ovulation using the synthetic estrogen stilbestrol dipropionate in combination with PMSG as a primer was not successful (Brandt and Schoonbee, 1980).

Androgens are present in female teleosts, can induce *in vitro* final maturation (Goetz and Bergman, 1978; Iwamatsu, 1978), and can lower the median effective dosage of gonadotropin required to induce *in vitro* final maturation in the rainbow trout (Jalabert, 1975, 1976). However, apart from an unsuccessful attempt using a very low dosage of methyltestosterone in the firetail gudgeon, *Hypseleotris galii* (MacKay, 1975), there appears to be little indication in the literature of attempts to use androgens to induce ovulation *in vivo*.

6. PROSTAGLANDINS

The role of prostaglandins in fish reproduction has been reviewed by Stacey and Goetz (1982). It is clear from *in vitro* studies that prostaglandins are involved in follicular rupture (Jalabert and Szollosi, 1975; Kagawa and Nagahama, 1981; Chapter 3, this volume). Of the various prostaglandins tested $PGF_{2\alpha}$ was usually the most effective (Fig. 12). *In vivo* studies in cyprinids have shown that indomethacin, an antiinflammatory agent which blocks prostaglandin synthesis from dihomo-γ-linolenic acid and arachidonic acid, inhibits ovulation, and that this inhibition can be overridden by administration of exogenous prostaglandin (Stacey and Pandey, 1975; Kapur, 1979). Further, in the pond loach, *Misgurnus anguillicaudatus*, ip injection of 100 IU HCG per fish increased the level of $PGF_{2\alpha}$ in the ovary at 24 hr. The increase in ovarian $PGF_{2\alpha}$ was completely inhibited by injection of indomethacin with HCG (Ogata *et al.*, 1979). In the goldfish plasma, PGF levels were reported to have increased sixfold after HCG-induced ovulation (Bouffard, 1979, in Stacey and Goetz, 1982). Jalabert (1976) has proposed that ovarian prostaglandin synthesis and/or release may be induced by stimulation of α-adrenergic sites in the ovary and follicle via the sympathetic nervous system in response to the end-of-maturation signal. Prostaglandins may also play a role in the modulation of gonadotropin release in fish, but the current data are inconclusive (Stacey and Goetz, 1982). However, there is conclusive evidence for the involvement of prostaglandins in reproductive behavior. In the goldfish, injection of $PGF_{2\alpha}$ (10 μg/kg) into the third ventricle induced female spawning behavior (Stacey and Peter, 1979).

Despite these promising experimental studies, only a few investigators

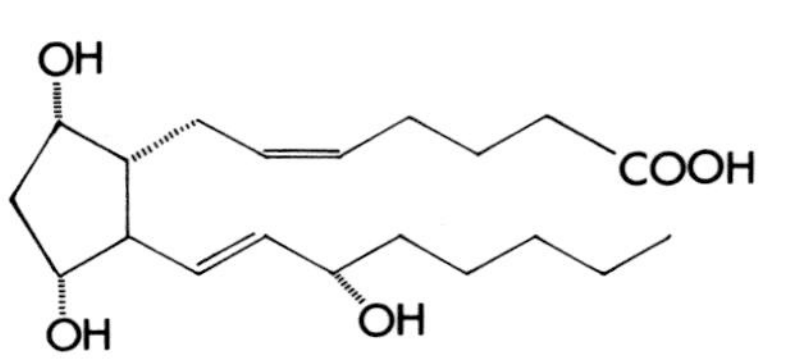

Fig. 12. Structure of prostaglandin $F_{2\alpha}$.

have attempted to use prostaglandins to induce ovulation on a practical basis. In the catfish, *Heteropneustes fossilis*, daily injection of PGE_1 or $PGF_{2\alpha}$ (100 μg/fish, body weight 41–47 gm) resulted in 90% ovulation after 5–6 days. Similar injections into catfish hypophysectomized 2 days previously failed to induce ovulation, suggesting possible pituitary mediation of the response to prostaglandin (Singh and Singh, 1976).

Prostaglandins clearly have potential as secondary injections after a final maturation-inducing primer. In the rainbow trout, oocytes in which final maturation had been induced *in vivo* have been ovulated *in vitro* using $PGF_{2\alpha}$ (Jalabert *et al.*, 1978b). This may open up the possibility of inducing ovulation *in vitro* in the mature ovaries of salmonids killed prematurely for egg take at culture facilities or harvested in terminal fisheries. Furthermore, the finding that prostaglandins are involved in the central regulation of spawning behavior (Stacey and Peter, 1979) may permit the stimulation of natural spawning under culture conditions.

7. OTHER FACTORS

A number of other hormones, besides those previously discussed, may play a role in reproduction (Fontaine, 1976); these include the catecholamines, calcitonin (Watts *et al.*, 1975), the thyroid hormones, the hormones of the urohypophysis (Lederis, 1972), growth hormone, and prolactin. However, there is a considerable difference between a hormone being known or suspected to play an indirect or direct role in one or more aspects of reproductive development, such as vitellogenesis, final maturation or ovulation, and the feasibility of using the hormone on a practical basis.

In the case of the catecholamines, they have been demonstrated to induce *in vitro* ovulation in trout follicles that had undergone final maturation *in vivo*. The potency of the tested catecholamines was in the order epinephrine > norepinephrine > phenylephrine > isoproterenol. The response time at 15°C was 20–24 hr, compared to 18–20 hr for $PGF_{2\alpha}$ (Jalabert, 1976). Inhibitors of α-adrenergic sites prevented the effect of the catecholamines on ovulation. Jalabert (1976) proposed that catecholamines either cause follicular contraction directly, or, more likely, by stimulation of prostaglandin synthesis and/or release. There have been few trials of catecholamines *in vivo*. In the brook trout, *Salvelinus fontinalis*, injection of 100 μg epinephrine every second day for almost 4 weeks accelerated sexual development in both females and males (Strand, 1958).

The thyroid hormones have long been regarded as having a role in fish reproduction (Fontaine, 1976). In the sturgeon, *Acipenser stellatus*, laboratory studies demonstrated that mature females fail to undergo final maturation and ovulation if held at low temperatures. Injection of triiodothyronine

(T_3) along with the pituitary extract facilitated maturation and ovulation. Initial attempts to utilize this technique on a production scale (300–1000 mg T_3 + 30–40 mg pituitary per fish) have met with mixed results, and in some cases the incidence of deformed embryos was higher in the T_3 groups (Detlaf and Davydova, 1979a,b).

IV. INDUCED SPERMIATION

A. Levels of Intervention

In the male, testicular development is regulated by the hypothalamic–pituitary–testicular axis. The structure, development, and endocrinology of the teleost testis have been recently reviewed by Grier (1981), Nagahama et al. (1982), Billard et al. (1982), and in Chapters 6 and 7 of Volume 9A, this series. The range of hormones that can be injected or implanted to induce spermatogenesis and/or spermiation is limited compared to the situation in the female; however, the culturist still has the option of intervening with hypothalamic, pituitary, or gonadal hormones. Because of the relative simplicity of the endocrine control of the testis compared to the ovary, it is easier to induce an immature male teleost to undergo spermatogenesis and spermiation than it is to induce an immature female teleost to undergo vitellogenesis, final maturation, and ovulation. In a number of cultured species, the male spermiates spontaneously; however, in others the captive male fails to spermiate or produces only a small volume of viscous milt. In most cases involving mature males, the volume and quality of the milt can be markedly improved by a single injection of an appropriate hormone. If the male is immature and fails to produce milt, a series of injections or one or more pellet implantations will stimulate spermatogenesis and subsequently spermiation.

B. Use of Specific Compounds

1. GONADOTROPIN RELEASING HORMONES

To date, the literature on induced spermiation in teleosts using LHRH or its analogues is very sparse. However, the literature on amphibians indicates that the deca- and nonapeptides show considerable promise. Spermiation was induced in the frog *Hyla regilla* by injection of LHRH at a level of 45 μg/kg (Licht, 1974). In the frog *Rana esculenta*, a single injection of LHRH at a level of 2–200 μg/kg induced 100% spermiation in 0.5–1 hr at 15°C. At the higher dosage, the duration of response was 2 days which compared to 5 days after 100 IU/kg HCG (Billard *et al.*, 1975). In the frog, *Rana sylvatica*, ova were fertilized after spermiation was induced by injection of LHRH at 50 μg/kg (Smith-Gill and Berven, 1980). In bull frogs, *Rana catesbeiana*, having a mean weight of 401 gm, the lowest dosage of LHRH which induced spermiation was 2.0 μg per individual, although the lowest dosage of the potent agonist (im Bzl-D-His[6], Pro[9]-NEt)-LHRH which induced spermiation was 0.4 μg per individual (McCreery *et al.*, 1982).

In the mature male goldfish, injection of LHRH or LHRHa D-Ala[6] induced an increase in plasma gonadotropin concentration, but spermiation was not reported (Peter, 1980). In mature male landlocked Atlantic salmon, *Salmo salar*, average body weight 90 gm, held at 8°C, spermiation was induced by ip injection of 0.75 mg/kg LHRHa D-Ala[6] in saline or 40% propylene glycol, or implantation of 1.5 mg/kg LHRHa D-Ala[6] in silastic tubing, or 25 μg per fish D-Nal(2)[6] LHRH in cholesterol pellets. Spermiation commenced in the ip-injected fish on day 1 postinjection, and the maximal response in most groups was noted after 8–12 days. In the control group, 75% were spermiating on day 12. The volume of sperm produced was correlated with plasma gonadotropin concentration (Weil and Crim, 1982).

In a recent study, E. M. Donaldson, G. A. Hunter and J. Stoss (unpublished) demonstrated that two ip injections of 0.05 mg/kg LHRHa D-Ala[6] 72 hr apart could be successfully used to induce spermiation in chum salmon (*Oncorhynchus keta*) in either salt or fresh water 24 hr after the second injection.

2. GONADOTROPINS

a. Fish Pituitary Extracts. In the European, Indian, and Chinese carps, where it is customary to induce ovulation in the female with two injections of a homologous or heterologous pituitary extract, it is usually also advisable to inject the male with a single dose of pituitary extract at the time that the female receives her second injection to ensure satisfactory spermiation in the male at the time of ovulation in the female. In the major Indian carps, rohu (*Labeo rohita*) and catla (*Catla catla*), for example, the males were given a single injection of 2–3 mg/kg of carp pituitary extract and two males were allowed to spawn with each female (Chaudhuri, 1976; Bhowmick *et al.*, 1977). However, toward the end of the spawning season, it is necessary to inject the male catla twice to obtain a satisfactory response (Bhowmick *et al.*, 1978). In the common carp injection of an aqueous extract of acetone-dried powder of carp pituitary glands at 2.2 or 22 mg/kg produced a dose-related increase in milt volume over a 24-hr period at 22°C from 0.12 ml in controls to 2.90 ml at the higher dosage level (Clemens and Grant, 1965). Recently, in Israel, male common carp have been found to spermiate in 24 hr at 16°C when injected with carp pituitary homogenate at a dosage of 1 gland per kg regardless of the month that the pituitary glands were collected (Rothbard and Rothbard, 1982). In an experimental study of the hypophysectomized goldfish, Billard (1977) induced spermiation with 50 mg/kg common carp, roach, or trout pituitary extract. When spermiation was quantified by measurement of milt production at the time of maximum response (1–2 days), the pituitary extract from the closely related carp gave the best response.

In the murrels, *Ophicephalus striatus*, *O. marutius*, *O. gachua*, and *O. punctatus*, spermiation and spawning in pools or aquaria has been induced with alcohol-preserved pituitary glands from the carp, *Labeo rohita*, or the catfish, *Wallago attu*, at a dose of 0–20 mg/kg for the first injection and 5–250 mg for the second injection (Parameswaran and Murugesan, 1976). In the northern pike, milt production was increased 3–7 times by a single injection of 1.2 or 14 mg wet wt/kg fresh pike pituitaries and 3–6 times by 0.5–3 mg/kg acetone-dried carp pituitary glands. The peak response was at 2 days and the duration was 10–14 days (Billard and Marcel, 1980).

Temperature can have a significant effect on the response of the testis to pituitary extracts. In the immature green sunfish, *Lepomis cyanellus*, the testis demonstrated little response to ip injection of 175 mg/kg carp pituitary acetone-dried powder at 10°C administered every second day. However, in fish injected at 21°C, spermiation occurred after 14 days (Kaya, 1973).

In the rainbow trout, Clemens and Grant (1965) induced a seminal hydration response by injection of 4.4 mg/kg bigmouth buffalo (*Ictiobus cyprinellus*) pituitary extract at 15°C. In the Pacific salmon, Donaldson *et al.* successfully induced a spermiation response using a single injection of 50 mg/kg homogenized frozen *Oncorhynchus* pituitary glands, or 6.5 mg/kg *Oncorhynchus* pituitary acetone-dried powder in saline. Other species in which spermiation has been induced with salmon pituitary homogenates include the Gulf croaker, *Bairdiella icistia* (Haydock, 1971), and the Pacific mackeral, *Scomber japonicus* (Leong, 1977).

   b. Purified Fish Gonadotropin Preparations. Although whole-pituitary extracts have been used most often at the practical level, a number of experimental studies on induced spermiation have been conducted using partially purified and purified fish gonadotropin. In the goldfish, Clemens and Reed (1967) induced both spermatogenesis and spermiation by injection every 3 days at 24°C with 10, 20, 60, or 120 mg/kg partially purified carp pituitary gonadotropin. At the highest dosage, spermiation occurred after 42 days. In hypophysectomized mature male goldfish, a single ip injection of partially purified salmon gonadotropin (SG-G100) over the range 0.1–3.0 mg/kg induced a dose-related spermiation response in 24 hr at 20°C (Yamazaki and Donaldson, 1968a). This technique was used as a bioassay during the purification of salmon gonadotropin, and fraction SG-DEAE-2 gave a significant spermiation response at 30 μg/kg (Donaldson et al., 1972b). Spermiation has also been induced in hypophysectomized catfish (*Heteropneustes fossilis*) with SG-G100 (Sundararaj et al., 1971). Injection of SG-G100 into juvenile pink salmon (*Oncorhynchus gorbuscha*) at 1 mg/kg 3 times per week resulted in spermiation after 10 weeks at 8°–11°C; however, juvenile chinook salmon responded more slowly (Funk and Donaldson, 1972). In rainbow trout, spermiation occurred after salmon gonadotropin injection two times per week for 12 weeks (Drance et al., 1976). In a subsequent study, spermiation was induced in juvenile pink salmon in 12 weeks by the aforementioned injection technique, and by implantation of a cholesterol pellet containing 9 mg/kg SG-G100 once every 3 weeks. Intraperitoneal injection of the same total dose of SG-G100 at intervals of 1 or 2 weeks, rather than three times per week, resulted in slower gonadal development and did not result in spermiation in 12 weeks (MacKinnon and Donaldson, 1978). These techniques have been used on a pilot scale for production of milt from pink salmon at exactly 1 year of age for cross-fertilization purposes (Donaldson et al., 1972a; MacKinnon, 1976). In the intact grey mullet (*Mugil cephalus*), spermiation has been induced with SG-G100 (Donaldson and Shehadeh, 1972; Shehadeh et al., 1973b), and the technique can be used to induce spermiation in mullet out of season to fertilize ova from photoperiod-regulated females (Nash and Shehadeh, 1980, p. 48). Spermiation has also been induced in the northern pike using a purified salmon gonadotropin PPSG at 5–100 μg/kg (Billard and Marcel, 1980).

c. *Mammalian Gonadotropins.* Of the various mammalian gonadotropins, HCG appears to the most effective in inducing spermiation in teleosts. For example, in the hypophysectomized goldfish, there was a dose-related response to HCG over the range of 100–10,000 IU/kg, but injection of up to 200 mg/kg LH and 300 mg/kg FSH failed to elicit a spermiation response (Yamazaki and Donaldson, 1968a). The spermiation response to HCG in goldfish was confirmed by Billard (1976) and has been utilized by Stacey and Peter (1979). Spermiation has also been induced with HCG in a number of other species including rainbow trout (Clemens and Grant, 1965), seabream, *Sparus aurata* (Arias, 1976), milkfish, *Chanos chanos* (Juario, 1981), the European eel, *Anguilla anguilla* [several workers including Boetius and Boetius (1967) and Bieniarz and Epler (1977)], and the Japanese eel, *Anguilla japonica*, using Synahorin (Yamamoto et al., 1972; Yamamoto and Yamauchi, 1974). The positive response of male *Puntius ticto* to a high dosage of ovine FSH (Kapoor, 1978) is an exception to the general failure of mammalian FSH to induce spermiation in teleosts. Billard (1977) investigated the ability of a variety of mammalian pituitary preparations to induce spermiation in hypophysectomized goldfish. Significant responses were obtained to sheep pituitary extract at 200 mg/kg, ovine LH at 2 gm/kg, but not at 200 mg/kg, ovine prolactin at 100 mg/kg and ACTH at 10,000 IU/kg. The responses to ovine FSH at 200 mg/kg and PMSG at 500 mg/kg were not significant.

3. STEROIDS

a. Androgens. Steroids were first demonstrated to induce spermiation in teleosts by Yamazaki and Donaldson (1969). In this study, mature male goldfish that had been hypophysectomized 3–7 days earlier were administered ip a single injection of 0.10–1 gm/kg methyltestosterone or dehydroepiandrosterone or 10–200 mg/kg 11-ketotestosterone. The minimal doses which elicited a spermiation response were 50–100 mg/kg, and maximal responses were only observed with the highest dosages. There were no differences in the dose-response relationships for these three androgens. In a subsequent study on induction of spermiation in hypophysectomized goldfish, Billard (1976) demonstrated that methyltestosterone was more effective than testosterone propionate. Using intact males, the response to 100 mg/kg methyltestosterone was similar to that obtained with 10,000 IU/kg HCG (Billard, 1976). However, in hypophysectomized goldfish, 1 gm/kg methyltestosterone (Yamazaki and Donaldson, 1969) was equivalent to 10,000 IU/kg HCG or 3 mg/kg SG-G100 (Yamazaki and Donaldson, 1968a).

In the green sunfish, injection of approximately 100 mg/kg testosterone propionate in corn oil every other day for 28 days at 21°C resulted in a marked spermatogenic response in seasonally regressed testes, but, in contrast to carp pituitary extract, did not result in spermiation within this period. However, injection of testosterone propionate at 400 mg/kg into mature northern pike resulted in spermiation after 24 hr (Quillier and Labat, 1977).

In the grey mullet, spermiation has been induced with methyltestosterone (Shehadeh *et al.*, 1973b); in the salmonids spermiation has been induced in the sockeye salmon (*Oncorhynchus nerka*) with 11-ketotestosterone (Idler *et al.*, 1961), and in the rainbow trout with testosterone but not 11-ketotesterone (Billard *et al.*, 1981). However, plasma 11-ketotesterone concentration has been correlated with sperm production in the latter species (Fostier *et al.*, 1982).

Recently, a commercial androgen preparation, Durandron Forte 250, has been used to induce spermiation in mature male milkfish. This preparation contains 30 mg testosterone propionate, 60 mg testosterone isocaproate, and 100 mg testosterone decanoate in 1 ml of vehicle, and when injected at the rate of 1 ml per fish (2–4 kg), it induced a spermiation response which lasted for up to 8 days compared to 3 days after a single injection of 5000 IU

HCG (Juario and Natividad, 1980; Juario, 1981). Therefore, a single high dosage injection of androgen in a suitable vehicle is an effective means of inducing spermiation on a practical basis.

b. Other Steroids. Although most investigators have emphasized the role and utility of androgens in induction of spermiation, Billard (1976) investigated the effect of a range of other steroids on intact and recently hypophysectomized mature male goldfish. Progesterone was found to be more potent than methyltestosterone in both intact and hypophysectomized goldfish over the dosage range of 1–100 mg/kg. In fact, when assessed by total volume of milt produced 10 mg/kg progesterone was equivalent to 100 mg/kg methyltestosterone. Furthermore, 100 mg/kg cortisone induced a response in intact fish which was greater than that induced by 10 mg/kg methyltestosterone. Even estradiol induced a small spermiation response at a low dosage of 1 mg/kg, but not at higher dosages.

The superiority of progesterone over methyltestosterone for induction of spermiation demonstrated by Billard (1976) has been confirmed in the northern pike where progesterone was more effective than testosterone (De Montalambert *et al.*, 1978b). In male rainbow trout, Scott and Baynes (1982) discovered that at the initiation of spermiation there is a rapid drop in androgen concentration and an increase in the plasma concentration of 17α-hydroxyprogesterone and 17α-hydroxy-20β-dihydroprogesterone. Progestogen levels remained high throughout spermiation and were positively correlated with sperm volume and the increase in the K^+-Na^+ ratio.

V. CONCLUSIONS AND FUTURE DEVELOPMENTS

It is clear from this discussion that major progress has occurred in recent years toward the development of reliable and economic procedures for the induction of ovulation and spermiation in economically important teleosts, and that "second generation" techniques are being applied for these purposes. However, this is not to say that all our goals have been reached and that no further progress will be made; on the contrary it is possible to forecast with a reasonable degree of certainty further progress in a number of areas.

1. One may expect that piscine GnRH will become available both in its natural form, which may vary between groups of teleosts, and in the form of high potency analogues, which are being produced by substituting amino acids in the basic fish GnRH structure in the same manner that has been achieved for mammalian LHRH. Large-scale production of the most effec-

tive GnRH analogues combined with further developments in administration techniques, e.g., oral or gill sprays analogous to the nasal sprays developed for administration of LHRH in the human, administration in the aquarium water or, for less mature fish, slow release preparations in pellets or silastic tubing.

2. Further research on the mode of action of antiestrogens in fish may reveal more effective compounds or better administration regimes.

3. In a number of species including salmonids it is currently difficult to accelerate the completion of vitellogenesis prior to induction of ovulation except by long-term environmental control and progress can be expected in this area with long acting or sequential hormone treatments.

4. One may expect to achieve a better understanding of the role of steroids, prostaglandins, and other substances on final maturation, ovulation, and spermiation, and it is possible that *in vitro* techniques may be developed to the point where mature ovaries could be harvested from valuable fish captured in terminal fisheries and ovulated *in vitro* for culture purposes.

ACKNOWLEDGMENTS

Thanks are extended to Drs. W. S. Hoar, R. E. Peter, J. Stoss, and Mr. Glen Van Der Kraak for their valuable comments on the manuscript. The assistance of Ms. Helen M. Dye in preparing a number of the figures and Morva Young for typing the manuscript is gratefully acknowledged.

REFERENCES

Adashi, E. Y., Hsueh, A. J. W., Bambino, T. H., and Yen, S. S. C. (1981). Disparate effect of clomiphene and tamoxifen on pituitary gonadotropin release in vitro. *Am. J. Physiol.: Endocrinol. Metab.* 3, E125–E130.

Aida, K., Iznmo, R. Z., Satah, H., and Hibiya, T. (1978). Induction of ovulation in plaice and goby with synthetic LH releasing hormone. *Bull. Jpn. Soc. Sci. Fish.* 44, 445–450.

Anand, T. C., and Sundararaj, B. I. (1974). Temporal effects of artificial induction of ovulation on the hypothalamo-hypophysial-ovarian system of the catfish, *Heteropneustes fossilis* (Bloch). *Neuroendocrinology* 15, 158–171.

Anonymous (1978). The role of LH-RH in induction of spawning in grass carp (*Ctenopharyngodon idellus*). 1. Cytological and histochemical studies of the pituitary and the ovary. *Sci. Sin.* 21, 390–398.

Arias, A. M. (1976). Reproduction artificielle de la Daurade *Sparus aurata* (L.). *Stud. Rev. Gen. Fish. Counc. Mediterr.* 55, 159–173.

Atz, J. W., and Pickford, G. E. (1959). The use of pituitary hormones in fish culture. *Endeavour* 18, 125–129.

Atz, J. W., and Pickford, G. E. (1964). The pituitary gland and its relation to the reproduction of fishes in nature and in captivity. An annotated bibiliography for the years 1956–1963. *FAO Fish. Biol. Tech. Pap.* No. 37 (FB/T37).

Babiker, M. M., and Ibrahim, H. (1979). Studies on the biology of reproduction in the cichlid *Tilapia nilotica* (L.): Effects of steroid and trophic hormones on ovulation and ovarian hydration. *J. Fish Biol.* **15**, 21–30.

Bhowmick, R. M. (1979). "Observations on the use of human chorionic gonadotropin prepared in the laboratory, in inducing spawning in major carp. *Labeo rohita* Ham. Tech. Pap. **16**, 11–12. Symposium on Inland Aquaculture. Barrackpore, India.

Bhowmick, R. M., and Chaudhuri, H. (1968). Preliminary observations on the use of tri-chloroacetic acid for extraction of gonadotropic hormones in induced spawning of carps. *J. Zool. Soc. India* **20**, 48–54.

Bhowmick, R. M., Kowtal, G. V., Jana, R. K., and Gupta, S. D. (1977). Experiments on second spawning of major Indian carps in the same season by hypophysation. *Aquaculture* **12**, 149–155.

Bhowmick, R. M., Kowtal, G. V., Jana, R. K., and Gupta, S. D. (1978). Large scale production of the seed of catla, *Catla catla* (Ham.) by hypophysation. *J. Inland Fish. Soc. India* **10**, 52–55.

Bieniarz, K., and Epler, P. (1977). Investigations on inducing sexual maturity in the male eel *Anguilla anguilla* L. *J. Fish Biol.* **10**, 555–559.

Bieniarz, K., Epler, P., Thuy, L. N., and Kogut, E. (1979). Changes in the ovaries of adult carp. *Aquaculture* **17**, 45–68.

Billard, R. (1976). Induction of sperm release in the goldfish by some steroids. *IRCS Med. Sci.: Libr. Compend.* **4**, 42.

Billard, R. (1977). Effect of various hormones on sperm release in the hypophysectomized goldfish. *IRCS Med. Sci.: Libr. Compend.* **5**, 188.

Billard, R. (1983). The control of fish reproduction in aquaculture. *Proc. World Conf. Aquacult., 1981.* In press.

Billard, R., and Marcel, J. (1980). Stimulation of spermiation and induction of ovulation in pike (*Esox lucius*). *Aquaculture* **21**, 181–195.

Billard, R., and Peter, R. E. (1977). Gonadotropin release after implantation of antiestrogens in the pituitary and hypothalamus of goldfish, *Carassius auratus*. *Gen. Comp. Endocrinol.* **32**, 213–220.

Billard, R., Lakomaa, E., and Kimlstrom, J. E. (1975). Induction of sperm release in *Rana esculenta* by LH/FSH releasing hormone. *IRCS Med. Sci.: Libr. Compend.* **3**, 17.

Billard, R., Breton, B., and Richard, M. (1981). On the inhibitory effect of some steroids on spermatogenesis in adult rainbow trout *Salmo gairdneri*. *Can. J. Zool.* **59**, 1479–1487.

Billard, R., Fostier, A., Weil, C., and Breton, B. (1982). Endocrine control of spermatogenesis in teleost fish. *Can. J. Fish. Aquat. Sci.* **39**, 65–79.

Boetius, I., and Boetius, J. (1967). Studies on the European eel, *Anguilla anguilla* (L.). Experimental induction of the male sexual cycle, its relation to temperature and other factors. *Medd. Dan. Fisk. Havunders.* **4**, 339–405.

Bouffard, R. E. (1979). The role of prostaglandins during sexual maturation, ovulation, and spermiation in the goldfish *Carassius auratus*. M.Sc. Thesis, University of British Columbia, Vancouver, B.C., Canada.

Brandt, F. De W., and Schoonbee, H. J. (1980). Further observations on the induced spawning of the phytophagous Chinese carp species *Ctenopharyngodon idella* (Val.) and *Hypophthalmichthys molitrix*. *Water SA* **6**, 27–30.

Breton, B., and Weil, C. (1973). Endocrinologie comparée—effets du LH/FSH-RH synthétique et d'extraits hypothalamiques de carpe sur la secretion d'hormone gonadotropic in vivo chez la carpe (*Cyprinus carpio* L.). *C. R. Hebd. Seances Acad. Sci.* **277**, 2061–2064.

Breton, B., Billard, R., Jalabert, B., and Kann, G. (1972). Dosage radioimmunologique des gonadotropines plasmatiques chez *Carassius auratus* au cours du nycthemère et de l'ovulation. *Gen. Comp. Endocrinol.* **18**, 463–468.

Breton, B., Jalabert, B., and Fostier, A. (1975). Induction de décharges gonadotropes hypophysaires chez la Carpe (*Cyprinus carpio* L.) à l'aide du citrate de cisclomiphène. *Gen. Comp. Endocrinol.* **25**, 400–404.

Breton, B., Jalabert, B., and Reinaud, P. (1976). Purification of gonadotropin from rainbow trout (*Salmo gairdneri* Richardson) pituitary glands. *Ann. Biol. Anim., Biochim., Biophys.* **16**, 25–36.

Bry, C. (1981). Temporal aspects of macroscopic changes in rainbow trout (*Salmo gairdneri*) oocytes before ovulation and of ova fertility during the post-ovulation period; effect of treatment with 17α-hydroxy-20β-dihydroprogesterone. *Aquaculture* **24**, 153–160.

Buckingham, J. C. (1978). The hypophysiotrophic hormones. *Prog. Med. Chem.* **15**, 165–198.

Burrows, R. E., Palmer, D. D., and Newman, H. W. (1952). Effects of injected pituitary material upon spawning of blueback salmon. *Prog. Fish-Cult.* **14**, 113–116.

Burzawa-Gerard, E. (1971). Purification d'une hormone gonadotrope hypophysaire du poisson teleosteen, la carpe (*Cyprinus carpio* L.). *Biochimie* **53**, 545–552.

Burzawa-Gerard, E. (1974). Etude biologique et biochemique de l'hormone gonadotrope d'un poisson téléostéen, la carpe (*Cyprinus carpio* L.). *Mem. Mus. Natl. Hist. Nat., Ser. A (Paris)* **86**, 1–77.

Burzawa-Gerard, E., Goncharov, B., Dumas, A., and Fontaine, Y. A. (1976). Further studies on carp gonadotropin (c-GtH), biochemical and biological comparison of cGtH and a gonadotropin from *Acipenser stellatus* Pall. (Chondrostei). *Gen. Comp. Endocrinol.* **29**, 498–505.

Carreon, J. A., Estocapio, F. A., and Enderez, F. M. (1976). Recommended procedures for induced spawning and fingerling production of *Clarias macrocephalus* Gunther. *Aquaculture* **8**, 269–281.

Cassifour, P., and Chambolle, P. (1975). Spawning induced by injection of progesterone into the teleost *Crenimugil labrosus* (Risso 1826) in a brackish milieu. *J. Physiol.* (Paris) **70**, 565–570.

Castagnolli, N., and Donaldson, E. M. (1981). Induced ovulation and rearing of the pacu (*Colossoma mitrei*). *Aquaculture* **25**, 275–279.

Chan, K. K. S. (1977). Effect of synthetic luteinizing hormone-releasing hormone (LH-RH) on ovarian development in Japanese medaka, *Oryzias latipes*. *Can. J. Zool.* **55**, 155–160.

Chang, J. P., and Peter, R. E. (1982). Actions of dopamine on gonadotropin release in goldfish, *Carassius auratus*. *Proc. Int. Symp. Reprod. Physiol. Fish, 1982* p. 51.

Chang, J. P., and Peter, R. E. (1983a). Effects of dopamine on gonadotropin release in female goldfish, *Carassius auratus*. *Neuroendocrinology* **36**, 351–357.

Chang, J. P,, and Peter, R. E. (1983b). Effects of pimozide and des-Gly[10][D-Ala[6]] luteinizing hormone ethylamide on serum gonadotropin concentrations, germinal vesicle migration and ovulation in female goldfish, *Carassius auratus*. *Gen. Comp. Endocrinol.* (in press).

Chang, J. P., Cook, A. F., and Peter, R. E. (1983). Influence of catecholamines on gonadotropin secretion in goldfish, *Carassius auratus*. *Gen. Comp. Endocrinol.* **49**, 22–31.

Chaudhuri, H. (1976). Use of hormones in induced spawning of carps. *J. Fish. Res. Board Can.* **33**, 940–947.

Chondar, S. L. (1980). "Hypophysation of Indian Major Carps." Satish Book Enterprise, India.

Clayton, R. N., and Shakespear, R. A. (1978). [D-Serine (TBu)[6]] luteinizing hormone releasing hormone 1-9 ethylamide: Enhanced biological activity because of resistance to degradation. *J. Endocrinol.* **77**, 34.

Clemens, H. P., and Grant, F. B. (1965). The seminal thinning response of carp (*Cyprinus carpio*) and rainbow trout (*Salmo gairdnerii*) after injections of pituitary extracts. *Copeia* pp. 174–177.

Clemens, H. P., and Reed, C. A. (1967). Long-term gonadal growth and maturation of goldfish (*Carassius auratus*) with pituitary injections. *Copeia* pp. 465–466.

Clemens, H. P., and Sneed, K. E. (1962). Bioassay and use of pituitary materials to spawn warm-water fishes. *Res. Rep.—U.S. Fish. Wildl. Serv.* **61**, 30 pp.

Colombo, L., Bern, H. A., Pieprzyk, J., and Johnson, D. W. (1973). Biosynthesis of 11-deoxycorticosteroids by teleost ovaries and discussion of their possible role in oocyte maturation and ovulation. *Gen. Comp. Endocrinol.* **21**, 168–178.

Conference on Application of Hormones to Economic Fish (1975). Experiments on inducement of spawning in domestic fish by injection of synthesized hypothalamic luteinizing hormone-releasing hormone (LH-RH). *Kexue Tongbao* **20**, 43–48.

Cook, A. F., Stacey, N. E., and Peter, R. E. (1980). Preovulatory changes in serum cortisol levels in the goldfish *Carassius auratus. Gen. Comp. Endocrinol.* **40**, 507–510.

Cooperative Team for Hormonal Application in Pisciculture (1977). A new highly effective ovulating agent for fish reproduction- practical application of LH-RH analogue for the induction of spawning of farm fishes. *Sci. Sin.* **20**, 469–474.

Coy, D. H., Coy, E. J., Schally, A. V., Vilchez-Martinez, J., Hirotsu, Y., and Arimura, A. (1974). Synthesis and biological properties of D-Ala[6],des Gly[10]-LH-RH ethylamide, a peptide with greatly enhanced LH and FSH releasing activity. *Biochem. Biophys. Res. Commun.* **57**, 335–340.

Coy, D. H., Vilchez-Martinez, J. A., and Meyers, C. A. (1979). Structure-function relationship of hypothalamic peptides. *In* "Clinical Neuroendocrinology: A Pathophysiological Approach" (G. Tolis *et al.*, eds.), pp. 83–88. Raven Press, New York.

Crim, L. W., Evans, D. M., and Vickery, B. H. (1983). Manipulation of the seasonal reproductive cycle of the landlocked salmon (*Salmo salar*) with LH RH administration at various stages of gonadal development. *Can. J. Fish. Aquat. Sci.* **40**, 61–67.

Davy, F. B., and Chouinard, A., eds. (1981). "Induced Fish Breeding in Southeast Asia," IDRC-178e. Int. Dev. Res. Cent., Ottawa.

de Azevedo, P., and de Oliveira, A. C. E. (1939). Sobre o emprego da hipofise conservada em alcool na des ova dos peixes. *In* "Livro de homenagem aos Professores Alvaro e Miguel Ozorio de Almeida," pp. 35–42. Rio de Janeiro.

De Montalembert, G., Jalabert, B., and Bry, C. (1978a). Precocious induction of maturation and ovulation in northern pike (*Esox lucius*). *Ann. Biol. Anim., Biochim., Biophys.* **18**, 969–975.

De Montalembert, G., Bry, C., and Billard, R. (1978b). Control of reproduction in northern pike. *Spec. Publ.—Am. Fish. Soc.* **11**, 217–225.

Dettlaf, T. A., and Davydova, S. I. (1979a). The effect of triiodothyronine on the maturation of the oocytes of the sevryuga, *Acipenser stellatus*, under the influence of the gonadotropic hormones. *J. Ichthyol.* **19**, 105–110.

Dettlaf, T. A., and Davydova, S. I. (1979b). Differential sensitivity of cells of follicular epithelium and oocytes in the stellate sturgeon to unfavourable conditions, and correlating influence of triiodothyronine. *Gen. Comp. Endocrinol.* **39**, 236–243.

Donaldson, E. M. (1973). Reproductive endocrinology of fishes. *Am. Zool.* **13**, 909–927.

Donaldson, E. M. (1975). Physiological and physiochemical factors associated with maturation and spawning. *EIFAC Tech. Pap.* **EIFAC/T25**, 53–71.

Donaldson, E. M. (1977). Bibliography of fish reproduction 1963–1974. Part 1. General, Non-teleostean species and Teleostei, Abramis to Clinostomus. Part 2. Teleostei, Clupea to Ompok. Part 3. Teleostei, Oncorhynchus to Zygonectes and Addendum. *Tech. Rep.—Fish. Mar. Serv. (Can.)* **732**, 1–572.

Donaldson, E. M. (1981). The pituitary-interrenal axis as an indicator of stress in fish. *In* "Stress and Fish" (A. D. Pickering, ed.), Chapter 2, pp. 11–48. Academic Press, New York.

Donaldson, E. M., and Fagerlund, U. H. M. (1970). Effect of sexual maturation and gonadectomy at sexual maturity on cortisol secretion rate in sockeye salmon (*Oncorhynchus nerka*). *J. Fish. Res. Board Can.* **27**, 2287–2296.

Donaldson, E. M., and Hunter, G. A. (1982). Sex control in fish with particular reference to salmonids. *Can. J. Fish. Aquat. Sci.* **39**, 99–110.

Donaldson, E. M., and Shehadeh, Z. H. (1972). Effect of salmon gonadotropin on ovarian and testicular development in immature grey mullet (*Mugil cephalus*). In "The Grey Mullet, Induced Breeding and Larval Rearing," Rep. No. 01-72-76-1, pp. 71–86. Oceanic Institute, Waimanalo, Hawaii.

Donaldson, E. M., and Yamazaki, F. (1968). Preparation of gonadotropic hormone from salmon pituitary glands. *51st Annual Conf. Chem. Inst. Can.* p. 64.

Donaldson, E. M., Funk, J. D., Withler, F. C., and Morley, R. B. (1972a). Fertilization of pink salmon (*Oncorhynchus gorbuscha*) ova by spermatozoa from gonadotropin injected juveniles. *J. Fish. Res. Board Can.* **29**, 13–18.

Donaldson, E. M., Yamazaki, F., Dye, H. M., and Philleo, W. W. (1972b). Preparation of gonadotropin from salmon (*Oncorhynchus tshawytscha*) pituitary glands. *Gen. Comp. Endocrinol.* **18**, 469–481.

Donaldson, E. M., Dye, H. M., and Wright, B. F. (1978a). The effect of storage conditions on the biological activity of salmon gonadotropin. *Ann. Biol. Anim., Biochim., Biophys.* **18**, 997–1000.

Donaldson, E. M., Hunter, G. A., and Dye, H. M. (1978b). Induced ovulation in the coho salmon (*Oncorhynchus kisutch*) using salmon pituitary preparations, gonadotropin releasing hormones and an antiestrogen. *West. Reg. Conf. Gen. Comp. Endocrinol., 1978* Abstract, p. 24.

Donaldson, E. M., Hunter, G. A., and Dye, H. M. (1979). Relative potency of gonadotropin releasing hormone and gonadotropin releasing hormone analogues for induced ovulation in the coho salmon *Oncorhynchus kisutch. West. Reg. Conf., Gen. Comp. Endocrinol., 1979* pp. 23–24.

Donaldson, E. M., Hunter, G. A., and Dye, H. M. (1981a). Induced ovulation in coho salmon (*Oncorhynchus kisutch*). II. Preliminary study of the use of LH-RH and two high potency LH-RH analogues. *Aquaculture* **26**, 129–141.

Donaldson, E. M., Hunter, G. A., and Dye, H. M. (1981b). Induced ovulation in coho salmon (*Oncorhynchus kisutch*). III. Preliminary study on the use of the antiestrogen tamoxifen. *Aquaculture* **26**, 143–154.

Donaldson, E. M., Hunter, G. A., Dye, H. M., and Van Der Kraak, G. (1984). Induced ovulation in Pacific salmon using LH-RH analogs and salmon gonadotropin. *Proc. Int. Symp. Comp. Endocrinol., 9th, 1981.* In press.

Donaldson, E. M., Hunter, G. A., Van Der Kraak, G., and Dye, H. M. (1982). Application of LH-RH and LH-RH analogues to the induced final maturation and ovulation of coho salmon (*Oncorhynchus kisutch*). *Proc. Int. Symp. Reprod. Physiol. Fish, 1982* pp. 177–180.

Donaldson, E. M., Van Der Kraak, G., Hunter, G. A., Dye, H. M., Rivier, J., and Vale, W. (1983). Teleost GnRH and Analogues: Effect on Plasma GtH Concentration and Ovulation in Coho Salmon (*Oncorhynchus kisutch*). XIIth Conf. Europ. Comp. Endocrinol., 1983. Sheffield, England, p. 122.

Drance, M. G., Hollenberg, M. J., Smith, M., and Wylie, V. (1976). Histological changes in trout testis produced by injections of salmon pituitary gonadotropin. *Can. J. Zool.* **54**, 1285–1293.

Eding, E. H., Janssen, J. A. L., Kleine Staarman, G. H. J., and Richter, C. J. J. (1982). Effects of human chorionic gonadotropin (hCG) on maturation and ovulation of oocytes in the catfish *Clarias lazera* (C & V). *Proc. Int. Symp. Reprod. Physiol. Fish, 1982* p. 195.

Fagerlund, U. H. M., and Donaldson, E. M. (1970). Dynamics of cortisone secretion in sockeye salmon (*Oncorhynchus nerka*) during sexual maturation and after gonadectomy. *J. Fish. Res. Board Can.* **27**, 2323–2331.

Farmer, S. W., and Papkoff, H. (1977). A teleost· (*Tilapia mossambica*) gonadotropin that resembles luteinizing hormone. *Life Sci.* **20**, 1227–1232.

Fedorov, K. Ye., and Smirnova, Zh. V. (1976). Partial purification of pink salmon (*Oncorhynchus gorbuscha*) gonadotropin. *J. Ichthyol.* **16**, 1024–1027.

Fontaine, M. (1976). Hormones and the control of reproduction in aquaculture. *J. Fish. Res. Board Can.* **33**, 922–939.

Fontenele, O. (1955). Injecting pituitary (hypophyseal) hormones into fish to induce spawning. *Prog. Fish-Cult.* **17**, 71–75.

Fostier, A., and Jalabert, B. (1982). Physiological basis of practical means to induce ovulation in fish. *Proc. Int. Symp. Reprod. Physiol. Fish, 1982* pp. 164–173.

Fostier, A., Billard, R., Breton, B., Legendre, M., and Marlot, S. (1982). Plasma 11-oxotestosterone and gonadotropin during the beginning of spermiation in rainbow trout *Salmo gairdneri* R. *Gen. Comp. Endocrinol.* **46**, 428–434.

Freshwater Commerical Artificial Propogation Work Groups in Fujian, Jiangsu, Zhejiang and Shanghai (1977). Further research into the effects of LRH-A on induced ovulation of farm fish. *Acta Biochim. Biophys. Sin.* **9**, 15–24; Fish. Mar. Serv. Transl. **4469.**

Fujino, M., Kobayashi, S., Obayashi, M., Shinagawa, S., Fukuda, T., Kitada, C., Nakayama, R., Yamazaki, I., White, W. F., and Rippel, R. H. (1972). Structure-activity relationships in the C-terminal part of luteinizing hormone releasing hormone. *Biochem. Biophys. Res. Commun.* **49**, 863–869.

Fujino, M., Yamazaki, I., Kobayashi, S., Fukuda, T., Shinagawa, S., Nakayama, R., White, W. F., and Rippel, R. H. (1974). Some analogs of luteinizing hormone releasing hormone having intense ovulation-inducing activity. *Biochem. Biophys. Res. Commun.* **57**, 1248–1256.

Funk, J. D., and Donaldson, E. M. (1972). Induction of precocious sexual maturity in male pink salmon (*Oncorhynchus gorbuscha*). *Can. J. Zool.* **50**, 1413–1419.

Gerbil'skii, N. L. (1938). Expedition for the study of the physiology of spawning. *Rybn. Khoz. (Moscow)* **18**(10), 33–36.

Goetz, F. W., and Bergman, H. C. (1978). The effects of steroids on final maturation and ovulation of oocytes from brook trout (*Salvelinus fontinalis*) and yellow perch (*Perca flavescens*). *Biol. Reprod.* **18**, 293–298.

Goswami, S. V., and Sundararaj, B. I. (1968). Effect of estradiol benzoate, human chorionic gonadotropin and follicle-stimulating hormone on unilateral ovariectomy-induced compensatory hypertrophy in catfish, *Heteropneustes fossilis* (Bloch). *Gen. Comp. Endocrinol.* **11**, 393–400.

Greenblatt, R. B., Barfield W. E., Jungck, E. C., and Ray, A. W. (1961). Induction of ovulation with MRL/41, preliminary report. JAMA, *J. Am. Med. Assoc.* **178**, 101–105.

Grier, H. J. (1981). Cellular organization of the testis and spermatogenesis in fishes. *Am. Zool.* **21**, 345–357.

Guillemin, R. (1978). Peptides in the brain: The new endocrinology of the neuron. *Science* **202**, 390–402.

Harvey, B. J., and Hoar, W. S. (1979). "The Theory and Practice of Induced Breeding in Fish," IDRC-TS21e. Int. Dev. Res. Cent., Ottawa.

Hasler, A. D., Meyer, R. K., and Field, H. M. (1939). Spawning induced prematurely in trout with the aid of pituitary glands of the carp. *Endocrinology* **25**, 978–983.

Hasler, A. D., Meyer, R. K., and Field, H. M. (1940). The use of hormones for the conservation of muskellunge, *Esox masquinongy immaculatus* Garrard. *Copeia* pp. 43–46.

Haydock, I. (1971). Gonad maturation and hormone-induced spawning in the gulf croaker, *Bairdiella icistia*. *Fish. Bull.* **69**, 157–180.

Hirose, K., and Ishida, R. (1974a). Induction of ovulation in the ayu (*Plecoglossus altivelis*) with LH releasing hormone (LH-RH). *Bull. Jpn. Soc. Sci. Fish.* **40**, 1235–1240.

Hirose, K., and Ishida, R. (1974b). Effects of cortisol and human chorionic gonadotropin (HCG) on ovulation in ayu *Plecoglossus altivelis* (Temminck and Schlegel) with special respect to water and ion balance. *J. Fish Biol.* **6**, 557–564.

Hirose, K., Machida, Y., and Donaldson, E. M. (1976). Induction of ovulation in the Japanese flounder (*Limanda yokohamae*) with human chorionic gonadotropin and salmon gonadotropin. *Bull. Jpn. Soc. Sci. Fish.* **42**, 13–20.

Hirose, K., Machida, Y., and Donaldson, E. M. (1979). Induced ovulation of Japanese flounder (*Limanda yokohamae*) with human chorionic gonadotropin and salmon gonadotropin, with special reference to changes in quality of eggs retained in the ovarian cavity after ovulation. *Bull. Jpn. Soc. Sci. Fish.* **45**, 31–36.

Hogendoorn, H. (1979). Controlled propagation of the African catfish, *Clarias lazera* (C. and V.). I. Reproductive biology and field experiments. *Aquaculture* **17**, 323–333.

Houssay, B. A. (1930). Accion sexual de la hipofisis en los perces y reptiles. *Rev. Soc. Argent. Biol.* **6**, 686–688.

Houssay, B. A. (1931). Action sexuelle de l'hypophyse sur les poissons et les reptiles. *C. R. Seances Soc. Biol. Ses Fil.* **106**, 377–378.

Huang, F. L., Huang, C.-J., Lin, S. H. Lo, T. B., and Papkoff, H. (1981). Isolation and characterization of gonadotropin isohormones from the pituitary gland of pike eel (*Muraenesox cinereus*). *Int. J. Pept. Res.* **18**, 69–78.

Hunter, G. A., Donaldson, E. M., Stone, E. T., and Dye, H. M. (1978). Induced ovulation of female chinook salmon (*Oncorhynchus tshawytscha*) at a production hatchery. *Aquaculture* **15**, 99–112.

Hunter, G. A., Donaldson, E. M., Dye, H. M., and Peterson, K. (1979). A preliminary study of induced ovulation in coho salmon (*Oncorhynchus kisutch*) at Robertson Creek Salmon Hatchery. *Tech. Rep.—Fish. Mar. Serv. (Can.)* **899**, 1–15.

Hunter, G. A., Donaldson, E. M., and Dye, H. M. (1981). Induced ovulation in coho salmon (*Oncorhynchus kisutch*). I. Further studies on the use of salmon pituitary preparations. *Aquaculture* **26**, 117–127.

Hyder, M., Shah, A. V., and Hartree, A. S. (1979). Methallibure studies on Tilapia. 3. Effects of tilapian partially purified pituitary gonadotropic fractions on the testes of methallibure-treated *Sarotherodon spirulus* (*Tilapia nigra*). *Gen. Comp. Endocrinol.* **39**, 475–480.

Ibrahim, K. H. (1969a). Techniques of collection, processing and storage of fish pituitary gland. *FAO/UNDP Reg. Semin. Induced Spawning Cultivated Fishes, 1969* pp. 1–13.

Ibrahim, K. H. (1969b). Preparation, preservation and ampouling of fish pituitary extract. *FAO/UNDP Reg. Semin. Induced Spawning Cultivated Fishes, 1969* pp. 1–12.

Ibrahim, K. H., and Chaudhuri, H. (1966). Preservation of fish pituitary extract in glycerine for induced breeding of fish. *Indian J. Exp. Biol.* **4**, 249–250.

Idler, D. R., and Hwang, S. J. (1978). Conditions for optimizing recovery of salmon gonadotropin activity from Con A Sepharose. *Gen. Comp. Endocrinol.* **34**, 642–644.

Idler, D. R., and Ng, T. B. (1979). Studies on two types of gonadotropins from both salmon and carp pituitaries. *Gen. Comp. Endocrinol.* **38**, 421–440.

Idler, D. R., Fagerlund, U. H. M., and Ronald, A. P. (1960). Isolation of pregn-4-ene-17α, 20β-diol-3-one from plasma of Pacific salmon. *Oncorhynchus nerka. Biochem. Biophys. Res. Commun.* **2**, 133–137.

Idler, D. R., Schmidt, P. J., and Biely, J. (1961). The androgenic activity of 11-ketotestosterone a steroid in salmon plasma. *Can. J. Biochem. Physiol.* **39**, 317–320.

Idler, D. R., Bazar, L. S., and Hwang, S. J. (1975a). Fish gonadotropin(s). II. Isolation of gonadotropin(s) from chum salmon pituitary glands using affinity chromatography. *Endocr. Res. Commun.* **2**, 215–235.

Idler, D. R., Bazar, L. S., and Hwang, S. J. (1975b). Fish gonadotropin in chum salmon pituitary glands. *Endocr. Res. Commun.* **2**, 237–249.

Igumnova, L. V. (1974). The time taken for beluga (*Huso huso*) females to mature at different temperatures following a pituitary injection. *J. Ichthyol.* **14**, 890–895.

Ishida, R., Hirose, K., and Donaldson, E. M. (1972). Induction of ovulation in ayu, *Plecoglossus altivelis*, with salmon pituitary gonadotropin. *Bull. Jpn. Soc. Sci. Fish.* **38**, 1007–1012.

Iwamatsu, T. (1978). Studies on oocyte maturation of the medaka *Oryzias latipes*. V. On the structure of steroids that induce maturation *in vitro*. *J. Exp. Zool.* **204**, 401–408.

Jalabert, B. (1975). Modulation par différents stéroides non maturants de l'efficacité de la 17α-hydroxy, 20β-dihydroprogestérone ou d'un extrait gonadotrope sur la maturation intra-folliculaire in vitro des ovocytes de la Truite arc-en-ciel *Salmo gairdnerii*. *C. R. Hebd. Seances Acad. Sci.* **281**, 811–814.

Jalabert, B. (1976). *In vitro* oocyte maturation and ovulation in rainbow trout (*Salmo gairdneri*), northern pike (*Esox lucius*) and goldfish (*Carassius auratus*). *J. Fish. Res. Board Can.* **33**, 974–988.

Jalabert, B., and Szollosi, D. (1975). *In vitro* ovulation of trout oocytes, effect of prostaglandins on smooth muscle like cells of the theca. *Prostaglandins* **9**, 765–778.

Jalabert, B., Bry, C., Breton, B., and Campbell, C. (1976). Action de la 17α-hydroxy-20β dihydroprogesterone et de la progestérone sur la maturation et l'ovulation in vivo et sur le niveau d'hormone gonadotrope plasmatique t-GtH chez la truite arc-en-ciel (*Salmo gairdneri*). *C. R. Hebd. Seances Acad. Sci., Ser. D* **283**, 1205–1208.

Jalabert, B., Breton, B., Brzuska, E., Fostier, A., and Wieniawski, J. (1977). A new tool for induced spawning; the use of 17α-hydroxy-20β-dihydroprogesterone to spawn carp at low temperature. *Aquaculture* **10**, 353–364.

Jalabert, B., Goetz, F. W., Breton, B., Fostier, A., and Donaldson, E. M. (1978a). Precocious induction of oocyte maturation and ovulation in coho salmon (*Oncorhynchus kisutch*). *J. Fish. Res. Board Can.* **35**, 1423–1429.

Jalabert, B., Breton, B., and Fostier, A. (1978b). Precocious induction of oocyte maturation and ovulation in rainbow trout (*Salmo gairdneri*): Problems when using 17α-hydroxy-20β-dihydroprogesterone. *Ann. Biol. Anim., Biochim., Biophys.* **18**, 977–984.

Jalabert, B., Breton, B., and Bry, C. (1980). Evolution de la gonadotropine plasmatique t-GtH après synchronisation des ovulations par injection de 17α-hydroxy-20βdihydroprogestérone chez la Truite arc-en-ciel (*Salmo gairdneri* R.). *C. R. Hebd. Seances Acad. Sci., Ser. D* **290**, 1431–1434.

Juario, J. V. (1981). *In* "Induced Fish Breeding in Southeast Asia" (F. B. Davy and A. Chouinard, eds.), IDRC-178e, Microfiche Annex 3/4 18 pp. Int. Dev. Res. Cent., Ottawa.

Juario, J. V., and Natividad, M. (1980). The induced spawning of captive milkfish. *Asian Aquacult.* **3**, 4.

Kagawa, H., and Nagahama, Y. (1981). In vitro effects of prostaglandins on ovulation in goldfish *Carassius auratus*. *Bull. Jpn. Soc. Sci. Fish.* **47**, 1119–1121.

Kapoor, C. P. (1976). Effect of luteinizing hormone on female *Puntius ticto* (Ham.). *Z. Mikrosk.-Anat. Forsch.* **90**, 514–520.

Kapoor, C. P. (1978). Effect of follicle stimulating hormone on the testes of *Puntius ticto* (Ham.). *Endokrinologie* **72**, 17–24.

Kapur, K. (1979). Effects of indomethacin on ovulation and spawning in exotic fish *Cyprinus carpio*. *Indian J. Exp. Biol.* **17**, 517–519.

Kapur, K., and Toor, H. S. (1979). The effect of clomiphene citrate on ovulation and spawning in indomethacin treated carp, *Cyprinus carpio*. *J. Fish Biol.* **14**, 59–66.

Katz, Y., and Eckstein, B. (1974). Changes in steroid concentration in blood of female *Tilapia aurea* (Teleostei, Cichlidae) during initiation of spawning. *Endocrinology* **95**, 963–967.

Katzman, P. A., and Doisy, E. A. (1932). Preparation of extracts of the anterior pituitary-like substance of urine of pregnancy. *J. Biol. Chem.* **98**, 739–754.

Kaya, C. M. (1973). Effects of temperature on responses of the gonads of green sunfish *Lepomis cyanellus* to treatment with carp pituitaries and testosterone propionate. *J. Fish. Res. Board Can.* **30**, 905–912.

Khalil, M. S. (1974). A study on the acceleration of maturation in *Tilapia mossambica* by the administration of mammalian hormones. *Bull. Fac. Sci., Cairo Univ.* **45**, 151–158.

Khoo, K. H. (1974). Steroidogenesis and the role of steroids in the endocrine control of oogenesis and vitellogenesis in the goldfish *Carassius auratus*. Ph.D. Thesis, University of British Columbia, Vancouver, B.C., Canada.

Kirshenblatt, J. D. (1952). The action of steroid hormones on the female vy'un. *Dokl. Akad. Nauk SSSR* **83**, 629–632.

Kirshenblatt, J. D. (1959). The effect of cortisone on ovaries of the loach. *Byull. Eksp. Biol. Med.* **47**, 108–112.

Klopper, A., and Hall, M. (1971). New synthetic agent for the induction of ovulation: Preliminary trials in women. *Br. Med. J.* **1**, 152–154.

Konig, W., Sandow, J., and Geiger, R. (1975). Structure-function relationship of LH-RH/FSH-RH. *In* "Chemistry, Structure and Biology" (R. Walter and J. Meienhofer, eds.), pp. 883–888. Ann Arbor Sci. Publ., Ann Arbor, Michigan.

Kouril, J., and Barth, T. (1981). Inducing fish egg ovulation by means of LH-RH in the artificial stripping of tench (*Tinca tinca* L.). *Bul. Vúrh Vodnany* **9**, 13–18.

Kuo, C. M., and Nash, C. E. (1975). Recent progress on the control of ovarian development and induced spawning of the grey mullet (*Mugil cephalus* L.). *Aquaculture* **5**, 19–29.

Kuo, C. M., Nash, C. E., and Shehadeh, Z. H. (1974). A procedural guide to induce spawning in grey mullet (*Mugil cephalus* L.). *Aquaculture* **3**, 1–14.

Lam, T. J. (1982). Applications of endocrinology to fish culture. *Can. J. Fish. Aquat. Sci.* **39**, 111–137.

Lam, T. J., Pandey, S., and Hoar, W. S. (1975). Induction of ovulation in goldfish by synthetic luteinizing hormone-releasing hormone (LH-RH). *Can. J. Zool.* **53**, 1139–1192.

Lam, T. J., Pandey, S., Nagahama, Y., and Hoar, W. S. (1976). Effect of synthetic luteinizing hormone-releasing hormone (LH-RH) on ovulation and pituitary cytology of the goldfish *Carassius auratus. Can. J. Zool.* **54**, 816–824.

Lam, T. J., Pandey, S., Nagahama, Y., and Hoar, W. S. (1978). Endocrine control of oogenesis, ovulation and oviposition in goldfish. *In* "Comparative Endocrinology" (P. J. Gaillard and H. H. Boer, eds.), pp. 55–64. Elsevier/North-Holland, Amsterdam.

Lederis, K. (1972). Recent progress in research on the urophysis. *Gen. Comp. Endocrinol., Suppl.* **3**, 339–344.

Leong, R. (1977). Maturation and induced spawning of captive Pacific mackeral, *Scomber japonicus. U.S., Natl. Mar. Fish. Serv., Fish. Bull.* **75**, 205–211.

Licht, P. (1974). Induction of sperm release in frogs by mammalian gonadotropin-releasing hormone. *Gen. Comp. Endocrinol.* **23**, 352–354.

Licht, P., Papkoff, H., Farmer, S. W., Muller, C. H., Tsui, H. W., and Crews, D. (1977). Evolution of gonadotropin structure and function. *Recent Prog. Horm. Res.* **33**, 169–248.

Lin, H.-R. (1982). Polycultural system of freshwater fish in China. *Can. J. Fish. Aquat. Sci.* **39**, 143–150.

Lin, H.-R., Peter, R. E., Nakorniak, C. S., and Bres, O. (1984). Actions of the superactive analogue Des-Gly[10][D-Ala[6]]LRH ethylamide (LRH-A) on gonadotropin secretion in goldfish. *Proc. Int. Symp. Comp. Endocrinol., 9th, 1981* In press.

Lo, T. B., Huang, F. L., and Chang, G. D. (1981). Separation of subunits of pike eel gonadotropin by hydrophobic interaction chromatography. *J. Chromatogr.* **215**, 229–233.

McBride, J. R., and van Overbeeke, A. P. (1969). Cytological changes in the pituitary gland of the adult sockeye salmon (*Oncorhynchus nerka*) after gonadectomy. *J. Fish. Res. Board Can.* **26**, 1147–1156.

McCreery, B. R., Licht, P., Barnes, R., Rivier, J. E., and Vale, W. W. (1982). Actions of agonistic and antagonistic analogs of gonadotropin releasing hormone (Gn-RH) in the bullfrog *Rana catesbeiana. Gen. Comp. Endocrinol.* **46**, 511–520.

Mackay, N. J. (1975). The reproductive cycle of the firetail gudgeon *Hypseleotris galii.* IV. Hormonal control of ovulation. *Aust. J. Zool.* **23**, 43–47.

MacKinnon, C. N. (1976). An investigation of the regulation of reproductive development in pink salmon (*Oncorhynchus gorbuscha*) in relation to the initiation of an off-year pink salmon run. M.Sc. Thesis, University of British Columbia.

MacKinnon, C. N., and Donaldson, E. M. (1978). Comparison of the effect of salmon gonadotropin administered by pellet implantation or injection on sexual development of juvenile male pink salmon (*Oncorhynchus gorbuscha*). *Can. J. Zool.* **56**, 86–89.

Monahan, M. W., Amoss, M. S., Anderson, H. A., and Vale, W. (1973). Synthetic analogs of the hypothalamic luteinizing hormone releasing factor with increased agonist or antagonist properties. *Biochemistry* **12**, 4616–4620.

Nagahama, Y., Kagawa, H., and Young, G. (1982). Cellular sources of sex steroids in teleost gonads. *Can. J. Fish. Aquat. Sci.* **39**, 56–64.

Nash, C. E., and Kuo, C. M. (1976). Preliminary capture husbandry and induced breeding results with the milkfish, *Chanos chanos. Proc. Int. Milkfish Workshop Conf. 1976* SEAFDEC, pp. 139–160.

Nash, C. E., and Shehadeh, Z. H., eds. (1980). Review of breeding and propagation techniques for grey mullet, *Mugil cephalus. ICLARM Stud. Rev.* **3**, 1–87.

Ng, T. B., and Idler, D. R. (1978a). Big and little forms of plaice vitellogenic and maturational hormones. *Gen. Comp. Endocrinol.* **34**, 408–420.

Ng, T. B., and Idler, D. R. (1978b). A vitellogenic hormone with a large and small form from salmon pituitaries. *Gen. Comp. Endocrinol.* **35**, 189–195.

Ng, T. B., and Idler, D. R. (1979). Studies on two types of gonadotropin from both American plaice and winter flounder pituitaries. *Gen. Comp. Endocrinol.* **38**, 410–420.

Ogata, H., Nomura, T., and Hata, M. (1979). Prostaglandin $F_2\alpha$ changes induced by ovulatory stimuli in the pond loach *Misgurnus anguillicaudatus. Bull. Jpn. Soc. Sci. Fish.* **45**, 929–931.

Palmer, D. D., Burrows, R. E., Robertson, O. H., and Newman, H. W. (1954). Further studies on the reactions of adult blueback salmon to injected salmon and mammalian gonadotropins. *Prog. Fish-Cult.* **16**, 99–107.

Pandey S., and Hoar, W. S. (1972). Induction of ovulation in goldfish by clomiphene citrate. *Can. J. Zool.* **50**, 1679–1680.

Pandey, S., and Stacey, N. (1975). Anti estrogenic action of clomiphene citrate in goldfish. *Can. J. Zool.* **53**, 102–103.

Pandey, S., Stacey, N., and Hoar, W. S. (1973). Mode of action of clomiphene citrate in inducing ovulation of goldfish. *Can. J. Zool.* **51**, 1315–1316.

Pandey, S., Lam, T. J., Nagahama, Y., and Hoar, W. S. (1977). Effect of dexamethasone and metopirone on ovulation in the goldfish, *Carassius auratus. Can. J. Zool.* **55**, 1342–1350.

Parameswaran, S., and Murugesan, V. K. (1976). Observations on the hypophysation of murrels (Ophicephalidae). *Hydrobiologia* **50**, 81–87.

Patterson, J. S. (1981). Clinical aspects and development of antioestrogen therapy: A review of the endocrine effects of tamoxifen in animals and man. *J. Endocrinol.* **89**, 67P–75P.

Peter, R. E. (1980). Serum gonadotropin levels in mature male goldfish in response to luteinizing hormone-releasing hormone (LH-RH) and desGly[10]-(D-Ala[6])- LH-RH ethylamide. *Can. J. Zool.* **58**, 1100–1104.

Peter, R. E. (1982). Neuroendocrine control of reproduction in teleosts. *Can. J. Fish. Aquat. Sci.* **39**, 48–55.

Peter, R. E., and Paulencu, C. R. (1980). Involvement of the preoptic region in gonadotropin release-inhibition in goldfish *Carassius auratus. Neuroendocrinology* **31**, 133–141.

Peter, R. E., Crim, L. W., Goos, H. J. T., and Crim, J. W. (1978). Lesioning studies on the gravid female goldfish: Neuroendocrine regulation of ovulation. *Gen. Comp. Endocrinol.* **35**, 391–401.

Pickford, G. E., and Atz, J. W. (1957). "The Physiology of the Pituitary Gland of Fishes." N.Y. Zool. Soc., New York.

Pierce, J. G., Faith, M. R., and Donaldson, E. M. (1976). Antibodies to reduced S-carboxymethylated alpha subunit of bovine luteinizing hormone and their application to the study of the purification of gonadotropin from salmon. *Gen. Comp. Endocrinol.* **30**, 47–60.

Pullin, R. (1975). Preliminary investigations into methods for controlling the reproduction of captive marine flatfish. *Pubbl. Stn. Zool. Napoli* **39**, Suppl., 282–296.

Pullin, R. S. V., and Kuo, C.-M. (1980). Developments in the breeding of cultured fishes. *Symp. Adv. Food-Prod. Syst. Arid Semiarid Lands, 1980* Manuscript, 35 pp.

Quillier, R., and Labat, R. (1977). Mise au point sur la reproduction des esocides; étude preliminaire. *Invest. Pesq.* **41**, 33–38.

Ramaswami, L. S. (1962). Endocrinology of reproduction in fish and frog. *Gen. Comp. Endocrinol., Suppl.* **1**, 286–299.

Ramaswami, L. S., and Lakshman, A. B. (1959). Action of mammalian hormones on the spawning of catfish. *J. Sci. Ind. Res., Sect. C* **18**, 185–191.

Refstie, T., Stoss, J., and Donaldson, E. M. (1982). Production of all female coho salmon (*Oncorhynchus kisutch*) by diploid gynogenesis using irradiated sperm and cold shock. *Aquaculture* **29**, 67–82.

Research Group of Eel Reproduction Xiamen (Amoy) Fisheries College and Fujian Fisheries Institute (1978). Preliminary studies on the induction of spawning in common eels. *Acta Zool. Sin.* **24**, 399–402.

Ribeiro, O. F., and Neto, J. F. T. (1944). Da ontencao de un extrato glicerinado para a hipofsacao de peixes. *Rev. Fac. Med. Vet., Univ. Sao Paulo* **2**, 227–232 (Engl. abstr.).

Rothbard, S. (1981). Induced reproduction in cultivated cyprinids—the common carp and the group of chinese carps. 1. The technique of induction, spawning and hatching. *Bamidgeh* **33**, 103–121.

Rothbard, S., and Rothbard, H. (1982). Spermiation response of carp to homologous pituitary gland. *Proc. Int. Symp. Reprod. Physiol. Fish, 1982* p. 201.

Schally, A. V. (1978). Aspects of hypothalamic regulation of the pituitary gland. *Science* **202**, 18–28.

Schally, A. V., and Kastin, A. J. (1972). Hypothalamic releasing and inhibiting hormones *Gen. Comp. Endocrinol., Suppl.* **3**, 76–85.

Schmidt, P. J., and Idler, D. R. (1962). Steroid hormones in the plasma of salmon at various states of maturation. *Gen. Comp. Endocrinol.* **2**, 204–214.

Schoonbee, H. J., Brandt, F. De W., and Bekker, C. A. L. (1978). Induced spawning of the two phytophagous Chinese carp species *Ctenopharyngodon idella* (Val.) and *Hypopthalmichthys molitrix* (Val.) with reference to the possible use of the grass carp in the control of aquatic weeds. *Water SA* **4**, 93–103.

Scott, A. P., and Baynes, S. M. (1982). Plasma levels of sex steroids in relation to ovulation and spermiation in rainbow trout (*Salmo gairdneri*). *Proc. Int. Symp. Reprod. Physiol. Fish,* *1982,* pp. 103–106.

Shehadeh, Z. H. (1975). Induced breeding techniques—a review of progress and problems. *EIFAC Tech. Pap.* **25,** 72–88.

Shehadeh, Z. H., and Ellis, J. N. (1970). Induced spawning of the striped mullet *Mugil cephalus* L. *J. Fish Biol.* **2,** 355–360.

Shehadeh, Z. H., Kuo, C. M., and Milisen, K. K. (1973a). Induced spawning of grey mullet *Mugil cephalus* L. with fractionated salmon pituitary extract. *J. Fish Biol.* **5,** 471–478.

Shehadeh, Z. H., Madden, W. D., and Dohl, T. P. (1973b). The effect of exogenous hormone treatment on spermiation and vitellogenesis in the grey mullet, *Mugil cephalus* L. *J. Fish Biol.* **5,** 479–487.

Sherwood, N. M., Eiden, L., Brownstein, M. J., Spiess, J., Rivier, J., and Vale, W. (1982). The sequence of teleost gonadotrophin releasing hormone (GnRH): Multiple forms of GnRH in the animal kingdom. *Soc. Neurosci. Abstr.* **8,** No. 8.4, p. 12.

Sherwood, N. M., Eiden, L., Brownstein, M., Spiess, J., Rivier, J., and Vale, W. (1983). Characterization of a teleost gonadotropin releasing hormone. *Proc. Natl. Acad. Sci., U.S.A.* **80,** pp. 2794–2798.

Singh, A. K., and Singh, T. P. (1976). Effect of Clomid, Sexovid and prostaglandins on induction of ovulation and gonadotropin secretion in a freshwater catfish, *Heteropneustes fossilis* (Bloch). *Endokrinologie* **68,** 129–136.

Smith-Gill, S. J., and Berven, K. A. (1980). In vitro fertilization and assessment of male reproductive potential using mammalian gonadotropin-releasing hormone to induce spermiation in *Rana sylvatica*. *Copeia* pp. 723–728.

Sokolowska, M., Popek, W., and Bieniarz, K. (1978). Synthetic releasing hormones LH/FSH-RH and LH-RH: Effect of intracerebral and intramuscular injections on female carp (*Cyprinus carpio* L.) maturation. *Ann. Biol. Anim., Biochim., Biophys.* **18,** 963–967.

Sower, S. A., Schreck, C. B., and Donaldson, E. M. (1982). Hormone induced ovulation of coho salmon (*Oncorhynchus kisutch*) held in sea water and fresh water. *Can. J. Fish. Aquat. Sci.* **39,** 627–632.

Stacey, N. E., and Goetz, F. W. (1982). Role of prostaglandins in fish reproduction. *Can. J. Fish. Aquat. Sci.* **39,** 92–98.

Stacey, N. E., and Pandey, S. (1975). Effects of indomethacin and prostaglandins on ovulation of goldfish. *Prostaglandins* **9,** 597–608.

Stacey, N. E., and Peter, R. E. (1979). Central action of prostaglandins in spawning behavior of female goldfish. *Physiol. Behav.* **22,** 1191–1196.

Stacey, N. E., Cook, A. F., and Peter, R. E. (1979a). Spontaneous and gonadotropin induced ovulation in the goldfish, *Carassius auratus* L.: Effects of external factors. *J. Fish Biol.* **15,** 349–361.

Stacey, N. E., Cook, A. F., and Peter, R. E. (1979b). Ovulatory surge of gonadotropin in the goldfish, *Carassius auratus*. *Gen. Comp. Endocrinol.* **37,** 246–249.

Strand, F. L. (1958). Effect of high carbohydrate diet and epinephrine in salmon and brook trout. *N.Y. Fish Game J.* **5,** 84–93.

Sundararaj, B. I. (1981). Reproductive physiology of teleost fishes. *FAO Rome* **ADCP/REP/ 81/16,** 1–82.

Sundararaj, B. I., and Goswami, S. V. (1966). Effects of mammalian hypophysial hormones, placental gonadotrophins, gonadal hormones and adrenal corticosteroids on ovulation and spawning in hypophysectomized catfish, *Heteropneustes fossilis* (Bloch). *J. Exp. Zool.* **161,** 287–296.

Sundararaj, B. I., and Goswami, S. V. (1968). Effect of estrogen, progesterone, and testosterone on the pituitary and ovary of catfish, *Heteropneustes fossilis* (Bloch). *J. Exp. Zool.* **169**, 211–228.

Sundararaj, B. I., and Goswami, S. V. (1969). Role of interrenal in luteinizing hormone-induced ovulation and spawning in the catfish, *Heteropneustes fossilis* (Bloch). *Gen. Comp. Endocrinol., Suppl.* **2**, 374–384.

Sundararaj, B. I., and Goswami, S. V. (1972). *In vivo* and *in vitro* induction of maturation and ovulation in oocytes of the catfish, *Heteropneustes fossilis* (Bloch), with protein and steroid hormones. *Int. Congr. Ser.—Excerpta Med.* **219**, 966–975.

Sundararaj, B. I., and Samy, T. S. A. (1974). Some aspects of chemistry and biology of piscine gonadotropins. *In* "Gonadotropins and Gonadal Function"(N. R. Mougdal, ed.), pp. 118–127. Academic Press, New York.

Sundararaj, B. I., Nayyar, S. K., Anand, T. C., and Donaldson, E. M. (1971). Effect of salmon pituitary gonadotropin, ovine luteinizing hormone and testosterone on the testes and seminal vesicles of hypophysectomized catfish, *Heteropneustes fossilis* (Bloch). *Gen. Comp. Endocrinol.* **17**, 73–82.

Sundararaj, B. I., Anand, T. C., and Donaldson, E. M. (1972). Effect of partially purified salmon pituitary gonadotropin on ovarian maintenance, ovulation and vitellogenesis in hypophysectomized catfish, *Heteropneustes fossilis* (Bloch). *Gen. Comp. Endocrinol.* **18**, 102–114.

Swann, C. G., and Donaldson, E. M. (1980). Bibliography of salmonid reproduction 1963–1970 for the family Salmonidae; Subfamilies Salmoninae, Coregoninae and Thymallinae. *Can. Tech. Rep. Fish. Aquat. Sci.* **970**, 1–221.

Tao, S.-K. (1973). Investigations of the use of gonadotropins to induce spawning in the white sucker (Catostomus commersoni (Lacepede) in North Dakota. *Natl. Taiwan Univ. Rep. Inst. Fish Biol.* **3**, 173–185.

Ueda, H., and Takahashi, H. (1976). Acceleration of ovulation in the loach *Misgurnus anguillicaudatus* by the treatment with clomiphene citrate. *Bull. Fac. Fish., Hokkaido Univ.* **27**, 1–5. *Fish. Mar. Serv. Transl.* **4070**, 1–12 (1977).

Ueda, H., and Takahashi, H. (1977a). Promotion of ovarian maturation accompanied with ovulation and changes of pituitary gonadotrophs after ovulation in the loach, *Misgurnus anguillicaudatus* treated with clomiphene citrate. *Bull. Fac. Fish., Hokkaido Univ.* **28**, 106–117.

Ueda, H., and Takahashi, H. (1977b). Promotion by clomiphene citrate of gonadal development and body growth in immature and maturing goldfish, *Carassius auratus. Bull. Fac. Fish., Hokkaido Univ.* **28**, 181–192.

Van Der Kraak, G., and Donaldson, E. M. (1983). Plasma levels of sex steroids in relation to oocyte maturation and ovulation in coho salmon (*Oncorhynchus kisutch*). *West Reg. Conf. Gen. Comp. Endocrinol.* Abst. 31, p. 23.

Van Der Kraak, G., Donaldson, E. M., Hunter, G. A., Dye, H. M., and Lin, H. R. (1982). The use of Des-Gly[10][D-Ala[6]]LH-RH ethylamide as an activator of the pituitary ovarian axis in adult coho salmon (*Oncorhynchus kisutch*)·*West. Reg. Conf. Gen. Comp. Endocrinol., 1982* Abstract, p. 3.

Van Der Kraak, G., Lin, H. R., Donaldson, E. M., Dye, H. M., and Hunter, G. A. (1983a). Effects of LH-RH and des-Gly[10][D-Ala[6]]LH-RH-ethylamide on plasma gonadotropin levels and oocyte maturaton in adult female coho salmon (*Oncorhynchus kisutch*). *Gen. Comp. Endocrinol.* **49**, 470–476.

Van Der Kraak, G., Dye, H. M., and Donaldson, E. M. (1983b). Effects of LH-RH and des-Gly[10] [D-Ala[6]]LH-RH-ethylamide on plasma sex steroid profiles in adult female coho salmon (*Oncorhynchus kisutch*). *Gen. Comp. Endocrinol.* (in press).

van Overbeek, A. P., and McBride, J. R. (1971). Histological effects of 11-ketotestoterone, 17α-methyltestosterone, estradiol, estradiol cypionate, and cortisol on the interrenal tissue, thyroid gland, and pituitary gland of gonadectomized sockeye salmon (*Oncorhynchus nerka*). *J. Fish. Res. Board Can.* **38**, 477–484.

Vanstone, W. E., Villaluz, A. C., and Tiro, L. B., Jr. (1976). Spawning of milkfish, *Chanos chanos*, in captivity. *Proc. Int. Milkfish Workshop Conf.*, 1976 SEAFDEC pp. 222–228.

von Ihering, R. (1935). Die Wirkung von Hypophyseninjektion auf den Laichakt von Fischen. *Zool. Anz.* **111**, 273–279.

von Ihering, R. (1937). A method for inducing fish to spawn. *Prog. Fish-Cult.* **34**, 15–16.

von Ihering, R., and de Azevedo, P. (1934). A curimata dos acudes nordestinos (*Prochilodus argenteus*). *Arch. Inst. Biol. (Sao Paulo)* **5**, 143–184.

von Ihering, R., and de Azevedo, P. (1936). As piabas dos acudes nordestinos (Characidae, Tetragonopterinae). *Arch. Inst. Biol. (Sao Paulo)* **7**, 75–106.

Watts, E. G., Copp, D. H., and Deftos, L. I. (1975). Changes in plasma calcitonin and calcium during the migration of salmon. *Endocrinology* **96**, 214–218.

Weil, C., and Crim, L. W. (1982). The effects of different methods of administration of LH-RH analogs on spermiation in the mature landlock salmon, *Salmo salar*. *Can. Soc. Zool. Bull.* **13**(2), 60 (abstr.).

Wingfield, J. C., and Grimm, A. S. (1977). Seasonal changes in plasma cortisol, testosterone and oestradiol-17β in the plaice, *Pleuronectes platessa* L. *Gen. Comp. Endocrinol.* **31**, 1–11.

Worthington, A. D., MacFarlane, N. A. A., and Easton, K. W. (1981). The induced spawning of roach (*Rutilus rutilus* L.) with carp pituitary extract and oestrogen—antagonists. *Symp. Fish Physiol.*, 3rd, 1981 Abstract.

Worthington, A. D., MacFarlane, N. A. A., and Easton, K. W. (1982). Controlled reproduction in the roach (*Rutilus rutilus* L.). *Proc. Int. Symp. Reprod. Physiol. Fish*, 1982 p. 220.

Woynarovich, E., and Horvath, L. (1980). The artificial propagation of warm-water finfishes; a manual for extension. *FAO Fish. Tech. Pap.* **201**, 183 pp.

Yamamoto, K., and Yamauchi, K. (1974). Sexual maturation of Japanese eel and production of eel larvae in the aquarium. *Nature (London)* **251**, 220–222.

Yamamoto, K., Hiroi, O., Hirano, T., and Morioka, T. (1972). Artificial maturation of cultivated male Japanese eels by Synahorin injection. *Bull. Jpn. Soc. Sci. Fish.* **38**, 1083–1090.

Yamazaki, F. (1965). Endocrinological studies on the reproduction of the female goldfish, *Carassius auratus* L., with special reference to the function of the pituitary gland. *Mem. Fac. Fish., Hokkaido Univ.* **13**, 64 pp.

Yamazaki, F., and Donaldson, E. M. (1968a). The spermiation of goldfish (*Carassius auratus*) as a bioassay for salmon (*Oncorhynchus tshawytscha*) gonadotropin. *Gen. Comp. Endocrinol.* **10**, 383–391.

Yamazaki, F., and Donaldson, E. M. (1968b). The effects of partially purified salmon pituitary gonadotropin on spermatogenesis, vitellogenesis, and ovulation in hypophysectomized goldfish, *Carassius auratus*. *Gen. Comp. Endocrinol.* **11**, 292–299.

Yamazaki, F., and Donaldson, E. M. (1969). Involvement of gonadotropin and steroid hormones in the spermiation of the goldfish (*Carassius auratus*). *Gen. Comp. Endocrinol.* **12**, 491–497.

Yaron, Z., Bogomolnaya, A., and Hilge, V. (1982). Application of an *in vitro* bioassay for fish GTH in improving harvest of carp pituitaries. *Proc. Int. Symp. Reprod. Physiol. Fish*, 1982 p. 27.

Yoneda, T., and Yamazaki, F. (1976). Purification of gonadotropin from chum salmon pituitary glands. *Bull. Jpn. Soc. Sci. Fish.* **42**, 343–350.

8

CHROMOSOME SET MANIPULATION AND SEX CONTROL IN FISH

GARY H. THORGAARD

Program in Genetics and Cell Biology
Washington State University
Pullman, Washington

I. INTRODUCTION

Techniques to control the sex of fish by manipulation of chromosomes sets with radiation, chemical, or physical treatments are practical because of the ease with which fish gametes can be handled and fertilized *in vitro*. Three broad classes of treatments are possible. In induced gynogenesis (all-maternal inheritance), the paternal chromosomes in the sperm are inactivated before fertilization. In induced androgenesis (all-paternal inheritance), the maternal chromosomes in the egg are inactivated before, or shortly after

FISH PHYSIOLOGY, VOL. IXB

fertilization. In induced polyploidy, early cell divisions in the fertilized egg are blocked to produce triploid or tetraploid individuals.

Gynogenesis and induced polyploidy in fish have been very actively investigated since the mid-1970s; various aspects of these "chromosome engineering" techniques have been reviewed (Purdom and Lincoln, 1973; Stanley and Sneed, 1974; Purdom, 1976; Stanley, 1981; Allen and Stanley, 1981; Cherfas, 1981; Chourrout, 1982a; Donaldson and Hunter, 1982; Purdom, 1983). The interest in these techniques arises because of their potential for sex control and genome manipulation. Gynogenesis could make it possible to produce all-female populations in species in which the female is the homogametic sex. It could also make the rapid production of inbred lines for breeding programs and research feasible. Androgenesis, although not as widely investigated to date, might also be useful in sex control and the production of inbred lines. Induced polyploidy has been studied because triploids are expected to be sterile as a result of disturbances in gonad development from having an odd number of chromosome sets pairing in meiosis. Sterile fish could be useful if they survive and grow better than normal fish at sexual maturation and, like monosex populations, do not overpopulate when natural reproduction is not desired. Tetraploids might be useful in the production of the sterile triploids if they are viable and fertile and can be successfully crossed with normal diploids. Extensive research in gynogenesis, androgenesis, and induced polyploidy has also been done with amphibians, and many of the techniques and results from these studies are relevant to investigators studying fish.

In this chapter, the techniques used in chromosome set manipulation are described. This is followed by a review of the results and prospects in the application of gynogenesis, androgenesis, and induced polyploidy to fish.

II. TECHNIQUES IN CHROMOSOME SET MANIPULATION

A. Sperm and Egg Chromosome Inactivation

Some populations of crucian carp (*Carassius auratus gibelio*) and of several species from the family Poeciliidae in Mexico reproduce naturally by gynogenesis (reviewed by Gold, 1979; Cherfas, 1981). In these all-female populations, the sperm from males of another species is used to trigger development, but the chromosomes from the sperm are not included in the embryo. Similar phenomena sometime take place in crosses involving species that are not naturally gynogenetic (e.g., Uyeno, 1972; Purdom and

Lincoln, 1974; Stanley, 1976a; Cherfas, 1981). This phenomenon of hybrid gynogenesis is interesting, but may not be reliable enough for induced gynogenesis studies. Consequently, treatments which inactivate the chromosomes in the sperm without destroying its ability to fertilize the egg and trigger development are needed.

The first successful example of artificial inactivation of sperm chromosomes without the impairment of sperm-fertilizing capacity was described by Hertwig (1911). In studies with frogs, Hertwig demonstrated that with increasing doses of radiation, survival of embryos decreased dramatically. However, with even higher doses, early survival was found to increase. This phenomenon, known as the "Hertwig effect," is attributed to total destruction of the chromosomes in the sperm at high levels of radiation resulting in production of haploid embryos which survive longer than diploid embryos expressing dominant lethal mutations induced by the radiation. Haploid amphibians and fish typically die early in development (Fankhauser, 1945; Purdom, 1969). Oppermann (1913), in studies with brown trout (*Salmo trutta*), was the first to use radiation treatments of sperm to produce haploid embryos in fish.

A variety of radiation and chemical treatments are available to inactivate sperm chromosomes. Radiation treatments that have been used successfully include irradiation with γ rays usually from ^{60}Co or ^{137}Cs sources (Purdom, 1969; Lincoln et al., 1974; Nagy et al., 1978; Chourrout et al., 1980; Ijiri, 1980; Onozato, 1982; Refstie et al., 1982), X rays (Romashov et al., 1963; Stanley and Sneed, 1974), and ultraviolet (UV) light (Nace et al., 1970; Stanley, 1976a, 1981, 1983; Ijiri and Egami, 1980; Hoornbeek and Burke, 1981; Streisinger et al., 1981; Chourrout, 1982b). The various radiation treatments have advantages and disadvantages.

Gamma radiation and X rays have good penetrating ability that facilitates treatment of large quantities of sperm. They act principally by inducing chromosome breaks. Stanley (1982) has found that sperm treated with X rays retain activity for a shorter time than normal sperm. However, Thompson et al. (1981) report that plaice sperm treated with ^{60}Co γ rays and kept at 0°C can retain activity for up to 5 days. Chourrout et al. (1980) found that γ-irradiated sperm was less effective at high dilution than normal sperm. The lower viability and fertilizing capacity observed in radiation-treated sperm are likely to be true for all types of radiation and chemical treatments. Residual paternal characteristics or chromosome fragments may sometimes be found in gynogenetic embryos even after high level γ irradiation and X irradiation of sperm (Ijiri, 1980; Chourrout and Quillet, 1982; Onozato, 1982).

Ultraviolet light is easier to work with and less dangerous than ^{60}Co, ^{137}Cs, or X-ray sources. Ultraviolet light damages chromosomes mainly by inducing thymine dimers. These may be repaired by the photoreactivation process which takes place in the presence of visible light (Ijiri and Egami, 1980). Photoreactivation may sometimes take place in small, transparent fish eggs exposed to visible light (Ijiri and Egami, 1980); investigators using UV light to inactivate sperm should be aware of this possibility.

Ultraviolet-light sources are inexpensive and can easily be set up in individual laboratories (Nace *et al.*, 1970; Stanley, 1976a, 1981). The low penetration of UV light, which makes it safer than X rays or γ rays, makes it important to have the sperm in a thin, relatively transparent film for treatment. Thick, opaque sperm samples may not be totally inactivated by UV light. Unfortunately, this makes the treatment of large volumes of sperm more difficult.

Chemicals which interact with the sperm DNA have also been used in induced-gynogenesis studies in fish and amphibians. Chemicals which have been used successfully include toluidine blue (Briggs, 1952; Uwa, 1965), ethyleneurea (Jones *et al.*, 1975), and dimethylsufate (Tsoi, 1969; Mantelman, 1980). Chemical mutagenesis to inactivate sperm chromosomes might be a convenient technique if proper treatment concentrations and times are identified.

Androgenesis in fish has not been as widely studied as gynogenesis. Spontaneous androgenesis sometimes has been observed in crosses of different fish species. Stanley (1976a) observed androgenetic grass carp (*Ctenopharyngodon idella*), in low frequency, from crosses of female carp (*Cyprinus carpio*) with male grass carp.

Much less research has been done on the inactivation of egg chromosomes to induce androgenesis than on the inactivation of sperm chromosomes to induce gynogenesis. A potential problem with the use of radiation to inactivate egg chromosomes is that mitochondrial DNA, messenger RNA (mRNA), and other constituents of the egg cytoplasm might be damaged as well as the chromosomal DNA. In spite of this problem, Romashov and Belyaeva (1964), Purdom (1969), and Arai *et al.* (1979) have successfully produced androgenetic haploids after treating eggs with ionizing radiation. Romashov and Belyaeva (1964) believed that the androgenetic haploid loach (*Misgurnus fossilis*) had more severe problems than the gynogenetic haploids, possibly because of radiation damage to cytoplasmic constituents. Arai *et al.* (1979) also believed that the radiation treatment may have damaged egg cytoplasmic constituents in a study with masu salmon (*Oncorhynchus masou*). However, Purdom (1969) found that androgenetic haploids were indistinguishable from gynogenetic haploids in plaice (*Pleuronectes platessa*).

Another approach to inactivation of egg chromosomes might be the use of UV-light treatments, as has been done in studies with amphibians (Gillespie and Armstrong, 1980, 1981). These studies have relied on the transparency of the amphibian egg and the fact that the egg pronucleus orients toward the animal pole after fertilization, facilitating treatment with UV light. The opacity of some fish eggs, and the failure of the egg nucleus to demonstrate any particular orientation before or after fertilization in some fish eggs (Devillers, 1961) could present problems for the use of UV light to successfully inactivate chromosomes in some species. However, because of its low penetration, UV light could have the advantage of minimizing damage to egg cytoplasmic constituents.

Physical treatments of fertilized eggs might be useful for inducing androgenesis. Gervai *et al.* (1980) found a high frequency of androgenetic haploid embryos in carp after cold-shock treatments of eggs shortly after fertilization. Cold (Fankhauser, 1945) or pressure (Elinson and Briedis, 1981; Briedis and Elinson, 1982) treatment of amphibian eggs shortly after fertilization also sometimes results in the production of androgenetic haploids. Yamazaki (1983) found evidence for an increased frequency of androgenetic haploid development in overripe eggs of rainbow trout. Physical treatments of eggs to induce androgenesis might damage egg constituents less than would radiation treatments.

B. Suppression of Cell Divisions

Production of diploid gynogenetic or androgenetic individuals or of polyploids requires either retention of the second polar body of the egg or suppression of an early mitotic division in the fertilized egg. Retention of the second polar body or suppression of the first mitosis may happen in some cases without special treatment of the fertilized egg, or may result from temperature, pressure, or chemical treatment of the fertilized egg.

Naturally occurring triploids in fish populations are believed to result from occassional failure of extrusion of the second polar body of the egg, which normally takes place after fertilization (Gold and Avise, 1976; Thorgaard and Gall, 1979). This probably also accounts for spontaneous gynogenetic diploids observed after fertilization of eggs with radiation-inactivated sperm (Thompson *et al.*, 1981). Spontaneous failure of second polar body extrusion in crosses within a species appears to be a relatively rare event.

Suppression of polar body extrusion can also occur in crosses between species. Stanley (1976a) has reviewed the numerous reports of spontaneous gynogenetic diploid fish resulting from crosses between species. Vasil'ev *et al.* (1975) have described triploid offspring observed in remote hybrids from crosses between carp and other Cyprinidae. A high frequency of triploid offspring are found in the hybrid between female grass carp and male bighead carp (*Aristichthys nobilis*) (Marian and Krasznai, 1978; Beck *et al.*,

1980). The high frequency of triploids observed in crosses between species might reflect increased viability of triploid hybrids (Section IV,A).

There have been several reports of an increased tendency to produce diploid eggs in cyprinid interspecific hybrids. The F_1 hybrids between crucian carp and carp have a high tendency to produce unreduced eggs, as measured by a tendency to produce spontaneous gynogenetic diploids (Cherfas and Ilyasova, 1980a) or triploids when backcrossed to the parental species (Cherfas et al., 1981). Ojima et al. (1975) described triploid offspring resulting from a backcross of funa (*Carassius auratus cuvieri*) × carp hybrids to male carp. The production of unreduced eggs in these cyprinid hybrids is probably related to the occurrence of gynogenetic reproduction in some crucian carp and funa populations.

Temperature treatments of fertilized eggs have been widely used to suppress the second meiotic division or second polar body extrusion in fish. Both cold-shock (e.g., Swarup, 1959a; Romashov and Belyaeva, 1965; Purdom, 1969, 1972; Valenti, 1975; Ojima and Makino, 1978; Nagy et al., 1978; Lemoine and Smith, 1980; Chourrout, 1980; Wolters et al., 1981a; Refstie et al., 1982) and heat-shock (e.g., Swarup, 1959a; Vasetskii, 1967; Chourrout, 1980; Thorgaard et al., 1981) treatments can be effective. Heat-shock treatments have also been used to block the first mitotic division in zebrafish (*Brachydanio rerio*) to produce homozygous diploids (Streisinger et al., 1981) and in rainbow trout (*Salmo gairdneri*) to produce tetraploids (Thorgaard et al., 1981; Chourrout, 1982c). The timing, duration, and temperature of treatment must be determined for each species. There are indications of differences in temperature-shock susceptibility related to genetic background (Streisinger et al., 1981) and egg-maturity stage (Refstie et al., 1982). A simple, practical approach for induction of polyploidy by heat shock may be the application of treatments shortly after fertilization (for induced triploidy) or shortly before first cleavage (for induced tetraploidy) at just below lethal temperatures. Heat shocks using longer treatment times at lower temperatures, where practical, are likely to provide greater uniformity of treatment among eggs. Temperature-shock treatments are inexpensive to apply and might be successfully adapted for mass production by fish farms or management agencies if polyploids prove valuable.

Hydrostatic pressure has been used in studies with amphibians to block second polar body extrusion (Dasgupta, 1962; Gillespie and Armstrong, 1979) or the first mitotic division (Reinschmidt et al., 1979; Gillespie and Armstrong, 1980). Streisinger et al. (1981) successfully used hydrostatic pressure to block second polar body extrusion and used such treatments combined with ether exposure to block the first mitotic division in zebrafish. Yamazaki (1983) reported the use of hydrostatic pressure to block second polar body extrusion or the first mitotic division in rainbow trout. Although application of hydrostatic pressure treatments requires more specialized equipment (pressure cell and hydraulic press) than temperature-shock treatments, the method deserves wide investigation because it may be less damaging to the embryo than temperature shock. Gillespie and Armstrong (1979), for example, found much better survival in triploid salamanders produced by hydrostatic pressure than in triploids produced by heat shock.

Chemicals may also be used to block polar body extrusion or mitotic division in fertilized eggs. Refstie *et al.* (1977) and Allen and Stanley (1979) reported producing mosaic polyploid–diploid Atlantic salmon (*Salmo salar*) after exposing fertilized eggs to Cytochalasin B. J. Kanka and P. Rab (personal communication) report observing diploid–triploid mosaic, triploid, and tetraploid *Tinca tinca* after treating fertilized eggs with Cytochalasin B. Smith and Lemoine (1979) reported producing similar mosacic polyploid–diploid brook trout (*Salvenlinus fontinalis*) after exposing fertilized eggs to colchicine. Polyethylene glycol is another potentially useful agent for inducing polyploidy; although it has not been used with fish, it has successfully induced tetraploidy in mouse embryos (Eglitis, 1980).

In view of the success and ease of temperature-shock and hydrostatic pressure treatments for blocking polar body extrusion and first mitotic division in fish, chemical treatments may not be the method of choice. They are probably also less adaptable to mass production than other methods.

C. Identification of Gynogenetic Diploids

When gynogenetic and androgenetic diploids are produced, it is important to have proof that the sperm (in gynogenesis) or the egg (in androgenesis) did not contribute genetically to the embryo. This may be obtained by several methods.

Perhaps the simplest method for gynogenesis studies is the use of irradiated sperm from a related species to trigger development (Nace *et al.*, 1970; Stanley, 1981). In this manner, any paternal inheritance might be recognized by (1) inviable hybrids, (2) morphologically recognizable hybrids, and (3) biochemically recognizable hybrids. Morphological (Stanley and Jones, 1976) and biochemical (Stanley *et al.*, 1976) studies were used to confirm the identity of androgenetic and gynogenetic grass carp in studies involving reciprocal hybridization of carp and grass carp. The use of foreign sperm in gynogenesis studies requires that suitable sperm-donor species be available, but eliminates the need for rearing genetically distinct strains of the test species as donors, or of biochemically testing members of the test species to identify suitable donors. The possibility of an incompatibility between egg cytoplasmic constituents and the paternal genome in androgenesis might limit the usefulness of foreign eggs for androgenesis studies. However, Stanley and Jones (1976), found that androgenetic grass carp, which developed from carp eggs fertilized by grass carp sperm, were morphologically normal. Study of nuclear–cytoplasmic compatability in androgenesis may provide some interesting insights into development.

Within a species, color, morphological, or biochemical markers can be conveniently used to provide proof of gynogenesis or androgenesis. Sperm carrying dominant color alleles were used to provide proof of all-maternal inheritance in rainbow trout (Chourrout, 1980) and zebrafish (Streisinger *et al.*, 1981). In carp, Nagy *et al.* (1978) used irradiated sperm from normal (scaled) carp to fertilize eggs of homozygous recessive individuals with a scattered (ss) scale pattern; the observation of 100% scattered progeny supported gynogenesis. Nagy *et al.* (1978) also used biochemical variation at the transferrin locus to provide proof of gynogenesis.

Although the second meiotic division and second polar body extrusion normally take place soon after fertilization, and the first mitotic division occurs shortly before the first cleavage furrow is visible, it is nevertheless important to obtain other evidence to clarify the origin of any gynogenetic diploids produced after treatments. Cytological studies of the fertilized egg can be used to pinpoint the timing of the early divisions (Streisinger *et al.* 1981). The ploidy of treated eggs fertilized with normal sperm gives indirect information about gynogenetic diploids treated at corresponding times. A treatment producing triploids presumably acts on the second meiotic division or second polar body extrusion, but a treatment producing tetraploids blocks the first mitotic division.

Genetic markers can also provide evidence of whether diploid gynogenesis results from a blockage of the first mitotic division or from a blockage of polar body extrusion. The progeny of a heterozygous female should all be homozygous if diploidy results from suppression of first mitosis. Suppression of first polar body extrusion, which is unlikely because this event is normally completed before ovulation (Cherfas, 1981), would result in almost 100% heterozygous progeny for genes near the centromere, decreasing to 66% heterozygotes for genes segregating randomly in relation to their centromere. Suppression of second polar body extrusion would result in predominantly homozygous progeny for genes near the centromere, with increasing proportions of heterozygotes for genes far from the centromere. For genes segregating randomly in relation to their centromere 66% of the progeny should be heterozygous (Nace *et al.*, 1970).

Streisinger *et al.* (1981), for example, investigated inheritance of an enzyme locus (*est*-3) in gynogenetic diploid zebrafish that they produced. Fertilized eggs of heterozygous females treated with hydrostatic pressure 1.5–6 min postfertilization gave rise to 14% heterozygous gynogenetic diploid offspring, a result consistent with suppression of the second meiotic division or second polar body extrusion. Fertilized eggs of heterozygous females treated with hydrostatic pressure 22.5–28 min postfertilization gave rise to 100% homozygous progeny, a result consistent with suppression of the first mitotic division. The absence of heterozygotes among the progeny that received the late pressure treatment also demonstrated that nondisjunction in meiosis I or meiosis II for the chromosome carrying the est-3 gene must be a rare event.

D. Identification of Polyploids

Polyploid fish have been identified by indirect methods such as measurement of nuclear or cellular volume, counting of nucleoli, electrophoresis of proteins, and examination of morphology, and direct methods such as chromosome counting and DNA content determination. Accurate methods for assessing ploidy are especially important because treatments are not always 100% successful and treated populations may be composed of mixtures of polyploids, diploids, and possibly polyploid–diploid mosaics.

Measurement of erythrocyte nuclear volume has been widely used to identify polyploid fish (e.g., Swarup, 1959b; Purdom, 1972; Valenti, 1975; Allen and Stanley, 1978, 1979; Lemoine and Smith, 1980; Beck and Biggers, 1983). Erythrocyte nuclear volume measurement is convenient and requires little specialized equipment. Unfortunately it does not always accurately reflect ploidy (Thorgaard and Gall, 1979; Wolters *et al.*, 1982a). Because of this, the method should be calibrated using direct ploidy measurement techniques. In the absence of other evidence, the potential inaccuracy of the method brings into question conclusions of polyploid–diploid mosaicism on the basis of erythrocyte nuclear volume (Thorgaard and Gall, 1979). However, erythrocyte volume measurement using a Coulter counter may be a practical method of determining ploidy (T. Benfey, personal communication).

The number of nucleoli in a cell can be used to estimate ploidy. Cherfas and Ilyasova (1980a) used nucleolar counts to identify triploid hybrids of the crucian carp and common carp.

Protein electrophoresis has also been used to identify polyploid fish. Liu *et al.* (1978) used muscle myogen and creatine kinase pattern differences to identify triploid and diploid ginbuna (*Carassius auratus langsdorfi*). Allen and Stanley (1981, 1983) and Magee and Philipp (1982) have successfully used electrophoresis to screen grass carp × bighead carp hybrids for polyploidy. The screening of large numbers of individuals for polyploidy in interspecific hybrids and in suitable intraspecific crosses should be feasible because of differences in relative dosages of alleles between diploids and triploids. However, the possibility of preferential expression of maternal or paternal alleles in interspecific hybrids (Whitt *et al.*, 1973; F. M. Utter, personal communication), emphasizes the importance of including control diploid hybrids, polyploid hybrids, and the parent species, in such studies.

Triploid fish do not appear to be strikingly morphologically different from diploids (Swarup 1959b; Gold and Avise, 1976; Thorgaard and Gall, 1979; Gervai *et al.*, 1980). Swarup (1959b) found that triploid sticklebacks had shorter trunks and longer tails than the diploid conrols. However, triploid interspecific hybrids may be readily distinguished from diploid hybrids in some cases because of differences in gene dosage from the parent species. Purdom (1972) was able to distinguish triploid plaice × flounder (*Platichthys flesus*) hybrids from diploids based on larval pigmentation patterns. The triploid grass carp × bighead carp hybrid is also morphologically distinguishable from the diploid hybrid (Allen and Stanley, 1981, 1983; Beck and Biggers, 1982).

Chromosome counting is an accurate, direct method of ploidy assessment that is relatively time consuming. Preparations of good quality are possible from embryos because of their high mitotic index (Hoornbeek and Burke, 1981; Thorgaard *et al.*, 1981; Yamazaki *et al.*, 1981). Preparations from most body tissues of larger fish are less successful, and require sacrifice of the fish. Chromosome preparations from regenerating fin (Streisinger *et al.*, 1981) or lymphocyte culture (Thorgaard and Gall, 1979; Gervai *et al.*, 1980; Wolters *et al.*, 1981b) allow ploidy determination of adults without sacrifice of the animal.

Cellular DNA-content determination is another effective, direct method of ploidy determination. Gervai *et al.* (1980) used microdensitometry of Feulgen-stained blood cell nuclei to identify triploid carp. Flow cytometry is a promising new technique in which the flourescence of large numbers of stained cells is rapidly measured to estimate DNA content. It has been used to identify triploid rainbow trout (Thorgaard *et al.*, 1982), triploid Pacific salmon (Utter *et al.*, 1983), triploid oysters (Allen, 1983), and diploid, triploid, and tetraploid grass carp × bighead carp hybrids (Allen and Stanley, 1983).

III. GYNOGENESIS AND ANDROGENESIS

A. Sex Control

One of the major rationales for the interest in gynogenesis has been the potential for producing monosex populations. In species with female homogamety, all the offspring should be female (XX) after gynogenesis. In species with female heterogamety (XY), both males and females might be produced. If the sex-determining locus is near the centromere or not recombined between the X and Y, and for all treatments suppressing the first mitotic division, equal proportions of XX male and YY female offspring would be anticipated. The YY individuals might or might not be viable. If the sex-determining locus is far from the centromere and recombined between the sex chromosomes, increasing proportions of females (as high as five-sixths) might be observed for treatments causing retention of the second polar body. Sex-determining loci could conceivably be recombined between sex chromosomes in fish because it is known that other loci sometimes are (Gold, 1979).

Therefore, the sex of offspring after gynogenesis can provide information about the sex-chromosome system prevalent among fish species. A program to produce monosex grass carp by gynogenesis succeeded in producing 100% females, a result consistent with a female homogametic (XX) system (Stanley, 1976b, 1981). Production of 100% females after gynogenesis in carp (Nagy *et al.*, 1978, 1981; Gomelsky *et al.*, 1979), coho salmon, *Oncorhynchus kisutch* (Refstie *et al.*, 1982), and pink salmon, *Oncorhynchus gorbuscha* (Maximovich and Petrova, 1980), and almost 100% females in rainbow trout (Chourrout and Quillet, 1982) is also consistent with female homogamety in these species. Several studies have also found male offspring after gynogenesis. Purdom and Lincoln (1973) found 24 female and 14 male offspring among gynogenetic plaice sexed at age 1 year, thereby suggesting female heterogamety in the plaice. Streisinger *et al.* (1981), in producing homozygous diploid clones of zebrafish by gynogenesis, found that there was considerable variation in the sex ratio characteristically found among various clones. Some clones that were derived from single homozygous females were predominantly male. These observations are not consistent with either a simple female homogametic or female heterogametic system, but they suggest possible effects of autosomal sex-determining genes or environmental influences.

Androgenesis should lead to the production of all-male offspring in species with male homogamety (XX). Gillespie and Armstrong (1981) produced androgenetic diploid males in the axolotl, a male homogametic salamander, by suppressing first cleavage after UV inactivation of egg chromosomes. In male heterogametic species, including carp and probably most salmonids, diploid androgenesis followed by suppresssion of first cleavage should lead to 50% XX and 50% YY offspring. The XX individuals would be homozygous females and YY individuals homozygous males with the potential of producing all-male offspring when crossed to normal females. On the basis of studies with hormonally sex-reversed fish, YY males are known to be viable in goldfish, *Carassius auratus* (Yamamoto, 1975), medaka, *Oryzias latipes* (Yamamoto, 1964), and coho salmon (Hunter *et al.*, 1982).

Although gynogenesis and androgenesis can be used to directly produce monosex fish populations, the resulting offspring are inbred and likely to have poorer survival and growth rates than normal fish. Consequently, in the absence of selective breeding, the monosex populations produced by gynogenesis or androgenesis are inferior to the outbred monosex populations produced genetically by crosses using YY males or sex-reversed XX males in male heterogametic species. Gynogenesis may not be the best method for sex control in populations to be released in nature because the offspring are inbred, are capable of reproduction (in contrast to triploid or hormonallly sterilized fish), and are genetically uniform and therefore able to establish an undesirable "monoculture" in nature (Streisinger *et al.*, 1981).

However, gynogenesis and androgenesis could be useful in the production of XX and YY males for creating outbred monosex populations. In male heterogametic species, gynogenesis followed by hormonal sex reversal of females into males should allow production of 100% sex-reversed XX males (see Yamazaki, 1983; Jensen *et al.*, 1983, Chapter 5, this volume). Androgenesis could be useful in the rapid production of YY males as previously discussed.

B. Gene-Centromere Mapping

Retention of the second polar body in eggs of heterozygous females results in varying proportions of heterozygous gynogenetic diploid offspring, depending on the distance between a gene and its centromere (Section II,C). When loci are near the centromere, virtually all offspring will be homozygous; as loci are located farther from the centromere, increasing proportions of heterozygous progeny will be observed. When genes assort randomly in relation to the centromere, two-thirds heterozygous progeny are expected.

Gene-centromere mapping has been done in many organisms. In *Neurospora*, gene-centromere distances have been mapped using tetrad analysis to estimate second-division segregation frequencies (Barratt *et al.*, 1954). In *Drosophila*, gene-centromere distances have been estimated in studies with attached-X females (Anderson, 1925; Beadle and Emerson, 1935). Gene-centromere distances have been estimated in amphibians by suppression of the second polar body extrusion with temperature shock or pressure and study of characteristics of triploid (Lindsley *et al.*, 1956) and gynogenetic diploid (Volpe and Dasgupta, 1962; Nace *et al.*, 1970; Volpe, 1970) offspring. A technique for gene-centromere mapping in humans involves the study of genetic markers in benign ovarian teratomas that apparently arise from fusion of the second polar body with the egg (Ott *et al.*, 1976; Ott, 1979). The extensive research of gene-centromere mapping in other organisms provides a theoretical base for such studies in fish using gynogenetic diploids.

Studies of inheritance of four enzyme loci have been conducted using gynogenetic diploid plaice (Purdom *et al.*, 1976; Thompson *et al.*, 1981). The proportion of heterozygous offspring for the loci were PGM (80%), MDH-A (38%), GPI-A (45%), and GPI-B (20%). The similarity of results for spontaneous- and induced-gynogenetic diploids suggested that both types arise as a result of retention of the chromosomes of the second polar body (Thompson *et al.*, 1981). The average proportion of heterozygous offspring among the four loci analyzed to date is 46%.

In carp, nine morphological and protein loci have been mapped using gynogenesis (Cherfas, 1977; Cherfas and Truweller, 1978; Nagy *et al.*, 1978, 1979). The proportion of heterozygous progeny ranges from about 8% for the genes for transferrin and scale pattern (s) to 97% for the gene N, which also is involved with scale pattern. The average proportion of heterozygous offspring among the nine loci is 35% (Nagy and Csanyi, 1982).

Streisinger *et al.* (1981), as previously described in Section II,C, found 14% heterozygotes among gynogenetic diploid zebrafish in families segregating for the esterase locus *est*-3.

Ten enzyme loci have been mapped in relation to their centromeres in gynogenetic diploid rainbow trout (Thorgaard *et al.*, 1983). The proportion of heterozygous progeny ranges from about 2% for the LDH–4 gene to close to 100% for the SOD and MDH–3,4 loci. The average proportion of heterozygotes among the loci was 55%.

Information from gene-centromere mapping is useful for several reasons. It provides basic information about the meiotic system in fish species. Observation of more than two-thirds heterozygous progeny among gynogenetic diploid offspring suggests that there are fewer double crossovers than expected if they occur randomly (positive interference). The very high proportion of heterozygous progeny reported for the N gene in carp (97%) and for several loci in rainbow trout suggests that virtually no double crossovers take place on the chromosomes carrying these loci.

Mapping information is also useful for comparative and evolutionary studies. Information in most fish species is difficult to obtain by traditional linkage analysis because of the large number of chromosomes in most fish species. Comparative mapping studies will provide information about the degree of conservation of gene arrangement among fish species and between fish and other vertebrates. The polyploid origin of many important fish species (Schultz, 1980) can also be examined with gene-centromere mapping. Duplicated loci originating from a tetraploid event might be expected to be unlinked, but at about equal distances from their centromeres if gene arrangements have been conserved since the tetraploid event. Nagy *et al.* (1979) studied the map locations of two genes involved with coloration in carp, and postulated on the basis of the results that the genes might represent duplicated loci that originated through a tetraploid event. Comparison of map locations for the LDH3 and LDH4 and the MDH3 and MDH4 loci, which were duplicated by a tetraploid event in the evolution of salmonid fish, revealed that they are located at similar distances from their centromeres in rainbow trout (Thorgaard *et al.*, 1983).

A third instance in which gene-centromere mapping information is useful is in the estimation of the degree of inbreeding obtained through gynogenesis (Nace *et al.*, 1970; Thompson *et al.*, 1981; Nagy and Csanyi, 1982; Thompson, 1983; Thorgaard *et al.*, 1983).

C. Inbreeding Depression

Inbreeding leads to an increased probability of homozygosity for harmful recessive alleles and a reduction in value for characters associated with reproduction or physiological efficiency known as inbreeding depression. Gynogenesis is a very rapid system of inbreeding (Section III,D), and, as expected, gynogenetic fish show evidence of inbreeding depression (Cherfas, 1981).

Survival of gynogenetic fish is reduced in sturgeon (Romashov *et al.*, 1963), loach (Romashov and Belyaeva, 1965), plaice (Purdom, 1969; Purdom and Lincoln, 1973), brown trout (*Salmo trutta*) (Purdom, 1969), carp (Cherfas, 1975; Nagy *et al.*, 1978), zebrafish (Streisinger *et al.*, 1981), coho salmon (Refstie *et al.*, 1982) and rainbow trout (Chourrout and Quillet, 1982). There is also evidence for a reduced growth rate in gynogenetic coho salmon (Refstie *et al.*, 1982). However, reports of studies of grass carp (Stanley and Sneed, 1974) and carp (Cherfas, 1975) have noted satisfactory growth in gynogenetic individuals. Growth rate should be less strongly influenced by inbreeding than "fitness" characters such as viability or reproductive efficiency (Lerner, 1954).

Several studies have also demonstrated an increased frequency of physical abnormalities in gynogenetic fish. In brown trout (Purdom, 1969) and rainbow trout (Chourrout, 1980), there was an increased incidence of gynogenetic fry with bent or curved spines. An increased frequency of abnormalities of gonad development has been found in gynogenetic carp (Gomelsky *et al.*, 1979).

Gynogenesis could potentially be used to estimate the degree of inbreeding depression for different characters in fish and to estimate the number of harmful recessive alleles normal individuals carry (genetic load). In doing this, it will be important to separate the treatment effects of gynogenesis (e.g., heat shock and pressure) from inbreeding effects. Improved survival and yield in offspring of gynogenetic females in carp (Cherfas and Ilyasova, 1980b) and zebrafish (Streisinger, *et al.*, 1981) demonstrate the importance of inbreeding effects in reducing survival initially. However, androgenesis studies with the axolotl (Gillespie and Armstrong, 1981), in which extremely poor survival in the offspring of androgenetic diploid males was attributed to the deleterious effects of hydrostatic pressure and heat shock, demonstrate the need for caution in using gynogenesis or androgenesis to numerically estimate inbreeding depression or genetic load.

D. Production of Inbred Lines

The main rationale for gynogenesis studies in fish, besides the possibility for producing monosex populations, has been the possibility of rapidly producing inbred lines. Such lines could be useful for selective breeding programs if lines that produce vigorous hybrids when crossed together can be identified (Stanley and Sneed, 1974; Purdom, 1976), as has been done with corn. Inbred lines of fish could also be useful for a variety of basic biological studies (Streisinger *et al.*, 1981; Simon *et al.*, 1981; Schultz and Schultz, 1982). Genetically uniform lines have been extremely useful tools that have made the laboratory mouse such a valuable experimental organism (Green, 1981).

There are two approaches that can be used to produce inbred lines by gynogenesis in fish. The approach most widely used to date has been gynogenesis by retention of the second polar body (pb gynogenesis). This leads to partial homozygosity among the offspring, with the degree depending on the amount of crossing over taking place in meiosis. Inheritance data collected for plaice, carp, and rainbow trout (Section III,B) allows one to estimate the proportion of homozygosity after one generation (equivalent to the inbreeding coefficient, F) as 0.54 in plaice, 0.65 in carp, and 0.45 in rainbow trout. This is higher than would be achieved through either brother–sister mating ($F = 0.25$) or (for plaice and carp) self-fertilization ($F = 0.5$). However, it may be difficult to attain complete homozygosity through pb gynogenesis. Loci with especially high gene-centromere recombination rates may remain heterozygous for many generations. Nagy and Csanyi (1982) have observed that although inbred lines generated by pb gynogenesis will not rapidly become totally homozygous, individuals within the lines will become similar to each other (isogenic) because they will remain heterozygous for the same loci.

The second approach for generating inbred lines by gynogenesis involves suppression of the first mitotic division (mitotic gynogenesis). Streisinger *et al.* (1981) used this approach to generate homozygous diploid zebrafish. The offspring of a single homozygous diploid female, produced by pb gynogenesis, were then all homozygous and genetically identical to each other. Mitotic gynogenesis is probably more difficult to achieve experimentally and results in a greater level of inbreeding depression than pb gynogenesis; however, it facilitates production of homozygous inbred lines of fish in two generations.

Homozygous diploid individuals have also been produced in mice by microsurgery (Hoppe and Illmensee, 1977), in amphibians by mitotic gynogenesis (Reinschmidt *et al.*, 1979) and androgenesis (Gillespie and Armstrong, 1980, 1981). and in plants by induction of chromosome doubling in haploid tissue (Kasha, 1974). The use of "doubled haploids" to generate inbred lines for plant breeding is basically analogous to gynogenesis, and some of the approaches and theory being developed by plant breeders may prove applicable to fish breeding (e.g., Chase, 1974; Kasha and Reinbergs, 1980; Snape and Simpson, 1981).

Androgenesis, followed by suppression of first mitosis, also leads to the production of homozygous diploid individuals. Induced diploid androgenesis has not yet been successful in fish, and it may not present any particular advantages over mitotic gynogenesis for species with short generation times. For species with longer generation times in which males may naturally mature earlier than females or may be induced to mature early (Donaldson *et al.*, 1972), androgenesis would permit much more rapid breeding progress with inbred lines. Homozygous diploid YY males, or hormonally sex-reversed XX males (see Chapter 5, this volume) could reach sexual maturity more rapidly than homozygous diploid females, and androgenesis would then allow "cloning" of selected individuals.

Streisinger *et al.* (1981) described the production of more vigorous clones of homozygous diploid zebrafish by crosses between clones (using hormonally sex-reversed males or occasional natural males) followed by generation of new clones from the hybrid by mitotic gynogenesis. In this manner, harmful alleles were selected out and improved inbred lines developed over time. Similar approaches are likely to be useful in breeding commercially important fish species.

IV. INDUCED POLYPLOIDY

A number of examples of fish species with a polyploid ancestry are known (reviewed by Schultz, 1980). Some large groups, such as the salmonids and the catostomids, apparently had a tetraploid event in an ancestral form.

Others, such as some unisexual goldfish and poeciliid fishes, and some loach, appear to be fairly recent polyploids. The polyploid ancestry of these fish demonstrates that new polyploids are sometimes viable and capable of reproduction. However, the experimental induction of polyploidy in fish may result in reduced viability and fertility. The major practical goal of induced polyploidy has been the production of sterile triploids.

A. Viability of Polyploids

The successful induction of triploidy in many fish species (e.g., Swarup, 1959a; Purdom, 1972; Valenti, 1975; Chourrout, 1980; Gervai et al., 1980; Wolters et al., 1981a; Thorgaard et al., 1981) and findings of spontaneous triploids (Cuellar and Uyeno, 1972; Gold and Avise, 1976; Thorgaard and Gall, 1979) support the belief that triploid fish have good viability. This is in contrast to the inviability of triploid mammals (Niebuhr, 1974), but similar to the good survival observed in triploid amphibians (Fankhauser, 1945).

Most studies of induced triploid fish have found that they have normal viability. Triploid stickleback, *Gasterosteus aculeatus* (Swarup, 1959b), plaice × flounder hybrids (Purdom, 1972; Purdom and Lincoln, 1973), blue tilapia, *Tilapia aurea* (Valenti, 1975), carp (Gervai et al., 1980), and zebrafish (G. Streisinger, personal communication) apparently survive as well as diploids. Studies with rainbow trout (Lincoln and Scott, 1983; Thorgaard et al., 1982) and coho salmon (Utter et al., 1983) suggest that induced triploids in these species may be somewhat less viable than diploids.

There has been interest in the possibility of producing tetraploid fish because, if fertile, they might be crossed to diploids to produce triploids (Refstie et al., 1977). Tetraploid fish might be expected to be less viable than triploids; viability decreased in tetraploid and pentaploid salamanders (Fankhauser, 1945). Tetraploid zebrafish have poor survival compared to diploids (G. Streisinger, personal communication). Tetraploid rainbow trout embryos are frequently abnormal in appearance (Thorgaard et al., 1981; Chourrout, 1982c). However, Refstie (1981) reports successful rearing of tetraploid rainbow trout to maturity. An increased frequency of abnormalities was found among fish he identified as tetraploid. Valenti (1975) found several probable tetraploids among blue tilapia treated as eggs with cold shock. These fish were larger than both controls and probable triploids at age 14 weeks. As yet, there is not enough information to draw general conclusions about the viability of tetraploid fish.

Triploidy may lead to increased viability in interspecies hybrids (Allen and Stanley, 1981; Chevassus et al., 1983; Scheerer and Thorgaard, 1983). In frogs, triploid interspecific hybrids are sometimes more viable than diploid hybrids (Bogart, 1980; Elinson and Briedis, 1981). Spontaneous triploids observed in hybrids of rainbow trout × brook trout (Capanna *et al.*, 1974), grass carp × common carp (Vasil'ev *et al.*, 1975), and grass carp × bighead carp (Marian and Krasznai, 1978; Beck *et al.*, 1980) may be related to better survival of triploid hybrids than diploid hybrids. The triploid grass carp × bighead carp hybrid is healthier than the frequently deformed diploid hybrid (Allen and Stanley, 1981, 1983; Beck and Biggers, 1982). The increased viability of triploid interspecific hybrids may be related to a "buffering" of the genetic incompatability between species by altering the ratio of parental chromosome sets from 1:1 to 2:1.

Interspecific triploid hybrids could prove useful in fish culture because hybrid vigor and desirable attributes of both species might be combined in a relatively healthy sterile hybrid (Allen and Stanley, 1981). For example, Refstie *et al.* (1982) found that coho salmon × chinook salmon (*Oncorhynchus tshawytscha*) hybrids grew much faster, but had a higher mortality rate than coho salmon. The hybrids also demonstrated high frequencies of deformities. Triploid coho × chinook hybrids might retain the hybrid vigor of the diploids, but have lower mortality rates and fewer deformities. Utter *et al.* (1983), with reciprocal pink salmon × chinook salmon hybrids, and W. R. Wolters (personal communication) with channel catfish (*Ictalurus punctatus*) × white catfish (*Ictalurus catus*) hybrids, are investigating the potential of triploid interspecific hybrids for fish culture.

B. Sex of Polyploids

Equal proportions of males and females are expected with induced triploidy in species with female homogamety and in many cases with female heterogametic species. In female homogametic species, retention of the second polar body results in an XX egg which is then fertilized with either an X or Y sperm. If the Y chromosome is male determining (Yamamoto, 1963; Thorgaard and Gall, 1979) then equal proportions of XXY-triploid males and XXX-triploid females are expected. In female heterogametic species, retention of the second polar body results in 50% XX and 50% YY eggs, if there is no crossing over between the sex-determining segment of the sex chromosome and its centromere. After fertilization with an X-bearing sperm, equal proportions of XXX (male) and XYY (female) offspring are expected. Further, some XXY (presumably female) offspring may be found if there was crossing over between the sex locus and the centromere. Lindsley *et al.* (1956) demonstrated that the sex locus was far from the centromere in the axolotl (a female heterogametic salamander) by studying the sex ratio in triploids induced by cold shock.

Studies of the sex of induced-triploid carp (Gervai *et al.*, 1980), induced-triploid channel catfish (Wolters *et al.*, 1982b), induced-triploid sticklebacks (Swarup, 1957, in Lincoln, 1981b), and spontaneous-triploid rainbow trout, which probably also result from retention of the second polar body of the egg (Thorgaard and Gall, 1979), have noted roughly equal proportions of males and females. These results are consistent with either male or female hetero-gamety in these groups, although both carp and rainbow trout are known to be male heterogametic from other lines of evidence (Thorgaard, 1977; Nagy *et al.*, 1978; Okada *et al.*, 1979; Johnstone *et al.*, 1979). Conflicting results have been obtained in studies with flatfish. Purdom (1972) found equal proportions of males and females among juvenile triploid plaice × flounder hybrids. However, Lincoln (1981b) found an excess of males among both triploid plaice × flounder hybrids and triploid plaice. Lincoln also found smaller excesses of males among the diploid controls. The excess of triploid males is not expected with either male or female heterogamety, although gynogenesis results suggest female heterogamety in plaice (Purdom and Lincoln, 1973). Lincoln (1981b) suggested that the deficit of females may have been caused by selective mortality during the larval stage.

The absence of intersexual characteristics among triploid carp, channel catfish, rainbow trout, and flatfish suggests that all these groups may have a "dominant Y" sex-determining mechanism. However, Swarup (1957, in Lincoln, 1981b) found some evidence of intersexuality in triploid sticklebacks. This suggests that sticklebacks, like *Drosophila* (Bridges, 1925) and chickens (Abdel-Hameed and Shoffner, 1971), may have a "genic balance" sex-deter-mining mechanism based on the ratio of X chromosomes to autosomes.

In tetraploids produced by suppression of the first mitotic division, equal proportions of males and females are expected. If the Y chromosome is dominant in sex determination then XXXY individuals, which might be produced by crossing XXYY and XXXX individuals, would be males in male heterogametic species and females in female heterogametic species. The dominant Y mechanism would therefore allow XXXY tetraploids to avoid being sterile intersexes. The possibility of producing XXXY sterile intersexes was one of Muller's (1925) explanations for the rarity of natural polyploids in animals.

C. Gonadal Development

The interest in producing triploid fish has been based on the assumption that they would be sterile and consequently might avoid overpopulation problems, and possibly grow faster or survive longer than normal fish. Tri-ploids are expected to be sterile because the odd number of chromosome sets will lead to disruption of meiosis and either a failure of gonad development or production of aneuploid gametes. The failure of gonad development might, in turn, prevent the appearance of undesirable side effects of sexual maturation, such as poor meat quality, slower growth, and high mortality. It appears that triploid fish are indeed functionally sterile, but that secondary sexual characteristics are not always suppressed.

Triploid males may generally show more gonadal development than triploid females, probably because triploidy does not interfere with the many mitotic divisions involved in bringing the testis to its mature size. The gonads of spontaneous-triploid rainbow trout showed considerable development, appearing similar to the testes of normal, developing males (Thorgaard and Gall, 1979). The testes of triploid channel catfish were slightly smaller than those of diploids at age 8 months and, unlike diploid testes, histologically evinced no sperm production (Wolters *et al.*, 1982b). Triploid plaice and plaice × flounder hybrids had testes that were similar in size and external appearance to those of diploids at age 3 years (Lincoln, 1981a). The testes of the triploid hybrids evinced histological abnormalities but did produce some sperm. However, the triploid plaice had histologically normal testes. Several triploid male plaice produced sperm that were used to fertilize normal eggs; the resulting embryos, apparently aneuploids, all died early in development. In contrast to the considerable gonad development observed in triploid male rainbow trout, channel catfish and flatfish, triploid male carp showed very little gonadal development (Gervai *et al.*, 1980).

The substantial gonad development in triploid males allows the expression of secondary sexual characters in some cases. Triploid male sticklebacks (Swarup, 1959b) and rainbow trout (Thorgaard and Gall, 1979) demonstrated the secondary characters of normal males at maturity. Triploid male plaice had testosterone levels which overlapped those of diploids at sexual maturity (Lincoln, 1981a). However, triploid male channel catfish did not develop the broad heads and dark coloration observed in diploid males at sexual maturity (Wolters *et al.*, 1982b).

Triploidy apparently inhibits gonadal development more in females than in males. Failure of meiosis may prevent oocyte development and the associated increase in size of the gonad. Triploid female rainbow trout at maturity had small, stringlike gonads with many cells arrested at the pachytene stage of meiosis (Thorgaard and Gall, 1979). Diploids have ovaries about four times as large at maturity as triploids in channel catfish (Wolters *et al.*, 1982b), plaice, and plaice × flounder hybrids (Lincoln, 1981b). In these three triploids, only occasional developing oocytes were observed histologically (Purdom, 1972; Lincoln, 1981b; Wolters *et al.*, 1982b). Gonadal development was also substantially inhibited in triploid female carp (Gervai *et al.*, 1980). Although gonadal development was inhibited in triploid female plaice × flounder hybrids, sex hormone levels were not significantly different from those in diploid controls (Lincoln, 1981c).

Tetraploids are not necessarily expected to be sterile. Refstie (1981) reported that some male rainbow trout he identified as tetraploid produced milt at age 2 years. Newly arisen autotetraploids may have fertility problems because multivalent pairing at meiosis might lead to failure of gonadal development or production of aneuploid gametes. Among tetraploid axolotls, males are sterile, but females produce a high frequency of aneuploid eggs (Fankhauser and Humphrey, 1959). Future studies on the fertility of induced-tetraploid fish should be interesting because fertile tetraploids, if they produce predominantly diploid gametes, might be conveniently crossed to diploids to produce sterile triploids (Refstie *et al.*, 1977).

D. Prospects

Induction of triploidy in fish might potentially be useful for the control of overpopulation, for increasing the growth rate in juveniles, and for extending survival and improving growth in mature fish.

Most of the interest in the use of triploidy for population control to date has focused on the grass carp and interspecific hybrids involving the grass carp (Stanley, 1979; Shireman *et al.*, 1983). Grass carp are useful for control of vegetation, but there is considerable concern about environmental damage if they become naturalized. Sterile triploid grass carp or sterile hybrids with the grass carp's ability to consume vegetation could be extremely valuable. Triploids might also be very useful for fisheries management in species such as brook trout and many warm water game fish that tend to overpopulate in lakes. For this purpose, the successful production of aneuploid sperm, as observed in triploid male plaice (Lincoln, 1981a), or aneuploid sperm and eggs, as observed in triploid axolotls (Fankhauser and Humphrey, 1959) might even be useful. Gamete-producing triploid fish might be used in population control if they mated with diploids and produced inviable offspring; similar genetic approaches to population control have been used in insects (Smith and von Borstel, 1972).

Superior growth and survival are not required for triploidy to be useful in population control. Highly efficient production of triploids, with few if any diploids in the treated groups, is necessary. This will require either very careful quality control in production and measurement of triploidy or the successful production of fertile tetraploids. A possible alternate approach to avoid these rigid requirements is the use of triploid interspecific hybrids in population control; if the diploid hybrid was sterile, it would not be essential to guarantee 100% induction of triploidy.

Most of the interest in induced triploidy has been from an aquaculture perspective with the hope that triploids might grow faster than diploids as juveniles or as mature fish. This might result from triploidy *per se* or as an indirect result of sterility of triploids.

Juvenile triploids have generally been found to grow no faster than diploids. Growth of juvenile triploids was similar to that of diploids in sticklebacks (Swarup, 1959b), plaice × flounder hybrids (Purdom, 1976), carp (Gervai *et al.*, 1980), and channel catfish (Wolters *et al.*, 1982b). In the blue tilapia, juvenile triploids were found to be larger than diploids (Valenti, 1975). However, in juvenile coho salmon (Utter *et al.*, 1983) triploids may grow slower than diploids.

Several studies have found that triploids may grow faster than diploids at sexual maturity, presumably because energy that is channeled to gonadal development in diploids is used for growth in triploids. Triploid female plaice × flounder hybrids (Lincoln, 1981c) and a spontaneous triploid female rainbow trout (Thorgaard and Gall, 1979) continued to gain weight during the spawning period of normal fish. Triploid channel catfish were significantly heavier than diploids at age 8 months and older (Wolters *et al.*, 1982b). However, in the female plaice × flounder hybrids, diploids compensated by rapid growth after spawning and were not significantly different in weight from triploids by 2 months after the end of the spawning period. However, mean fillet weight remained significantly higher in the triploid hybrids than in the diploids (Lincoln, 1981c). This illustrates that quality factors (e.g., fillet weight and meat quality) of triploids, as well as overall size, must be considered in assessing their value. For example, Wolters *et al.*, (1982b) found that triploid male channel catfish had smaller heads than diploid males, which could lead to less wastage in processing.

Sterile triploids might also survive longer than diploids in species which experience losses at sexual maturity. For example, in salmonids, rainbow trout experience high mortality in lakes that lack spawning tributaries (Purdom and Lincoln, 1973), and Pacific salmon routinely die after spawning. The possibility of extending survival and producing "trophy" fish for sport fishing in these and other species has obvious appeal and deserves investigation.

The potential advantages of triploidy (e.g., population control, improved growth, and survival) are primarily consequences of triploid sterility and not of triploidy *per se*. Because sterility may also be induced with hormone treatments (see Chapter 5, this volume), the relative merits of both approaches will need to be considered in the future. In addition to growth and survival considerations, a number of variables such as ease and efficiency of treatment, government regulations, consumer acceptance, and residual gonadal development in triploids will be important. In any case, induced triploidy does have the capability of producing unique new forms of sterile hybrids that hormonal sterilization lacks.

V. SUMMARY

Chromosome-set manipulation techniques of sperm chromosome inactivation (with radiation or chemicals) and suppression of cell divisions (with heat shock, cold shock, or pressure) can be readily applied to fish to produce gynogenetic and polyploid individuals. Gynogenetic individuals have all their chromosomes from the female parent and should all be females in species with XX females. Polyploids include triploids, which are expected to be sterile, and tetraploids, which have the potential of being fertile and producing sterile triploids when crossed to normal diploids. Partially inbred gynogenetic diploids and triploids may be produced by treatments causing retention of the second polar body of the egg. Completely homozygous gynogenetic diploids and tetraploids may be produced by treatments blocking the first mitotic division.

Gynogenesis has been shown to be useful for gene mapping and for the rapid generation of homozygous inbred lines. These lines are likely to be useful for basic biological studies and may be useful for selective breeding. Gynogenesis may not be as advantageous as some other techniques for the purpose of sex control because of inbreeding depression in the offspring. Androgenesis, a technique in which the egg chromosomes are inactivated and offspring have all their chromosomes from the male parent, has not yet been widely applied in fish, but has potential as a complementary technique to gynogenesis for the generation of inbred lines.

Induced-triploid fish have generally good viability and are sterile because of disturbances in gonad development resulting from disruptions in meiotic pairing. Males show more residual gonadal development than females. Several studies have found that triploids may maintain growth better at sexual maturity than diploids. Triploid hybrids between species may have greater viability and fewer abnormalities in many cases than diploid hybrids. There is not yet enough information to draw general conclusions about the viability and fertility of tetraploid fish.

ACKNOWLEDGMENTS

Thanks are extended to Drs. M. Beck, N. Cherfas, V. Csanyi, F. Hoornbeek, J. Kanka, R. Lincoln, A. Maximovich, P. Rab, T. Refstie, L. Smith, J. Stanley, G. Streisinger, D. Thompson, F. Utter, W. Wolters, F. Yamazaki, Mr. S. Allen, Jr., and Mr. T. Beufey for providing unpublished data during preparation of this chapter. During its preparation, I was supported by National Science Foundation grant PCM 8108787.

REFERENCES

Abdel-Hameed, F., and Shoffner, R. N. (1971). Intersexes and sex determination in chickens. *Science* **172**, 962–964.

Allen, S. K., Jr. (1983). Flow cytometry:Assaying experimental polyploid fish and shellfish. *Aquaculture* **33**, 317–328.

Allen, S. K., Jr., and Stanley, J. G. (1978). Reproductive sterility in polyploid brook trout, *Salvelinus fontinalis*. *Trans. Am. Fish. Soc.* **107**, 473–478.

Allen, S. K., Jr., and Stanley, J. G. (1979). Polyploid mosaics induced by cytochalasin B in landlocked Atlantic salmon *Salmo salar*. *Trans. Am. Fish. Soc.* **108**, 462–466.

Allen, S. K., Jr., and Stanley, J. G. (1981). Polyploidy and gynogenesis in the culture of fish and shellfish. *Coop. Res. Rep., Int. Counc. Explor. Sea Ser. B* **28**, 1–18.

Allen, S. K., Jr., and Stanley, J. G. (1983). Ploidy of hybrid grass carp × bighead carp determined by flow cytometry. *Trans. Am. Fish. Soc.* **112**, 431–435.

Anderson, E. G. (1925). Crossing over in a case of attached X chromosomes in *Drosophila melanogaster*. *Genetics* **10**, 403–417.

Arai, K., Onozato, H., and Yamazaki, F. (1979). Artificial androgenesis induced with gamma irradiation in masu salmon, *Oncorhynchus masou*. *Bull. Fac. Fish., Hokkaido Univ.* **30**, 181–186.

Barratt, R. W., Newmeyer, D., Perkins, D. D., and Garnjobst, L. (1954). Map construction in *Neurospora crassa*. *Adv. Genet.* **6**, 1–93.

Beadle, G. W., and Emerson, S. (1935). Further studies of crossing over in attached X chromosomes of *Drosophila melanogaster*. *Genetics* **20**, 192–206.

Beck, M. L., and Biggers, C. J. (1982). Chromosomal investigation of *Ctenopharyngodon idella* × *Aristichthys nobilis* hybrids. *Experientia* **38**, 319.

Beck, M. L., and Biggers, C. J. (1983). Erythrocyte measurements of diploid and triploid *Ctenopharyngodon idella* × *Hypophthalmichthys nobilis* hybrids. *J. Fish Biol.* **22**, 497–502.

Beck, M. L., Biggers, C. J., and Dupree, H. K. (1980). Karyological analysis of *Ctenopharyngodon idella*, *Aristichthys nobilis*, and their F_1 hybrid. *Trans. Am. Fish. Soc.* **109**, 433–438.

Bogart, J. P. (1980). Evolutionary implications of polyploidy in amphibians and reptiles. *In* "Polyploidy: Biological Relevance" (W. H. Lewis, ed.), pp. 341–378. Plenum, New York.

Bridges, C. B. (1925). Sex in relation to chromosomes and genes. *Am. Nat.* **59**, 127–137.

Briedis, A., and Elinson, R. P. (1982). Suppression of male pronuclear movement in frog eggs by hydrostatic pressure and deuterium oxide yields androgenetic haploids. *J. Exp. Zool.* **222**, 45–57.

Briggs, R. (1952). An analysis of the inactivation of the frog sperm nucleus by toluidine blue. *J. Gen. Physiol.* **35**, 761–780.

Capanna, E., Cataudella, S., and Volpe, R. (1974). Un ibrido intergenerico tra trota iridea e salmerino di fonte (*Salmo gairdneri* × *Salvelinus fontinalis*). *Boll. Pesca, Piscic. Idrobiol.* **29**, 101–106.

Chase, S. S. (1974). Utilization of haploids in plant breeding: Breeding diploid species. *In* "Haploids in Higher Plants: Advances and Potential" (K. J. Kasha, ed.) pp. 211–230. University of Guelph, Guelph, Ontario, Canada.

Cherfas, N. B. (1975). Investigation of radiation-induced diploid gynogenesis in the carp

(*Cyprinus carpio* L.). I. Experiments on obtaining the diploid gynogenetic progeny in mass quantities. *Genetika (Moscow)* **11**(7), 78–86.

Cherfas, N. B. (1977). Investigation of radiation-induced diploid gynogenesis in the carp (*Cyprinus carpio* L.). II. Segregation with respect to certain morphological characters in gynogenetic progenies. *Genetika (Moscow)* **13**(5), 811–820.

Cherfas, N. B. (1981). Gynogenesis in fishes. *In* "Genetic Bases of Fish Selection" (V. S. Kirpichnikov, ed.), pp. 255–273. Springer-Verlag, Berlin and New York.

Cherfas, N. B., and Ilyasova, V. A. (1980a). Induced gynogenesis in silver crucian carp and carp hybrids. *Genetika (Moscow)* **16**(7), 1260–1269.

Cherfas, N. B., and Ilyasova, V. A. (1980b). Some results of studies on diploid radiation gynogenesis in carp *Cyprinus carpio* L. *In* "Karyological Variability, Mutagenesis and Gynogenesis in Fishes," pp. 74–81. Inst. Cytol. USSR Acad. Sci., Leningrad.

Cherfas, N. B., and Truweller, K. A. (1978). Investigation of radiation-induced diploid gynogenesis in carp (*Cyprinus carpio* L.). III. Gynogenetic offspring analysis by biochemical markers. *Genetika (Moscow)* **14**(4), 599–604.

Cherfas, N. B., Gomelsky, B. I., Emeljanova, O. V. and Rekoubratsky, A. V. (1981). Triploidy in reciprocal hybrids obtained from crucian carp and carp. *Genetika (Moscow)* **17**(6), 1136–1139.

Chevassus, B., Guyomard, R., Chourrout, D., and Quillet, E. (1983). Production of viable hybrids in salmonids by triploidization. *Génet. Sel. Evol.*, (in press).

Chourrout, D. (1980). Thermal induction of diploid gynogenesis and triploidy in the eggs of the rainbow trout (*Salmo gairdneri* Richardson). *Reprod. Nutr. Dev.* **20**, 727–733.

Chourrout, D. (1982a). La gynogenèse chez les vertébrés. *Reprod. Nutr. Dévelop.* **22**, 713–724.

Chourrout, D. (1982b). Gynogenesis caused by ultraviolet irradiation of salmonid sperm. *J. Exp. Zool.* **223**, 175–181.

Chourrout, D. (1982c). Tetraploidy induced by heat shocks in the rainbow trout (*Salmo gairdneri* R.). *Reprod. Nutr. Dev.* **20**, 727–733.

Chourrout, D. and Quillet, E. (1982). Induced gynogenesis in the rainbow trout: Sex and survival of progenies. Production of all-triploid populations. *Theor. Appl. Genet.* **63**, 201–205.

Chourrout, D., Chevassus, B., and Herioux, F. (1980). Analysis of an Hertwig effect in the rainbow trout (*Salmo gairdneri* Richardson) after fertilization with gamma-irradiated sperm. *Reprod. Nutr. Dev.* **20**, 719–726.

Cuellar, O., and Uyeno, T. (1972). Triploidy in rainbow trout. *Cytogenetics* **11**, 508–515.

Dasgupta, S. (1962). Induction of triploidy by hydrostatic pressure in the leopard frog, *Rana pipiens. J. Exp. Zool.* **151**, 105–121.

Devillers, C. (1961). Structural and dynamic aspects of the development of the teleostean egg. *Adv. Morphog.* **1**, 379–428.

Donaldson, E. M., and Hunter, G. A. (1982). Sex control in fish with particular reference to salmonids. *Can. J. Fish Aquat. Sci.* **39**, 99–110.

Donaldson, E. M., Funk, J. D., Withler, F. C. and Morley, R. B. (1972). Fertilization of pink salmon (*Oncorhynchus gorbuscha*) ova by spermatozoa from gonadotropin-injected juveniles. *J. Fish. Res. Board Can.* **29**, 13–18.

Eglitis, M. A. (1980). Formation of tetraploid mouse blastocysts following blastomere fusion with polyethylene glycol. *J. Exp. Zool.* **213**, 309–313.

Elinson, R. P., and Briedis, A. (1981). Triploidy permits survival of an inviable amphibian hybrid. *Dev. Genet.* **2**, 357–367.

Fankhauser, G. (1945). The effect of changes in chromosome number on amphibian development. *Q. Rev. Biol.* **20**, 20–78.

Fankhauser, G., and Humphrey, R. R. (1959). The origin of spontaneous heteroploids in the progeny of diploid, triploid and tetraploid axolotl females. *J. Exp. Zool.* **142**, 379–421.

Gervai, J., Peter, S., Nagy, A., Horvath, L., and Csanyi, V. (1980). Induced triploidy in carp, *Cyprinus carpio* L. *J. Fish Biol.* **17**, 667–671.

Gillespie, L. L., and Armstrong, J. B. (1979). Induction of triploid and gynogenetic diploid axolotls (*Ambystoma mexicanum*) by hydrostatic pressure. *J. Exp. Zool.* **210**, 117–122.

Gillespie, L. L., and Armstrong, J. B. (1980). Production of androgenetic diploid axolotls by suppression of first cleavage. *J. Exp. Zool.* **213**, 423–425.

Gillespie, L. L., and Armstrong, J. B. (1981). Suppression of first cleavage in the Mexican axolotl (*Ambystoma mexicanum*) by heat shock or hydrostatic pressure. *J. Exp. Zool.* **218**, 441–445.

Gold, J. R. (1979). Cytogenetics. In "Fish Physiology" (W. S. Hoar, D. J. Randall, and J. R. Brett, eds.) Vol. 8, pp. 353–405. Academic Press, New York.

Gold, J. R., and Avise, J. C. (1976). Spontaneous triploidy in the California roach, *Hesperoleucus symmetricus* (Pisces:Cyprinidae). *Cytogenet. Cell Genet.* **17**, 144–149.

Gomelsky, B. I., Ilyasova, V. A., and Cherfas, N. B. (1979). Investigation of radiation-induced gynogenesis in carp (*Cyprinus carpio* L.). IV. Gonad state and evaluation of reproductive ability in carp of gynogenetic origin. *Genetika (Moscow)* **15**(9), 1643–1650.

Green, M. C., ed. (1981). "Genetic Variants and Strains of the Laboratory Mouse." Gustav Fischer Verlag, New York.

Hertwig, O. (1911). Die Radiumkrankheit tierischen Kiemzellen. *Arch. Mikrosk. Anat.* **77**, 1–97.

Hoornbeek, F. K., and Burke, P. M. (1981). Induced chromosome number variation in the winter flounder. *J. Hered.* **72**, 189–192.

Hoppe, P. C., and Illmensee, K. (1977). Microsurgically produced homozygous-diploid uniparental mice. *Proc. Natl. Acad. Sci. U.S.A.* **74**, 5657–5661.

Hunter, G. A., Donaldson, E. M., Goetz, F. W., and Edgell, P. F. (1982). Production of all female and sterile groups of coho salmon (*Oncorhynchus kisutch*) and experimental evidence for male heterogamety. *Trans. Am. Fish. Soc.* **111**, 367–372.

Ijiri, K. (1980). Gamma-ray irradiation of the sperm of the fish *Oryzias latipes* and induction of gynogenesis. *J. Radiat. Res.* **21**, 263–270.

Ijiri, K., and Egami, N. (1980). Hertwig effect caused by UV-irradiation of sperm of *Oryzias latipes* (Teleost) and its photoreactivation. *Mutat. Res.* **69**, 241–248.

Jensen, G. L., Shelton, W. L., Yang, S-L., and Wilken, L. O. (1983). Sex reversal of gynogenetic grass carp by implantation of methyltestosterone. *Trans. Am. Fish. Soc.* **112**, 79–85.

Johnstone, R., Simpson, T. H., Youngson, A. F., and Whitehead, C. (1979). Sex reversal in salmonid culture. Part II. The progeny of sex-reversed rainbow trout. *Aquaculture* **18**, 13–19.

Jones, P., Jackson, H., and Whiting, M. H. S. (1975). Parthenogenetic development after chemical treatment of *Xenopus laevis* spermatozoa. *J. Exp. Zool.* **192**, 73–82.

Kasha, K. J., ed. (1974). "Haploids in Higher Plants: Advances and Potential." University of Guelph, Guelph, Ontario, Canada.

Kasha, K. J., and Reinbergs, E. (1980). Achievements with haploids in barley research and breeding. In "The Plant Genome" (D. R. Davies and D. A. Hopwood, eds.), pp. 215–230. John Innes Charity, Norwich, England.

Lemoine, H. L., Jr., and Smith, L. T. (1980). Polyploidy induced in brook trout by cold shock. *Trans Am. Fish. Soc.* **109**, 626–631.

Lerner, I. M. (1954). "Genetic Homeostasis." Wiley, New York.

Lincoln, R. F. (1981a). Sexual maturation in triploid male plaice (*Pleuronectes platessa*) and plaice × flounder (*Platichythys flesus*) hybrids. *J. Fish Biol.* **19**, 415–426.

Lincoln, R. F. (1981b). Sexual maturation in female triploid plaice, *Pleuronectes platessa*, and plaice × flounder, *Platichthys flesus*, hybrids. *J. Fish Biol.* **19**, 499–507.

Lincoln, R. F. (1981c). The growth of female diploid and triploid plaice (*Pleuronectes platessa*) × flounder (*Platichthys flesus*) hybrids over one spawning season. *Aquaculture* **25**, 259–268.

Lincoln, R. F., and Scott, A. P. (1983). Production of all-female triploid rainbow trout. *Aquaculture* **30**, 375–380.

Lincoln, R. F., Aulstad, D., and Grammeltvedt, A. (1974). Attempted triploid induction in Atlantic salmon (*Salmo salar*) using cold shocks. *Aquaculture* **4**, 287–297.

Lindsley, D. L., Fankhauser, G., and Humphrey, R. R. (1956). Mapping centromeres in the axolotl. *Genetics* **41**, 58–64.

Liu, S., Sezaki, D., Hashimoto, K., Kobayasi, H., and Nakamura, M. (1978). Simplified techniques for determination of polyploidy in ginbuna *Carassius auratus langsdorfi*. *Bull. Jpn. Soc. Sci. Fish.* **44**, 601–606.

Magee, S. M., and Philipp, D. P. (1982). Biochemical genetic analyses of the grass carp ♀ × bighead carp ♂ F_1 hybrid and the parental species. *Trans. Am. Fish. Soc.* **111**, 593–602.

Mantelman, I. I. (1980). First results obtained from usage of chemical mutagens for producing gynogenetic progeny in the Siberian white fish, *Coregonus peled* Gm. *In* "Karyological Variability, Mutagenesis and Gynogenesis in Fishes," pp. 82–85. Inst. Cytol., USSR Acad. Sci., Leningrad.

Marian, T., and Krasznai, Z. (1978). Kariological investigation on *Ctenopharyngodon idella* and *Hypophthalmichthys nobilis* and their cross-breeding. *Aquacult. Hung.* **1**, 44–50.

Maximovich, A. A., and Petrova, G. A. (1980). Production of radiation-induced diploid gynogenetic pink salmon. *In* "Proceedings of the First International Conference on Biology of Pacific Salmon," pp. 151–155. Moscow.

Muller, H J. (1925). Why polyploidy is rarer in animals than in plants. *Am. Nat.* **59**, 346–353.

Nace, G. W., Richards, C. M., and Asher, J. H., Jr. (1970). Parthenogenesis and genetic variability. I. Linkage and inbreeding estimations in the frog, *Rana pipiens*. *Genetics* **66**, 349–368.

Nagy, A. and Csanyi, V. (1982). Changes of genetic parameters in successive gynogenetic generations and some calculations for carp gynogenesis. *Theor. Appl. Genet.* **63**, 105–110.

Nagy, A. Rajki, K., Horvath, L., and Csanyi, V. (1978). Investigation on carp *Cyprinus carpio* L. gynogenesis. *J. Fish Biol.* **13**, 215–224.

Nagy, A., Rajki, K., Bakos, J., and Csanyi, V. (1979). Genetic analysis in carp (*Cyprinus carpio*) using gynogenesis. *Heredity* **43**, 35–40.

Nagy, A., Bercsenyi, M., and Csanyi, V. (1981). Sex reversal in carp (*Cyprinus carpio*) by oral administration of methyltestosterone. *Can. J. Fish. Aquat. Sci.* **38**, 725–728.

Niebuhr, D. (1974). Triploidy in man: Cytogenetical and clinical aspects. *Humangenetik* **21**, 103–125.

Ojima, Y., and Makino, S. (1978). Triploidy induced by cold shock in fertilized eggs of the carp. A Preliminary study. *Proc. Jpn. Acad.* **54**, 359–362.

Ojima, Y., Hayashi, M., and Ueno, K. (1975). Triploidy appeared in the backcross offspring from funa-carp crossings. *Proc. Jpn. Acad.* **51**, 702–706.

Okada, H., Matumoto, H., and Yamazaki, F. (1979). Functional masculinization of genetic females in rainbow trout. *Bull. Jpn. Soc. Sci. Fish.* **45**, 413–419.

Oppermann, K. (1913). Die Entwicklung von Forelleneiern nach Befruchtung mit radium bestrahlten Samemfaden. *Arch. Mikrosk. Anat.* **83**, 141–189.

Onozato, H. (1982). The "Hertwig effect" and gynogenesis in chum salmon *Oncorhynchus keta* eggs fertilized with [60]Co γ-ray irradiated milt. *Bull. Jpn. Soc. Sci. Fish.* **48**, 1237–1244.

Ott, J. (1979). Human gene mapping by postreduction and recombination frequencies under complete interference. *Clin. Genet.* **15**, 11–16.

Ott, J., Linder, D., McCaw, B. K., Lovrien, E. W., and Hecht, F. (1976). Estimating distances from the centromere by means of benign ovarian teratomas in man. *Ann. Hum. Genet.* **40**, 191–196.

Purdom, C. E. (1969). Radiation-induced gynogenesis and androgenesis in fish. *Heredity* **24**, 431–444.

Purdom, C. E. (1972). Induced polyploidy in plaice (*Pleuronectes platessa*) and its hybrid with the flounder (*Platichthys flesus*). *Heredity* **29**, 11–24.

Purdom, C. E. (1976). Genetic techniques in flatfish culture. *J. Fish. Res. Board Can.* **33**, 1088–1093.

Purdom, C. E. (1983). Genetic engineering by the manipulation of chromosomes. *Aquaculture* **33**, 287–300.

Purdom, C. E., and Lincoln, R. F. (1973). Chromosome manipulation in fish. *In* "Genetics and Mutagenesis of Fish" (J. H. Schroder, ed.), pp. 83–89. Springer-Verlag, Berlin and New York.

Purdom, C. E., and Lincoln, R. F. (1974). Gynogenesis in hybrids within the Pleuronectidae. *In* "The Early Life History of Fish" (J. H. S. Blaxter, ed.), pp. 537–544. Springer-Verlag, Berlin and New York.

Purdom, C. E., Thompson, D., and Dando, P. R. (1976). Genetic analysis of enzyme polymorphisms in plaice (*Pleuronectes platessa*). *Heredity* **37**, 193–206.

Refstie, T. (1981). Tetraploid rainbow trout produced by Cytochalasin B. *Aquaculture* **25**, 51–58.

Refstie, T., Vassvik, V., and Gjedrem, T. (1977). Induction of polyploidy in salmonids by Cytochalasin B. *Aquaculture* **10**, 65–74.

Refstie, T., Stoss, J., and Donaldson, E. M. (1982). Production of all female coho salmon (*Oncorhynchus kisutch*) by diploid gynogenesis using irradiated sperm and cold shock. *Aquaculture* **29**, 67–82.

Reinschmidt, D. C., Simon, S. J., Volpe, E. P., and Tompkins, R. (1979). Production of tetraploid and homozygous diploid amphibians by suppression of first cleavage. *J. Exp. Zool.* **210**, 137–143.

Romashov, D. D., and Belyaeva, V. N. (1964). Cytology of radiation gynogenesis and androgenesis in the loach (*Misgurnus fossilis* L.). *Dokl. Akad. Nauk SSSR* **157**(4), 964–967.

Romashov, D. D., and Belyaeva, V. N. (1965). Analysis of diploidization induced by low temperature during radiation gynogenesis in loach. *Tsitologiya* **7**(5), 607–615.

Romashov, D. D., Nikolyukin, N. I., Belyaeva, V. N., and Timofeeva, N. A. (1963). Possibility of producing diploid radiation-induced gynogenesis in sturgeons. *Radiobiologiya* **3**, 104–110.

Scheerer, P. D., and Thorgaard, G. H. (1983). Increased survival in salmonid hybrids by induced triploidy. *Can. J. Fish. Aquat. Sci.* (in press).

Schultz, M. E., and Schultz, R. J. (1982). Diethylnitrosamine-induced hepatic tumors in wild vs. inbred strains of a viviparous fish. *J. Hered.* **73**, 43–48.

Schultz, R. J. (1980). Role of polyploidy in the evolution of fishes. *In* "Polyploidy: Biological Relevance" (W. H. Lewis, ed.), pp. 313–340. Plenum, New York.

Shireman, J. V., Rottmann, R. W., and Aldridge, F. J. (1983). Consumption and growth of hybrid grass carp fed four vegetation diets and trout chow in circular tanks. *J. Fish Biol.* **22**, 685–693.

Simon, R. C., Schill, W. B., and Kincaid, H. L. (1981). Causes and importance of genetic differences between groups of test fish. *Dev. Biol. Stand.* **49**, 267–272.

Smith, L. T., and Lemoine, H. L. (1979). Colchicine-induced polyploidy in brook trout. *Prog. Fish-Cult.* **41**, 86–88.

Smith, R. H., and von Borstel, R. C. (1972). Genetic control of insect populations. *Science* **178**, 1164–1174.

Snape, J. W., and Simpson, E. (1981). The genetical expectations of doubled haploid lines derived from different filial generations. *Theor. Appl. Genet.* **60**, 123–128.

Stanley, J. G. (1976a). Production of hybrid, androgenetic, and gynogenetic grass carp and carp. *Trans. Am. Fish. Soc.* **105**, 10–16.

Stanley, J. G. (1976b). Female homogamety in grass carp (*Ctenopharyngodon idella*) determined by gynogenesis. *J. Fish. Res. Board Can.* **33**, 1372–1374.

Stanley, J. G. (1979). Control of sex in fishes, with special reference to the grass carp. *In* "Proceedings of the Grass Carp Conference" (J. V. Shireman, ed.), pp. 201–242. University of Florida, Gainesville.

Stanley, J. G. (1981). Manipulation of developmental events to produce monosex and sterile fish. *Rapp. P.-V. Reun, Cons. Int. Explor. Mer* **178**, 485–491.

Stanley, J. G. (1983). Gene expression in haploid embryos of Atlantic salmon. *J. Hered.* **74**, 19–22.

Stanley, J. G., and Jones, J. B. (1976). Morphology of androgenetic and gynogenetic grass carp, *Ctenopharyngodon idella* (Valenciennes). *J. Fish Biol.* **9**, 523–528.

Stanley, J. G., and Sneed, K. E. (1974). Artificial gynogenesis and its application in genetics and selective breeding of fishes. *In* "The Early Life History of Fish" (J. H. S. Blaxter, ed.), pp. 527–536. Springer-Verlag, Berlin and New York.

Stanley, J. G., Biggers, C. J., and Schultz, D. E. (1976). Isozymes in androgenetic and gynogenetic white amur, gynogenetic carp, and carp-amur hybrids. *J. Hered.* **67**, 129–134.

Streisinger, G., Walker, C., Dower, N., Knauber, D., and Singer, F. (1981). Production of clones of homozygous diploid zebra fish (*Brachydanio rerio*). *Nature (London)* **291**, 293–296.

Swarup, H. (1957). The production and effects of triploidy in the 3-spined stickleback *Gasterosteus aculaetus* (L.). Ph.D. Thesis, Oxford University.

Swarup, H. (1959a). Production of triploidy in *Gasterosteus aculeatus* (L.). *J. Genet.* **56**, 129–142.

Swarup, H. (1959b). Effect of triploidy on the body size, general organization and cellular structure in *Gasterosteus aculeatus* (L.). *J. Genet.* **56**, 143–155.

Thompson, D. (1983). The efficiency of induced diploid gynogenesis in inbreeding. *Aquaculture* **33**, 237–244.

Thompson, D., Purdom, C. E., and Jones, B. W. (1981). Genetic analysis of spontaneous gynogenetic diploids in the plaice *Pleuronectes platessa*. *Heredity* **47**, 269–274.

Thorgaard, G. H. (1977). Heteromorphic sex chromosomes in male rainbow trout. *Science* **196**, 900–902.

Thorgaard, G. H., and Gall, G. A. E. (1979). Adult triploids in a rainbow trout family. *Genetics* **93**, 961–973.

Thorgaard, G. H., Jazwin, M. E., and Stier, A. R. (1981). Polyploidy induced by heat shock in rainbow trout. *Trans. Am. Fish. Soc.* **110**, 546–550.

Thorgaard, G. H., Rabinovitch, P. S., Shen, M. W., Gall, G. A. E., Propp, J., and Utter, F. M. (1982). Triploid rainbow trout identified by flow cytometry. *Aquaculture* **29**, 305–309.

Thorgaard, G. H., Allendorf, F. W., and Knudsen, K. L. (1983). Gene-centromere mapping in rainbow trout: High interference over long map distances. *Genetics* **103**, 771–783.

Tsoi, R. M. (1969). Action of nitrosomethylurea and dimethylsulfate on the sperm cells of the rainbow trout and the peled. *Dokl. Akad. Nauk SSSR* **189**(1), 411–414.

Utter, F. M., Johnson, O. W., Thorgaard, G. H., and Rabinovitch, P. S. (1983). Measurement and potential applications of induced triploidy in Pacific salmon. *Aquaculture* **35**, (in press).

Uwa, H. (1965). Gynogenetic haploid embryos of the medaka (*Oryzias latipes*). *Embryologia* **9**, 40–48.

Uyeno, T. (1972). Chromosomes of offspring resulting from crossing coho salmon and brook trout. *Jpn. J. Ichtyol.* **19**, 166–171.

Valenti, R. J. (1975). Induced polyploidy in *Tilapia aurea* (Steindachner) by means of temperature shock treatment. *J. Fish Biol.* **7**, 519–528.

Vasetskii, S. G. (1967). Changes in the ploidy of sturgeon larvae induced by heat treatment of eggs at different stages of development. *Dokl. Akad. Nauk SSSR* **172**(5), 1234–1237.

Vasil'ev, V. P., Makeeva, A. P., and Ryabov, I. N. (1975). On the triploidy of remote hybrids of carp (*Cyprinus carpio* L.) with other representatives of Cyprinidae. *Genetika (Moscow)* **11**(8), 49–56.

Volpe, E. P. (1970). Chromosome mapping in the leopard frog. *Genetics* **64**, 11–21.

Volpe, E. P., and Dasgupta, S. (1962). Gynogenetic diploids of mutant leopard frogs. *J. Exp. Zool.* **151**, 287–301.

Whitt, G. S., Childers, W. F., and Cho, P. L. (1973). Allelic expression at enzyme loci in an intertribal hybrid sunfish. *J. Hered.* **64**, 54–61.

Wolters, W. R., Libey, G. S., and Chrisman, C. L. (1981a). Induction of triploidy in channel catfish. *Trans. Am. Fish. Soc.* **110**, 310–312.

Wolters, W. R., Chrisman, C. L., and Libey, G. S. (1981b). Lymphocyte culture for chromosomal analyses of channel catfish, *Ictalurus punctatus. Copeia* pp. 503–504.

Wolters, W. R., Chrisman, C. L., and Libey, G. S. (1982a). Erythrocyte nuclear measurements of diploid and triploid channel catfish, *Ictalurus punctatus. J. Fish Biol.* **20**, 253–258.

Wolters, W. R., Libey, G. S., and Chrisman, C. L. (1982b). Effect of triploidy on growth and gonad development of channel catfish. *Trans. Am. Fish. Soc.* **111**, 102–105.

Yamamoto, T. (1963). Induction of reversal in sex differentiation of YY zygotes in the medaka, *Oryzias latipes. Genetics* **48**, 293–306.

Yamamoto, T. (1964). The problem of viability of YY zygotes in the medaka, *Oryzias latipes. Genetics* **50**, 45–48.

Yamamoto, T. (1975). A YY male goldfish from mating estrone-induced XY female and normal male. *J. Hered.* **66**, 2–4.

Yamazaki, F. (1983). Sex control and manipulation in fish. *Aquaculture* **33**, 329–354.

Yamazaki, F., Onozato, H., and Arai, K. (1981). The chopping method for obtaining permanent chromosome preparation from embryos of teleost fishes. *Bull. Jpn. Soc. Sci. Fish.* **47**, 963.

AUTHOR INDEX

Numbers in *italics* refer to the pages on which the complete references are listed.

SYSTEMATIC INDEX

Note: Names listed are those used by the authors of the various chapters. No attempt has been made to provide the current nomenclature where taxonomic changes have occurred. Boldface letters refer to Parts A and B of Volume 9.

A

Acanthias vulgaris, see Squalus acanthias
Acanthogobius flavimanus, **A**, 112, 363
Acara, brown, *see Aequidens portalegrensis*
Acheilognathus
   A. lanceolata, **B**, 316
   A. tabira, **B**, 316
Acipenser
   A. güldenstadti, **B**, 327
   A. stellatus, **A**, 194, 202; **B**, 123, 133, 137, 383
Aequidens
   A. latifrons, **B**, 19, 30
   A. portalegrensis, **B**, 38, 43, 88
   A. pulcher, **A**, 323
Alewife, *see Alosa pseudoharengus*
Alosa pseudoharengus, **B**, 308
Ameiurus nebulosus, **A**, 381
Amphiprion, **B**, 176
   A. alkallopisos, **B**, 206, 208
   A. bicinctus, **B**, 206, 208
   A. melanopus, **B**, 86, 87, 89, 206
Amur, black, **B**, 365, 366
Anabas testudineus, **A**, 320, 322, 324, 393
Anchovy, *see Stolepholus*
Anemonefish, *see Amphiprion melanopus*
Angelfish, **B**, 30, *see also Pterophyllum scalare*
Anguilliformes, **B**, 307
   Anguilla
      A. anguilla, **A**, 123, 151, 157, 165, 172, 199, 249, 285, 286, 297, 304, 320, 321, 325, 333, 338, 390, 392; **B**, 121, 123, 133, 137, 174, 201, 204, 205, 266, 387
      A. japonica, **A**, 103, 166, 248, 249, 300, 320, 339; **B**, 121, 366, 387
Anolis carolinesis, **B**, 45, 46

Anthias squamipinnis, **B**, 179, 206–208, 210
Aphanius dispar, **A**, 237
Apode, **A**, 138, 139
Aristichthys noblis, **A**, 113; **B**, 331, 362, 409
Astronotus ocellatus, **B**, 30
Asyntanax
   A. bimaculatus, **B**, 373
   A. mexicanus, **B**, 13, 93
   A. taeniatus, **B**, 373
Atheriniformes, **A**, 226, 227, 281; **B**, 175
   Belone belone, **A**, 301
   Dermogenys pusillus, **A**, 301
   Fundulus
      F. confluentus, **B**, 73, 83
      F. heteroclitus, **A**, 106, 249, 251, 260, 281, 300, 301, 303, 318, 320, 323, 333, 342, 344, 387; **B**, 28, 29, 48, 70, 73, 83, 86, 87, 91, 93, 120, 124, 133, 136, 137, 312, 315, 321, 327
      F. similis, **A**, 118
   Jenynsia lineata, **A**, 285, 303, 308
   Oryzias latipes, **A**, 111, 117, 161, 171, 197, 200, 201, 224, 281, 283, 284, 300, 301, 328, 331–333, 339; **B**, 4, 26, 28, 29, 40, 67, 69, 77, 78, 89, 92, 94, 95, 120, 126, 134, 137, 183, 191, 206, 224, 227, 228, 231, 233, 236–240, 242, 245, 249, 251, 258, 259, 263, 275, 326, 327, 363, 416
   Poecilia
      P. caudofasciata, **B**, 183
      P. latipinna, **A**, 107, 142, 143, 145, 147, 148, 151, 162, 165, 170–172, 247, 281, 300, 301, 303, 308, 314, 320, 328, 333, 340, 342
      P. reticulata, **A**, 143, 145, 147, 226, 231–233, 254, 281, 283, 285, 290, 297, 300, 301, 303, 328, 333,

SUBJECT INDEX

Note: Boldface **A** refers to entries in Volume IXA; **B** refers to entries in Volume IXB.

A

ACTH, *see* Corticotropin
Actinomycin D, **B,** 132, 138, 149, 235
Adenohypophysis, *see* Pituitary gland
Aggressive behavior, **B,** 16–33
Ambosexual (Amphisexual) fishes, **B,** 173
Androgenesis, induced, **B,** 405
Androgenine, **B,** 183–185
Androgens
    biochemistry of, **A,** 303–315
    biosynthesis in testis, **A,** 304–315
    conjugates of, **A,** 315
    conversion to estrogen, **A,** 255
    in cyclostomes, **A,** 15–17
    induced breeding and, **B,** 381–382
    oocyte maturation and, **B,** 122–133
    sex control of cichlids, **B,** 270–273
    sex control of cyprinids, **B,** 288–290
    sex control of salmonids, **B,** 277–284
    in sex determination, **B,** 191–193
    in sperm, **A,** 316
    in teleost ovary, **A,** 286–297
    in teleost testis, **A,** 304–313, 339–344
    vitellogenesis and, **A,** 337–338, 391
Androtermone, **B,** 183–233
Antiestrogens, *see also* Clomiphene,
    Tamoxifen
    in fish culture, **B,** 355–361
    structure of, **B,** 356
Ammocoete stage, **A,** 2–4, 7
Anti-Müllerian hormone (AMH), **B,** 172,
    185
Apomorphine, **B,** 370
Aquaculture
    chromosome manipulation in, **B,** 405–427
    environmental control in, **B,** 96–99
    hormonal sex control, **B,** 243–291
    induced maturation, **B,** 352–384
    induced spermiation, **B,** 384–390
Aromatase, brain distribution, **A,** 123
Atresia, *see* Corpora atretica

B

Balbiani bodies, **A,** 238, 387
Barr bodies, **B,** 186
Behavior, *see* Reproductive behavior
Blood–testis barrier, **A,** 237
Brain, *see also* Hypothalamus
    aromatase activity, **A,** 317
    hormonal action on, **B,** 47–48
    hormones of, **A,** 97–135
    sex determination, *see* Sex determination
Breeding cycles, *see also* Reproductive
    cycles
    androgens and, **B,** 3–5, 6
    in Chondrichthyes, **A,** 51, 75–85
    corticosteroids and, **B,** 4, 6
    estrogens and, **B,** 5–7
    GtH regulation of, **A,** 124–127
    of lampreys, **A,** 3–4, 23–25
    of myxinoids, **A,** 4
    progestins and, **B,** 4, 6
Broodstock management, photothermal ma-
    nipulations in, **B,** 96–98
Buccal lobe (BL), **A,** 66

C

Candle, **A,** 83
cAMP, *see* Cyclic AMP
Castration
    behavior effects, **B,** 18–28
    effect on GtH, **A,** 120–121, 123, 210
    pituitary cytology and, **A,** 160
Catecholamines
    gonadotropin release and, **A,** 115–116; **B,**
    370
    male behavior and, **B,** 23
    ovulation and, **B,** 154
Central nervous system, *see* Brain
Centromere mapping, **B,** 416–418
Chorionic gonadotropin, *see* Human
    chorionic gonadotropin

471